PRINCIPLES OF
Virology
THIRD EDITION

VOLUME II *Pathogenesis and Control*

PRINCIPLES OF
Virology
THIRD EDITION

S. J. FLINT
Department of Molecular Biology
Princeton University
Princeton, New Jersey

L. W. ENQUIST
Department of Molecular Biology
Princeton University
Princeton, New Jersey

V. R. RACANIELLO
Department of Microbiology
College of Physicians and Surgeons
Columbia University
New York, New York

A. M. SKALKA
Fox Chase Cancer Center
Philadelphia, Pennsylvania

ASM
PRESS
WASHINGTON, DC

Front cover illustration: A model of the atomic structure of the poliovirus type 1 Mahoney strain. The model has been highlighted by radial depth cuing so that the portions of the model that are farthest from the center are bright. Prominent surface features include a star-shaped mesa at each of the fivefold axes and a propeller-shaped feature at each of the threefold axes. A deep cleft or canyon surrounds the star-shaped feature. This canyon is the receptor-binding site. Courtesy of Robert Grant, Stéphane Crainic, and James Hogle (Harvard Medical School).

Back cover illustration: Progress in the global eradication of poliomyelitis has been striking, as illustrated by maps showing areas of known or probable circulation of wild-type poliovirus in 1988, 1998, and 2008. Dark red indicates the presence of virus. In 1988, the virus was present on all continents except Australia. By 1998, the Americas were free of wild-type poliovirus, and transmission was interrupted in the western Pacific region (including the People's Republic of China) and in the European region (with the exception of southeastern Turkey). By 2008, the number of countries reporting endemic circulation of poliovirus had been reduced to four: Afghanistan, Pakistan, India, and Nigeria.

Address editorial correspondence to ASM Press, 1752 N St. NW, Washington, DC 20036-2904, USA

Send orders to ASM Press, P.O. Box 605, Herndon, VA 20172, USA
Phone: (800) 546-2416 or (703) 661-1593
Fax: (703) 661-1501
E-mail: books@asmusa.org
Online: estore.asm.org

Library of Congress Cataloging-in-Publication Data

Principles of virology / S.J. Flint ... [et al.]. — 3rd ed.
    p. ; cm.
  Includes bibliographical references and index.
   ISBN 978-1-55581-443-4 (pbk. : set) — ISBN 978-1-55581-479-3
(pbk. : v. 1) — ISBN 978-1-55581-480-9 (pbk. : v. 2)
   1. Virology. I. Flint, S. Jane. II. American Society for Microbiology.
  [DNLM: 1. Viruses. 2. Genetics, Microbial. 3. Molecular Biology.
4. Virology—methods. QW 160 P957 2009]

QR360.P697 2009
579.2—dc22
                                            2008030964

10 9 8 7 6 5 4 3 2 1

ISBN 978-1-55581-480-9

Illustrations and illustration concepting: Patrick Lane, ScEYEnce Studios
Cover and interior design: Susan Brown Schmidler

*We dedicate this book to the students, current and future scientists
and physicians, for whom it was written.
We kept them ever in mind.*

*We also dedicate it to our families:
Jonn, Gethyn, and Amy Leedham
Kathy and Brian
Doris, Aidan, Devin, and Nadia
Rudy, Jeanne, and Chris*

*Oh, be wiser thou!
Instructed that true knowledge leads to love.*

WILLIAM WORDSWORTH
Lines left upon a Seat in a Yew-tree
1888

Contents

Preface xiii
Acknowledgments xvii

1 Infection of a Susceptible Host 2

Introduction 3

A Brief History of Viral Pathogenesis 3

Microbes as Infectious Agents 3
The First Human Viruses 4
The Golden Age of Viral Pathogenesis 5
The New Millennium and Viral Pathogenesis 6

Infection Basics 6

A Series of Unfortunate Events 6
Initiating an Infection 6
Viral Entry 9
Successful Infections Must Modulate or Bypass Host Defenses 14
Viral Spread 16
Organ Invasion 21
Tropism 23

Perspectives 26

References 26

2 Infection of Populations 28

Introduction 29

Principles of Viral Pathogenesis 29

Statistics 30
Epidemiology 31
Shedding of Virions 34

Transmission of Viral Infection 36
Geography and Season 37
Viral Virulence 40
Host Susceptibility to Viral Disease 48
Other Determinants of Susceptibility 48

Perspectives 50

References 51

3   Virus Offense Meets Host Defense: Early Actions 52

Introduction 53
Primary Physical and Chemical Defenses 54
The First Critical Moments of Infection 54

Intrinsic Cellular Defenses 55
How Do Individual Cells Detect Foreign Invaders? 55
Receptor-Mediated Recognition of Pathogen-Associated Molecules 55
Cytokines, the Primary Output of Intrinsic Cell Defense 59
Interferons, Cytokines of Early Warning and Action 61
Apoptosis (Programmed Cell Death) 72

The Hostile Cytoplasm: Other Intrinsic Defenses 78
Autophagy 78
Epigenetic Silencing 78
RNA Silencing 78
Cytosine Deamination (Apobec, [Apolipoprotein B Editing Complex]) 79
Trim (Tripartite Interaction Motif) Proteins 79

Perspectives 80

References 82

4   Immune Defenses 86

Introduction 87
Innate and Adaptive Immune Defenses 87

The Innate Immune Response 89
General Features 89
Sentinel Cells 89
Natural Killer Cells 91
Complement 93
The Inflammatory Response 97

The Adaptive Immune Response 99
General Features 99
Cells of the Adaptive Immune System 101
Adaptive Immunity: the Action of Lymphocytes
That Carry Distinct Antigen Receptors 102
Antigen Presentation and Activation of Immune Cells 107
The Cell-Mediated Adaptive Response 110
The Antibody Response 116
The Immune System and the Brain 120

Immunopathology: Too Much of a Good Thing 121
Immunopathological Lesions 121
Viral Infection-Induced Immunosuppression 124
Systemic Inflammatory Response Syndrome 124

Autoimmune Diseases   124

Heterologous T-Cell Immunity   125

Superantigens "Short-Circuit" the Immune System   126

Mechanisms Mediated by Free Radicals   127

Perspectives   127

References   131

5   Patterns of Infection   134

Introduction   135

Life Cycles and Host Defenses   135

**Mathematics of Growth Correlate with Patterns
of Infection   136**

Acute Infections   138

Definition and Requirements   138

Acute Infections Tend To Be Efficiently Contained and Cleared   138

Antigenic Variation Provides a Selective Advantage in Acute Infections   140

Acute Infections Present Common Public Health Problems   141

Persistent Infections   142

Definition and Requirements   142

An Ineffective Intrinsic or Innate Immune Response Can
Promote a Persistent Infection   143

Modulation of the Adaptive Immune Response Perpetuates
a Persistent Infection   143

Persistent Infections May Be Established in Tissues with
Reduced Immune Surveillance   147

Persistent Infections May Occur When Cells of the
Immune System Are Infected   147

Two Viruses That Cause Persistent Infections   148

Measles Virus   148

Lymphocytic Choriomeningitis Virus   149

Latent Infections   150

General Properties   150

Herpes Simplex Virus   150

Epstein-Barr Virus   156

Slow Infections: Sigurdsson's Legacy   160

Abortive Infections   160

Transforming Infections   161

Perspectives   161

References   162

6   Human Immunodeficiency Virus Pathogenesis   164

Introduction   165

Worldwide Scope of the Problem   165

HIV Is a Lentivirus   166

Discovery and Characterization   166

Distinctive Features of the HIV Replication Cycle and the Roles
of Auxiliary Proteins   169

Cellular Targets   176

Routes of Transmission   177
Sources of Virus Infection   177
Modes of Transmission   177
Mechanics of Spread   179

The Course of Infection   180
Patterns of Virus Appearance and Immune Cell
Indicators of Infection   180
Variability of Response to Infection   181

Origins of Cellular Immune Dysfunction   182
CD4$^+$ T Lymphocytes   182
Cytotoxic T Lymphocytes   182
Monocytes and Macrophages   182
B Cells   183
Natural Killer Cells   183
Autoimmunity   183

Immune Responses to HIV   184
Humoral Responses   184
The Cellular Immune Response   186
Summary: the Critical Balance   186

Dynamics of HIV-1 Replication in AIDS Patients   186

Effects of HIV on Different Tissues and Organ Systems   188
Lymphoid Organs   188
The Nervous System   188
The Gastrointestinal System   190
Other Organ Systems   190

HIV and Cancer   191
Kaposi's Sarcoma   191
B-Cell Lymphomas   193
Anogenital Carcinomas   194

Prospects for Treatment and Prevention   194
Antiviral Drugs and Therapies   194
Highly Active Antiretroviral Therapy   194
Prophylactic Vaccine Development To Prevent Infection   195

Perspectives   196

References   197

7   Transformation and Oncogenesis   200

Introduction   201
Properties of Transformed Cells   202
Control of Cell Proliferation   204

Oncogenic Viruses   207
Discovery of Oncogenic Viruses   208
Viral Genetic Information in Transformed Cells   212
The Origin and Nature of Viral Transforming Genes   217
Functions of Viral Transforming Proteins   218

Activation of Cellular Signal Transduction Pathways by Viral Oncogene Products   221
Viral Mimics of Cellular Signaling Molecules   221
Alteration of the Production or Activity of Cellular Signal Transduction Proteins   224

Disruption of Cell Cycle Control Pathways by Viral Oncogene Products 230
Abrogation of Restriction Point Control Exerted by the Rb Protein 230
Production of Virus-Specific Cyclins 233
Inactivation of Cyclin-Dependent Kinase Inhibitors 233

Transformed Cells Must Also Grow and Survive 234
Integration of Mitogenic and Growth-Promoting Signals 234
Mechanisms That Permit Survival of Transformed Cells 234

Tumorigenesis Requires Additional Changes in the Properties of Transformed Cells 239
Inhibition of Immune Defenses 240

Other Mechanisms of Transformation and Oncogenesis by Human Tumor Viruses 241
Nontransducing, Complex Oncogenic Retroviruses: Tumorigenesis with Very Long Latency 241
Oncogenesis by Hepatitis Viruses 242

Perspectives 246

References 247

8 Vaccines 250

Introduction 251

The Historical Origins of Vaccination 251
Smallpox: a Historical Perspective 251
Large-Scale Vaccination Programs Can Be Dramatically Effective 253

Vaccine Basics 256
Immunization Can Be Active or Passive 256
Active Vaccines Stimulate Immune Memory 256
The Fundamental Challenge 260

The Science and Art of Making Vaccines 261
Basic Approaches 261

Vaccine Technology 271
Most Killed and Subunit Vaccines Rely on Adjuvants To Stimulate an Immune Response 271
Delivery 272
Immunotherapy 273

The Quest for an AIDS Vaccine 274
Formidable Challenges 274
The Central Issues 275

Perspectives 275

References 276

9 Antiviral Drugs 278

Introduction 279
Paradox? So Much Knowledge, So Few Antivirals 279
Historical Perspective 281

Discovering Antiviral Compounds 281
The New Lexicon of Antiviral Discovery 281
Screening for Antiviral Compounds 282
Designer Antivirals and Computer-Based Searching 285

The Difference between "R" and "D"   287
Examples of Some Approved Antiviral Drugs   289
The Search for New Antiviral Targets   293
Antiviral Gene Therapy and Transdominant Inhibitors   295
Resistance to Antiviral Drugs   298

Human Immunodeficiency Virus and AIDS   299
Examples of Anti-HIV Drugs   299
The Combined Problems of Treating a Persistent Infection and Emergence of Drug Resistance   303
Combination Therapy   305
Strategic Treatment Interruption   307
Challenges and Lessons Learned   307

Perspectives   307

References   308

10 Evolution and Emergence   310

Virus Evolution   311
The Classic Theory of Host-Parasite Interactions   311
How Do Viral Populations Evolve?   312
The Origin of Viruses   321
The Fundamental Properties of Viruses Constrain and Drive Evolution   330

Emerging Viruses   333
The Spectrum of Host-Virus Interactions   333
Encountering New Hosts: Fundamental Problems in Ecology   339
Expanding Viral Niches: Snapshots of Selected Emerging Viruses   341
Host Range Can Be Expanded by Mutation, Recombination, or Reassortment   345
Some Emergent Viruses Are Truly Novel   349
A Revolution in Diagnostic Virology   350

Perceptions and Possibilities   350
Infectious Agents and Public Perceptions   350
What Next?   351

Perspectives   353

References   354

APPENDIX A Diseases, Epidemiology, and Disease Mechanisms of Selected Animal Viruses Discussed in This Book   357

APPENDIX B Unusual Infectious Agents   385

Glossary   393

Index   399

Preface

The enduring goal of scientific endeavor, as of all human enterprise, I imagine, is to achieve an intelligible view of the universe. One of the great discoveries of modern science is that its goal cannot be achieved piecemeal, certainly not by the accumulation of facts. To understand a phenomenon is to understand a category of phenomena or it is nothing. Understanding is reached through creative acts.

A. D. HERSHEY
Carnegie Institution Yearbook 65

The major goal of all three editions of this book has been to define and illustrate the basic principles of animal virus biology. In this information-rich age, the quantity of data describing any given virus can be overwhelming, if not indigestible, for student and expert alike. Furthermore, the urge to write more and more about less and less is the curse of reductionist science and the bane of those who write textbooks meant to be used by students. Consequently, in the third edition, we have continued to distill information with the intent of extracting essential principles, while retaining some descriptions of how the work is done. Our goal is to illuminate process and strategy as opposed to listing facts and figures. We continue to be selective in our choice of topics, viruses, and examples in an effort to make the book readable, rather than comprehensive. Detailed encyclopedic works like *Fields Virology* (2007) have made the best attempt to be all-inclusive, and *Fields* is recommended as a resource for detailed reviews of specific virus families.

What's New

The major change in the third edition is the separation of material into two volumes, each with its unique appendix(es) and general glossary. Volume I covers molecular aspects of the biology of viruses, and Volume II focuses on viral pathogenesis, control of virus infections, and virus evolution. The organization into two volumes follows a natural break in pedagogy and provides considerable flexibility and utility for students and teachers alike. The smaller size and soft covers of the two volumes make them easier for students to carry and work with than the single hardcover volume of earlier editions. The volumes can be used for two courses, or as parts I and II of a one-semester course. While differing in content, the volumes are integrated in style and presentation. In addition to updating the material for both volumes, we have used the new format to organize the material more efficiently and to keep chapter size manageable.

As in our previous edition, we have tested ideas for inclusion in the text in our own classes. We have also received constructive comments and suggestions from other virology instructors and their students. Feedback from students was particularly useful in finding typographical errors, clarifying confusing or complicated illustrations, and pointing out inconsistencies in content.

For purposes of readability, references again are generally omitted from the text, but each chapter ends with an updated and expanded list of relevant books, review articles, and selected research papers for readers who wish to pursue specific topics. In general, if an experiment is featured in a chapter, one or more references are listed to provide more detailed information.

Principles Taught in Two Distinct, but Integrated Volumes

These two volumes outline and illustrate the strategies by which all viruses are propagated in cells, how these infections spread within a host, and how such infections are maintained in populations. The principles established in Volume I enable understanding of the topics of Volume II: viral disease, its control, and the evolution of viruses.

Volume I: the Science of Virology and the Molecular Biology of Viruses

This volume features the molecular processes that take place in an infected host cell. Chapters 1 and 2 discuss the foundations of virology. A general introduction with historical perspectives as well as definitions of the unique properties of viruses is provided first. The unifying principles that are the foundations of virology, including the concept of a common strategy for viral propagation, are then described. Chapter 2 establishes the principle of the infectious cycle with an introduction to cell biology. The basic techniques for cultivating and assaying viruses are outlined, and the concept of the single-step growth cycle is presented.

Chapter 3 introduces the fundamentals of viral genomes and genetics, and it provides an overview of the perhaps surprisingly limited repertoire of viral strategies for genome replication and mRNA synthesis. Chapter 4 describes the architecture of extracellular virus particles in the context of providing both protection and delivery of the viral genome in a single vehicle. In Chapters 5 through 13, we describe the broad spectrum of molecular processes that characterize the common steps of the reproductive cycle of viruses in a single cell, from decoding genetic information to genome replication and production of progeny virions. We describe how these common steps are accomplished in cells infected by diverse but representative viruses, while emphasizing principles applicable to all.

The appendix in Volume I provides concise illustrations of viral life cycles for the main virus families discussed in the text. It is intended to be a reference resource when one is reading individual chapters and a convenient visual means by which specific topics may be related to the overall infectious cycles of the selected viruses.

Volume II: Pathogenesis, Control, and Evolution

This volume addresses the interplay between viruses and their host organisms. Chapters 1 to 7 focus on principles of virus replication and pathogenesis. Chapter 1 provides a brief history of viral pathogenesis and addresses the basic concepts of how an infection is established in a host as opposed to infection of single cells in the laboratory. In Chapter 2, we focus on how viral infections spread in populations. Chapter 3 presents our growing understanding of crucial autonomous reactions of cells to infection and describes how these actions influence the eventual outcome for the host. Chapter 4 provides a virologist's view of immune defenses and their integration with events that occur when single cells are infected. Chapter 5 describes how a particular virus replication strategy and the ensuing host response influence the outcome of infection such that some are short and others are of long duration. Chapter 6 is devoted entirely to the AIDS virus, not only because it is the causative agent of the most serious current worldwide epidemic, but also because of its unique and informative interactions with the human immune defenses. In Chapter 7, we discuss virus infections that transform cells in culture and promote oncogenesis (the formation of tumors) in animals.

Chapters 8 and 9 outline the principles involved in treatment and control of infection. Chapter 8 focuses on vaccines, and chapter 9 discusses the approaches and challenges of antiviral drug discovery. In Chapter 10, the final chapter, we present a foray into the past and future, providing an introduction to viral evolution. We illustrate important principles taught by zoonotic infections, emerging infections, and humankind's experiences with epidemic and pandemic viral infections.

Appendix A summarizes the pathogenesis of common viruses that infect humans in three "slides" (viruses and diseases, epidemiology, and disease mechanisms) for each virus or virus group. This information is intended to provide a simple snapshot of pathogenesis and epidemiology. Appendix B provides a concise discussion of unusual infectious agents, such as viroids, satellites, and prions, that are not viruses but that (like viruses) are molecular parasites of the cells in which they replicate.

Reference

Knipe, D. M., and P. M. Howley (ed. in chief). 2007. *Fields Virology,* 5th ed. Lippincott Williams & Wilkins, Philadelphia, PA.

Acknowledgments

These two volumes of *Principles* could not have been composed and revised without help and contributions from many individuals. We are most grateful for the continuing encouragement from our colleagues in virology and the students who use the text. Our sincere thanks also go to colleagues who have taken considerable time and effort to review the text in its evolving manifestations. Their expert knowledge and advice on issues ranging from teaching virology to organization of individual chapters and style were invaluable, even when orthogonal to our approach, and are inextricably woven into the final form of the book.

We thank the following individuals for their reviews and comments on multiple chapters in both volumes: Nicholas Acheson (McGill University), Karen Beemon and her virology students (Johns Hopkins University), Clifford W. Bond (Montana State University), Martha Brown (University of Toronto Medical School), Teresa Compton (University of Wisconsin), Stephen Dewhurst (University of Rochester Medical Center), Mary K. Estes (Baylor College of Medicine), Ronald Javier (Baylor College of Medicine), Richard Kuhn (Purdue University), Muriel Lederman (Virginia Polytechnic Institute and State University), Richard Moyer (University of Florida College of Medicine), Leonard Norkin (University of Massachusetts), Martin Petric (University of Toronto Medical School), Marie Pizzorno (Bucknell University), Nancy Roseman (Williams College), David Sanders (Purdue University), Dorothea Sawicki (Medical College of Ohio), Bert Semler (University of California, Irvine), and Bill Sugden (University of Wisconsin).

We also are grateful to those who gave so generously of their time to serve as expert reviewers of these or earlier individual chapters or specific topics: James Alwine (University of Pennsylvania), Edward Arnold (Center for Advanced Biotechnology and Medicine, Rutgers University), Carl Baker (National Institutes of Health), Amiya Banerjee (Cleveland Clinic Foundation), Silvia Barabino (University of Basel), Albert Bendelac (University of Chicago), Susan Berget (Baylor College of Medicine), Kenneth I. Berns (University of Florida),

John Blaho (MDL Corporation), Sheida Bonyadi (Concordia University), Jim Broach (Princeton University), Michael J. Buchmeier (The Scripps Research Institute), Hans-Gerhard Burgert (University of Warwick), Allan Campbell (Stanford University), Jim Champoux (University of Washington), Bruce Chesebro (Rocky Mountain Laboratories, National Institute of Allergy and Infectious Diseases), Marie Chow (University of Arkansas Medical Center), Barclay Clements (University of Glasgow), Don Coen (Harvard Medical School), Richard Condit (University of Florida), David Coombs (University of New Brunswick), Michael Cordingley (Bio-Mega/Boehringer Ingelheim), Ted Cox (Princeton University), Andrew Davison (Institute of Virology, MRC Virology Unit), Ron Desrosiers (Harvard Medical School), Robert Doms (University of Pennsylvania), Emilio Emini (Merck Sharp & Dohme Research Laboratories), Alan Engelman (Dana-Farber Cancer Center), Ellen Fanning (Vanderbilt University), Bert Flanagan (University of Florida), Nigel Fraser (University of Pennsylvania Medical School), Huub Gelderblom (University of Amsterdam), Charles Grose (Iowa University Hospital), Samuel Gunderson (European Molecular Biology Laboratory), Pryce Haddix (Auburn University at Montgomery), Peter Howley (Harvard Medical School), James Hoxie (University of Pennsylvania), Frederick Hughson (Princeton University), Clinton Jones (University of Nebraska), Christopher Kearney (Baylor University), Walter Keller (University of Basel), Tom Kelly (Memorial Sloan-Kettering Cancer Center), Elliott Kieff (Harvard Medical School), Elizabeth Kutter (Evergreen State College), Robert Lamb (Northwestern University), Ihor Lemischka (Mount Sinai School of Medicine), Arnold Levine (Institute for Advanced Study), Michael Linden (Mount Sinai School of Medicine), Daniel Loeb (University of Wisconsin), Adel Mahmoud (Princeton University), Michael Malim (King's College London), James Manley (Columbia University), Philip Marcus (University of Connecticut), Malcolm Martin (National Institutes of Health), William Mason (Fox Chase Cancer Center), Loyda Melendez (University of Puerto Rico Medical Sciences Campus), Baozhong Meng (University of Guelph), Edward Mocarski (Emory University), Bernard Moss (Laboratory of Viral Diseases, National Institutes of Health), Peter O'Hare (Marie Curie Research Institute), Radhakris Padmanabhan (University of Kansas Medical Center), Peter Palese (Mount Sinai School of Medicine), Philip Pellett (Cleveland Clinic and Case Western Reserve University), Stuart Peltz (University of Medicine and Dentistry of New Jersey, Robert Wood Johnson Medical School), Roger Pomerantz (Thomas Jefferson University), Glenn Rall (Fox Chase Cancer Center), Charles Rice (Rockefeller University), Jack Rose (Yale University), Barry Rouse (University of Tennessee College of Veterinary Medicine), Rozanne Sandri-Goldin (University of California, Irvine), Nancy Sawtell (Childrens Hospital Medical Center), Priscilla Schaffer (University of Arizona), Robert Schneider (New York University School of Medicine), Christoph Seeger (Fox Chase Cancer Center), Aaron Shatkin (Center for Advanced Biotechnology and Medicine, Rutgers University), Thomas Shenk (Princeton University), Geoff Smith (Wright-Fleming Institute), Greg Smith (Northwestern University), Kathryn Spink (Illinois Institute of Technology), Joan Steitz (Yale University), Victor Stollar (University of Medicine and Dentistry of New Jersey), Wesley Sundquist (University of Utah), John M. Taylor (Fox Chase Cancer Center), Alice Telesnitsky (University of Michigan Medical School), Heinz-Jürgen Thiel (Institut für Virologie, Giessen, Germany), Adri Thomas (University of Utrecht), Gerald Thrush (Western University of Health Sciences), Paula Traktman (Medical College of Wisconsin), James van

Etten (University of Nebraska, Lincoln), Chris Upton (University of Victoria), Luis Villarreal (University of California, Irvine), Herbert Virgin (Washington University School of Medicine), Peter Vogt (The Scripps Research Institute), Simon Wain-Hobson (Institut Pasteur), Gerry Waters (TB Alliance), Robin Weiss (University College London), Sandra Weller (University of Connecticut Health Center), Michael Whitt (University of Tennessee), Lindsay Whitton (The Scripps Research Institute), and Eckard Wimmer (State University of New York at Stony Brook). Their rapid responses to our requests for details and checks on accuracy, as well as their assistance in simplifying complex concepts, were invaluable. All remaining errors or inconsistencies are entirely ours.

Since the inception of this work, our belief has been that the illustrations must complement and enrich the text. Execution of this plan would not have been possible without the support of Jeff Holtmeier (Director, ASM Press) and the technical expertise and craft of our illustrator. The illustrations are an integral part of the exposition of the information and ideas discussed, and credit for their execution goes to the knowledge, insight, and artistic talent of Patrick Lane of ScEYEnce Studios. As noted in the figure legends, many of the figures could not have been completed without the help and generosity of our many colleagues who provided original images. Special thanks go to those who crafted figures tailored specifically to our needs or provided multiple pieces: Mark Andrake (Fox Chase Cancer Center), Edward Arnold (Rutgers University), Bruce Banfield (The University of Colorado), Christopher Basler and Peter Palese (Mount Sinai School of Medicine), Amy Brideau (Peregrine Pharmaceuticals), Roger Burnett (Wistar Institute), Rajiv Chopra and Stephen Harrison (Harvard University), Marie Chow (University of Arkansas Medical Center), Bob Craigie (NIDDK, National Institutes of Health), Richard Compans (Emory University), Friedrich Frischknecht (European Molecular Biology Laboratory), Wade Gibson (Johns Hopkins University School of Medicine), Ramón González (Universidad Autónoma del Estado de Morelos), David Knipe (Harvard Medical School), Thomas Leitner (Los Alamos National Laboratory), Maxine Linial (Fred Hutchinson Cancer Center), Pedro Lowenstein (University of California, Los Angeles), Paul Masters (New York State Department of Health), Rolf Menzel (National Institutes of Health), Thomas Mettenleiter (Federal Institute for Animal Diseases, Insel Reims, Germany), Heather Ongley and Michael Chapman (Oregon Health and Science University), B. V. Venkataram Prasad (Baylor College of Medicine), Botond Roska (Friedrich Miescher Institute, Basel, Switzerland), Michael Rossmann (Purdue University), Alasdair Steven (National Institutes of Health), Phoebe Stewart (Vanderbilt University), Wesley Sundquist (University of Utah), Jose Varghese (Commonwealth Scientific and Industrial Research Organization), Robert Webster (St. Jude's Children's Research Hospital), Thomas Wilk (European Molecular Biology Laboratory), Alexander Wlodawer (National Cancer Institute), and Li Wu (Medical College of Wisconsin).

The collaborative work undertaken to prepare the third edition was facilitated greatly by an authors' retreat at The Institute for Advanced Study, Princeton, NJ, in August 2007. We thank Arnold Levine for making the Biology Library available to us. ASM Press generously provided financial support for this retreat as well as for our many other meetings

We thank all those who guided and assisted in the preparation and production of the book: Jeff Holtmeier (Director, ASM Press) for steering us through the complexities inherent in a team effort, Ken April (Production Manager, ASM Press) for keeping us on track during production, and Susan Schmidler

(Susan Schmidler Graphic Design) for her elegant and creative designs for the layout and cover. We are also grateful for the expert secretarial and administrative support from Trisha Barney and Ellen Brindle-Clark (Princeton University) and Mary Estes and Rose Walsh (Fox Chase Cancer Center) that facilitated preparation of this text. Special thanks go to Ellen Brindle-Clark for obtaining the permissions required for many of the figures.

This often-consuming enterprise was made possible by the emotional, intellectual, and logistical support of our families, to whom the two volumes are dedicated.

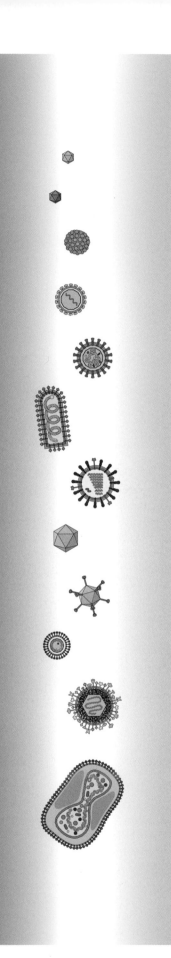

1

Introduction

A Brief History of Viral Pathogenesis

Microbes as Infectious Agents

The First Human Viruses

The Golden Age of Viral Pathogenesis

The New Millennium and Viral
Pathogenesis

Infection Basics

A Series of Unfortunate Events

Initiating an Infection

Viral Entry

Successful Infections Must Modulate
or Bypass Host Defenses

Viral Spread

Organ Invasion

Tropism

Perspectives

References

Infection of a Susceptible Host

Introduction

While the field of viral pathogenesis is almost 100 years old, the viral and host genes that control this process are only now being enumerated and analyzed. Even though many such genes have been identified, this information rarely provides more than a glimpse of the molecular mechanisms responsible. For example, knowing that a gene controls the spread of infection via the blood or the nervous system does not answer the question of how the gene product determines which route will be taken. Although it is difficult to obtain such mechanistic knowledge, many investigators believe that viral pathogenesis is the most exciting field in virology today, simply because more fundamental questions remain to be answered than in any other area. The challenge is to build on our understanding of viral replication that was established in Volume I of this book to provide a comprehensive molecular description of how viruses cause disease. One obvious consequence of such knowledge will be new approaches to preventing, treating, and curing viral disease. In this chapter, we begin our analysis of viral pathogenesis and control, by outlining some history and the basic principles of how a viral infection is established in a single host.

A Brief History of Viral Pathogenesis

Microbes as Infectious Agents

From the earliest times, poisonous air (miasma) was generally invoked to account for **epidemics** of contagious diseases, and there was little recognition of the differences among their causative agents. The association of particular microorganisms, initially bacteria, with specific diseases can be attributed to the ideas of the German physician Robert Koch. He developed and applied a set of criteria for identification of the agent responsible for a specific disease (a **pathogen**). These criteria, **Koch's postulates**, are still applied in the identification of pathogens that can be propagated in the laboratory and tested in an appropriate animal model. The postulates are as follows.

- The organism must be associated regularly with the disease and its characteristic lesions.
- The organism must be isolated from the diseased host and grown in culture.
- The disease must be reproduced when a pure preparation of the organism is introduced into a healthy, susceptible host.
- The same organism must be reisolated from the experimentally infected host.

Guided by these postulates, and the methods for the sterile culture and isolation of pure preparations of bacteria developed by Pasteur, researchers identified and classified many pathogenic bacteria (as well as yeasts and fungi) during the last part of the 19th century. From these beginnings, investigation into the causes of infectious disease was placed on a secure scientific foundation, the first step toward rational treatment and ultimately control. During the last decade of the 19th century, failures of the paradigm that bacterial or fungal agents are responsible for all diseases led to the identification of a new class of infectious agents—submicroscopic pathogens that came to be called **viruses** (see Volume I, Chapter 1).

The First Human Viruses

The first human virus to be identified, in 1901, was that responsible for yellow fever. The high lethality of this disease, the lack of effective treatments, and the complicated transmission and life cycle of the virus made this achievement all the more remarkable. Yellow fever is known to have been widespread in tropical countries since the 15th century. It was responsible for devastating epidemics associated with such high rates of mortality (e.g., 28% in the New Orleans epidemic of 1853) that normal life became impossible (Volume I, Appendix; also see Box 10.15 in this volume). However, this disease is not directly contagious, and an infectious agent could not be demonstrated in yellow fever patients. These puzzling properties defeated efforts to establish the origin of yellow fever until 1880, when the Cuban physician Carlos Juan Finlay proposed that a bloodsucking insect, most likely a mosquito, played a part in the transmission of the disease. A commission to study the etiology of yellow fever was established in 1899 by the U.S. Army under Colonel Walter Reed, in part because of the high incidence of the disease among soldiers who were occupying Cuba. Jesse Lazear, a member of Reed's commission, provided leadership and ultimately gave his life to demonstrate that mosquitoes transmitted yellow fever. Lazear was the first experimentally infected person who died from the disease. The members of this courageous team are depicted in a dramatic 1939 painting by Dean Cornwell (Fig. 1.1). The results of the Reed Commission's

Figure 1.1 *Conquerors of Yellow Fever.* This painting by Dean Cornwell (1939) depicts the experimental infection of a soldier volunteer with yellow fever virus, via transmission through mosquitoes. Standing (left to right) are Carlos Finlay, in a dark suit; Aristedes Agramonte, holding a hat; Jesse Lazear, applying a cage with mosquitos to the arm of the soldier; and Walter Reed, in a white uniform. Reproduced with the generous permission of Wyeth Laboratories.

study proved conclusively that mosquitoes are the vectors for this disease. In 1901, Reed and James Carroll injected diluted, filtered serum from an experimentally infected yellow fever patient into three nonimmune individuals. Two subsequently developed yellow fever. Reed and Carroll concluded that a filterable virus was the cause of the disease. In the same year, Juan Guiteras, a professor of pathology and tropical medicine at the University of Havana, attempted to produce immunity to yellow fever by exposing volunteers to mosquitoes carrying yellow fever virus. Of 19 volunteers, 8 contracted yellow fever and 3 died. One of the dead was Clara Louise Maass, a U.S. Army nurse from New Jersey. Yellow fever had been a constant scourge of Havana for 150 years, but the conclusions of Reed and his colleagues were a revelation. Rapid introduction of mosquito control by the mayor of Havana, William Gorgas, dramatically reduced the incidence of disease within a year. To this day, mosquito control remains an important method for reducing the incidence of yellow fever.

Other human viruses were identified during the early decades of the 20th century (Fig. 1.2). However, the pace of discovery was slow, not least because of the dangers and difficulties associated with experimental manipulation of human viruses so amply illustrated by the experience with yellow fever virus. Consequently, agents of some

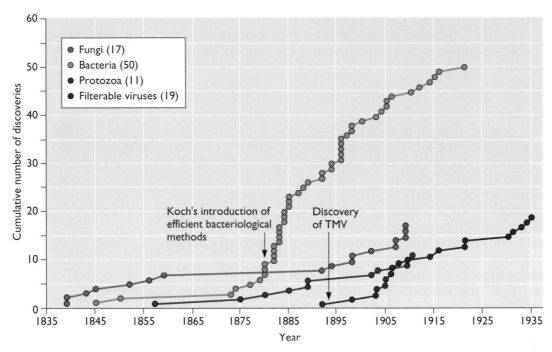

Figure 1.2 The pace of early discovery of new infectious agents. Koch's introduction of efficient bacteriological techniques spawned an explosion of new discoveries of bacterial agents in the early 1880s. Similarly, the discovery of filterable agents launched the field of virology in the early 1900s. Despite an early surge of virus discovery, only 19 distinct human viruses had been reported by 1935. TMV, tobacco mosaic virus. Adapted from K. L. Burdon, *Medical Microbiology* (MacMillan Co., New York, NY, 1939), with permission.

important human diseases were not identified for many years, and then only with some good luck. A classic case in point is the virus responsible for influenza, a name derived in the mid-1700s from Italian because of the belief that the disease resulted from the "influence" of miasma (bad air) and adverse astrological signs. The human disease is now thought to have arisen as a result of the transfer of virus among humans and livestock following domestication of animals about 10,000 years ago. Worldwide epidemics, called **pandemics**, of influenza have been documented in humans for well over 100 years. These pandemics were typically associated with mortality among the very young and the very old, but the 1918 to 1919 pandemic following the end of World War I was especially devastating. Over 40 million people died, more than were killed in the preceding war. Despite many efforts, a human influenza virus was not isolated until 1933. This virus was first identified by Wilson Smith, Christopher Andrewes, and Patrick Laidlaw only because they found a host suitable for its propagation. They infected ferrets with human throat washings and isolated the virus now known as influenza A virus. Ferrets may seem to be an exotic animal host, and, in fact, the success achieved with these animals was serendipitous: Laidlaw was using ferrets in studies of canine distemper

virus and therefore they were available in his laboratory. Subsequently, influenza A virus was shown to infect adult mice and chicken embryos. The latter proved to be an especially valuable host system, for vast quantities of the virus are produced in the allantoic sac. Indeed, chicken eggs are still used today to produce influenza vaccines.

The Golden Age of Viral Pathogenesis

While the first 50 years of virology saw ground-breaking work in describing the viral etiology of diseases, the study of viral pathogenesis was placed on a firm foundation during the next 25 years. In the first half of the 20th century, methods were developed to isolate and identify viruses and use embryonated chicken eggs in quantitative studies of viral infection. However, in the 1950s, several new technologies were developed to usher in what some now call the golden age of **viral pathogenesis**. Indeed, the basic principles of this field were established in the 25 years after 1950. The development of seemingly simple techniques was critical. For example, the plaque assay for lytic viruses and the focus-forming assay for viral transformation brought methods and concepts developed by phage biologists to analysis of animal virus infections. Synchronous infection and analysis of a single reproductive cycle, as pioneered by Max Delbrück and colleagues for bacteriophage, were now applicable to animal viruses. Tissue culture techniques and methods for centrifugation enabled early forays into biochemistry. The revolutionary use of fluorescent antibodies to identify viral proteins enabled scientists to observe viral infection and spread in cells and tissues. Electron microscopy became routine, and the amazing subcellular changes in virus-infected cells could be catalogued and analyzed. The development of methods to measure the immune response was probably the most significant accelerant of the golden age. In particular, the ability to identify antibodies and their activities transformed the study of viral pathogenesis. Complement fixation, neutralization, and hemagglutination inhibition are but three of the many innovative assays developed in the 1950s. With these techniques in hand, scientists performed the classic studies of the pathogenesis of poliovirus, mousepox virus, rabies virus, and lymphocytic choriomeningitis virus, which stand to this day as models of elegance and careful analyses.

The New Millennium and Viral Pathogenesis

The time from the mid-1970s to the end of the 20th century saw a true revolution in biology. Recombinant DNA technology enabled the cloning, sequencing, and manipulation of host and viral genomes. The polymerase chain reaction (PCR) was first among the many new offshoots of recombinant DNA technology that transformed biology. One can point to the Nobel Prizes of the 1980s and 1990s to see the transformative power of recombinant DNA technology: transgenic animals, gene targeting, and RNA interference are prime examples.

The concepts and methods of molecular and cell biology marked a transition from the descriptive phase of virology to the reductionist phase. Genomes were isolated, proteins were identified, functions were deduced by genetic and biochemical methods, and new models of disease were established. This reductionist approach was remarkably fruitful, not only for mechanistic understanding, but also for practical applications including the development of diagnostic reagents, antiviral drugs, and vaccines. As the 20th century came to a close, many scientists moved from reductionism to a more holistic philosophy of analysis. They embraced the concept of "systems biology," the idea that by using appropriate technologies, one could know all the molecules or reactions of a biological system, monitor them during an infection, and discover new mechanisms of host and viral biology missed by the classical "one gene at a time" methods. These ideas were first developed using DNA microarray technology, with which it is possible to measure the global transcriptional response of both host and viruses after infection of single cells and tissues (Box 1.1). The systematic compilation of data has revealed common and cell-specific responses to infection, as well as host-specific responses to a given infectious agent. Other massive data-gathering technologies enable similar profiling of proteins and even small-molecule metabolites of cells and tissues (Box 1.2). Computer programs capable of handling and integrating these massive databases are available to most scientists. Not surprisingly, this veritable gold mine of information can be overwhelming. It is likely that reductionist approaches will be required to test the many ideas that are emerging from systems biology. Furthermore, despite the new approaches, some fundamental tenets of viral pathogenesis remain unchallenged. These principles are considered in the next sections.

Infection Basics

A Series of Unfortunate Events

Infection of a susceptible host can be viewed as a sequence of individual events (Fig. 1.3A). The pathogenesis of mousepox is a classic example of such a sequence (Fig. 1.4). After local viral multiplication in the foot, the virus spreads via the bloodstream to the spleen, liver, and skin. The steps involved in infection have been defined for many viral pathogens by using animal models of infection. A contrary view is that infection comprises a series of random (stochastic) events that permit infection to bypass bottlenecks imposed by the host (Fig. 1.3B). The architecture of tissues and the immune system are examples of such bottlenecks. Selection of viruses that can bypass the bottleneck occurs because of the the diversity of viral populations (Chapter 10). These two views of infection are not as difficult to reconcile as they appear; the stochastic view does not exclude the idea that infection is a series of defined steps, but rather adds random elements to the outcome of each step.

Initiating an Infection

Three requirements must be satisfied to ensure successful infection in an individual host: sufficient **virions** must be available to initiate infection; the cells at the site of infection must be physically accessible to virions, **susceptible** (bear receptors for entry), and **permissive** (contain intracellular gene products needed for viral replication); and the local host antiviral defense systems must be absent or at least initially ineffective.

The first requirement erects a substantial barrier to any infection, and represents a significant weak link in the transmission of infection from host to host. Free virus particles face both a harsh environment and rapid dilution that can reduce their concentration. Viruses that are spread in contaminated water and sewage must be stable in the presence of osmotic shock, pH changes, and sunlight, and must not adsorb irreversibly to debris. Aerosol-dispersed virus particles must remain hydrated and highly concentrated to infect

EXPERIMENTS
*Determining the host response to infection by
transcriptional profiling*

Since the first paper in 1998 demonstrating the use of DNA microarrays to monitor the host transcriptional response to human cytomegalovirus infection, hundreds of similar and more advanced studies have been published covering most of the well-known viral infections.

The results of these studies are always complex. For example, it is difficult to determine if an RNA changes in abundance because of new synthesis or changes in stability or even if the transcript levels produce corresponding changes in protein abundance. However, it is possible to define some common host transcriptional responses to infection.

One such study provided a systematic cluster analysis of 32 studies that compared 785 experiments and 77 different host-pathogen interactions including a variety of viruses and bacteria. The authors found a cluster of 511 coregulated RNAs that they designated the "common host response." The authors postulated that some of these gene products constituted a general alarm signal for infection as well as the common intrinsic, cell-autonomous defenses. They also suggest that pathogens can modulate these responses to enhance virulence.

The products of these 511 genes fall into five general groups:

1. Inflammation mediators (proinflammatory cytokines, chemokines, prostaglandin synthesis)
2. Interferon stimulated
3. Activators of intrinsic and immune defenses (transcription proteins such as Nf-κB, activators of cytoplasmic nucleic acid detectors)
4. Negative regulators of cellular and immune defenses (IκBα, IκBε, various kinases, antiapoptosis genes)
5. Host response (lymphocyte activation, cell adhesion, tissue invasion)

However, despite being clustered by the computer, these responses are cell type and pathogen specific, and vary in degree and temporal appearance. The challenge is to understand the relationship of transcriptional changes in cultured cells to those found in infected tissues and hosts.

Jenner, R., and R. Young. 2005. Insights into host responses against pathogens from transcriptional profiling. *Nat. Rev. Microbiol.* **3**:281–294.

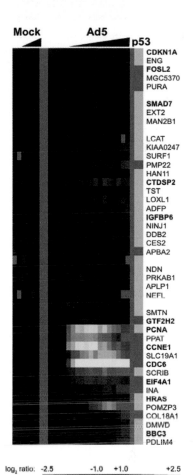

Changes in expression of cellular genes during adenovirus type 5 infection. Human foreskin fibroblasts were infected or mock infected, and RNA was prepared at different times after infection and analyzed by microarray. Shown are the results for ~50 genes that are targeted by p53. Blue indicates a reduction in RNA; yellow indicates an increase. The column labeled p53 summarizes genes that are transcriptionally activated (yellow) or repressed (blue) by the p53 protein. Ramps above columns indicate time after infection. Reprinted from D. L. Miller et al., *Genome Biol.* **8**:R58, 2007, with permission.

the next host. Viruses that are spread in this way do best in populations in which individuals are in close contact. In contrast, viruses that are spread via biting insects, contact with mucosal surfaces, or other means of direct contact, including contaminated needles, have little environmental exposure.

Even if one virus particle survives the passage from one host to another, infection may fail simply because the concentration is not sufficient. In principle, a single virion should be able to initiate an infection, but host physical and immune defenses, coupled with the complexity

DISCUSSION
Virus infection markedly affects cellular metabolic pathways: a genomics and metabolomics approach

The biochemical infrastructure of host cells is essential for virus propagation. An underlying, but poorly understood, fact is that this infrastructure includes the cellular metabolic machinery that provides the energy and building blocks necessary for their replication. Liquid chromatography-mass spectrometry has been used to quantitate directly the concentrations of a large number of metabolic compounds (energy molecules and biochemical building blocks) during human cytomegalovirus infection of cultured human cells. In addition, changes in cellular RNA were measured by microarray analysis in parallel. After infection, the concentration of many metabolites increased dramatically, far more than is seen when cells switch from resting to growing states. Often the change

in metabolite levels coincided with an apparent increase in the level of RNA associated with the production of that metabolite. One striking conclusion was that virus-infected cells produced a characteristic metabolic program.

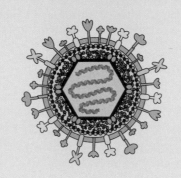

We can expect more studies like this one to provide a comprehensive characterization of the metabolic environment of virus-infected cells. It is likely that these substantial metabolic changes will affect the global host response to infection. Moreover, it may be that these changes are required for virus reproduction and spread and may therefore be targets for antiviral therapies.

Munger J., S. Bajad, H. Coller, T. Shenk, and J. Rabinowitz. 2006. Dynamics of the cellular metabolome during human cytomegalovirus infection. *PLoS Pathog.* **2:**1165–1175.

of the infection process itself, demand the presence of many particles. The number of particles required to initiate and maintain an infection depends on the particular virus, the site of infection, and the physiology and age of the host. However, some basic facts help to guide us.

Statistical analysis of infections in cultured cells demonstrates that on average a single virus particle can initiate an infection, but that many perfectly competent virions fail to do so. Such failure can be explained in part by the complexity of the infectious cycle: there are many distinct reactions, and the probability of a single virus particle

Figure 1.3 Views of viral pathogenesis. (A) Infection viewed as a series of steps with a predictable outcome. **(B)** Infection seen as a series of stochastic events. Viruses are selected that can pass through bottlenecks in cells, organs, or hosts. Adapted from H. W. Virgin, *Nat. Immunol.* **8:**1143–1147, 2007, with permission.

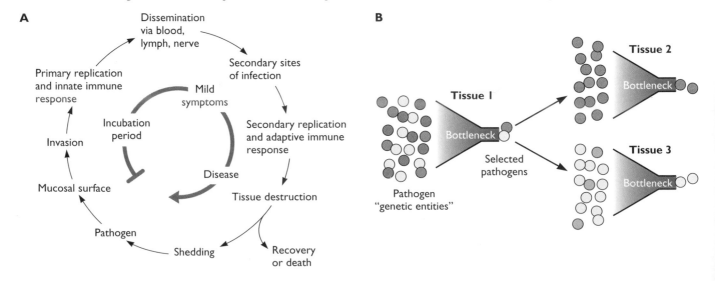

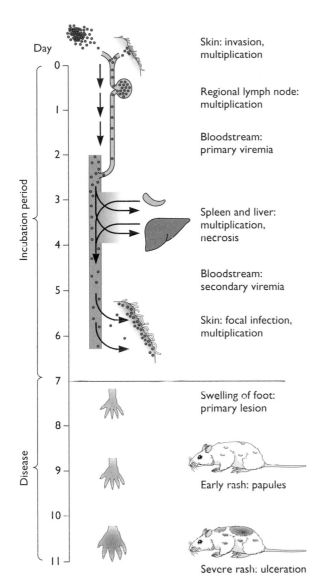

Day

Skin: invasion, multiplication

Regional lymph node: multiplication

Bloodstream: primary viremia

Spleen and liver: multiplication, necrosis

Bloodstream: secondary viremia

Skin: focal infection, multiplication

Swelling of foot: primary lesion

Early rash: papules

Severe rash: ulceration

Incubation period

Disease

Figure 1.4 Pathogenesis of mousepox. Sequence of events in the pathogenesis of mousepox after inoculation of virus into the footpad. The figure is based on the classic studies of Frank Fenner, which were the first to demonstrate how disseminated viral infections develop from local multiplication to primary and secondary viremia. In the case of mousepox, after local multiplication in the foot, the host response leads to swelling at the site of inoculation; after viremia, the host response to replication in the skin results in a rash. Adapted from F. Fenner et al., *The Biology of Animal Viruses* (Academic Press, New York, NY, 1974), with permission.

completing any one is not 100%. For example, there are many potentially nonproductive interactions of virus particles with debris and extracellular material during their initial encounter with the cell surface. Even if a virus attaches successfully to a permissive cell, it may be delivered to a digestive lysosome upon entry. Many of these

false starts or inappropriate interactions are irreversible, aborting infection by the virion.

In addition, populations of viruses often contain particles that are not capable of completing an infectious cycle. For example, defective particles can be produced by mistakes during virus replication or from interaction with inhibitory compounds in the environment. In the laboratory, a quantitative measure of the proportion of infectious viruses is the particle–to–**plaque-forming-unit** (PFU) ratio. As described in Volume I, Chapter 2, the number of physical particles in a given preparation are counted, usually with an electron microscope, and compared with the number of infectious units, or PFU, per unit volume. This ratio is a useful indicator of the quality of a virus preparation, as it should be constant for a given virus prepared by identical or comparable procedures.

Viral Entry

In general, virions must first enter cells at a body surface. Common sites of entry include the mucosal linings of the respiratory, alimentary, and urogenital tracts, the outer surface of the eyes (conjunctival membranes or cornea), and the skin (Fig. 1.5).

Respiratory Tract

Probably the most common route of viral entry is through the respiratory tract. In a human lung there are about 300 million terminal sacs, called alveoli, which function in gaseous exchange between inspired air and the blood. To accomplish this function, each sac is in close contact with capillary and lymphatic vessels. The combined absorptive area of the human lung is almost 140 m². Humans have a resting ventilation rate of 6 liters of air per min. Consequently, large numbers of foreign particles and aerosolized droplets are introduced into the lungs with every breath. Many of these particles and droplets contain virions. Fortunately, there are numerous host defense mechanisms to block respiratory tract infection. Mechanical barriers play a significant role in antiviral defense. For example, the tract is lined with a mucociliary blanket consisting of ciliated cells, mucus-secreting goblet cells, and subepithelial mucus-secreting glands (Fig. 1.6). Foreign particles deposited in the nasal cavity or upper respiratory tract are trapped in mucus, carried to the back of the throat, and swallowed. In the lower respiratory tract, particles trapped in mucus are brought up from the lungs to the throat by ciliary action. The lowest portions of the tract, the alveoli, lack cilia or mucus, but macrophages lining the alveoli are responsible for ingesting and destroying particles. Other cellular and humoral immune responses also intervene.

Many viruses enter the respiratory tract in the form of aerosolized droplets expelled by an infected individual by

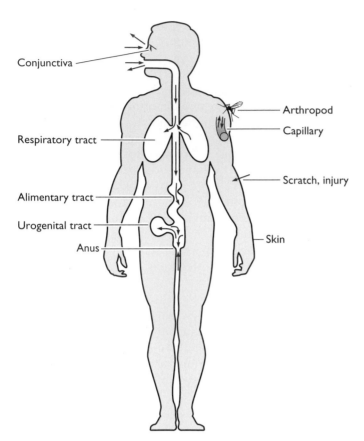

Figure 1.5 Sites of viral entry into the host. A representation of the human host is shown, with sites of virus entry and shedding indicated. The body is covered with skin, which has a relatively impermeable (dead) outer layer. However, there are accessible layers of living cells to absorb food, exchange gases, and release urine and other fluids. These layers offer easier pathways for the entry of viruses than the skin. Virions can be introduced through the skin by a scratch or injury, a vector bite, or inoculation with a needle.

coughing or sneezing (Table 1.1). Infection can also spread through contact with saliva from an infected individual. Larger virus-containing droplets are deposited in the nose, while smaller droplets find their way into the airways or the alveoli. To infect the respiratory tract successfully, viruses must not be swept away by mucus, neutralized by antibody, or destroyed by alveolar macrophages.

Alimentary Tract

The alimentary tract is a common route of infection and dispersal (Table 1.1). This tube, which connects the oral cavity to the anus, is always in motion. Eating, drinking, and some social activities routinely place viruses in the alimentary tract. It mixes, digests, and absorbs food, providing a good opportunity for viruses to encounter a susceptible cell and to interact with cells of the circulatory, lymphatic, and immune systems. However, it is an extremely hostile environment for virions. The stomach is acidic, the intestine is alkaline, digestive enzymes and bile detergents abound, mucus lines the epithelium, and the lumenal surfaces of intestines carry antibodies and phagocytic cells.

Virions that infect by the intestinal route must, at a minimum, be resistant to extremes of pH, proteases, and bile detergents. Indeed, virions that lack these features are destroyed when exposed to the alimentary tract, and must infect at other sites. The family *Picornaviridae* comprises both acid-labile viruses (e.g., rhinoviruses) and acid-resistant viruses (e.g., poliovirus) (Box 1.3). Rhinoviruses are respiratory pathogens and cannot infect the upper intestine. They spread in a population as inhaled aerosols. In contrast, poliovirus can survive ingestion and establish an infection of the upper intestine, and spreads by the fecal-oral route. The hostile environment of the alimentary tract actually facilitates infection by some viruses. For example, reovirus particles are converted by host proteases in the intestinal lumen into infectious subviral particles, the forms that subsequently infect intestinal cells. As might be expected, most enveloped viruses do not initiate infection in the alimentary tract, because viral envelopes are susceptible to dissociation by detergents such as bile salts. Enteric coronaviruses are notable exceptions, but it is not known why these enveloped viruses can withstand the harsh conditions in the alimentary tract.

Nearly the entire intestinal surface is covered with columnar villous epithelial cells with apical surfaces that are densely packed with microvilli (Fig. 1.7). This brush border, together with a surface coat of glycoproteins and glycolipids, and the overlying mucus layer, is permeable to electrolytes and nutrients but presents a formidable barrier to microorganisms. Nevertheless, viruses such as enteric adenoviruses and Norwalk virus, a calicivirus, replicate extensively in intestinal epithelial cells. The mechanisms by which they bypass physical barriers and enter susceptible cells are beginning to be understood (Volume I, Chapter 5). Scattered throughout the intestinal mucosa are lymphoid follicles that are covered on the lumenal side with a specialized follicle-associated epithelium consisting mainly of columnar absorptive cells and M (membranous epithelial) cells. The M cell cytoplasm is very thin, resulting in a membrane-like bridge that separates the lumen from the subepithelial space. As discussed in Chapter 4, M cells ingest and deliver antigens to the underlying lymphoid tissue by **transcytosis**. In this process, material taken up on the lumenal side of the M cell traverses the cytoplasm virtually intact, and is delivered to the underlying basal membranes and extracellular space (Fig. 1.7). It is thought that M cell transcytosis provides the mechanism by which some enteric viruses gain entry

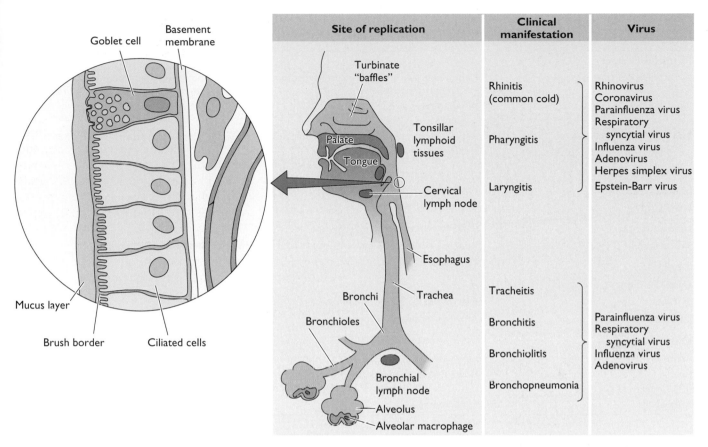

Figure 1.6 Sites of viral entry in the respiratory tract. (Left) A detailed view of the respiratory epithelium. A layer of mucus, produced by goblet cells, is a formidable barrier to virion attachment. Virions that pass through this layer may multiply in the ciliated cells or pass between them, reaching another physical barrier, the basement membrane. Beyond this extracellular matrix are tissue fluids from which particles may be taken into lymphatic capillaries and reach the blood. Local macrophages patrol the tissue fluids in search of foreign particles. Adapted from C. A. Mims et al., *Mims' Pathogenesis of Infectious Disease* (Academic Press, Orlando, FL, 1995), with permission. **(Right)** Viruses that replicate at different levels of the respiratory tract, with the associated clinical syndromes.

to deeper tissues of the host from the intestinal lumen. After crossing the mucosal epithelium in this manner, a virus particle could enter lymphatic vessels and capillaries of the circulatory system, facilitating spread within the host. A particularly well studied example is transcytosis of reovirus. After attaching to the M cell surface, reovirus subviral particles are transported to cells underlying the lymphoid follicle, where they replicate and spread to other tissues. Rather than spread by transcytosis across the M cell, some viruses actively replicate in them and do not spread to underlying tissues. For example, infection by human rotavirus and the coronavirus transmissible gastroenteritis virus destroys M cells, resulting in mucosal inflammation and diarrhea.

It is possible for virions to enter the body through the lower gastrointestinal tract without passing through the upper tract and its defensive barriers. For example, human immunodeficiency virus can be introduced into the body by anal intercourse. As M cells abound in the lower colon, these cells are likely to provide a portal of entry for this virus into susceptible lymphocytes in the underlying lymphoid follicles. Once in the follicle, the virus can infect migratory lymphoid cells and spread throughout the body.

Urogenital Tract

Some viruses enter the urogenital tract as a result of sexual activities (Table 1.1). The urogenital tract is well protected by physical barriers, including mucus and low pH (in the case of the vagina). Normal sexual activity can result in minute tears or abrasions in the vaginal epithelium or the urethra, allowing virions to enter. Some viruses infect the epithelium and produce local lesions (certain human papillomaviruses, which cause genital warts). Other viruses gain access to cells in the underlying tissues and infect cells of the

Table 1.1   Different routes of viral entry into the host

Location	Virus(es)
Respiratory tract	
Localized upper tract	Rhinovirus; coxsackievirus; coronavirus; arenaviruses; hantavirus; parainfluenza virus types 1–4; respiratory syncytial virus; influenza A and B viruses; human adenovirus types 1–7, 14, 21
Localized lower tract	Respiratory syncytial virus; parainfluenza virus types 1–3; influenza A and B viruses; human adenovirus types 1–7, 14, 21
Entry via respiratory tract followed by systemic spread	Rubella virus, arenaviruses, hantavirus, mumps virus, measles virus, varicella-zoster virus, poxviruses
Alimentary tract	
Systemic	Enterovirus, reovirus, adenovirus
Localized	Coronavirus, rotavirus
Urogenital tract	
Systemic	Human immunodeficiency virus type 1, hepatitis B virus, herpes simplex virus
Localized	Papillomavirus
Eyes	
Systemic	Enterovirus 70, herpes simplex virus
Localized	Adenovirus types 8, 22
Skin	
Arthropod bite	Bunyavirus, flavivirus, poxvirus, reovirus, togavirus
Needle puncture, sexual contact	Hepatitis C and D viruses, cytomegalovirus, Epstein-Barr virus, hepatitis B virus, human immunodeficiency virus, papillomavirus (localized)
Animal bite	Rhabdovirus

immune system (human immunodeficiency virus type 1), or sensory and autonomic neurons (herpes simplex virus). Infection by these two viruses invariably spreads from the urogenital tract to establish lifelong persistent or latent infections, respectively.

Eyes

The epithelium covering the exposed part of the sclera (the outer fibrocollagenous coat of the globe of the eye) and the inner surfaces of the eyelids (conjunctivae) is the route of entry for several viruses. Every few seconds the eyelid

BOX 1.3

DISCUSSION
Why is rhinovirus but not poliovirus sensitive to low pH?

Low pH induces irreversible disassembly of the rhinovirus capsid, but the changes in the poliovirus capsid that occur under acidic conditions are fully reversible. Although high-resolution crystallographic structures of rhinovirus and poliovirus virions have been determined, they do not provide information about the basis for the difference in pH sensitivity. Acid-resistant mutants of rhinovirus contain amino acid changes near the fivefold axes of symmetry. Such mutants behave similarly to poliovirus, in that low pH induces reversible conformational changes in the capsid. The

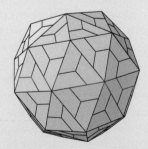

mutations responsible for acid resistance probably result in stabilization of the particle, so that it does not dissociate at low

pH. Such mutations are not selected during natural infections, because they are not necessary for the virus to replicate in the respiratory tract.

Giranda, V. L., B. A. Heinz, M. A. Oliveira, I. Minor, K. H. Kim, P. R. Kolatkar, M. G. Rossmann, and R. R. Rueckert. 1992. Acid-induced structural changes in human rhinovirus 14: possible role in uncoating. *Proc. Natl. Acad. Sci. USA* **89:**10213–10217.

Skern, T., H. Torgersen, H. Auer, E. Kuechler, and D. Blaas. 1991. Human rhinovirus mutants resistant to low pH. *Virology* **183:**757–763.

A

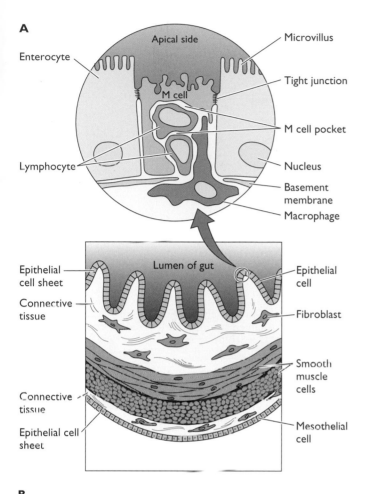

B

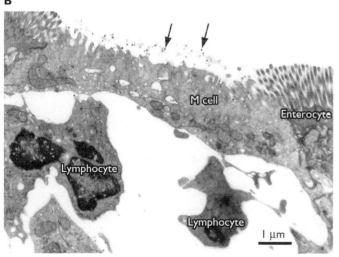

Figure 1.7 Viral entry through M cells in the intestine. **(A)** Schematic drawing of the intestinal wall. This organ is made up of epithelial, connective, and muscle tissues. Each is formed by different cell types that are organized by cell-cell adhesion within an extracellular matrix. A section of the epithelium has been enlarged, and a typical M cell is shown, surrounded by two enterocytes. Lymphocytes and macrophages move in and out of

passes over the sclera, bathing it in secretions that wash away foreign particles. There is usually little opportunity for viral infection of the eye, unless it is injured by abrasion. Direct inoculation into the eye may occur during ophthalmologic procedures or from environmental contamination, such as improperly sanitized swimming pools and hot tubs. In most cases, replication is localized and results in inflammation of the conjunctiva, a condition called conjunctivitis. Systemic spread of the virus from the eye is rare, although it does occur; paralytic illness after enterovirus 70 conjunctivitis is one example. Herpesviruses, in particular herpes simplex virus type 1, can also infect the cornea, mainly at the site of a scratch or other injury. Such an infection may lead to immune destruction of the cornea and blindness. Inevitably, herpes simplex virus infection of the cornea is followed by spread of the virus to sensory neurons, and then to neuronal cell bodies in the sensory ganglia, where a latent infection is established.

Skin

The skin protects the body yet provides sensory contact with the environment. The external surface of the skin, or **epidermis**, is composed of several layers, including a basal germinal layer of proliferating cells, a granular layer of dying cells, and an outer layer of dead, keratinized cells (Fig. 1.8). The skin of most animals is an effective barrier against viral infections, as the dead outer layer cannot support viral replication. Replication is usually limited to the site of entry, because the epidermis is devoid of blood or lymphatic vessels that could provide pathways for further spread. Entry through this organ occurs primarily when its integrity is breached by breaks or punctures. Examples of viruses that can gain entry in this manner are some human papillomaviruses and certain poxviruses (e.g., myxoma virus) that are transmitted mechanically by insect vectors such as arthropods (Fig. 1.5; Table 1.1). However, the epidermis is supported by the highly vascularized **dermis**. Other viruses can gain entry to the dermis through the bites of arthropod vectors such as mosquitoes, mites, ticks, and sand flies. Even deeper inoculation, into the tissue and muscle below the dermis, can occur by hypodermic

invaginations on the basolateral side of the M cell. Adapted from A. Siebers and B. B. Finlay, *Trends Microbiol.* **4:**22–28, 1996, and B. Alberts et al., *Molecular Biology of the Cell* (Garland Publishing, New York, NY, 1994), with permission. **(B)** Reovirus attached to, and within vesicles of, an M cell. An electron micrograph of the gut epithelium shows reovirus (small black arrows) attached to the surface of an M cell and also within intracellular vesicles. Reprinted from J. L. Wolf and W. A. Bye, *Annu. Rev. Med.* **35:**95–112, 1984, with permission. Photo courtesy of R. Finberg, Harvard Medical School.

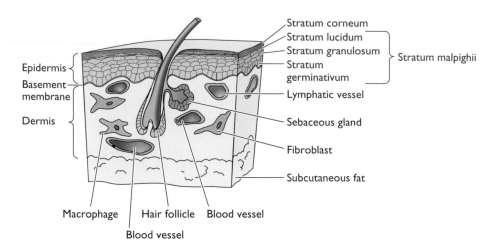

Figure 1.8 Schematic diagram of the skin. The epidermis consists of a layer of dead, keratinized cells (stratum corneum) over the stratum malpighii. The latter may have two layers of cells with increasing numbers of keratin granules (stratum granulosum and stratum lucidum) and a basal layer of dividing epidermal cells (stratum germinativum). Below this is the basement membrane. The dermis contains blood vessels, lymphatic vessels, fibroblasts, and macrophages. A hair follicle and a sebaceous gland are shown. Adapted from F. Fenner et al., *The Biology of Animal Viruses* (Academic Press, New York, NY, 1974), with permission.

needle punctures, body piercing or tattooing, or sexual contact when body fluids are mingled through skin abrasions or ulcerations (Table 1.1). Animal bites can introduce rabies virus into tissue and muscle rich with nerve endings, through which virions can invade motor neurons. In contrast to the strictly localized replication of viruses in the epidermis, viruses that initiate infection in dermal or subdermal tissues can reach nearby blood vessels, lymphatic tissues, and cells of the nervous system. As a consequence, they may spread to other sites in the body.

Successful Infections Must Modulate or Bypass Host Defenses

To initiate any infection, there must be viral mechanisms for countering the host defenses: these mechanisms may be active or passive, or a combination of the two. The actual pattern of infection that ensues is determined by the kinetics of virus replication in the face of host defenses. The interplay between virus offense and host defense is dynamic: there will be different consequences of a fast-replicating or a slow-replicating virus, in combination with strong or weak host defenses. We discuss many such mechanisms in Chapters 3 and 4.

Some host defenses may be overcome passively by an overwhelming inoculum of virus particles. Single droplets found in the aerosol produced by sneezing can contain as many as 100 million rhinovirus particles; a similarly large number of hepatitis B virus particles can be found in 1 ml of blood from a patient with hepatitis. At these concentrations, it may be impossible for physical and intrinsic

Figure 1.9 Polarized release of viruses from cultured epithelial cells visualized by electron microscopy. (A) Influenza virus released by budding from the apical surface of canine kidney cells. **(B)** Budding of measles virus on the apical surface of human colon carcinoma cells. **(C)** Release of vesicular stomatitis virus at the basal surface of canine kidney cells. Arrows indicate virus particles. Magnification, ×324,000. Reprinted from D. M. Blau and R. W. Compans, *Semin. Virol.* **7:**245–253, 1996, with permission. Courtesy of D. M. Blau and R. W. Compans, Emory University School of Medicine.

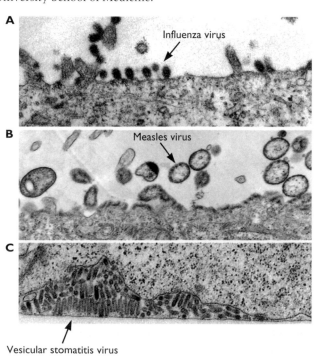

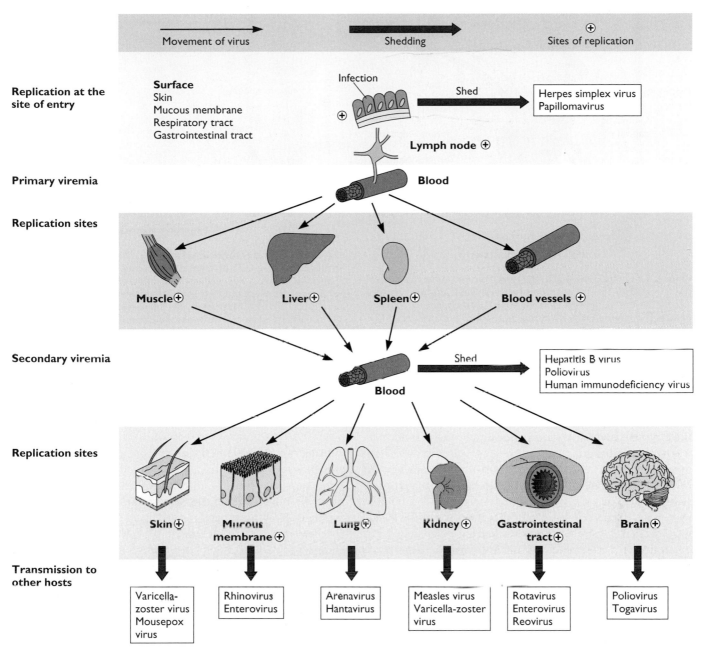

Figure 1.10 Entry, dissemination, and shedding of blood-borne viruses. Shown are the target organs for some viruses that enter at epithelial surfaces and spread via the blood. The sites of virus shedding (red arrows), which may lead to transmission to other hosts, are shown. Adapted from N. Nathanson (ed.), *Viral Pathogenesis* (Lippincott-Raven Publishers, Philadelphia, PA, 1997), with permission.

defenses to block every infecting virus particle. Free passage of virus through the primary physical barriers of skin and mucus layers, made possible by a cut, abrasion, or needle stick, may also allow passive evasion of defenses. Some viruses, including herpesviruses, papillomaviruses, and rabies virus, establish unique infections because they infect organs or cells not exposed to antibodies or cytotoxic lymphocytes. A more egregious breach of both primary and secondary defenses may occur during organ transplantation, which places potentially infected tissues in direct contact with potentially susceptible cells in immunosuppressed patients.

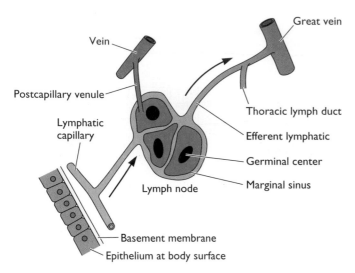

Figure 1.11 The lymphatic system. Lymphocytes flow from the blood into the lymph node through postcapillary venules. Adapted from C. A. Mims et al., *Mims' Pathogenesis of Infectious Disease* (Academic Press, Orlando, FL, 1995), with permission.

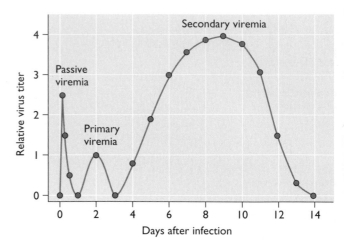

Figure 1.12 Characteristics of viremia. The graph was produced using data from different viral infections. For passive viremia, La Crosse virus, a bunyavirus, was injected into weanling mice, and virus titers in plasma, brain, and muscle were determined at different times afterward. No virus can be detected in the blood after 1 day. For primary viremia, ectromelia virus was inoculated into the footpad of mice; after local multiplication, the virus enters the blood. For secondary viremia, viral progeny produced by multiplication of the ectromelia virus in the target organs were counted. Virus reaches these organs during primary viremia Adapted from N. Nathanson (ed.), *Viral Pathogenesis and Immunity* (Academic Press, London, United Kingdom, 2007), with permission.

Viral Spread

Following replication at the site of entry, virus particles can remain localized or can spread to other tissues (Table 1.1). Local infections in the epithelium are usually contained by the physical constraints of the tissue and brought under control by the intrinsic and immune defenses discussed in Chapters 3 and 4. In general, an infection that spreads beyond the primary site of infection is said to be **disseminated**. If many organs become infected, the infection is described as **systemic**. Spread of an infection beyond the primary site requires that physical and immune barriers be breached. For example, after crossing an epithelium, virus particles reach the basement membrane (Fig. 1.7). The integrity of that structure may be compromised by epithelial cell destruction and inflammation. Below the basement membrane are subepithelial tissues, where virions encounter tissue fluids, the lymphatic system, and phagocytes. All three play significant roles in clearing foreign particles, but also may disseminate infectious virus from the primary site of infection.

One important mechanism for avoiding local host defenses and facilitating spread within the body is the directional release of virions from polarized cells at a mucosal surface (Volume I, Chapter 12). Virions can be released from the apical surface, from the basolateral surface, or from both (Fig. 1.9). After replication, particles released from the apical surface are back where they started, "outside" the host. Such directional release facilitates the dispersal of many newly replicated enteric viruses in the feces (e.g., poliovirus). In general, virions released at apical membranes establish a localized or limited infection.

In this case, local lateral spread from cell to cell occurs in the infected epithelium, but the underlying lymphatic and circulatory vessels are rarely invaded. In contrast, virus particles released from the basolateral surfaces of polarized epithelial cells have been moved away from the defenses of the lumenal surface. Release of particles at the basal membrane provides access to the underlying tissues and may facilitate systemic spread. Directional release is therefore a major determinant of the infection pattern.

The consequences of directional release are striking. Sendai virus, which is normally released from the apical surfaces of polarized epithelial cells, causes only a localized infection of the respiratory tract. In stark contrast, a mutant virus, which is released from both apical and basal surfaces, is disseminated.

Hematogenous Spread

Virions that escape from local defenses to produce a disseminated infection often do so by entering the bloodstream (**hematogenous spread**). Virus particles may enter the blood directly through capillaries, by replicating in endothelial cells, or through inoculation by a vector bite. Once in the blood, virions have access to almost every tissue in the host (Fig. 1.10). Hematogenous spread begins when newly replicated particles produced at the entry site

BOX 1.4

TERMINOLOGY
Infection of the nervous system: definitions and distinctions

A **neurotropic virus** can infect neurons; infection may occur by neural or hematogenous spread initiating from a peripheral site.

A **neuroinvasive virus** can enter the central nervous system (spinal cord and brain) after infection of a peripheral site.

A **neurovirulent** virus can cause disease of nervous tissue, manifested by neurological symptoms and often death.

Examples:

Herpes simplex virus has low neuroinvasiveness of the central nervous system, but high neurovirulence. It always enters the peripheral nervous system but rarely enters the central nervous system. When it does, the consequences are almost always severe, often fatal.

Mumps virus has high neuroinvasiveness but low neurovirulence. Most infections lead to invasion of the central nervous system, but neurological disease is mild.

Rabies virus has high neuroinvasiveness and high neurovirulence. It readily infects the peripheral nervous system and spreads to the central nervous system with 100% lethality unless antiviral therapy is administered shortly after infection.

Figure 1.13 Possible pathways for the spread of infection in nerves. Virus particles may enter the sensory or motor neuron endings. They may be transported within axons, in which case viruses taken up at sensory endings reach dorsal root ganglion cells. Those taken up at motor endings reach motor neurons. Viruses may also travel in the endoneural space, in perineural lymphatics, or in infected Schwann cells. Directional transport of virus particles inside the sensory neuron is defined as anterograde [movement from the (−) to the (+) ends of microtubules] or retrograde [movement from the (+) to the (−) ends of microtubules]. Adapted from R. T. Johnson, *Viral Infections of the Nervous System* (Raven Press, New York, NY, 1982), with permission.

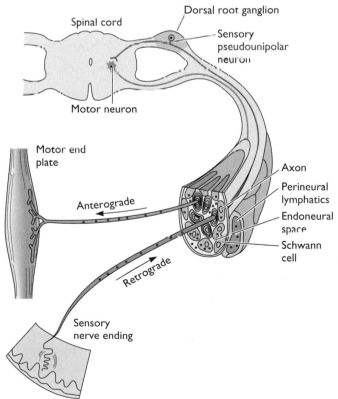

arc released into the extracellular fluids, which can be taken up by the local lymphatic vascular system (Fig. 1.11). Lymphatic capillaries are considerably more permeable than circulatory system capillaries, facilitating virus entry. As lymphatic vessels ultimately drain into the circulatory system, virus particles in lymph have free access to the bloodstream. In the lymphatic system, virions pass through lymph nodes, where they encounter migratory cells of the immune system. Viral pathogenesis resulting from the direct infection of immune system cells is initiated in this fashion (e.g., human immunodeficiency virus and measles virus). Some viruses replicate in the infected lymphoid cells, and progeny are released into the blood plasma. The infected lymphoid cell may also migrate away from the local lymph node to distant parts of the circulatory system. There may be little viral replication while the cell is in the bloodstream. However, when the infected cell receives chemotactic signals that direct it to enter other tissues, new virus particles may be produced in the activated cell.

The term **viremia** describes the presence of infectious virus particles in the blood. These virions may be free in the blood or contained within infected cells such as lymphocytes. **Active viremia** is produced by replication, while **passive viremia** results when particles are introduced into the blood without viral replication at the site of entry (injection of a virion suspension into a vein) (Fig. 1.12). Progeny virions released into the blood after initial replication at the site of entry constitute **primary viremia**. The concentration of particles during primary viremia is usually low. However, the subsequent disseminated infections that result are often extensive, releasing considerably more virions. Such delayed appearance of a high concentration of infectious virus in the blood is termed **secondary viremia**. The two phases of viremia were first described in classic studies of mousepox (Fig. 1.4).

BOX 1.5

TERMINOLOGY
Which direction, anterograde or retrograde?

Those who study virus spread in the nervous system often use the words **retrograde** and **anterograde** to describe direction. Unfortunately, confusion arises because the terms can be used to describe directional movement of virus particles inside a cell as well as spread between synaptically connected neurons. Spread from the primary neuron to the second-order neuron in the direction of the nerve impulse is said to be anterograde spread (see figure). Spread in the opposite direction is said to be retrograde. Spread inside a neuron is defined by microtubule polarity. Transport on microtubules from (−) to (+) ends is said to be anterograde, while transport on microtubules from (+) to (−) ends is said to be retrograde.

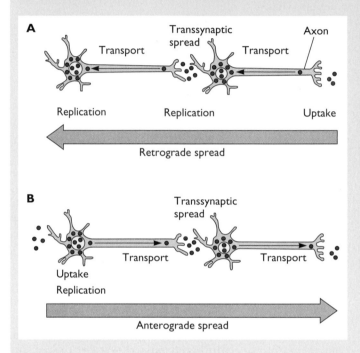

Retrograde and anterograde spread of virus in nerves.
(A) Retrograde spread of infection. Virus invades at axon terminals and spreads to the cell body, where replication ensues. Newly replicated virus particles spread to a neuron at sites of synaptic contact. Particles enter the axon terminal of the second neuron to initiate a second cycle of replication and spread. **(B)** Anterograde spread of infection. Virus invades at dendrites or cell bodies and replicates. Virus particles then spread to axon terminals, where virions cross synaptic contacts to invade dendrites or cell bodies of the second neuron.

The concentration of virions in blood is determined by the rates of their synthesis in permissive tissues, and by how quickly they are released into, and removed from, the blood. Circulating particles are removed by phagocytic cells of the reticuloendothelial system in the liver, lungs, spleen, and lymph nodes. When serum antibodies appear, virions in the blood may bind these antibodies and be neutralized, as described in Chapter 4. Formation of a complex of antibodies and virus particles facilitates uptake by Fc receptors carried by macrophages lining the circulatory vessels. Virion-antibody complexes can be sequestered in significant quantities in the kidneys, spleen, and liver. The time a virion is present in the blood usually varies from 1 to 60 min, depending on such parameters as the physiology of the host (e.g., age and health) and the size of the virus (large particles are cleared more rapidly than small particles). Some viral infections are noteworthy for the long-lasting presence of infectious particles in the blood.

Hosts infected with hepatitis B and C viruses or lymphocytic choriomeningitis virus may have viremia that lasts for years.

Viremia obviously is of diagnostic value and can be used to monitor the course of infection, but it also presents practical problems. Infections can be spread inadvertently in the population, when pooled blood from thousands of individuals is used directly for therapeutic purposes (transfusions) or as a source of therapeutic proteins (e.g., gamma globulin or blood-clotting factors). We have learned from unfortunate experience that hepatitis and acquired immunodeficiency syndrome can be spread by contaminated blood and blood products. Obviously, sensitive detection methods and stringent purification protocols are required to protect those who use and dispense these products. Frequently, it may be difficult, or technically impossible, to quantify infectious particles in the blood, as is currently the case for hepatitis B virus. Accordingly, the presence

BOX
1.6

BACKGROUND
*The path rarely taken: direct entry into the central
nervous system by olfactory routes*

Olfactory neurons are unusual in that
their cell bodies are present in the olfactory epithelia and their axon termini are
in synaptic contact with olfactory bulb
neurons. Literally, these direct conduits

to the brain project from cells that are in
direct contact with the environment. The
olfactory nerve fiber passes through the
skull via an opening called the arachnoid.
Remarkably, few viral infections enter the

brain by the olfactory route, despite significant replication of many in the nasopharyngeal cavity.

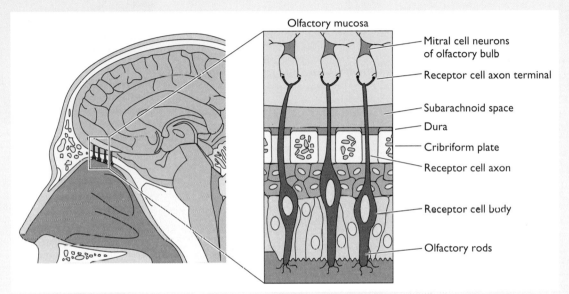

Adapted from R. T. Johnson, *Viral Infections of the Nervous System* (Raven Press, New York, NY, 1982), with permission.

of characteristic viral proteins provides surrogate markers
for viremia.

Neural Spread

Many viruses spread from the primary site of infection
by entering local nerve endings. In some cases, neuronal
spread is the definitive characteristic of their pathogenesis (e.g., rabies virus and alphaherpesviruses). For others, invasion of the nervous system is a rare but important
diversion from their normal site of replication (e.g., poliovirus and reovirus). Mumps virus, human immunodeficiency virus, and measles virus replicate in the brain, but
spread by the hematogenous route. The molecular mechanisms that dictate spread by neural or hematogenous pathways are generally not well understood. While viruses that
infect the nervous system are often said to be **neurotropic**
(Box 1.4), they are generally capable of infecting a variety
of cell types. Viral replication usually occurs first in nonneuronal cells, with virions subsequently spreading into
afferent (e.g., sensory) or efferent (e.g., motor) nerve fibers
innervating the infected tissue (Fig. 1.13).

Neurons are polarized cells with functionally and structurally distinct processes (axons and dendrites) that can be
separated by enormous distances. For example, in adult
humans the axon terminals of motor neurons that control stomach muscles can be 50 cm away from the cell
bodies and dendrites in the brain stem. We currently have
a limited understanding of how viral particles move in and
among cells of the nervous system. It is likely that virions enter neurons by the same mechanisms used to enter
other cells. Virus particles must be transported over relatively long distances to the site of viral replication in the
neuronal cell body. All evidence indicates that virions are
carried in the infected neuron by cellular systems, but viral
proteins may facilitate the direction of spread.

Directionality of movement within a neuron is most
likely to be mediated by microtubules and their attendant motor proteins, including kinesin and dynein (Box
1.5). Drugs, such as colchicines, that disrupt microtubules
efficiently block the spread of many neurotropic viruses
from the site of peripheral inoculation to the central nervous system. The precise intracellular form of any virus

BOX 1.7

DISCUSSION

Tracing neuronal connections in the nervous system with viruses

The identification and characterization of synaptically linked multineuronal pathways in the brain are important in understanding the functional organization of neuronal circuits. Conventional tracing methodologies have relied on the use of markers such as wheat germ agglutinin-horseradish peroxidase or fluorochrome dyes. The main limitations of these tracers are their low specificity and sensitivity. During experimental manipulation, it is difficult to restrict the diffusion of tracers to a particular cell group or nucleus, and so uptake occurs in neighboring neurons or adjacent unconnected axon fibers, producing false-positive labeling of a circuit. Neurons located one or more synapses away from the injection site receive progressively less label, because the tracer is diluted at each stage of transneuronal transfer.

Some alphaherpesviruses and rhabdoviruses have considerable promise as self-amplifying tracers of synaptically connected neurons. Under proper conditions, second- and third-order neurons show the same labeling intensity as those infected initially. Moreover, the specific pattern of infected neurons observed in tracing studies is consistent with transsynaptic passage of virus rather than lytic spread through the extracellular space.

The detection of viruses typically involves immunohistochemical localization of viral antigens by light microscopy. More recently, reporter genes such as that encoding green fluorescent protein gene from *Aequorea victoria* have been introduced into the genomes of neurotropic viruses for simpler visualization of viral infection.

Ekstrand, M., L. Pomeranz, and L. W. Enquist. 2008. The alpha-herpesviruses: molecular pathfinders in nervous system circuits. *Trends Mol. Med.* **14**:134–140.

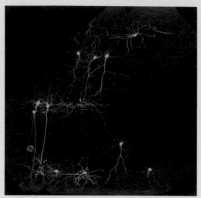

Identification of a possible microcircuit in the rodent visual cortex (V2) after injection of a GFP-expressing strain of pseudorabies virus into the synaptically connected, but distant V1 region. Infection spread via V1 axons (V1 cell bodies are located far out of the field of view) in a retrograde manner to a subset of V2 cell bodies seen here. Confocal microscopy and image reconstruction by Botond Roska, Friedrich Miescher Institute, Basel, Switzerland.

---

particle that spreads in the nervous system has not been established. Both mature virions and nucleocapsids have been observed in axons and dendrites of animals infected by rabies virus and alphaherpesviruses.

In general, viral replication in neurons is preceded by replication in epithelial cells. Virions must enter the neuron at axon terminals and virions or subviral particles (e.g., the nucleocapsid) then must be transported to the cell body of the neuron where replication may occur. After replication, virions or subviral particles are assembled, and may be released directly from the neuronal cell body, may spread directly to glia or support cells in contact with the neuron, or may be transported within the axon, to axon terminals for release. Under some circumstances, virions can enter neurons directly with no prior replication in nonneuronal cells. However, this event is probably infrequent, as nerve endings are rarely exposed to an infecting virus.

With rare exceptions (Box 1.6), cells of the peripheral nervous system are the first to be infected. These neurons represent the first cells in circuits connecting the innervated peripheral tissue with the spinal cord and brain. Once in the peripheral nervous system, alphaherpesviruses, some rhabdoviruses (e.g., rabies virus), and some flaviviruses (e.g., West Nile virus) can spread among neurons connected by synapses (Box 1.7). Nonneuronal support cells and satellite cells in ganglia may also be infected. Virus spread by this mode can continue through chains of connected neurons of the peripheral nervous system to the spinal cord and brain, with devastating results (Fig. 1.14).

An important component of neuronal infection is movement and release of infectious particles in polarized neurons. As with polarized epithelial cells discussed earlier, directional release of infectious virus from neurons affects the outcome of infection. For example, alphaherpesviruses become latent in peripheral neurons that innervate the site of infection (Fig. 1.14). Reactivation from the latent state results in viral replication in the primary neuron and subsequent transport of progeny virus particles from the neuron cell body back to the innervated peripheral tissue. Virions can then spread from the peripheral to the central nervous system, or it can spread back to the peripheral site serviced by that particular group of neurons in the ganglion (Fig. 1.14). The direction taken is the difference

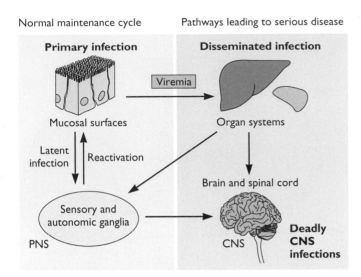

Figure 1.14 Outline of the spread of alphaherpesviruses and relationship to disease. Abbreviations: PNS, peripheral nervous system; CNS, central nervous system.

between a minor local infection (a cold sore) or a lethal viral encephalitis. Luckily, spread back to the peripheral site is by far more common.

Organ Invasion

Once virions enter the blood and are dispersed from the primary site, any subsequent replication requires invasion of new cells and tissues. There are three main types of blood vessel-tissue junctions that serve as routes for

Figure 1.15 Blood-tissue junction in a capillary, venule, and sinusoid. (Left) Continuous endothelium and basement membrane found in the central nervous system, connective tissue, skeletal and cardiac muscle, skin, and lungs. **(Center)** Fenestrated endothelium found in the choroid plexus, villi of the intestine, renal glomerulus, pancreas, and endocrine glands. **(Right)** Sinusoid, lined with macrophages of the reticuloendothelial system, as found in the adrenal glands, liver, spleen, and bone marrow. Adapted from C. A. Mims et al., *Mims' Pathogenesis of Infectious Disease* (Academic Press, Orlando, FL, 1995), with permission.

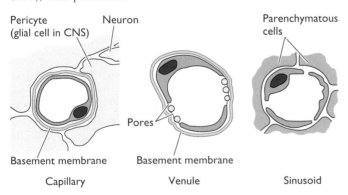

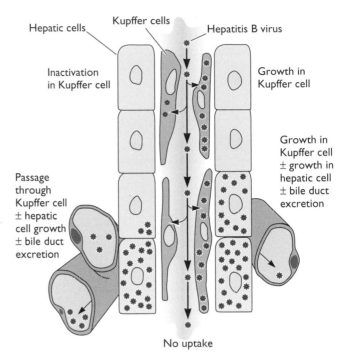

Figure 1.16 Routes of viral entry into the liver. Two layers of hepatocytes are shown, with the sinusoid at the center lined with Kupffer cells. Endothelial cells are not shown. Adapted from C. A. Mims et al., *Mims' Pathogenesis of Infectious Disease* (Academic Press, Orlando, FL, 1995), with permission.

tissue invasion (Fig. 1.15). In some tissues the endothelial cells are continuous with a dense basement membrane. At other sites, the endothelium contains gaps, and at still others there may be **sinusoids**, in which macrophages form part of the blood-tissue junction.

The Liver, Spleen, Bone Marrow, and Adrenal Glands

The liver, spleen, bone marrow, and adrenal glands are characterized by the presence of sinusoids lined with macrophages. Such macrophages, known as the reticuloendothelial system, function to filter the blood and remove foreign particles. They often provide a portal of entry into tissues. For example, viruses that infect the liver, the major filtering and detoxifying organ of the body, usually enter from the blood. The presence of virus particles in the blood invariably leads to the infection of **Kupffer cells**, the macrophages that line liver sinusoids (Fig. 1.16). Virions may be transcytosed across Kupffer and endothelial cells without replication to reach the underlying hepatic cells. Alternatively, viruses may multiply in these cells and then infect underlying hepatocytes. Either mechanism may induce inflammation and necrosis of liver tissue, a condition termed **hepatitis**.

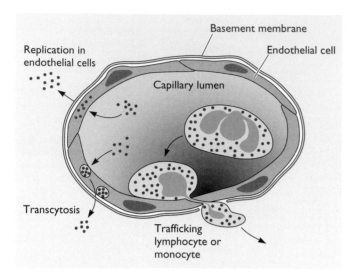

Figure 1.17 How viruses travel from blood to tissues. Schematic of a capillary illustrating different pathways by which viruses may leave the blood and enter underlying tissues. Adapted from N. Nathanson (ed.), *Viral Pathogenesis and Immunity* (Academic Press, London, United Kingdom, 2007), with permission.

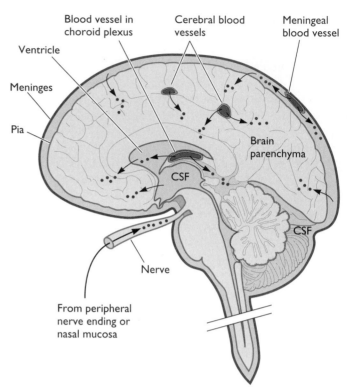

Figure 1.18 How viruses gain access to the central nervous system. A summary of the mechanisms by which viruses can enter the brain is shown. CSF, cerebrospinal fluid. Adapted from C. A. Mims et al., *Mims' Pathogenesis of Infectious Disease* (Academic Press, Orlando, FL, 1995), with permission.

Central Nervous System, Connective Tissue, and Skeletal and Cardiac Muscle

In the central nervous system, connective tissue, and skeletal and cardiac muscle, capillary endothelial cells are backed by a dense basement membrane (Fig. 1.15 and 1.17). In the central nervous system the basement membrane is the foundation of the blood-brain barrier. However, it is not so much a barrier as a selective permeability system. Much study has been devoted to determining how infection spreads to the brain from the blood. Routes of invasion into the central nervous system are summarized in Fig. 1.18.

In several well-defined parts of the brain, the capillary epithelium is fenestrated (with "windows" between cells; loosely joined together), and the basement membrane is sparse. These highly vascularized sites include the choroid plexus, a tissue that produces more than 70% of the cerebrospinal fluid that bathes the spinal cord and ventricles of the brain. Some viruses (mumps virus and certain togaviruses) pass through the capillary endothelium and enter the stroma of the choroid plexus, where they may cross the epithelium into the cerebrospinal fluid either by transcytosis or by replication and directed release. Once in the cerebrospinal fluid, infection spreads to the ependymal cells lining the ventricles and the underlying brain tissue (Fig. 1.18). Other viruses (picornaviruses and togaviruses) may infect directly, or be transported across, the capillary endothelium. Some viruses (human immunodeficiency virus and measles virus) cross the endothelium within infected

monocytes or lymphocytes. Increased local permeability of the capillary endothelium, caused, for example, by certain hormones, may also permit virus entry into the brain and spinal cord.

Coxsackieviruses, members of the *Picornaviridae*, multiply in skeletal and cardiac muscle. The mechanisms by which these viruses enter muscles from the blood have not been adequately studied. It is known that certain coxsackieviruses can replicate in cultured endothelial cells. Virions may be released into muscle cells after multiplying within cells of the capillary walls. Some coxsackieviruses infect B lymphocytes, which might carry the infection from the blood into muscle.

The Renal Glomerulus, Pancreas, Ileum, and Colon

To enter tissues that lack sinusoids (Fig. 1.15), virions must first adhere to the endothelial cells lining capillaries or venules, where the blood flow is slowest and the walls are thinnest. Endothelial cells are not highly phagocytic, and therefore adhesion of virus particles to these cells requires the presence of cellular receptors. To increase the chances of adhesion, virions must be present in a high concentration and circulate for a sufficient period. Clearly there is a "race" between adhesion and removal of virions by the reticuloendothelial system. Once blood-borne virus particles have adhered to the vessel wall, they can readily invade the renal glomerulus, pancreas, ileum, or colon, because the endothelial cells that make up the capillaries are fenestrated, permitting virions or virus-infected cells to cross into the underlying tissues. Some viruses cross the endothelium while being carried by infected monocytes or lymphocytes, in a process called **diapedesis**.

Skin

In a number of systemic viral infections, rashes are produced when virions leave blood vessels (Table 1.2). Different types of skin lesions are recognized. **Macules** and **papules** develop when inflammation occurs in the dermis, with the infection confined in or near the vascular bed. **Vesicles** and **pustules** occur when viruses spread from the capillaries to the superficial layers of the skin. Destruction of cells by virus replication is the primary cause of lesions.

During some viral infections, lesions may also occur in mucosal tissues such as those in the mouth and throat. Because these surfaces are wet, vesicles break down more rapidly than on the skin. During measles infection, vesicles in the mouth become ulcers before the appearance of skin lesions. Such Koplik spots are diagnostic for measles virus infection. In the respiratory tract, after virions leave the subepithelial capillaries, only a single layer of cells must be traversed before particles reach the exterior. Hence, during infections with measles virus and varicella-zoster virus, virions appear in respiratory tract secretions before the skin

Table 1.2 Viruses that cause skin rashes in humans

Virus	Disease	Features
Coxsackievirus A16	Hand-foot-and-mouth disease	Maculopapular rash
Measles virus	Measles	Maculopapular rash
Parvovirus	Erythema infectiosum	Maculopapular rash
Rubella virus	German measles	Maculopapular rash
Varicella-zoster virus	Chickenpox, zoster	Vesicular rash

rash appears. By the time that the infection is recognized from the skin rash, viral transmission to other persons may already have occurred.

The Fetus

The basement membrane is less well developed in the fetus, and infection can occur by invasion of the placental tissues and subsequent invasion of fetal tissue. Infected circulating cells such as monocytes may enter the fetal bloodstream directly. In a pregnant female, viremia may result in infection of the developing fetus. The risk of fetal infection in infants whose mothers have been infected with rubella virus during the first trimester is approximately 80%. Similarly, intrauterine transmission of human cytomegalovirus occurs in approximately 40% of pregnant women with primary infection. Human immunodeficiency virus can be transmitted from mother to infant.

Tropism

Most viruses do not infect all the cells of a host, but are restricted to specific cell types of certain organs. **Tropism** is the predilection of viruses to infect certain tissues and

---

BOX 1.8 BACKGROUND
JC virus, a ubiquitous human polyomavirus

JC virus is widespread in the human population. Most humans experience inapparent childhood infections, but the virus then persists for life in the kidneys or brain with little consequence. If the immune system is compromised by pregnancy or chemotherapy, virus often reactivates from kidney tissues, and infectious virus particles can be found in the urine. On rare occasions, JC virus reactivates in the brain, causing more serious problems. The ensuing disease is called pro-

gressive multifocal leukoencephalopathy, a demyelinating disease affecting oligodendrocytes. This disease is often seen in patients with acquired immunodeficiency syndrome and after immunosuppressive therapy for organ transplant procedures. Given the rarity of the disease and the lack of suitable animal models, it has been difficult to determine how the genome is maintained and virus replication is reactivated.

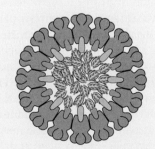

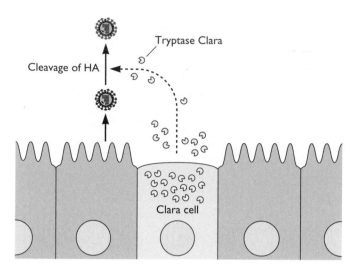

Figure 1.19 Cleavage of influenza virus HA0 by tryptase Clara. Influenza viruses replicate in respiratory epithelial cells in humans. These virus particles contain the uncleaved form of HA (HA0) and are noninfectious. Clara cells secrete the protease tryptase Clara, which cleaves the HA0 of extracellular particles, thereby rendering them infectious. Adapted from M. Tashiro and R. Rott, *Semin. Virol.* **7:**237–243, 1996, with permission.

not others. For example, an **enterotropic** virus replicates in the gut, whereas a **neurotropic virus** replicates in cells of the nervous system. Some viruses are **pantropic**, infecting and replicating in many cell types and tissues. Tropism is governed by at least four parameters. It can be determined by the distribution of receptors for entry (**susceptibility**), or by a requirement for differentially expressed intracellular gene products to complete the infection (**permissivity**). However, even if the cell is susceptible and permissive, infection may not occur because virus particles are physically prevented from interacting with the tissue (**accessibility**). Finally, an infection may not occur even when the tissue is accessible and the cells are susceptible and permissive, because of local intrinsic and innate immune defenses. In most cases, tropism is determined by a combination of two or more of these parameters.

Tropism influences the pattern of infection, pathogenesis, and long-term virus survival. Human herpes simplex virus is often said to be neurotropic because of its noteworthy ability to infect, and be reactivated from, the nervous system. But in fact, herpes simplex virus is pantropic and replicates in many cells and tissues in the host. By infecting neurons, it may establish a stable latent infection, but, because it is pantropic, infection may spread to other tissues and host cells. One serious consequence is that, if an infection is not contained by host defenses at the site of infection, it spreads widely, causing disseminated disease, as can occur when herpes simplex virus infects infants and immunocompromised adults (Fig. 1.14). Herpes simplex

virus neurotropism leads to yet another serious result of infection. On rare occasions, this virus can enter the central nervous system and cause encephalitis that is often fatal. This lethal excursion is not due to a breach in immune defenses because immunodeficient individuals experience the same low rate of brain infections as do immunocompetent individuals.

Cell Receptors for Viruses

A cell may be susceptible to infection if the viral receptor(s) is present and functional. If the proper form of viral receptor is not made, the tissue cannot become infected. The location of the receptor might also be a determinant of cellular susceptibility. If the cellular receptor is present only on the basal cell membrane of polarized epithelial cells, a virus cannot replicate unless it first reaches that location by some means. In some cases, virus particles bound to antibodies can be taken up by Fc receptors on nonsusceptible cells (see "Immunopathological lesions caused by B cells" in Chapter 4).

As cell receptors for viruses have been identified and studied, it has become evident that they can be determinants of tropism. Despite the simplicity of this mechanism, the distribution of cellular receptors in host tissues generally is more widespread than the observed tropism of the virus (Table 1.3). Clearly, receptors are necessary for infection but are not sufficient to explain viral tropism.

Host Cell Proteins That Regulate Viral Transcription

Sequences in viral genomes that control transcription of viral genes such as enhancer regions may be determinants of viral tropism. In the brain, JC polyomavirus replicates only in oligodendrocytes (Box 1.8). The results of *in vitro* and *in vivo* experiments indicate that the JC virus enhancer is active only in this cell type. Other examples include the liver-specific enhancer of hepatitis B virus, the keratinocyte-specific enhancer of human papillomavirus type 11, and the enhancers in the long terminal repeat of human immunodeficiency virus type 1 that are specific for T cells.

Table 1.3 Viral receptors and viral tropism

Virus	Tropism	Receptor (distribution)
Major-group rhinovirus	Respiratory tract	Icam-1 (ubiquitous)
Influenza virus	Respiratory tract	Sialic acid (ubiquitous)
Poliovirus	Oropharyngeal and intestinal mucosa, motor neurons	CD155 (most organs)
Herpes simplex virus	Mucoepithelial cells, neurons	Glycosaminoglycans (ubiquitous)

BOX
1.9

DISCUSSION
A mechanism for expanding the tropism of influenza virus is revealed by analyzing infections that occurred in 1940

Until the isolation of the H5N1 virus from 16 persons in Hong Kong, viruses with the HA0 cleavage site change that permits cleavage by intracellular furin proteases had not been found in humans. However, the WSN/33 strain of influenza virus, produced in 1940 by passage of a human isolate in mouse brain, is pantropic in mice. Unlike most human influenza virus strains, WSN/33 can replicate in cell culture in the absence of added trypsin, because its HA can be cleaved by serum plasmin. Surprisingly, it was found that the NA of WSN/33 is necessary for HA cleavage by serum

plasmin. This altered NA protein can bind plasminogen, sequestering it on the cell surface, where it can be converted to the active form, plasmin (see figure, panel A). Plasmin then cleaves HA into HA1 and HA2. Therefore, a change in NA, not in HA, allowed cleavage of HA by a ubiquitous cellular protease. This property may, in part, explain the pantropic nature of WSN/33.

WSN/33 can replicate in the mouse brain, causing encephalitis. Interestingly, recombinant viruses containing only NA from WSN/33 did not cause encephalitis;

replication in the brain requires the presence of the M and NS genes in addition to the NA from WSN/33. The virulence of WSN/33 is polygenic and cannot be explained solely by the expanded tropism of this strain.

Goto, H., and Y. Kawaoka. 1998. A novel mechanism for the acquisition of virulence by a human influenza A virus. *Proc. Natl. Acad. Sci. USA* **95:**10224–10228.

Taubenberger, J. K. 1998. Influenza virus hemagglutinin cleavage into HA1 and HA2: no laughing matter. *Proc. Natl. Acad. Sci. USA* **95:**9713–9715.

Proposed mechanism for activation of plasminogen and cleavage of HA.
(A) Plasminogen binds to NA, which has a lysine at the carboxyl terminus. A cellular protein converts plasminogen to the active form, plasmin. Plasmin then cleaves HA0 into HA1 and HA2. **(B)** When NA does not contain a lysine at the carboxyl terminus, plasminogen cannot interact with NA and is not activated to plasmin. Therefore, HA is not cleaved. Adapted from H. Goto and Y. Kawaoka, *Proc. Natl. Acad. Sci. USA* **95:**10224–10228, 1998, with permission.

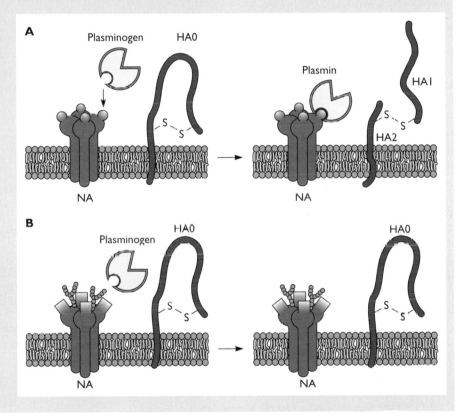

Cellular Proteases

Cellular proteases are often required to cleave viral proteins to form the mature infectious virus particle. A cellular protease cleaves the influenza virus HA0 precursor into two subunits so that fusion of the viral envelope and cell membrane can proceed. In mammals, the replication of influenza virus is restricted to epithelial cells of the upper and lower respiratory tract. The tropism of this virus is thought to be influenced by the limited production of the protease that

processes HA0. This serine protease, called tryptase Clara, is secreted by nonciliated Clara cells of the bronchial and bronchiolar epithelia (Fig. 1.19). The purified enzyme can cleave HA0 and activate HA0 in virions *in vitro*. Alteration of the HA cleavage site so that it can be recognized by other cellular proteases dramatically influences the tropism of the virus and its pathogenicity: some highly virulent avian influenza virus strains contain an insertion of multiple basic amino acids at the cleavage site of HA0. This new sequence permits processing by ubiquitous intracellular proteases such as furins. As a result, these variant viruses are released in an active form and are able to infect many organs of the bird, including the spleen, liver, lungs, kidneys, and brain. Naturally occurring mutants of this type cause high mortality in poultry farms. Avian influenza viruses isolated from 16 people in Hong Kong contained similar amino acid substitutions at the HA cleavage site. Indeed, many of these individuals had gastrointestinal, hepatic, and renal symptoms as well as respiratory symptoms. A virus with such an HA site alteration had not been previously identified in humans, and its isolation led to fears that an influenza pandemic was imminent. To prevent the virus from spreading, all chickens in Hong Kong were slaughtered. The concern was based on the knowledge that the HA of a pantropic influenza virus strain isolated over 50 years ago is processed by ubiquitous proteases (Box 1.9).

The Site of Entry Often Establishes the Pathway of Spread

For viruses that spread by neural pathways, the innervation at the primary site of inoculation determines the neuronal circuits that become infected. The only areas in the brain or spinal cord that become infected by herpes simplex virus are those that contain neurons with axon terminals or dendrites connected to the site of inoculation. After peripheral infection, poliovirus never reaches certain areas of the spinal cord and brain. However, replication occurs if the virus is placed directly into these sites. The conclusion is that accessibility of susceptible cells can determine the tropism of infection.

As we have discussed, the brain can be infected by either hematogenous or neural spread. When the neurotropic NWS strain of influenza virus is inoculated intraperitoneally into mice, it is disseminated to the brain by hematogenous spread and replicates in the meninges, choroid plexus, and ependymal cells lining the brain and spinal cord. However, when the virus inoculum is placed in the nose, there is no viremia. Instead, virus enters the brain by neural spread. It first replicates in olfactory epithelia, and subsequently enters the sensory neurons that richly endow the nasopharyngeal cavity (Box 1.6). Virions then can be found in the trigeminal ganglia, neurons in the brain stem that are in contact with trigeminal neurons, and the olfactory bulb in the brain. The experiment with the NWS influenza virus strain makes a significant point: the site of inoculation can determine the pathway of spread of a particular virus.

Perspectives

In this chapter, we begin discussion of the transition from the well-controlled environment of single cells in the tissue culture dish to the constantly changing "real world."

Infection of a single susceptible host is the first step in the complicated process by which a viral infection is established and maintained in a host population. Every viral pathogen has a distinctive route of infection, which leads ultimately to shedding of infectious particles. Primary sites of replication are often the mucosal membranes of the nasopharyngeal tract, the respiratory system, the gastrointestinal tract, and the genital tract. Injection of virions or infected cells into the bloodstream by insect vectors, needles, or wounds provides another common route of infection. Remarkably, some viral infections remain localized and do not spread throughout the infected host. Others show characteristic spread through the blood or nervous system. Subsequent rounds of viral replication during systemic spreading lead to characteristic features of viral pathogenesis (e.g., the rash of measles, herpes simplex virus encephalitis, or viral hepatitis). After infection and replication, virions are released from the infected individual in a form that can be passed on to other susceptible hosts. Events in a single infected host set the stage for an essential, and even more complicated, process, serial transmission of infection in a susceptible host population, which is the topic of the next chapter.

References

Books

Brock, T. (ed.). 1961. *Milestones in Microbiology*. American Society for Microbiology, Washington, DC.

Johnson, R. T. 1982. *Viral Infections of the Nervous System*. Raven Press, New York, NY.

Mims, C. A., A. Nash, and J. Stephen. 2001. *Mims' Pathogenesis of Infectious Disease*, 5th ed. Academic Press, Orlando, FL.

Nathanson, N. (ed.). 2007. *Viral Pathogenesis and Immunity*, 2nd ed. Academic Press, London, United Kingdom.

Notkins, A. L., and M. B. A. Oldstone (ed.). 1984. *Concepts in Viral Pathogenesis*. Springer-Verlag, New York, NY.

Richman, D. D., R. J. Whitley, and F. G. Hayden (ed.). 2009. *Clinical Virology*, 3rd ed. ASM Press, Washington, DC.

Review Articles

Collins, P., and B. Graham. 2008. Viral and host factors in human respiratory syncytial virus pathogenesis. *J. Virol.* **82:**2040–2055.

Diefenbach, R., M. Miranda-Saksena, M. Douglas, and A. Cunningham. 2008. Transport and egress of herpes simplex virus in neurons. *Rev. Med. Virol.* **18:**35–51.

Enquist, L. W., P. J. Husak, B. W. Banfield, and G. A. Smith. 1999. Infection and spread of alphaherpesviruses in the nervous system. *Adv. Virus Res.* **51:**237–347.

Kennedy, P. G. E. 1992. Molecular studies of viral pathogenesis in the central nervous system: the Linacre lecture 1991. *J. R. Coll. Physicians Lond.* **26:**204–214.

Pereira, L., E. Maidji, S. McDonagh, and T. Yamamoto-Tabata. 2006. Pathogenesis of human cytomegalovirus infection at the maternal-fetal interface. *In* M. J. Reddehase (ed.), *Cytomegalovirus: Molecular Biology and Immunology*. Caister Academic Press, Wymondham, United Kingdom.

Takada, A., and Y. Kawaoka. 2002. The pathogenesis of Ebola hemorrhagic fever. *Trends Microbiol.* **9:**506–511.

Tellinghuisen, T., M. Evans, T. von Hahn, S. You, and C. Rice. 2007. Studying hepatitis C virus: making the best of a bad virus. *J. Virol.* **81:**8853–8867.

Virgin, H. W. 2007. *In vivo* veritas: pathogenesis of infection as it actually happens. *Nat. Immunol.* **8:**1143–1147.

Zampieri, C., N. Sullivan, and G. Nabel. 2007. Immunopathology of highly virulent pathogens: insights from Ebola virus. *Nat. Immunol.* **8:**1159–1164.

Papers of Special Interest

Billam, P., F. Huang, Z. Sun, F. Pierson, R. Duncan, F. Elvinger, D. Guenette, T. Toth, X. Meng. 2005. Systematic pathogenesis and replication of avian hepatitis E virus in specific-pathogen-free adult chickens. *J. Virol.* **79:**3429–3437.

Bodian, D. 1955. Emerging concept of poliomyelitis infection. *Science* **122:**105–108.

Cheung, C., L. Poon, I. Ng, W. Luk, S.-F. Sia, M. Wu, K.-H. Chan, K.-Y. Yuen, S. Gordon, Y. Guan, and J. Peiris. 2005. Cytokine responses in severe acute respiratory syndrome coronavirus-infected macrophages in vitro: possible relevance to pathogenesis. *J. Virol.* **79:**7819–7826.

Crawford, S., D. Patel, E. Cheng, Z. Berkova, J. Hyser, M. Ciarlet, M. Finegold, M. Conner, and M. Estes. 2006. Rotavirus viremia and extraintestinal viral infection in the neonatal rat model. *J. Virol.* **80:**4820–4832.

Croxford, J., J. Olson, H. Anger, and S. Miller. 2005. Initiation and exacerbation of autoimmune demyelination of the central nervous system via virus-induced molecular mimicry: implications for the pathogenesis of multiple sclerosis. *J. Virol.* **79:**8581–8590.

Hamelin, M.-E., K. Yim, K. Kuhn, R. Cragin, M. Boukhvalova, J. Blanco, G. Prince, and G. Boivin. 2005. Pathogenesis of human metapneumovirus lung infection in BALB/c mice and cotton rats. *J. Virol.* **79:**8894–8903.

Ida-Hosonuma, M., T. Iwasaki, T. Yoshikawa, N. Nagata, Y. Sato, T. Sata, M. Yoneyama, T. Fujita, C. Taya, H. Yonekawa, and S. Koike. 2005. The alpha/beta interferon response controls tissue tropism and pathogenicity of poliovirus. *J. Virol.* **79:**4460–4469.

Jung, K., K. Alekseev, X. Zhang, D.-S. Cheon, A. Vlasova, and L. Saif. 2007. Altered pathogenesis of porcine respiratory coronavirus in pigs due to immunosuppressive effects of dexamethasone: implications for corticosteroid use in treatment of severe acute respiratory syndrome coronavirus. *J. Virol.* **81:**13681–13693.

Ku, C.-C., J. Besser, A. Abendroth, C. Grose, and A. Arvin. 2005. Varicella-zoster virus pathogenesis and immunobiology: new concepts emerging from investigations with the SCIDhu mouse model. *J. Virol.* **79:**2651–2658.

Maidji, E., O. Genbacev, H.-T. Chang, and L. Pereira. 2007. Developmental regulation of human cytomegalovirus receptors in cytotrophoblasts correlates with distinct replication sites in the placenta. *J. Virol.* **81:**4701–4712.

Morrison, L. A., and B. N. Fields. 1991. Parallel mechanisms in neuropathogenesis of enteric virus infections. *J. Virol.* **65:**2767–2772.

Publicover, J., E. Ramsburg, M. Robek, and J. Rose. 2006. Rapid pathogenesis induced by a vesicular stomatitis virus matrix protein mutant: viral pathogenesis is linked to induction of tumor necrosis factor alpha. *J. Virol.* **80:**7028–7036.

Rambaut, A., O. Pybus, M. Nelson, C. Viboud, J. Taubenberger, and E. Holmes. 2008. The genomic and epidemiological dynamics of human influenza A virus. *Nature* **453:**615–619.

Rivers, T. 1937. Viruses and Koch's postulates. *J. Bacteriol.* **33:**1–12.

Sacher, T., J. Podlech, C. Mohr, S. Jordan, Z. Ruzsics, M. Reddehase, and U. Koszinowski. 2008. The major virus-producing cell type during murine cytomegalovirus infection, the hepatocyte, is not the source of virus dissemination in the host. *Cell Host Microbe* **3:**263–272.

Samuel M., and M. Diamond. 2006. Pathogenesis of West Nile virus infection: a balance between virulence, innate and adaptive immunity, and viral evasion. *J. Virol.* **80:**9349–9360.

Smith, G. A., S. P. Gross, and L. W. Enquist. 2001. Herpesviruses use bidirectional fast-axonal transport to spread in sensory neurons. *Proc. Natl. Acad. Sci. USA* **98:**3466–3470.

Smith, G. A., L. Pomeranz, S. P. Gross, and L. W. Enquist. 2004. Local modulation of plus-end transport targets herpesvirus entry and egress in sensory neurons. *Proc. Natl. Acad. Sci. USA* **101:**16034–16039.

Souza, M., M. Azevedo, K. Jung, S. Cheetham, and L. Saif. 2008. Pathogenesis and immune responses in gnotobiotic calves after infection with the genogroup II.4-HS66 strain of human norovirus. *J. Virol.* **82:**1777–1786.

Tyler, K., D. McPhee, and B. Fields. 1986. Distinct pathways of virus spread in the host determined by reovirus S1 gene segment. *Science* **233:**770–774.

Tyler, K. L., H. W. Virgin IV, R. Bassel-Duby, and B. N. Fields. 1989. Antibody inhibits defined stages in the pathogenesis of reovirus serotype-3 infection of the central nervous system. *J. Exp. Med.* **170:**887–900.

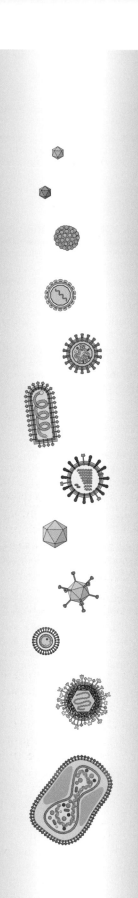

2

Introduction

Principles of Viral Pathogenesis

    Statistics

    Epidemiology

    Shedding of Virions

    Transmission of Viral Infection

    Geography and Season

    Viral Virulence

    Host Susceptibility to Viral Disease

    Other Determinants of Susceptibility

Perspectives

References

Infection of Populations

Introduction

Uncountable numbers of viral particles are released constantly into the environment from all living things. The number of virions that impinge on any one individual is unknowable, yet most encounters are of no consequence for many reasons (discussed in Chapter 10). However, in some instances, not only do we become infected, we become ill and may even die. Furthermore, the disease may spread rampantly to many other individuals for a time. Perhaps surprisingly, there is no one answer to the seemingly straightforward questions: how does a viral infection cause disease in its hosts and how is disease transmitted and maintained in populations? In this chapter, we discuss the complicated problem of viral disease and its transmission in populations.

Principles of Viral Pathogenesis

Viral pathogenesis refers to the series of events that occur during viral infection of a host. It is the sum of the effects on the host of virus replication and of the immune response (Table 2.1). The simple fact is that interaction between virus-infected cells and host defense systems determines the severity of disease. However, these interactions are complex, multifactorial processes that are difficult to control and study (Table 2.2). Conclusions derived from reductionist approaches to studying pathogenesis, such as focusing on the function of a viral receptor protein in cultured cells, frequently are called into question when tested in animals (Box 2.1). Studying pathogenesis in living animals is thorny because so many variables come into play that it is often impossible to prove mechanisms. Consequently, viral pathogenesis has often been called a phenomenological discipline. With the recent development of new experimental tools, pathogenesis is becoming a mechanistic science.

Some human viruses have a broad host range and can infect different animals such as monkeys, ferrets, and guinea pigs. These various animal models have proven invaluable for understanding viral diseases. The mouse has become a particularly fruitful host for studying viral pathogenesis. Because the

29

Table 2.1 Determinants of viral pathogenesis

Interaction with target tissue

Access to target tissue

Presence of receptors

Stability of virus particles in body

  Temperature

  Acid and bile of gastrointestinal tract

Capacity to establish viremia

Capacity to spread through the reticuloendothelial system

Ability to kill cells (cause cytopathology)

Efficiency of viral replication in the cell

  Best temperature for replication

  Cell permissivity

Cytotoxic viral proteins

Inhibition of macromolecular synthesis

Production of viral proteins and structures (inclusion bodies)

Altered cell metabolism

Host response to infection

Intrinsic cell response

Innate immune response

Acquired immune response

Viral immune escape mechanisms

Immunopathology

Interferon: systemic symptoms

T-cell responses: delayed-type hypersensitivity

Antibody: complement, antibody-dependent cellular cytotoxicity,
  immune complexes

Table 2.2 Determinants of viral disease

Nature of the disease

Target tissue

  Site of entry

  Ability of virus to gain access to target tissue

  Viral tropism

  Permissivity of cells

Strain of virus

Severity of disease

Ability to kill cells (cytopathic effect)

Immunity to virus

Intact immune response

Immunopathology

Quantity of virions inoculated

Duration of infection

General health of the host

Host nutritional status

Other infections which might affect immune response

Host genotype

Age of the host

an understanding of pathogenesis mechanisms but also information about transmission and propagation. Progress in understanding the basis for these processes has been complicated, because the molecular correlates of pathogenesis are often defined in animal models, while the understanding of disease spread has come from analyzing human populations.

Statistics

When studying viral infections *in vivo*, scientists rarely obtain results that are so clear and obvious that everyone agrees with the conclusions. Often the effects are subtle, or the data are noisy with wide variation from sample to sample or study to study. Indeed, authors sometimes are stunned to realize that their colleagues do not accept their conclusions. Statistical methods (Table 2.3), properly employed, provide the common language of critical analysis to determine whether the differences between groups are significant. Unfortunately, surveys of articles published in virology journals indicate that errors in statistical analyses abound, which makes it even more difficult for the reader to interpret results. In fact, the term "significant difference" may be one of the most misused phrases in scientific papers, because the actual statistical support for the statement is often absent. While a detailed presentation of basic statistical considerations for virology experiments is beyond the scope of this textbook, critical principles are provided.

It is essential to consider experimental design carefully **before** going to the bench or to the field (Fig. 2.1). The

mouse genome can be manipulated readily, it is possible to engineer this host to allow susceptibility to some human viruses (Box 2.2). The ability to disrupt specific genes in mice enables assessment of the role of individual proteins in pathogenesis. In some cases, insights into human disease are gleaned by studying close relatives of human viruses. An example is simian immunodeficiency virus, which has proven invaluable as a model for human immunodeficiency virus infection. While the knowledge obtained from animal models is essential for understanding how viruses cause disease in humans, the results of such studies must be interpreted with caution. No human disease is completely reproduced in an animal model: what is true for a mouse is not always true for a human. For example, simple differences in size, metabolism, and development can have substantial effects on pathogenesis. Nevertheless, principles and mechanisms obtained from the study of animal models of virus infections often apply to human infections.

Interest in viral pathogenesis stems in large part from the desire to treat or eliminate viral diseases that affect humans. However, treatment of disease requires not only

BOX 2.1

EXPERIMENTS
Of mice and humans

The conclusion that human influenza virus strains are preferentially bound by sialic acids attached to galactose via an α(2,6) linkage was derived by studying the binding of virus particles to cultured cells and to purified sugars. This is the major sialic acid present on human respiratory epithelium, suggesting that it is the receptor bound by virus during infection of animals. This hypothesis was tested using mice that lack the gene encoding ST6Gal I sialyltransferase, the main enzyme used for linking of α(2,6) sialic acid to glycoproteins. Such mice have no detectable α(2,6) sialic acid in the respiratory tract. Nevertheless, human influenza viruses efficiently replicate in the lung and trachea of these mice, indicating that α(2,6) sialic acid is not essential for influenza virus infection of mice.

The lesson to be learned from this experiment is clear: the findings of reductionist experiments must always be validated by experiments with animals.

Glaser, L., G. Conenello, J. Paulson, and P. Palese. 2007. Effective replication of human influenza viruses in mice lacking a major α(2,6) sialyltransferase. *Virus Res.* **126:**9–18.

fundamental problem in study design is to understand the number of observations required to detect a significant difference. The significance level is defined as the probability of mistakenly saying that a difference is meaningful; typically this probability is set at 0.05. An important concept is **power**, the probability of detecting a difference that truly is significant. In the simplest case, power can be increased by having a larger sample size. As an example, consider a study of vaccine efficacy in which laboratory animals are used. Animals are injected with a placebo or with a vaccine, and then challenged with a pathogenic virus. How many animals do you need before you can be confident that the vaccine is effective? Suppose that three animals are in each of the control groups and three animals are in each of the vaccine groups. After the challenge, all control animals die and all vaccinated animals survive. The result seems to be clear cut; the vaccine works—or does it? In fact, with this number of animals, the result will never be significant ($P < 0.05$) (Table 2.4). There is simply insufficient power in a study with so few animals to make a statistically meaningful conclusion.

It is critical to have a detailed description of how statistical analyses were performed (preferably in the Methods section of the paper). These facts are just as important as a description of laboratory methods. More complex data, study design issues, and analyses may require consultation with a statistician. The modern world of virology is incorporating more quantitative and data-rich components into

Table 2.3 Statistical terms[a]

Term	Definition
Alternative hypothesis	Hypothesis that contradicts the null hypothesis
Binary data	Data that consist of only two values (e.g., positive, negative)
Cardinal data	Data that are on a scale in which common arithmetic is meaningful
Confidence interval	Likely range of the true value of a parameter of interest
Hypothesis testing	Use of statistical testing to objectively assess whether results seen in experiments are real or due to random chance
Nonparametric test	Statistical test that requires no assumptions regarding the underlying distribution of the data
Normally distributed data	Data which, when plotted in a histogram, look approximately like a bell-shaped curve
Null hypothesis	Hypothesis which presumes that there are no differences between treated and untreated groups; if hypothesis testing results in a statistically significant difference, the null hypothesis is rejected
P value	Probability of getting a result as extreme as or more extreme than the value obtained in one's sample, given that the null hypothesis is true
Parametric test	Statistical test that assumes the data follow a particular distribution (e.g., normal)
Power	Probability of detecting a statistically significant difference that truly exists
Sample size	Number of experimental units in a study
Significance level	Probability of falsely finding a statistically significant difference

[a]Reprinted from B. A. Richardson and J. Overbaugh, *J. Virol.* **79:**669–676, 2005, with permission.

BOX 2.2

BACKGROUND

Transgenic and knockout mice for studying viral pathogenesis

Mice have always played an important role in the study of viral pathogenesis (see figure). Because it is possible to manipulate this animal genetically, a wealth of new information about how viruses cause disease is emerging. Introducing a gene into the mouse germ line to produce a transgenic mouse and ablating specific genes (gene knockouts) both have wide use in virology.

New mouse models have been established for poliomyelitis and measles by producing transgenic mice that synthesize the human viral receptors. When viral receptors have not been identified, or are not sufficient for infection, an alternative approach is to express either the entire viral genome or a selected viral gene in mice. For example, transgenic mice expressing the hepatitis B virus genome have been used to study interactions between the virus and the host immune response. Transgenic mice that express T-cell-receptor transgenes or genes encoding soluble immune mediators have also been produced. Such mice have been used to study the effect of immune cells on virus clearance, and the protective and deleterious effects of cytokines.

Mice lacking specific components of the immune response have proven invaluable for studying immunity and immunopathogenesis. For example, mice lacking the gene encoding perforin, a molecule essential for the ability of cytotoxic T lymphocytes to lyse target cells, cannot clear infection by lymphocytic choriomeningitis virus, despite the presence of an otherwise intact immune response. Studies of mice with disruptions in genes encoding components of the immune response have led to the identification of cells that are important for mediating recovery from a variety of viral infections, including measles, influenza, and lymphocytic choriomeningitis.

Rall, G. F., D. M. P. Lawrence, and C. E. Patterson. 2000. The application of transgenic and knockout mouse technology for the study of viral pathogenesis. *Virology* **271:**220–226.

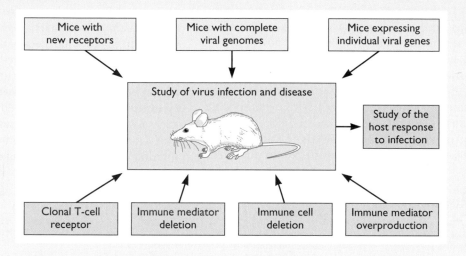

analyses. Consequently, statistical methods will be even more important for interpreting results and drawing conclusions. The objective is to be precise and consistent in determining whether observed biological differences are real, or due to random chance.

Epidemiology

While all viral infections begin with events in single cells, the expansion of infection within an individual must lead to the subsequent spread of infection in many individuals, if the virus population is to survive. **Epidemiology** is the study of the events and actions that affect the health and illness of populations. It is the cornerstone of public health research, providing the rationale for intervention and control.

A viral epidemiologist is an expert in communicable disease who investigates disease outbreaks by undertaking careful data collection and statistical analysis. An epidemiologist specializing in viral diseases must be knowledgeable about not only viral biology and pathogenesis, but also social science disciplines. Social actions and group dynamics are an integral part of the consideration of mechanisms of viral transmission, risk factors for infection, size of the population needed for virus transmission, geography, season, and means of control (Box 2.3; Table 2.5).

Incidence and Prevalence

The quantitation of disease occurrence is the primary result of epidemiology studies. **Incidence** (attack rate)

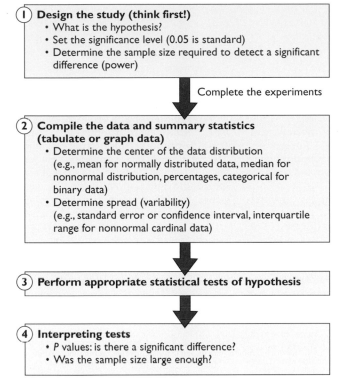

1. **Design the study (think first!)**
   - What is the hypothesis?
   - Set the significance level (0.05 is standard)
   - Determine the sample size required to detect a significant difference (power)

Complete the experiments

2. **Compile the data and summary statistics (tabulate or graph data)**
   - Determine the center of the data distribution (e.g., mean for normally distributed data, median for nonnormal distribution, percentages, categorical for binary data)
   - Determine spread (variability) (e.g., standard error or confidence interval, interquartile range for nonnormal cardinal data)

3. **Perform appropriate statistical tests of hypothesis**

4. **Interpreting tests**
   - *P* values: is there a significant difference?
   - Was the sample size large enough?

Figure 2.1 Experimental design, execution, and interpretation.

is computed as a ratio of the number of cases of disease divided by some measure of population size and time frame. Incidence is of use mainly for acute infections. For example, the incidence of influenza in New York City is expressed as the number of reported cases per year. Disease **prevalence**, an alternative way to express attack rate, is often used for persistent infections where disease onset is not easily determined. In this case, a particular date is selected and the number of cases of a particular disease on that day is divided by an appropriate measure of population. Prevalence is often expressed as cases per million at a particular time.

Prospective and Retrospective Studies

Infections of natural populations obviously differ from those under controlled conditions in the laboratory. Nevertheless, it is possible to determine if one or more variables affect disease incidence and spread in nature. Two general experimental protocols are used: **prospective** (cohort or longitudinal) and **retrospective** (case-control). In prospective studies, the population is divided into two groups, such that one variable is present in one group but not the other. The incidence of disease or side effect is determined. Protocols of this type are often used for drug or vaccine trials where one group is treated and the other is not. Experimental design is critical in these studies. Placebo controls as well as single- or double-blind analyses are essential to remove investigator bias and patient expectations. Prospective studies require a large number of subjects who often are studied for months or years. The number of subjects and time required depend on the incidence of the disease or side effect under study, and the statistical significance required for decision-making.

In contrast, retrospective studies are not burdened by the need for large numbers of subjects and long study times. The protocol simply is to choose a number of subjects with the disease or side effect and an equal number who do not have the malady and classify them as to the variable to be analyzed. For example, in one retrospective study of measles vaccine, a group of 100 children with an adverse side effect and 100 age-matched controls were chosen randomly and classified as vaccinated or not vaccinated. The variable was a particular adverse side effect that may or may not be associated with vaccination. The incidence of the side effect in each group can be computed,

Table 2.4   *P* values for the differences in infection rates between experimental and control groups[a]

No. of animals per group (*n*)	P value for indicated group[b]		
	All control animals infected and no experimental animals infected	All control animals and one experimental animal infected or one control animal infected and no experimental animal infected	One control animal infected and one experimental animal infected
3	0.1	0.4	1.0
4	0.03	0.1	0.5
5	0.008	0.05	0.2
6	0.002	0.02	0.08
7	<0.001	0.005	0.03
8	<0.001	0.001	0.01

[a]Reprinted from B. A. Richardson and J. Overbaugh, *J. Virol.* **79**:669–676, 2005, with permission.

[b]Determined by Fisher's exact test, using a two-sided hypothesis test with the significance level fixed at 0.05. Fisher's exact test is used because it is appropriate for experiments with small numbers of observations.

DISCUSSION
Video games model infectious-disease epidemics

The hugely popular online video game *World of Warcraft* recently became a model for the transmission of virus infections. In this game, players adventure in a fantasy world populated by humans, elves, and orcs and other exotic beasts. In late 2005, a dungeon was added in which players could confront and kill a powerful creature called Hakkar. In his death throes, Hakkar hits foes with "corrupted blood" that contains a virus and causes a fatal infection. The infection was meant to affect only those in the immediate vicinity of Hakkar's corpse, but the virus spread as players and their virtual pets traveled to other cities in the game. Within hours after the software update that installed the new dungeon, a full-blown epidemic ensued as millions of characters became infected.

Although such games are meant only for entertainment, they do model disease spread in a realistic manner. For example, the spread of the virus depended on the ease of travel within the game, interspecies transmission by pets, and transmission via asymptomatic carriers. These aspects of the game world mirrored real-world epidemiology, except for how the disease was halted: the game developers removed Hakkar's dungeon and rebooted their computers.

Epidemiologists are limited to observational and retrospective studies when studying human infectious diseases. Computer models of epidemics have been developed, but they lack the variability and unexpected outcomes found in real-world epidemics. Massively multiplayer online role-playing games have large numbers of participants (10 million for *World of Warcraft*) and therefore are excellent pools for experimental study of infectious diseases. While enjoyment and entertainment are the central focus of such games, the players are serious and devoted, and their responses to situations of danger approximate real-world reactions. For example, during the "corrupted-blood" epidemic, players with healing ability were the first to attempt to help the infected players. This action probably affected the dynamics of the epidemic since infected players survived longer and were able to travel and spread the infection. Multiplayer video games provide an excellent opportunity to examine the consequences of human actions within a statistically significant and controlled computer simulation.

The computer game disease model differs from a real-word virus infection in one significant way: the death of a character in *World of Warcraft* is not permanent. Resurrection is as simple as the click of a mouse.

Lofgren, E. T., and N. H. Fefferman. 2007. The untapped potential of virtual game worlds to shed light on real world epidemics. *Lancet Infect. Dis.* **7:**625–629.

and the ratio of these values yields the relative risk associated with vaccination.

Shedding of Virions

The release of infectious virus particles from an infected host is called **shedding** (see also Fig. 1.10). Shedding is usually an absolute requirement for viral propagation in the host population. The exceptions are direct transmission of viral genomes in the germ lines of their hosts and infections transmitted in the blood supply or by organ transplantation such as acquired immunodeficiency syndrome or hepatitis. During localized infections, shedding takes place from the primary site of replication at one of the body openings. In contrast, release of virions that cause disseminated infections can occur at many sites. Effective transmission of virions from one host to another depends directly on the concentration of released particles, and the mechanisms by which the virions are introduced into the next host. The shedding of small quantities of virus particles may be irrelevant to transmission, while the shedding of high concentrations may facilitate transmission with minute quantities of tissue or fluid. For example, the concentration of hepatitis B virus in blood can be so high that as little as a few microliters can be sufficient to initiate an infection. How well virions survive in the environment also influences the efficiency of transmission.

Respiratory Secretions

Respiratory transmission depends on the production of airborne particles, or aerosols, that contain viruses. Aerosols are produced during speaking, singing, and normal breathing, while coughing produces even more forceful expulsion. Shedding from the nasal cavity requires sneezing and is much more effective if infection induces the production of nasal secretions. A sneeze produces up to 20,000 droplets (in contrast to several hundred expelled by coughing), and all may contain rhinovirus if the individual has a common cold. The largest droplets fall to the ground within a few meters. Many virus particles are inactivated by drying (e.g., measles virus, influenza virus, and rhinovirus), and therefore close proximity is required for transmission. The remaining droplets travel a distance determined by their size. Droplet nuclei, which are 1 to 4 μm in diameter, may remain suspended indefinitely, because air is continually in motion. Such particles may reach the lower respiratory tract. Nasal secretions also frequently contaminate hands or tissues. The infection may be transmitted when these objects contact another person's fingers and

Table 2.5 The many components of epidemiology

Mechanisms of transmission
Aerosol
Food and water
Fomites
Body secretions
Sexual activity
Birth
Transfusion or transplantation
Zoonoses (animals, insects)

Factors that promote transmission
Virion stability
Presence in aerosols and secretions
Asymptomatic shedding
Ineffective immune response

Geography and season
Vector ecology (habitat and season)
School year
Home-heating season

Risk factors
Age
Health
Immunity
Occupation
Travel
Lifestyle
Children (school, day care centers)
Sexual activity

Critical population size
Numbers of seronegative susceptible individuals

Means of control
Quarantine
Vector elimination
Immunization
Antivirals

---

that person in turn touches his or her nose or conjunctiva. In today's crowded society, the physical proximity of people may select for viruses that spread efficiently by this route.

Saliva

Some viruses that replicate in the lungs, nasal mucosa, or salivary glands are shed into the oral cavity. Transmission may occur through aerosols, as discussed above, via contaminated fingers, or by kissing or spitting. Animals that lick, nibble, and groom may also transmit infections in saliva. Human cytomegalovirus, mumps virus, and some retroviruses are known to be transmitted by this route.

Feces

Enteric and hepatic virus particles are shed in the feces and are generally more resistant to inactivation by environmental conditions than those released from other sites. An important exception is hepatitis B virus, which is shed in bile into the intestine, but is inactivated as a consequence and not transmitted in feces. Instead, hepatitis B virus is transmitted through the blood. Viruses transmitted by fecal spread usually survive dilution in water, as well as drying. Inefficient or no sewage treatment, contaminated irrigation systems, and the use of animal manures are prime sources of fecal contamination of food, water supplies, and living areas. Any one of these conditions provides an efficient mode for continual reentry of these viruses into the alimentary canal of their hosts. Two hundred years ago, such contamination was inevitable in most of the world, as disposal of human feces in the streets was a common practice. With modern sanitation, the flow of feces into human mouths has been largely interrupted in developed countries, but is still common throughout the rest of the world. Even so, the closing of beaches because of high coliform counts and the contamination of clam and oyster beds by sewage outflows provide regular reminders of our continuous exposure to enteric and hepatic viruses.

Blood

Viremias are a common feature of many viral infections, and viremic blood is a prime vehicle of virus transmission. Arthropods acquire virions when they bite viremic hosts. Hepatitis and acquired immunodeficiency syndrome can be transmitted by virus-laden blood during transfusions and injections. Infections may be transmitted from viremic blood during coitus or childbirth, and eating raw meat may place viremic blood in contact with the alimentary and respiratory tracts. Health care and emergency rescue workers and dentists are exposed routinely to viremic blood. Indeed, for many of the fatal hemorrhagic fevers caused by viruses (such as members of the *Bunyaviridae* and *Filoviridae*), their only mode of transmission to humans is via blood and body fluids. Consequently, health care workers often are the first people to die in an outbreak of such viral diseases.

Urine, Semen, and Milk

Virus-containing urine is a common contaminant of food and water supplies. The presence of virus particles in the urine is called **viruria**. Hantaviruses and arenaviruses that infect rodents cause persistent viruria. Consequently, humans may be infected by exposure to dust that contains dried urine of infected rodents. A few human viruses replicate in the kidneys and are shed in urine. However, viruria is not important for transmission of most human viruses.

Some retroviruses, including human immunodeficiency virus type 1, herpesviruses, and hepatitis B virus, are shed in semen and transmitted during coitus. Herpesviruses that infect the genital mucosa are also shed from lesions and transmitted by genital secretions, as are papillomaviruses.

Mouse mammary tumor virus is spread primarily by milk, as are some tick-borne encephalitis viruses. Mumps virus and cytomegalovirus are shed in human milk, but are probably not often transmitted by this route.

Skin Lesions

Many viruses replicate in the skin, and the lesions that form contain infectious virus particles that can be transmitted to other hosts. Spread of infection is usually by direct body contact. For example, herpes simplex virus causes a common rash in wrestlers, known as herpes gladiatorum. Warts caused by certain poxviruses and papillomavirus may also be transmitted by contact.

Transmission of Viral Infection

The chain of infection can be maintained only by spreading from one susceptible host to another (**transmission**). There are two general patterns of viral transmission: (i) the perpetuation of infection in one species and (ii) alternate infection of insect and vertebrate hosts. Some viruses such as rabies virus and influenza viruses spread across species, but transmission within each species is relatively self-contained. Most human viruses are transmitted from human to human, although there are some for which an extrahuman cycle is needed for maintenance. Measles and hepatitis A viruses are transmitted from human to human and are maintained solely in the human population. Humans are therefore the **reservoir** for these viruses. In contrast, rabies virus is transmitted from animal to human, but is maintained in animal-to-animal cycles. Some arboviruses are transmitted and maintained in vector-to-human cycles (dengue virus and urban yellow fever virus), while others are maintained in vector-to-vertebrate cycles (St. Louis encephalitis virus and western equine encephalitis virus). Viral diseases shared by humans and animals or insects are called **zoonoses** (see also Chapter 10).

Viral infections are transmitted among hosts in specific ways (Table 2.6). The site of virion excretion and the physical stability of the virion determine the route of transmission. The presence or absence of a lipid envelope is a major determinant of the mode of transmission. Enveloped virions are fragile and sensitive to low pH. Consequently, they are most often transmitted by aerosols or secretions, by

Table 2.6   Viral transmission

Route of transmission	Examples
Respiratory	Paramyxoviruses, influenza viruses, picornaviruses, varicella-zoster virus
Fecal-oral	Picornaviruses, rotavirus, adenovirus
Contact: lesions, saliva, fomites	Herpes simplex virus, rhinovirus, poxvirus, adenovirus
Zoonoses: insects, animals	Togaviruses (arthropod bite), flaviviruses (arthropod bite), bunyaviruses (urine, arthropod bite), arenaviruses (urine), rabies virus (animal bite)
Blood	Human immunodeficiency virus, human T-lymphotropic virus, hepatitis B virus, hepatitis C virus, cytomegalovirus
Sexual contact	Herpes simplex virus, human papillomavirus
Maternal-neonatal	Rubella virus, cytomegalovirus, echovirus, herpes simplex virus, varicella-zoster virus
Germ line	Retroviruses

injection, or by organ transplantation. In contrast, non-enveloped virions can withstand drying, detergents, low pH, and higher temperatures. These virions can be transmitted by the respiratory and fecal-oral routes, and are often acquired via contaminated objects (**fomites**). Various descriptions are used to characterize transmission (Box 2.4). Vertical transmission may occur during gestation, when infection crosses the placenta, during birth, or by close physical contact (Table 2.7). Most maternal infections have little effect on the fetus, but some viral infections that spread by viremia can infect the placenta and even reach the fetal circulation. As a result, the fetus may die and be aborted. The immature state of the fetal immune system may contribute to infections that cause congenital birth defects, including deafness, blindness, and abnormalities of the heart and nervous system. Other symptoms of fetal infection are obvious upon birth. For example, congenital rubella syndrome is recognized in the newborn by liver and spleen enlargement, jaundice, and skin discoloration. Human cytomegalovirus infection has been implicated in many fetal and neonatal birth defects.

Acute viral infections are transmitted efficiently to new hosts and are usually transmitted by the respiratory or fecal-oral route (e.g., common cold virus or poliovirus infections). Such infections often result in excretion of large numbers of virus particles, another property that ensures efficient transmission. Persistent infections present more complicated issues for transmission in populations. In some cases, large quantities of virions are found in the blood (e.g., hepatitis B virus infections), and transmission occurs

BOX 2.4

T E R M I N O L O G Y
Types of viral transmission

Iatrogenic transmission occurs when some activity of a health care worker leads to infection of the patient. Such transmission can occur when nonsterile instruments and needles are used or if a health care worker is infectious. **Nosocomial** transmission occurs when an individual is infected while in a hospital or health care facility. **Vertical transmission** refers to the transfer of infection between parent and offspring, while **horizontal transmission** includes all other forms. In **germ line transmission** the agent is transmitted as part of the host genome (e.g., integrated proviral DNA).

via shared needles or blood transfusions. In other cases, viremia is absent and no visible lesions can be detected (e.g., latent herpes simplex virus infections). Transmission is probable only for a short time after reactivation and only then by close contact.

Most arthropod-borne viruses can replicate in both an insect and a vertebrate host, and infections are maintained by cycling transmission between these diverse animals. Such viruses have a major or exclusive insect vector, and these insects in turn prefer to feed on particular vertebrate species. Humans are the major vertebrate host for dengue virus and urban yellow fever virus. In most other arthropod-mediated viral infections, humans are incidental hosts and are accidentally infected during vector bites (Chapter 10). In these so-called dead-end host infections, the insects have a preferred nonhuman target such as birds or woodland rodents. Examples of the dead-end host interaction are rabies, hantavirus pneumonia, and Korean hemorrhagic fever.

Geography and Season
Some viruses are found only in specific geographical locations. Such restrictions may reflect the requirement for a specific vector or animal reservoir and, in turn, the subsequent interactions of host and virus. For example, migratory animals may carry potential zoonotic infections and therefore the disease is confined to areas encompassing the highest concentrations of humans and other animals. Serial transmission of some acute viral infections occurs only if the host population is quite large and interactive (e.g., measles virus can be maintained only in populations that exceed 200,000). These infections are rarely found in isolated small groups that might populate islands or areas with extreme climates. Before global travel was possible, isolated host populations were the norm and the distribution of viruses was far more limited. Now, viral infections are transported routinely around the globe in planes, trains, and automobiles. A striking example of how the vector can affect localization of viral infection is the global spread of the hitherto exotic chikungunya virus caused by a viral mutation leading to a change in the mosquito vector (Box 2.5).

Despite the increasing propensity for global mixing of hosts and viruses, most acute viral infections have a striking seasonal variation in incidence (Fig. 2.2). Respiratory virus infections are more frequent in winter months, and enteric virus infections predominate in the summer. Seasonal differences in diseases caused by arthropod-borne viruses are clearly a consequence of the life cycle of the vector or the animal reservoir. However, the basis for the seasonal nature of non-arthropod-borne virus infections is less

Table 2.7 Some congenital viral infections

Syndrome	Virus	Fetus
Fetal death and abortion	Smallpox virus	Human
	Parvovirus	Human
	Various alphaherpesviruses	Swine, horses, cattle
Congenital defects	Cytomegalovirus	Human
	Rubella virus	Human
Immunodeficiency	Human immunodeficiency virus type 1	Human
Inapparent (lifelong carrier)	Lymphocytic choriomeningitis virus	Mouse
	Noncytopathic bovine viral diarrhea virus	Cattle
	Murine leukemia virus	Mouse
	Avian leukosis virus	Chicken

An exotic virus on the move

Chikungunya virus is a togavirus in the alphavirus genus. The infection is spread by mosquitoes (primarily the notorious *Aedes aegypti*). The viral disease has been known for more than 50 years in the tropics and savannahs of developing countries of Asia and Africa, but had never been a problem of the developed countries in Europe. The disease is uncomfortable (rashes and joint pains), but not fatal and certainly nothing out of the ordinary for an alphavirus disease. In the last 5 years, something changed dramatically and brought this once third-world viral disease into the forefront of public concern.

In 2004, outbreaks of chikungunya disease spread rapidly from Kenya to islands in the Indian Ocean and then to India (where it had not been reported in over 30 years). In 2007, there was an outbreak in Italy, the first ever in Europe. In some of the Indian Ocean islands, more than 40% of the population fell ill (e.g., the island of Réunion, population 785,000). What had happened to change the pattern of infection?

An alarming finding for the developed world was that the Asian tiger mosquito (*A. albopictus*) apparently is an efficient new vector for the virus. A point mutation in the viral genome appears to be the cause of this vector switch and, perhaps, for the epidemic spread of the disease where it had been unknown. *A. albopictus* is spreading across the globe from eastern Asia and is now found in mainland Europe and the United States. *A. albopictus* is a maintenance (occasionally epidemic) vector of dengue viruses in parts of Asia, and is a competent vector of several other viral diseases. Since its discovery in the United States, five arboviruses (eastern equine encephalomyelitis, Keystone, Tensaw, Cache Valley, and Potosi viruses) have been isolated from this mosquito.

Enserink, M. 2007. Chikungunya: no longer a third world disease. *Science* **318**:1860–1861.

Distribution of *Aedes albopictus* mosquitoes in the United States, 2000. *A. albopictus*, an Asian mosquito, is thought to have been introduced into Hawaii in the late 19th century. The mosquito was not found in the New World until 1985, when it was isolated in Houston, TX. As of 2000, it had been detected in 26 states in the continental United States. INT, intermediate; NEG, negative; POS, positive; UNK, unknown. From http://www.cdc.gov/ncidod/dvbid/arbor/albopic_new.htm.

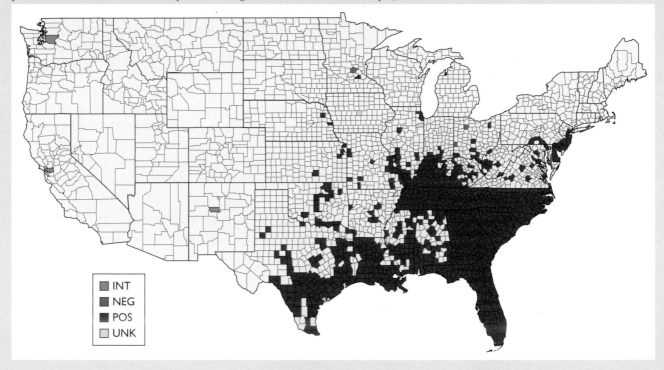

obvious. Variations in disease incidence correlate with changes in climate. For example, poliomyelitis was seasonal in New England but not in Hawaii. It has been suggested that seasonality of infections is due to differences in sensitivity to humidity or to stability of virions to temperature.

According to this hypothesis, during winter months when humidity is low, poliovirus is inactivated but influenza virus remains infectious. The results of recent experiments demonstrate that transmission of influenza A virions is more efficient at low temperature and humidity (Box 2.6).

A Rubella, 1963–1968

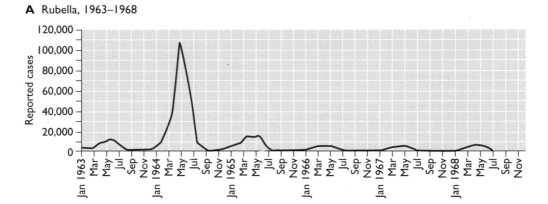

B Influenza, 1994–1999

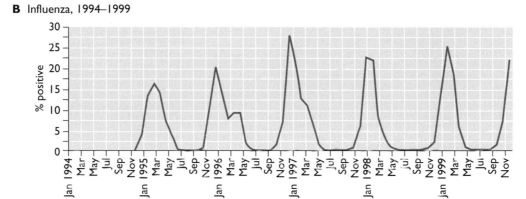

C Poliomyelitis, 1956–1957

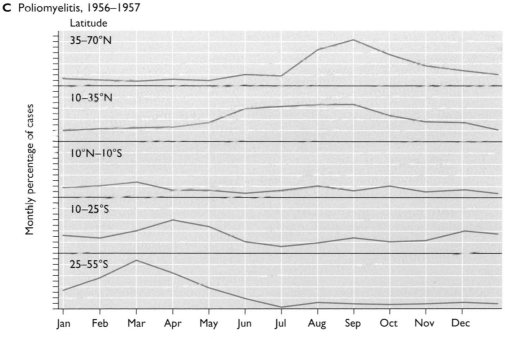

Figure 2.2 Seasonal variation in disease caused by three human pathogens in the United States. (A) Annual cycles of rubella between larger epidemics, which occurred every 6 to 9 years. **(B)** Percentage of specimens testing positive for influenza viruses. **(C)** Monthly incidence of poliomyelitis at different latitudes. Adapted from S. F. Dowell, *Emerg. Infect. Dis.* **7:**369–374, 2001, with permission.

BOX
2.6

EXPERIMENTS
Seasonal factors that affect transmission of influenza virus

Seasonality is a familiar feature of influenza: in temperate climates the infection occurs largely from November to March in the northern hemisphere and from May to September in the southern hemisphere. There have been many hypotheses to explain this seasonality, but none have been supported by experimental data. Recently a guinea pig model was used to show that spread of the virus in aerosols is dependent upon both temperature and relative humidity.

Transmission experiments were conducted by housing infected and uninfected guinea pigs together in an environmental chamber. Transmission of infection was most effective at humidities of 20 to 35% and blocked at a humidity of 80%. In addition, transmission occurred with greater frequency when guinea pigs were housed at 5°C than at 20°C. The authors conclude that low temperature and humidity, conditions found during winter, favor

influenza virus spread. The dependence of influenza virus transmission on low humidity might be related to the nature of the droplets produced by coughing and sneezing (see figure).

Curiously, at 30°C, no transmission of infection took place, an observation at odds with the fact that influenza occurs all year in tropical climates. One explanation of these findings is that transmission in the tropics might occur by contact.

Model for the effect of humidity on transmission of influenza virus.
Transmission efficiency at 20°C (dashed line) or 5°C (solid line) is shown as a function of percent humidity. At 20°C transmission is highest at low humidity, conditions which would favor conversion of exhaled droplets into droplet nuclei (defined as droplets less than 5 μm in diameter and which remain airborne). Reduced virion stability at intermediate humidity is the cause of poor transmission. At high humidity, the conversion from droplets to droplet nuclei is inhibited, and the heavier droplets fall from the air, reducing transmission. At 5°C transmission is more efficient than at 20°C, but there is a gradual loss of transmission with increasing humidity, presumably also as a consequence of reduced formation of droplet nuclei. Adapted from A. C. Lowen et al., *PLoS Pathog.* **3:**1470–1476, 2007, with permission.

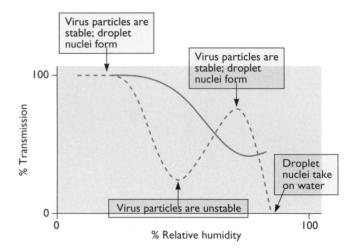

Annual variations in viral disease may also be caused by changes in the susceptibility of the host. Such changes might be linked to circadian rhythms, and could be governed by alterations in mucosal surfaces, epithelial receptors, immune-cell numbers, and responsiveness. If annual changes in host resistance contribute to the seasonality of viral disease, then the gene products that regulate such changes should be identified to provide therapeutic targets for intervention.

Viral Virulence

Once infected, a host may develop a wide range of disease symptoms depending on several variables. **Virulence** refers to the capacity of infection to cause disease. It is a quantitative statement of the degree or extent of pathogenesis. In general, a **virulent virus** causes disease whereas an **avirulent virus** does not. In populations, viral virulence may be manifested as efficiency of spread from individual to individual.

From the earliest days of experimental virology, it was recognized that viral strains often differ in virulence. Virologists thought that the study of viruses with reduced virulence (**attenuated**) would answer the question of how viruses cause disease. This understanding has stood the test of time; indeed, the study of attenuated viruses is still a common strategy. We have learned how to alter viral virulence by direct and indirect methods, and have produced viruses of such limited virulence that they can be used as live vaccines (Chapter 8). Today, the methods of recombinant DNA technology permit a more systematic analysis, by introducing defined mutations into viral genomes so that virulence genes can be identified. The goal of these studies is to understand the mechanisms by which viral and cellular gene products control virulence.

Measuring Viral Virulence

Virulence can be quantified in a number of ways. One approach is to determine the amount of virus that causes death or disease in 50% of the infected animals. This parameter is called the 50% lethal dose (LD_{50}), the 50% paralytic dose (PD_{50}), or the 50% infectious dose (ID_{50}), depending on the parameter that is measured. Other measurements of virulence include time to death (Fig. 2.3A) or appearance of symptoms, and degree of fever or weight loss. Virus-induced tissue damage can be measured directly by examining histological sections or blood (Fig. 2.3B). The safety of live, attenuated poliovirus vaccine is determined by assessing the extent of pathological lesions in the central nervous system in experimentally inoculated monkeys. The reduction in the concentration of $CD4^+$ lymphocytes in blood as a result of human immunodeficiency virus type 1 infection is another example. Indirect measures of virulence include assays for levels of liver enzymes (alanine or aspartate aminotransferases) that are released into the blood following virus-induced liver damage.

It is important to recognize that the virulence of a single virus strain may vary dramatically depending on the dose and the route of infection, as well as on the species, age, gender, and susceptibility of the host. The effect of inoculation route on virulence is illustrated in Table 2.8. Clearly, virulence is a relative property. Consequently, when the degree of virulence of two very similar viruses are compared, the assays must be identical. Quantitative terms such as LD_{50} cannot be used to compare virulence among different viruses.

Genetic Determinants of Virulence

A major goal of animal virology is to identify viral and host genes that control virulence. Once such genes have been identified, defined mutations can be made, gene products can be purified, and mechanistic hypotheses can be tested. If the molecular mechanisms by which viruses cause disease are known, drugs that block these processes may be synthesized. In addition, the contribution of virulence genes to the characteristic patterns of viral infections can be determined. It may be possible to test the hypothesis that virulence influences the long-term survival of viruses in nature.

Alteration of Viral Virulence

To identify viral virulence genes, it is necessary to compare viruses that differ only in their degree of virulence. Before the era of modern virology, several approaches were used to attain this goal. Occasionally, avirulent viruses were isolated from clinical specimens. For example, although wild-type strains of poliovirus type 2 readily cause paralysis after intracerebral inoculation into monkeys, an isolate from the feces of healthy children was shown to be completely avirulent after inoculation by the same route. A second approach to isolate viruses with reduced virulence was to serially passage viruses either in animal hosts or in cell culture (Chapter 8).

Although these approaches were useful, they were unpredictable. To overcome this limitation, viral genomes were often mutagenized, as described in Volume I, Chapter 2,

Figure 2.3 Two methods for measuring viral virulence. (A) Measurement of survival. Mice (5 per virus) were inoculated intracerebrally with either type 1 or type 2 poliovirus, and observed daily for survival. **(B)** Measurement of pathological lesions. Monkeys were inoculated intracerebrally with different viruses, and lesions in different areas of the central nervous system were assigned numerical values. C, cerebrum; B, brain stem; S, spinal cord. (A) Adapted from V. Racaniello, *Virus Res.* **1**:669–675, 1984, with permission. (B) Adapted from N. Nathanson (ed.), *Viral Pathogenesis and Immunity* (Academic Press, London, United Kingdom, 2007) with permission.

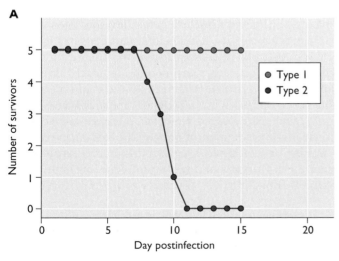

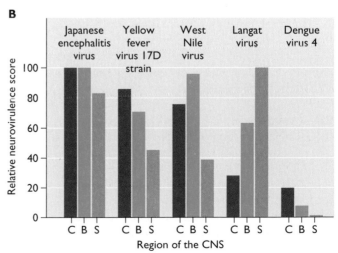

Table 2.8 Effect of route of inoculation on viral virulence[a]

Virus	No. of virions needed to kill 50% of animals			
	Suckling mice		Adult mice	
	Intracerebral infection	Subcutaneous infection	Intracerebral infection	Subcutaneous infection
Wild-type La Crosse virus	~1	~1	~1	~10
Attenuated La Crosse virus mutant	~1	>10^5	>10^6	>10^7

[a]Adapted from Table 9.1 of N. Nathanson, *Viral Pathogenesis and Immunity*, 2nd ed. (Academic Press, London, United Kingdom, 2007), with permission.

and the altered viruses were assayed for virulence in animals. However, controlling the degree of mutagenesis was difficult, and multiple mutations were often introduced. Until the advent of recombinant DNA technology, the ability to identify mutations in a specific gene was limited. More recently, rapid sequencing of entire viral genomes, polymerase chain reaction (PCR) amplification of selected genomic segments, and site-directed mutagenesis have become routine procedures in the quest to identify viral virulence genes and their products (Fig. 2.4).

Viral Virulence Genes

Despite modern technological advances, the identification and analysis of virulence genes in a systematic way

Figure 2.4 Attenuation of viral virulence by a point mutation. Mice were inoculated intracerebrally with two strains of poliovirus which differ by a single base change at nucleotide 472. **(A)** The dose of virus causing death in 50% of the animals (LD$_{50}$) was determined. The change from C to U is accompanied by a large increase in LD$_{50}$. **(B)** Viral replication in mice was determind by plaque assay of spinal cord homogenates. The change from C to U decreases viral replication in the spinal cord. Adapted from N. La Monica, J. W. Almond, and V. R. Racaniello, *J. Virol.* **61:**2917–2920, 1987, with permission.

Virus	Base at 472	LD$_{50}$
PRV7.3	U	>2×10^7
PRV8.4	C	9×10^3

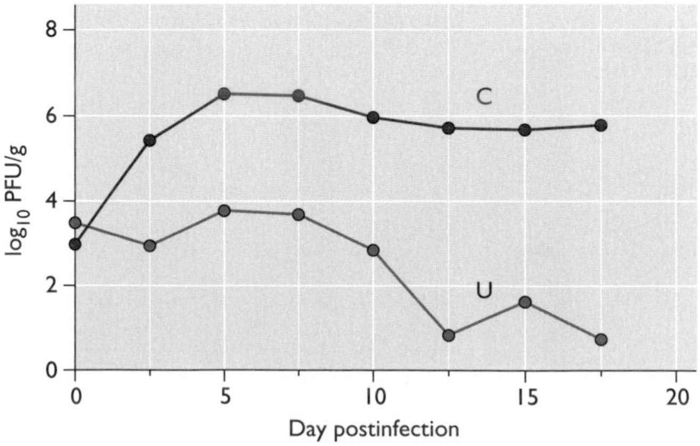

have not been straightforward. Part of the problem is that there are no simple tissue culture assays for virulence. Many of the pathogenic effects promoted by viruses are a result of action of the intrinsic immune defense systems, and it is not possible to reproduce their complex actions in a tissue culture dish. Another problem confronting investigators is the simple fact that it is not obvious *a priori* by inspection of viral genomes what comprises a virulence gene. Consequently, most studies begin with the premise that a virus that causes reduced or no disease in an animal host harbors a defective virulence gene. The genomes of attenuated viruses obtained by empirical efforts (Chapter 8) were often found to harbor multiple mutations, and the contributions of the individual mutations to the attenuation phenotype was often difficult to ascertain. However, by reversion of point mutations, repair of deletions, and crossing of mutants and wild-type strains, many relevant defects were identified. Some mutations reduced, eliminated, or augmented protein function, while others affected the binding of transcription, translation, or replication proteins.

A significant drawback to studies of virulence phenotypes of viruses that infect humans is that relevant animal models of disease are not always readily available. Nevertheless, considerable progress has been made in recent years. In the following sections, we discuss examples of viral virulence genes that can be placed in one of four general classes (Box 2.7). It should be understood, however, that the vast majority of known virulence genes have not been studied sufficiently to be placed in one of these categories.

Although this discussion focuses on producing viruses that are less virulent, the opposite approach, producing viruses that are **more** virulent than the wild type, is possible. The approach is rarely used simply because of the unknown risks involved. However, there are some cases of inadvertent production of a more virulent pathogen. Perhaps the best example is the production of a recombinant ectromelia virus containing the gene encoding interleukin-4 (IL-4) (Box 2.8).

Gene products that alter virus replication. Genes that encode proteins affecting both viral replication and virulence can be placed in one of two subclasses (Fig. 2.5).

BOX 2.7

TERMINOLOGY
Four classes of viral virulence genes

In general, virulence genes are defined by mutations that reduce virulence. For reasons of safety and ethics, experiments to increase virulence are rarely done.

The viral genes affecting virulence can be sorted into four general classes (and some may be included in more than one). The genes in these classes specify proteins that

- affect the ability of the virus to replicate
- modify the host's defense mechanisms

- facilitate virus spread in and among hosts
- are directly toxic

As might be expected, mutations in these genes often have minimal or no effect on replication in cell culture and, as a consequence, are often called "nonessential genes," an exceedingly misleading appellation.

Virulence genes require careful definition, as exemplified by the first general

class listed above (ability of the virus to replicate). **Any** defect that impairs virus reproduction or propagation often results in reduced virulence. In many cases, this observation is not particularly insightful or useful. The difficulty in distinguishing an indirect effect due to inefficient replication from an effect directly relevant to disease has plagued the study of viral pathogenesis for years. There is an adage in genetics that says, "You always get what you select, but you may not get what you want."

Classification is accomplished by analyzing mutants defective in the gene of interest. Viral mutants with alterations in one subclass of genes exhibit reduced or no replication in the animal host and in many cultured cell types. Reduced virulence results from failure to produce sufficient numbers of virus particles to cause disease. Such a phenotype may be caused by mutations in any viral gene.

Mutants of the second subclass exhibit impaired virulence in animals, but no replication defects in cells in culture (except perhaps in cell types representative of the tissue in which disease develops). Such host range mutants should provide valuable insight into the basis of viral virulence, because they identify genes specifically required for disease. Host range mutations in a wide variety of viral

BOX 2.8

EXPERIMENTS
Inadvertent creation of a more virulent poxvirus

Australia had a wild-rodent infestation, and scientists were attempting to attack this problem with a genetically engineered ectromelia virus, a member of the *Poxviridae*. The idea was to introduce the gene for the mouse egg shell protein zona pellucida 3 into a recombinant ectromelia virus. When the virus infects mice, the animals would mount an antibody response that would destroy eggs in female mice. Unfortunately, the strategy did not work in all the mouse strains that were tested. It was decided to incorporate the gene for IL-4 into the recombinant virus. This strategy was based on the previous observation that incorporation of this gene into vaccinia virus boosts antibody production in mice. The presence of IL-4 was therefore expected to increase the immune response against zona pellucida.

To the researchers' great surprise, the recombinant virus replicated out of control in inoculated mice, destroying their livers and killing them. Moreover, mice that were vaccinated against ectromelia could

not survive infection with the recombinant virus; half of them died. "This was a complete shock to us," said one researcher. Essentially, they had shown that the common laboratory technique of recombinant DNA technology could be used to overcome the host immune response and create a more virulent poxvirus.

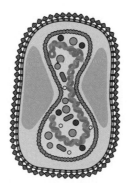

Those who conducted this work debated whether to publish their findings, but eventually did so. Their findings raised alarms about whether such technology could be used to produce biological weapons, and the incident was widely reported in the press. Although the result was a surprise to the investigators, analysis of previously published data suggests that increased virulence of the recombinant virus could have been predicted. This incident emphasizes the need to consider carefully one's experimental design, and to be aware of possible dangers that might arise from the inappropriate use of genetic engineering.

Jackson, R. J., A. J. Ramsay, C. D. Christensen, S. Beaton, D. F. Hall, and I. A. Ramshaw. 2001. Expression of mouse interleukin-4 by a recombinant ectromelia virus suppresses cytolytic lymphocyte responses and overcomes genetic resistance to mousepox. *J. Virol.* **75:**1205–1210.

Müllbacher, A., and M. Lobigs. 2001. Creation of killer poxvirus could have been predicted. *J. Virol.* **75:**8353–8355.

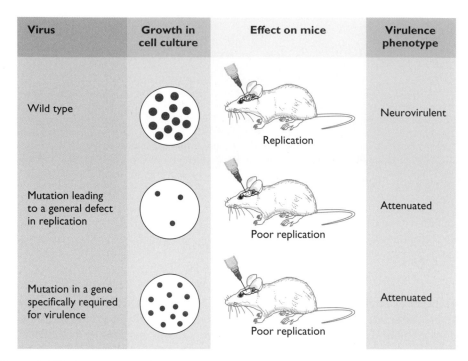

Virus	Growth in cell culture	Effect on mice	Virulence phenotype
Wild type		Replication	Neurovirulent
Mutation leading to a general defect in replication		Poor replication	Attenuated
Mutation in a gene specifically required for virulence		Poor replication	Attenuated

Figure 2.5 Different types of virulence genes. Examples of virulence genes that affect viral growth, using intracerebral neurovirulence in adult mice as an example. Wild-type viruses grow well in cell culture; after inoculation into the mouse brain, they replicate and are virulent. Mutants with defects in replication do not grow well in cultured cells, or in mouse brain, and are attenuated. Mutants with a defect in a gene specifically required for virulence replicate well in certain cultured cells, but not in the mouse brain, and are attenuated. Adapted from N. Nathanson (ed.), *Viral Pathogenesis* (Lippincott-Raven Publishers, Philadelphia, PA, 1997), with permission.

genes encoding proteins that participate in many of the steps in viral replication have been described. We discuss three gene families found in alphaherpesviruses as specific examples of this type of mutant.

A primary requirement for the replication of DNA viruses is access to large pools of deoxyribonucleoside triphosphates. This need poses a significant problem for viruses that replicate in terminally differentiated, nonreplicating cells such as neurons. The genomes of many small DNA viruses encode proteins that alter the cell cycle, such that the cellular substrates for DNA synthesis are produced. Another solution, exemplified by alphaherpesviruses, is to encode enzymes that function in nucleotide metabolism, such as thymidine kinase and ribonucleotide reductase. Mutations in these genes often reduce the neurovirulence of herpes simplex virus because the mutants cannot replicate in neurons or in any other cell unable to complement the deficiency.

Several attenuated herpesvirus strains harbor viral DNA polymerase gene mutations that alter virus replication in neurons but not in other cell types. Such mutants are attenuated after direct inoculation into the brains of mice,

yet replicate well if introduced in the periphery. In another example, point mutations that affect the helicase activity of the helicase-primase complex result in a virus that cannot replicate in neurons and so is attenuated. These observations imply that neuron-specific proteins cooperate with the viral DNA replication machinery to promote DNA synthesis. Analogous attenuating mutations are also found in the genomes of members of other virus families.

Deletion of the herpes simplex virus gene encoding ICP34.5 protein produces a mutant virus so dramatically attenuated that it is difficult to determine an LD_{50}, even when it is injected directly into the brain. Such mutants can replicate in some, but not all, cell types in culture and in the brain (Box 2.9). Notably, they are unable to grow in postmitotic neurons. The molecular basis for the cell type specificity of ICP34.5 mutants has yet to be determined. Their lack of virulence is related to the interaction of this protein with components of the interferon-activated Pkr pathway (Chapter 3).

Noncoding sequences that affect virus replication. The attenuated strains that comprise the live Sabin

**BOX
2.9**

DISCUSSION
The use of attenuated herpes simplex viruses to clear human brain tumors

Malignant glioma, a common brain tumor, is almost universally fatal, despite advances in surgery, radiation, and chemotherapy. Patients rarely survive longer than a year after diagnosis. Several groups have proposed the use of cell-specific replication mutants of herpes simplex virus to kill glioma cells *in situ*. One such virus under study carries a deletion of the ICP34.5 gene and the gene encoding the large subunit of ribonucleotide reductase. These mutant viruses replicate well in dividing cells, such as glioma cells, but not in nondividing cells, such as neurons. The theory is that attenuated virus injected into the glioma will replicate and kill the dividing tumor cells, but will not replicate or spread in the nondividing neurons.

This idea works in principle: studies of mice have indicated that direct injection of this mutant virus into human gliomas transplanted into mice causes clearing of the tumor. The virus is attenuated and safe: injection of 1 billion virus particles into the brain of *Aotus nancymai* (a monkey highly sensitive to herpes simplex virus brain infections) had no pathogenic effect on the animal. This degree of attenuation is remarkable.

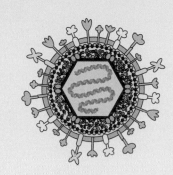

Several human trials are in progress to test safety and dosage. In one study, up to 10^5 PFU was inoculated directly into the brain tumors of nine patients. No encephalitis, adverse clinical symptoms, or reactivation of latent herpes simplex virus were observed. Higher concentrations will be used until a therapeutic effect is attained.

Mineta, T., S. D. Rabkin, T. Yazaki, W. D. Hunter, and R. L. Martuza. 1995. Attenuated multi-mutated herpes simplex virus-1 for the treatment of malignant gliomas. *Nat. Med.* **1:** 938–943.

Rampling, R., G. Cruickshank, V. Papanastassiou, J. Nicoll, D. Hadley, D. Brennan, R. Petty, A. MacLean, J. Harland, E. McKie, R. Mabbs, and M. Brown. 2000. Toxicity evaluation of replication-competent herpes simplex virus (ICP 34.5 null mutant 1716) in patients with recurrent malignant glioma. *Gene Ther.* **7:**859–866.

poliovirus vaccine are examples of viruses with mutations that are not in protein-coding sequences (Chapter 8). Each of the three serotypes in the vaccine contains a mutation in the 5′ noncoding region of the viral RNA that impairs replication in the brain (Fig. 2.4). They also reduce translation of viral messenger RNA (mRNA) in cultured cells of neuronal origin, but not in certain other cell types. An interesting finding is that attenuated viruses bearing these mutations apparently do not replicate efficiently at the primary site of infection in the gut. Consequently, many fewer virus particles are available for hematogenous or neural spread to the brain. Mutations in the 5′ noncoding regions of other picornaviruses also affect virulence in animal models. For example, deletions within the long poly(C) tract within the 5′ noncoding region of mengovirus reduce virulence in mice without affecting viral replication in cell culture.

Gene products that modify host defense mechanisms. The study of viral virulence genes has identified a diverse array of viral proteins that sabotage the body's intrinsic, innate, and adaptive defenses. Some of these viral proteins are called **virokines** (secreted proteins that mimic cytokines, growth factors, or similar extracellular immune regulators) or **viroceptors** (homologs of host receptors). Mutations in genes encoding either class of protein affect

virulence, but these genes are **not** required for growth in cell culture (Fig. 2.6). Most virokines and viroceptors have been discovered in the genomes of large DNA viruses (Box 2.10).

As discussed in Chapter 3, many viral infections induce apoptosis, an intrinsic response that contributes significantly to viral pathogenesis. For example, the pattern of infection of Sindbis virus changes from nonlethal persistent to lethal acute, depending on whether the infected cell can mount an apoptotic response. After infection, apoptosis can be either indirect (uninfected cells die) or direct (infected cells die), and both responses influence subsequent pathogenesis. African swine fever, a highly contagious disease of pigs that is caused by a double-stranded DNA virus transmitted by insects, is an example of a disease caused by indirect apoptosis. The severe lesions and hemorrhages are striking, but intense destruction of lymphoid tissue is characteristic of the disease. This destruction is caused by apoptosis of uninfected lymphocytes induced by cytokines and apoptotic mediators released from infected macrophages. Direct pathogenic effects of virus-induced apoptosis have been suggested in a model of herpes simplex virus-mediated fulminant hepatitis, in which massive hepatocyte death occurs within 24 h of virus injection into the bloodstream. In this disease model, virions in the blood are

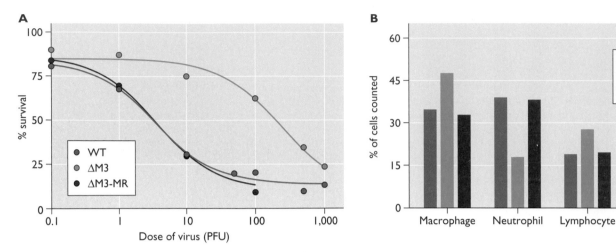

Figure 2.6 Role of a herpesviral chemokine in pathogenesis. (A) Survival of mice after intracerebral inoculation with wild-type gammaherpesvirus type 68, with a mutant virus lacking the M3 gene (ΔM3), which encodes a protein that binds CC chemokines, or with the mutant virus to which the M3 gene has been restored (ΔM3-MR). **(B)** Cellular infiltration in meninges of mice after intracerebral inoculation with the three viruses described in panel A. Adapted from V. van Berkel et al., *J. Clin. Investig.* **109:**905–914, 2002, with permission.

removed rapidly by cells of the reticuloendothelial system, primarily those found in the liver. If the phagocytic activity of macrophages is impaired, virus replication ensues in the liver and is followed by massive apoptotic death throughout the organ.

Gene products that enable the virus to spread in the host. The mutation of some viral genes disrupts spread from peripheral sites of inoculation to the organ in which disease is manifested. For example, after intramuscular inoculation in mice, reovirus type 1 spreads to the central

---

BOX
2.10

DISCUSSION
Variola virus virulence: a highly efficient inhibitor of complement encoded in the genome

Variola virus, which causes the human disease smallpox, is the most virulent member of the *Orthopoxvirus* genus. The prototype poxvirus, vaccinia virus, does not cause disease in immunocompetent humans, and is used to vaccinate against smallpox. Both viral genomes encode inhibitors of the complement pathway. The vaccinia virus complement control protein is secreted from infected cells and functions as a cofactor for the serine protease factor I. The variola virus homolog, called smallpox inhibitor of complement, differs from the vaccinia virus protein by 11 amino acid substitutions. Because the variola virus protein had not been studied, it was produced by changing the 11 codons in DNA encoding the vaccinia virus homolog. The variola virus protein produced in this way was found to be 100 times more potent than the vaccinia virus

protein at inactivating human complement. This finding provides an explanation for the virulence of variola virus.

These findings suggest that the virulence of variola virus, and the avirulence

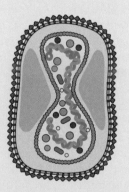

of vaccinia virus, might be controlled in part by complement inhibitors encoded in the viral genome. Furthermore, if smallpox should reemerge, the smallpox inhibitor of complement might be a useful therapeutic target.

Rosengard, A. M., Y. Liu, N. Zhiping, and R. Jimenez. 2002. Variola virus immune evasion design: expression of a highly efficient inhibitor of human complement. *Proc. Natl. Acad. Sci. USA* **99:**8808–8813.

nervous system through the blood, while type 3 spreads by neural routes. Studies of viral recombinants between types 1 and 3 indicate that the gene encoding the viral outer capsid protein s1, which recognizes the cell receptor, determines the route of spread. Only 1 plaque-forming unit (PFU) of La Crosse virus (a bunyavirus) causes lethal encephalitis after intracerebral injection into mice. However, subcutaneous inoculation of over 10^7 PFU causes no disease because it cannot produce a viremia and spread to the brain (Table 2.8).

Viral membrane proteins have been implicated in neuroinvasiveness. For example, the change of a single amino acid in the gD glycoprotein of herpes simplex virus type 1 blocks spread to the central nervous system via nerves after footpad inoculation. Similarly, studies of neuroinvasive and nonneuroinvasive strains of bunyaviruses indicate that the G1 glycoprotein is an important determinant of entry into the central nervous system from the periphery. Although it is tempting to speculate that these viral glycoproteins, which participate in entry, facilitate direct ingress into nerve termini, the mechanisms by which they govern neuroinvasiveness are unknown. These glycoproteins may also influence the ability of host antibodies to clear virus from the primary site of infection.

Toxic viral proteins. Some viral gene products cause cell injury directly, and alterations in these genes reduce viral virulence. Evidence of their intrinsic activity is usually obtained by adding purified proteins to cultured cells, or by synthesis of the proteins from plasmids or viral vectors. The most convincing example of a viral protein with intrinsic toxicity relevant to the viral disease is the nsP4 protein of rotaviruses, which cause gastroenteritis and diarrhea. nsP4 is a nonstructural glycoprotein that participates in the formation of a transient envelope as the particles bud into the endoplasmic reticulum. When nsP4 is fed to young mice, it causes diarrhea. It is thought that in cultured cells, the protein induces a phospholipase C-dependent calcium signaling pathway that leads to chloride secretion (Fig. 2.7). nsP4 therefore acts as a viral enterotoxin, and triggers a signal transduction pathway in the intestinal mucosa.

The SU and TM glycoproteins of human immunodeficiency virus are toxic to cultured cells. TM causes death of cultured cells, most probably as a result of alterations in membrane permeability. The addition of SU to cells results in a high influx of calcium. The contribution of this toxicity to pathogenesis in humans remains untested.

Targets of viral virulence gene products. Mutagenesis can also be used to identify the cellular target of a viral virulence gene product. Mutation of the ICP34.5 gene of herpes simplex virus dramatically reduces neurovirulence

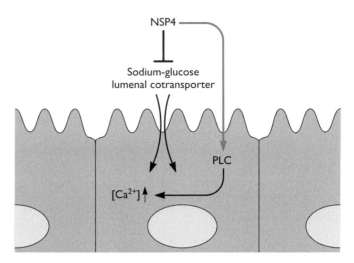

Figure 2.7 Model for rotavirus-induced diarrhea. nsP4, produced during rotavirus replication in intestinal epithelial cells, inhibits the sodium-glucose lumenal cotransporter. Because this transporter is required for water reabsorption in the intestine, its inhibition by nsP4 could be one mechanism of diarrhea induction. nsP4 also induces a phospholipase C (PLC)-dependent calcium signaling pathway. The increase in the concentration of intracellular calcium could induce calcium-dependent chloride secretion. Adapted from M. Lorrot and M. Vasseur, *Virol. J.* **4:**31–36, 2007, with permission.

in mice. It was hypothesized that the protein product of this gene inhibits the antiviral effects of the cellular protein Pkr (Chapter 3). In support of this hypothesis, the ICP34.5 mutant virus was found to be neurovirulent in mice lacking the gene encoding Pkr. Similar experiments have been conducted with a variety of viruses, allowing identification of specific cellular pathways that are altered by viral virulence gene products.

Cellular Virulence Genes

Disruption of cellular genes that encode proteins required for innate and adaptive immune responses may have enormous effects on viral infection. In some cases viral disease becomes more severe, while in others the disease severity is lessened. Such cellular genes can therefore be considered virulence determinants. A consideration of immune response genes can be found in Chapter 4.

Host proteins required for viral translation, genome replication, and mRNA synthesis are prime candidates for virulence determinants. Tissue-specific differences in such proteins could in principle influence virus replication. A common approach in studying viral pathogenesis is to construct mice with specific gene disruptions (Box 2.1). It may be possible to derive mice lacking such proteins as further tools with which to study viral virulence. Few studies have been done in this largely uncharted territory of animal virology, an area that may be one of the most interesting yet to be explored in viral pathogenesis.

Host Susceptibility to Viral Disease

When populations of humans or other animals are infected, many different responses are possible. Some hosts may be highly resistant to infection, some may become infected, and still others may fall in the spectrum between the two extremes. Those who become infected may develop disease ranging from asymptomatic to fatal. Susceptibility to infection and susceptibility to disease vary independently. Understanding the basis for such variation is important, because it may suggest methods for preventing viral disease. The results of such studies demonstrate that both host genes and nongenetic parameters control how populations respond to viral infections (see further discussion in Chapter 10 on emerging viral diseases).

Intrinsic and Immune Defenses

Epidemiologists divide human populations into two groups: susceptible and immune (or resistant). Individuals who have been infected in the past are immune and are not likely to transmit infection. Susceptible individuals can develop disease and spread the virus to others. Persistence of a virus in a population depends on the presence of a sufficient number of susceptible individuals. How efficiently infection occurs determines this number. Immunization against viral infection, by natural infection or vaccination, reduces the number of susceptible people, and therefore limits viral persistence and spread (discussed in more detail in Chapters 5 and 10). For example, epidemics of poliomyelitis were self-limiting, because the asymptomatic spread of the virus immunized the population. The competence of the immune response also determines the speed and efficiency with which the infection is resolved, and the severity of symptoms.

Genetic Determinants of Susceptibility

Several examples of susceptibility genes were first identified in animal systems. The *mx* gene of mice, which confers resistance to infection with influenza A virus, is a well-characterized example. The *mx* gene encodes a guanosine triphosphatase (GTPase) that inhibits influenza virus replication (Chapter 3). Resistance of mice to flavivirus disease has been mapped to the *flv* gene. Flavivirus titers in mice with an *flv* mutation are 1,000- to 10,000-fold lower than in susceptible animals, and the infection is cleared before disease symptoms develop. The product of the *flv* locus is 2′-5′-oligo(A) synthetase, an interferon-induced enzyme that activates ribonuclease L (RNase L), leading to degradation of host and viral mRNAs (Chapter 3). *mx* and *flv* genes have not been shown to play roles in human infections.

It is rare to find single human genes that influence susceptibility to viral infections. The few that do influence susceptibility have been found to encode components of the intrinsic and innate immune systems, e.g., the Toll-like receptors and the chemokine receptors that serve as cofactors for entry of human immunodeficiency virus type 1 into cells. Some genes may influence susceptibility to more than one virus infection. Remarkably, a mutation in the gene encoding the CCr5 chemokine that is protective against infection with human immunodeficiency virus type 1 increases susceptibility to lethal encephalitis caused by West Nile virus.

Recently, researchers found two different mutations in humans that predispose carriers to herpes simplex virus encephalitis. The mutations were in the gene encoding Tlr3 or in the gene expressing the protein Unc-93B. Both gene products affect the production of alpha/beta interferon. The *TLR3* mutation is autosomal dominant (a single copy of the mutant gene increases susceptibility to herpes simplex virus encephalitis), while the *UNC-93B* mutation is recessive (two copies of the mutated gene are required for the susceptibility phenotype). One might expect these mutations to result in broad sensitivity to many viral pathogens, because similar defects in mice are not pathogen specific. As far as could be determined, these patients did not have increased susceptibility to other microbial pathogens. The implication is that in addition to being a general response to all pathogens, intrinsic and innate defenses may evolve to be targeted directly toward a single pathogenic process in a particular species.

Proteins that mediate the humoral and cellular immune responses are other well-known determinants of susceptibility to viral infections. The class I and class II major histocompatibility complex proteins, for which there are many genes, present foreign peptides to T cells. The ability of these proteins to interact with peptides derived from viral proteins determines, in part, how efficiently the infection is cleared. Diversity in major histocompatibility protein genes makes it more likely that there will be a suitable molecule to present peptides for any given infectious agent. Populations from isolated islands have less polymorphism in these genes, a property that may account for their greater susceptibility to infection.

Other Determinants of Susceptibility

The age of the host plays an important role in determining the result of viral infections. Very young and very old humans are most susceptible to disease (Fig. 2.8). The increased susceptibility of infants and young children is likely to be explained by the immaturity of their immune responses. Although young animals may have an immature immune response and become infected more frequently,

A

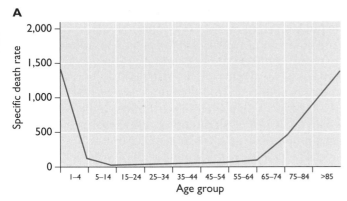

B

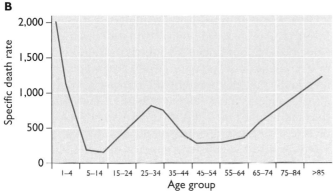

Figure 2.8 Age dependence of influenza pneumonia. Influenza pneumonia deaths per 100,000 in the United States from 1911 to 1915 **(A)** and 1918 **(B)** are shown. Adapted from R. Ahmed et al., *Nat. Immunol.* **8:**1188–1193, 2007, with permission.

they also have greater freedom from immunopathology. Intracerebral inoculation of lymphocytic choriomeningitis virus in adult mice is lethal, because of the T-cell response, while infant mice survive because of their weaker response (Box 2.11).

Other physiological differences may explain age-dependent variation in susceptibility. Infection of human infants with enteric coronaviruses is severe because the alimentary canal is not fully active and presents a particularly hospitable niche for infection. In these individuals, the gastric pH tends toward neutrality, and digestive enzymes are not available. Infection with rubella virus during the first 11 weeks of human gestation results in severe damage to the fetus. Infants that survive have abnormalities in the heart, eyes, and central nervous system. Infection at later times during gestation results in fewer congenital abnormalities, but babies who survive may have hearing loss, mental retardation, and growth deficits.

A major reason for the increased susceptibility of the very old to infection is immune senescence. Furthermore, as animals age, their alveoli become less elastic, the respiratory muscles weaken, and the cough reflex is diminished. These changes may explain in part why elderly people have increased susceptibility to respiratory infections. However, there are notable exceptions to this general trend. Respiratory syncytial virus causes severe lower respiratory tract infections in infants, but only mild upper tract infections in adults. It is not known whether the difference is due to

---

BOX 2.11

DISCUSSION
Congenital brain infections: the lymphocytic choriomeningitis virus model

The fetal brain is one of the most vulnerable organs during development: viral infections of the fetus often result in severe brain injury. Unfortunately, many animal models of congenital brain infection do not mimic human disease, for a variety of poorly understood reasons.

In contrast, the neonatal rat model for congenital lymphocytic choriomeningitis virus (LCMV) infection reproduces virtually all of the neuropathological changes observed in congenitally infected humans.

Within the developing rat brain, the virus selectively infects mitotically active neuronal precursors, a fact that explains the variation in pathology with time of infection during gestation.

Importantly, LCMV infection results in delayed-onset neuronal loss after the virus has been cleared by the immune system. Accordingly, many researchers think that this model can be used to study neurodegenerative or psychiatric diseases

associated with loss of neurons or their function.

Bonthius, D., and S. Perlman. 2007. Congenital viral infection of the brain: lessons learned from lymphocytic choriomeningitis virus in the neonatal rat. *PLoS Pathog.* **3:**1541–1550.

host defenses or to variations in the susceptibility of cells to viral infection. The 1918 influenza pandemic was particularly lethal, not only for the very young and the very old, but unexpectedly also for young adults, 18 to 30 years of age (Fig. 2.8B). It has been suggested that the increased lethality in young adults occurred because they lacked protective immunity that would be conferred by previous infection with a related virus.

Some viral infections, including those caused by poliovirus, mumps virus, and measles virus, are less severe in children than in adults. The basis for this property is not known, but one possibility is that the protective and pathogenic immune responses are better balanced in children.

Males are more susceptible to viral infections than females, but the difference is slight and the reasons are not understood. Hormonal differences, which affect the immune system, may be partly responsible. Pregnant women are more susceptible to infectious disease than nonpregnant women, probably for similar reasons. Hepatitis A, B, and E are more lethal, and paralytic poliomyelitis was more common, in pregnant women than in others.

Malnutrition increases susceptibility to infection because the physical barriers to infection, as well as the immune response, are compromised. An example is the increased susceptibility to measles in children with protein deficiency. For this reason, measles is 300 times more lethal in developing countries than in Europe and North America. When children are malnourished, the small red spots in the buccal mucosa that are pathognomic for measles (Koplik spots) become massive ulcers, the skin rash is much worse, and lethality may approach 10 to 50% (Chapter 5). Such severe measles infections are observed in children in tropical Africa and in aboriginal children in Australia.

Many other parameters influence susceptibility to infection. Corticosteroid hormones have large effects, because they are essential for the body's response to the stress of infection. These hormones have an anti-inflammatory effect, which is thought to limit tissue damage. Cigarette smoking increases susceptibility to respiratory infections in some situations. Increased susceptibility is correlated with a poor mental state, such as occurs in stressful life situations (e.g., a death in the family, injury, or the loss of a job). Epidemiological studies indicate that there is a link between air pollution and hospital admissions for viral respiratory diseases. Our mothers told us that exposure to changes in temperature increases susceptibility to infection. However, comprehensive studies with rhinovirus have failed to reveal any relationship between exposure to low temperature and the common cold. Fortunately for some mothers, hot chicken soup still works wonders.

Perspectives

A fundamental principle of virology is that to perpetuate and maintain any given viral population, virions must be released from one infected host and infect another. This process of serial infection, while simple in principle, is surprisingly difficult to study in natural systems. Nevertheless, local clusters of infection in our homes, more widespread epidemics in our communities, and global pandemics of epic proportions all arise from serial transmission of virions. Epidemiology, the study of this process, is evolving rapidly as new technology for working with populations, as well as tracking and identifying infectious agents, progresses. Advances in genetics have now opened doors to identification of human susceptibility genes that were inconceivable a decade ago. Similarly, rapid progress is being made in our understanding of the molecular biology of viral replication and pathogenesis. Despite all this progress, predicting and derailing the transmission of **any** viral disease remains a challenge (Table 2.9). The simple fact is that individual hosts and their viruses are exceptionally diverse. Moreover, the environment in the real world is a kaleidoscope of variability compared to the stable confines of the virology laboratory.

Our current understanding of the fundamental principles of viral pathogenesis comes largely from studies with animal models. For example, the large number of genetically identical strains of mice has led to the identification of at least 25 loci conferring susceptibility to particular viral infections. It is noteworthy that most of these gene products affect a single virus or virus family and many target a particular step in virus replication (Chapter 3). Some gene products that have broad effects on susceptibility to infection affect intrinsic and innate immune pathways (e.g., interferon).

At one time, it seemed impossible to do similar studies with humans, because genetic techniques were not powerful enough to detect rare susceptibility mutations in outbred human populations. However, recent studies have identified several human genes whose products are involved in intrinsic or innate defense for particular viral

Table 2.9 Fundamental questions of viral pathogenesis

How does a virion enter the host?
What is the initial host response?
Where does primary replication occur?
How does the infection spread in the host?
What organs and tissues are infected?
Is the infection cleared from the host or is a persistent infection established?
How is the virus transmitted to other hosts?

infections. It will be exciting to read about how the modern techniques of genetics and genomics reveal the genes and perhaps selection pressures that drove the evolution of these remarkable viral defenses in different species.

References

Books

Crosby, M. 2006. *The American Plague: the Untold Story of Yellow Fever, the Epidemic That Shaped Our History.* Berkley Books, New York, NY.

Krauss, H., A. Weber, M. Appel, B. Enders, H. D. Isenberg, H. G. Schiefer, W. Slenczka, A. von Graevenitz, and H. Zahner. 2003. *Zoonoses: Infectious Diseases Transmissible from Animals to Humans,* 3rd ed. ASM Press, Washington, DC.

Mims, C. A., A. Nash, and J. Stephen. 2001. *Mims' Pathogenesis of Infectious Disease,* 5th ed. Academic Press, Orlando, FL.

Nathanson, N. (ed.). 2007. *Viral Pathogenesis and Immunity,* 2nd ed. Academic Press, London, United Kingdom.

Review Articles

Ahmed, R., M. B. A. Oldstone, and P. Palese. 2007. Protective immunity and susceptibility to infectious diseases: lessons from the 1918 influenza pandemic. *Nat. Immunol.* **8:**1188–1193.

Biondi, M., and L. Zannino. 1997. Psychological stress neuro-immunomodulation and susceptibility to infectious diseases in animals and man: a review. *Psychother. Psychosom.* **66:**3–26.

Casanova, J., and L. Abel. 2007. Primary immunodeficiencies: a field in its infancy. *Science* **317:**617–619.

Collins, P. L., and B. S. Graham. 2008. Viral and host factors in human respiratory syncytial virus pathogenesis. *J. Virol.* **82:**2040–2055.

Hoenen, T., A. Groseth, D. Falzarano, and Feldmann H. 2006. Ebola virus: unraveling pathogenesis to combat a deadly disease. *Trends Mol. Med.* **12:**206–215.

Lorrot, M., and M. Vasseur. 2007. How do the rotavirus NSP4 and bacterial endotoxins lead differently to diarrhea? *Virol. J.* **4:**31–36.

Rall, G. F., D. M. P. Lawrence, and C. E. Patterson. 2000. The application of transgenic and knockout mouse technology for the study of viral pathogenesis. *Virology* **271:**220–226.

Sancho-Shimizu, V., S. Y. Zhang, L. Abel, M. Tardieu, F. Rozenberg, E. Jouanguy, and J. L. Casanova. 2007. Genetic susceptibility to herpes simplex encephalitis in mice and humans. *Curr. Opin. Allergy Clin. Immunol.* **7:**495–505.

Virgin, H. W. 2002. Host and viral genes that control herpesvirus vasculitis. *Clevel. Clin. J. Med.* **69:**SII7–SII12.

Virgin, H. W. 2007. In vivo veritas: pathogenesis as it actually happens. *Nat. Immunol.* **8:**1143–1147.

Papers of Special Interest

Casrouge, A., S. Zhang, C. Eidenschenk, E. Jouanguy, A. Puel, K. Yang, A. Alcais, C. Picard, N. Mahfoufi, N. Nicolas, L. Lorenzo, S. Plancoulaine, B. Sénéchal, F. Geissmann, K. Tabeta, K. Hoebe, X. Du, R. L. Miller, B. Héron, C. Mignot, T. B. de Villemeur, P. Lebon, O. Dulac, F. Rotenberg, B. Beutler, M. Tardieu, L. Abel, and J. L. Casanova. 2006. Herpes simplex virus encephalitis in human UNC-93B deficiency. *Science* **314:**308–312.

Evans, A. 1978. Causation and disease: a chronological journey. *Am. J. Epidemiol.* **108:**249–258.

Gibbs, S. E., M. C. Wimberly, M. Madden, J. Masour, M. J. Yabsley, and D. E. Stallknecht. 2006. Factors affecting the geographic distribution of West Nile virus in Georgia, USA: 2002–2004. *Vector Borne Zoonotic Dis.* **6:**73–82.

Glass, W. G., D. H. McDermott, J. K. Lim, S. Lekhong, S. F. Yu, W. A. Frank, J. Pape, R. C. Cheshier, and P. M. Murphy. 2006. CCR5 deficiency increases risk of symptomatic West Nile virus infection. *J. Exp. Med.* **203:**35–40.

Goodman, L. B., A. Loregian, G. A. Perkins, J. Nugent, E. L. Buckles, B. Mercorelli, J. H. Kydd, G. Palù, K. C. Smith, N. Osterriedcr, and N. Davis-Poynter. 2007. A point mutation in a herpesvirus polymerase determines neuropathogenicity. *PLoS Pathog.* **3:**1583–1592.

Jackson, D., M. J. Hossain, D. Hickman, D. R. Perez, R. A. Lamb. 2008. A new influenza virus virulence determinant: the NS1 protein four C-terminal residues modulate pathogenicity. *Proc. Natl. Acad. Sci. USA* **105:**4381–4386.

Leib, D. A., M. A. Machalek, B. R. Williams, R. II. Silverman, and H. W. Virgin. 2000. Specific phenotypic restoration of an attenuated virus by knockout of a host resistance gene. *Proc. Natl. Acad. Sci. USA* **97:**6097–6101.

Lindesmith, L. C., E. F. Donaldson, A. D. Lobue, J. L. Cannon, D. P. Zheng, J. Vinje, and R. S. Baric. 2008. Mechanisms of GII.4 norovirus persistence in human populations. *PLoS Med.* **5:**269–290.

Lowen, A. C., S. Mubareka, J. Steel, and P. Palese. 2007. Influenza virus transmission is dependent on relative humidity and temperature. *PLoS Pathog.* **3:**1470–1476.

Perelygin, A. A., S. V. Scherbik, I. B. Zhulin, B. M. Stockman, Y. Li, and M. A. Brinton. 2002. Positional cloning of the murine flavivirus resistance gene. *Proc. Natl. Acad. Sci. USA* **99:**9322–9327.

Virgin, H. W. 2002. Host and viral genes that control herpesvirus vasculitis. *Clevel. Clin. J. Med.* **69:**SII7–SII12.

Zhang, S., E. Jouanguy, S. Ugolini, A. Smahi, G. Elain, P. Romero, D. Segal, V. Sancho-Shimizu, L. Lorenzo, A. Puel, C. Picard, A. Chapgier, S. Plancoulaine, M. Titeux, C. Cognet, H. von Bernuth, C. L. Ku, A. Casrouge, X. X. Zhang, L. Barreiro, J. Leonard, C. Hamilton, P. Lebon, B. Héron, L. Vallée, L. Quintana-Murci, A. Hovanian, F. Rozenberg, E. Vivier, F. Geissmann, M. Tardieu, L. Abel, and J. L. Casanova. 2007. TLR3 deficiency in patients with herpes simplex encephalitis. *Science* **317:**1522–1527.

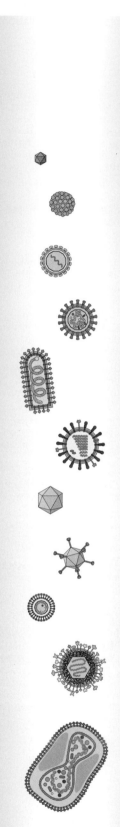

3

Introduction
  Primary Physical and Chemical Defenses
  The First Critical Moments of Infection

Intrinsic Cellular Defenses
  How Do Individual Cells Detect
  Foreign Invaders?
  Receptor-Mediated Recognition of
  Pathogen-Associated Molecules
  Cytokines, the Primary Output of
  Intrinsic Cell Defense
  Interferons, Cytokines of Early Warning
  and Action
  Apoptosis (Programmed Cell Death)

**The Hostile Cytoplasm: Other
Intrinsic Defenses**
  Autophagy
  Epigenetic Silencing
  RNA Silencing
  Cytosine Deamination (Apobec
  [Apolipoprotein B Editing Complex])
  Trim (Tripartite Interaction Motif)
  Proteins

Perspectives

References

Virus Offense Meets Host Defense: Early Actions

Keep a-knockin' but you can't come in!
Come back tomorrow night and try again!
Louis Jordan and His Tympany Five
Vocal by Louis Jordan, recorded 29 March 1939

Introduction

We live and prosper in a literal cloud of viruses. The numbers of potentially infectious particles that impinge on us daily are astronomical. Seasonal "colds," "flu," childhood rashes, measles, chicken pox, and mumps, as well as acquired immunodeficiency syndrome (AIDS) and Ebola fever, all serve notice of our vulnerability. Despite the fact that multicellular organisms evolved in close association with microbes, our understanding of how the defense systems of metazoans recognize and deal with these organisms is incomplete. In this and the next two chapters, we consider the powerful primary and secondary defense systems that ensure our survival.

The labyrinthine complexity of the host defense system is bewildering, and the details can be overwhelming initially. We offer this piece of advice: in the beginning, do not be sidetracked by all the details, but rather try to grasp the big picture of host defense against viral infections (Table 3.1). The apparently distracting nuances of each viral infection can be appreciated **only** when taken in this admittedly oversimplified context. When a cell is infected, intrinsic defensive actions initiate almost immediately. The defensive actions, while initially cell autonomous, escalate in complexity and intensity, recruiting more and more cells if viral replication is not stopped. Control of this response is critical, as the final effector actions usually include the destruction of infected cells. When improperly stimulated or controlled, defensive responses can lead to collateral cellular damage or even host death.

Defense mechanisms, even in healthy hosts, are imperfect despite millions of years of evolution in the face of microbial infections, in large part because the genomes of successful pathogens encode gene products to modify, redirect, or block every step of host defense. Indeed, for every host defense, there will be a viral offense. As many viral family members express a small number of proteins (some genomes have only one or two open reading frames), we can be both amazed and daunted by the realization that the genome of every virus on the planet today **must** encode countermeasures to modulate the defenses of its

Table 3.1 Basic concepts: host defense systems against viral infections are multistep, sequential, and intercommunicating

First: physical and chemical defenses

The skin, surface coatings of tissues such as mucous secretions, tears, acid pH, and surface-cleansing mechanisms

Second: frontline defense

Cell-autonomous, intrinsic defense systems

Detection of altered cell metabolism

Detection of unusual macromolecules made only by invading parasites

Production of cytokines, induction of apoptosis, interference with early steps of viral replication

Third: attack and clean up

Innate and adaptive immune defense

Direct, amplified response by coordinated action of cytokines and lymphocytes.

Infection cleared by pathogen-specific antibodies, helper T cells, and cytotoxic T cells

Production and maintenance of B-cell and T-cell "memory" cells

"Immune" host, ready to respond instantly to the same infection that induced the memory response

host (Box 3.1). The challenge is to identify the key nodes in host defense that are targeted by viral proteins. Only then will we appreciate the foibles of our own vulnerabilities.

Primary Physical and Chemical Defenses

Most virions have no chance in the real world; almost all that impinge on prospective hosts never encounter a susceptible cell. This is so simply because of primary, but often underappreciated, host defenses. These defenses are extracellular and seemingly crude and primitive. They are the physical and chemical barriers of the body: the dead skin, surface coatings such as mucous secretions, tears, and surface-cleansing mechanisms (discussed in more detail in Chapter 1).

The skin is the largest organ of the body, weighing more than 5 kg in an average adult, and is a strong barrier to infection. It is impervious, unless broken by cuts, abrasions, or punctures (e.g., insect bites and needle sticks). Many virus particles that land on the skin are inactivated by desiccation, acids, or other inhibitors formed by indigenous commensal microorganisms. Surfaces exposed to the environment but not covered by skin are lined by living cells and therefore depend on other primary defenses for immediate protection. These surfaces are at risk for infection, despite the remarkable and continuous action of cleansing mechanisms.

The First Critical Moments of Infection

An infection can be initiated only when physical and chemical barriers are breached and virions encounter living cells that are susceptible and permissive. A central dictum for those who study viral pathogenesis is the following: "What happens early dictates what happens late." Indeed, the critical time is within the first hour or so when single cells are infected and spread of infection is local or nonexistent. During this time, various outcomes are in the balance, depending on which molecular switch is activated: will defensive actions be initiated? Which actions are appropriate? If the decisions are too late, the host may die. If the response is too strong or otherwise inappropriate, the host also may suffer. The decisions are based on two coupled, primary processes: intracellular recognition of the invader and subsequent, appropriate control responses. Once the invader is identified, the response must match the invader. Is it a DNA or RNA invader, an intracellular bacterium, a virus, or none of the above? The molecular coupling between the invader detectors and the response effectors is understood only in broad outline, but more details are emerging every day. As we will discuss, these two key processes are played out at each subsequent step in immune defense; the processes of recognition and control are monitored, adapted, and directed (or redirected) as the infection proceeds. Remarkably, the whole process begins with events initiated in a single infected cell.

BOX 3.1

TERMINOLOGY
Is it evasion or modulation?

From the online *Merriam-Webster Dictionary:*

Evade: to elude by dexterity or stratagem

Modulate: to adjust to or keep in proper measure or proportion

The phrase "immune evasion" is popular in the virology literature. It is meant to describe the viral mechanisms that thwart the host immune defense systems. However, in many cases, the phrase can be inaccurate, imprecise, and even misleading. The term "evasion" implies that host defenses are ineffective. In reality, defense and offense are matters of degree, not absolutes. If viruses really could evade the immune system, we might not be here discussing semantic issues.

Perhaps a more accurate term to describe viral gene products that engage immune defenses is "immune modulators." The strategic point is that given the speed of viral replication, an infection can be successful if defenses are only transiently suppressed.

Intrinsic Cellular Defenses

All cells have genetic programs that respond to various stresses, such as starvation, temperature extremes, irradiation, and infection. Some of these programs are designed to maintain homeostasis, and others have evolved to repel cellular invaders. These latter programs recognize and respond to alien nucleic acids and other microbial products, such as viral membrane proteins, bacterial cell wall and flagellum components, or other foreign protein modifications and lipids.

We define these cell-autonomous, protective programs as **intrinsic cellular defenses** to distinguish them from **immune defenses** (Chapter 4). Immune defenses tend to be more global (organism-wide) responses, whereas intrinsic defenses begin with a single cell and tend to be local. Intrinsic defenses arose very early in the evolution of cells. In contrast, immune defenses appeared later during the evolution of complicated multicellular organisms and depend on mobile lymphocytes and antibodies released in blood and secretions. Below, we consider several primary intrinsic defenses that are relevant to viral infection. It is important to note that while we make a clear distinction between intrinsic defense and innate immune defense, these two early-action arms of host defense are coupled by the action of cytokines (Fig. 3.1).

We first discuss what we understand about how an invader is recognized by a cell and how the coordination of recognition and appropriate defensive response is achieved. We then discuss cytokines, with a focus on the interferons, secreted proteins of early warning. Finally, we review six widely conserved processes that function in cellular defense against viral infections.

How Do Individual Cells Detect Foreign Invaders?

A major conundrum for many years was to understand how cells recognize microbial invaders as not "self." The problem in the case of viruses is obvious: the basic building materials are derived from self cells, the only difference being the way the materials are put together. We now understand that an infected cell differs from an uninfected one in at least two general properties. First, as soon as virions engage their receptors, new signals may flow through cellular signal transduction pathways. In these first minutes of infection, no nonhost proteins are made, but the cell may respond in many ways. For example, the dynamics of ion flow, membrane permeability, protein modification and localization, and even transcription of host genes may change. Second, soon after virions engage receptors, viral nucleic acid appears in the cytoplasm or the nucleus. As we shall see, the cytoplasm is a hostile environment for foreign nucleic acids. It is likely that the same holds true for the nucleus,

but much less is known about this organelle's defenses. As soon as viral genomes are exposed to the cytoplasm, host proteins can bind to the foreign nucleic acid (recognizing structures that are not commonly found in cellular nucleic acids). These protein-nucleic acid complexes engage signal transduction pathways that stimulate synthesis of cytokines such as the type I interferons (IFN-α and IFN-β) (Fig. 3.2).

We have only a minimal understanding of nuclear defenses against foreign nucleic acids. For example, it is clear that if viral DNA or RNA enters the nucleus, a DNA damage response may ensue as a result of detection of single-stranded nucleic acid or double-strand ends of DNA. There is some evidence that nuclear proteins may bind the incoming foreign DNA and block transcription. After infection by retroviruses, the host genome invariably suffers direct damage: proviral DNA is always integrated in the cell genome. The process of integration depends on activation of DNA repair pathways, and the integrated proviral DNA may affect the expression or integrity of a cellular gene, a process known as **insertional mutagenesis** (Chapter 6; also see Volume I, Chapter 7).

As soon as viral proteins are produced, the cell is permanently changed, sometimes in dramatic ways (Table 3.2). For example, infection may result in visible changes in the cells, collectively called **cytopathic effect**. The consequences of viral infection also may include cessation of essential host processes such as translation, DNA and RNA synthesis, and vesicular transport. These deviations from the norm are detected by two general systems that monitor cell health. First, intracellular molecules monitor cell integrity and homeostasis. Second, extracellular proteins (cytokines) are produced and secreted in response to infection and bind to surface receptors on nearby cells. These cytokines announce the infection to neighboring uninfected cells. Cells that bind the cytokines initiate defensive actions. When homeostasis is altered or when signaling cytokines bind to their receptors, a common output pathway that results in cell death is activated. The integration of detection and response is one of the crucial features of intrinsic cellular defense. Viral evolution in the face of these responses has resulted in viral gene products that counter, modulate, or even bypass intrinsic cell defenses.

Receptor-Mediated Recognition of Pathogen-Associated Molecules

Pathogens are distinguished by the presence of unique molecules including bacterial and fungal cell wall and membrane materials (lipopolysaccharide, peptidoglycans, lipotechoic acids, glucans), bacterial and viral RNA and DNA, N-formylmethionine, and lipoproteins. Such pathogen-associated molecules are recognized by certain cellular proteins called **pattern recognition receptors**. These

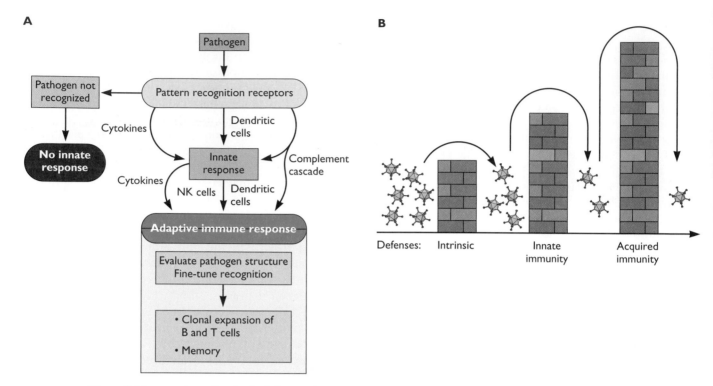

Figure 3.1 Integration of intrinsic defense with the innate and adaptive immune response. (A) An invading pathogen is first detected by molecular interactions that depend on pattern recognition receptors of the intrinsic defense system (see also Fig. 3.10). Molecules of microbial origin are usually detected, resulting in a variety of responses including cytokine production and release of stimulators of inflammation. As infection proceeds in the face of intrinsic defenses, the innate immune response comes into play. Complement proteins, natural killer cells (NK cells), dendritic cells, and other phagocytic cells act to contain the infection. During this phase, migratory sentinel cells (e.g., dendritic cells) take packets of ingested proteins to lymph nodes, where they contact and stimulate cells of the adaptive immune system. Molecules of the invading microbe and the effector molecules of the innate immune defenses then interact further with the adaptive immune system, often causing clonal expansion of distinct classes of lymphocytes. Highly specific effectors such as antibodies produced by B cells, and cytotoxic T cells, are released into the circulation to recognize the invading microbe and the infected cells. This adaptive response enables recognition of foreign proteins with a high degree of structural specificity. The intrinsic and innate immune defenses are essential, not only for immediate deployment, but also for reconnaissance and transfer of information to the adaptive immune system. **(B)** The sequential nature of host defenses depicted as the breaching of successive barriers by viral infection. Most infections are blocked by intrinsic defenses. If intrinsic defenses are breached, then innate defenses come into play to contain the infection. Activation of acquired immune defenses is usually sufficient to contain and clear any infections that escape intrinsic and innate defense. In rare instances, host defenses may be absent or inefficient and severe or lethal pathogenesis occurs. Adapted from D. T. Fearon and R. M. Locksley, *Science* **272:**50–54, 1996, with permission.

receptors also detect endogenous stress signals such as uric acid and some heat shock proteins. Pattern recognition receptors have been selected over evolutionary time to be highly pathogen specific. Our first insights into the nature of these receptors came from *Drosophila* developmental genetics (Box 3.2). We now understand that all intrinsic and innate defense systems arose early in the evolution of multicellular organisms, and remain absolutely essential for survival.

Several receptors detect specific motifs characteristic of invading microbial pathogens in single cells (Table 3.3).

Four of these are the DEXD/H box RNA helicases (e.g., RigI and Mda5) capable of detecting foreign RNA in the cytoplasm (Box 3.3), the Dai protein (also known as the Z-DNA-binding protein) that binds foreign DNA in the cytoplasm, the Toll-like receptors (Tlrs) (Box 3.4) that detect a wide variety of microbial products including RNA and DNA, and the complement lectin C1q family members (see "Complement" below) (Table 3.4). All these receptors likely detect not only microbial but also nonmicrobial danger signals. Different cell types and tissues vary in the distribution and concentration of these receptors. Most cells

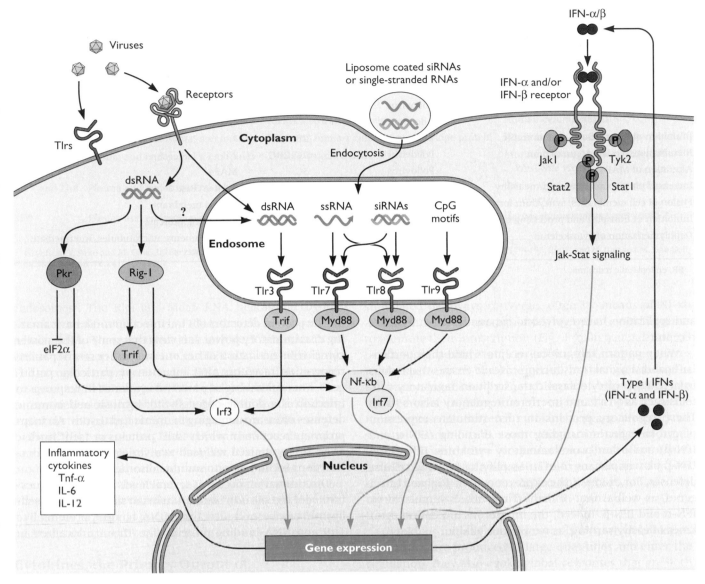

Figure 3.2 Recognition of foreign nucleic acids in mammalian cells. The Toll-like receptors (Tlrs), RigI, and protein kinase R (Pkr) all contribute to detection of pathogen-specific nucleic acids including single-stranded RNA, dsRNA, RNA nucleotides, siRNAs, and unmethylated CpG-containing oligonucleotides. As the receptor's cognate nucleic acid is bound on the cell surface, in the cytoplasm, or in the lumen of endosomes, signal transduction events lead to activation of Nf-κb, Irf3, or Irf7 to induce expression of inflammatory cytokines and IFN-α/β. Important cytoplasmic proteins in the signal transduction cascade, including Trif and Myd88, bind the cytoplasmic tails of endosomal Tlr proteins after they have engaged their cognate ligand. Viral RNA and DNA may be exposed in the lumen of endosomes after degradation or uncoating events. Pkr is autophosphorylated when dsRNA is bound, leading to phosphorylation of its substrates. One such substrate is the a subunit of the eukaryotic translation initiation factor 2α. Phosphorylation of this protein blocks protein synthesis. Many IFN-inducible genes, including Pkr, are induced when IFN-α and IFN-β bind the IFN receptor (autocrine pathway).

synthesize all these, but the Tlrs are critical for the function of cells in the immune system, particularly dendritic cells and macrophages (Table 3.3).

The pattern recognition receptor proteins are localized strategically to the sites at which the earliest stages of

viral infections begin. For example, while the RNA helicase receptors tend to be found in the cytoplasm, certain Tlrs are present on the cell surface or in endosomes, where entering viral proteins and nucleic acids first appear (Figure 3.2). It is likely that all processes of viral uncoating

BOX
3.3

DISCUSSION
Detecting viral invaders

A fundamental problem solved over eons of evolution is the detection by individual cells of invading viral RNA or DNA. How an invading nucleic acid is distinguished from cellular RNA and DNA remained an enigma for decades. However, in the last few years, scientists have found a veritable treasure trove of powerful systems that detect alien nucleic acid and activate cellular alarm systems like the IFN response. The Toll-like receptors (Tlrs) are one class of frontline microbial sensors that bind to a variety of unique microbial products, while retinoic acid-inducible protein I (RigI) and melanoma differentiation-associated protein (Mda5) represent another class that recognizes RNA in the cytoplasm. RigI and Mda5 are RNA helicases of the DEXD/H box family. A distinguishing property of these two helicases is their tandem caspase activation and recruitment domains (CARDs). These modular domains interact with other CARD-containing proteins during the apoptotic response. After binding their ligand, RigI and Mda5 signal using a unique interaction of their CARD domains with an adapter protein. This adapter is an outer mitochondrial membrane protein (Mav) and, when bound to the CARD domain of either RigI or Mda5, activates Irf3 and Nf-κb by the pathways shown. Mav binds to Traf6 and induces its polyubiquitination by E1 ligase and Ubc12/Uve1A. In turn, Tak1 kinase and Tab2 adapter protein bind and phosphorylate Jnk kinases and the Iκκ complex. The common adaptor Mav integrates two different interactions with a common output. The coordination of these three signal transduction pathways leads to the assembly of a multiprotein enhancer complex in the nucleus, which drives expression of the IFN-β gene.

A crucial question is how RigI and Mda5 can distinguish viral RNA from cellular RNA. It was clear for some time that the two receptors had different specificities and actions *in vivo*. Use of mice that lacked either of these two RNA detectors showed that RigI is required for the *in vivo* response to paramyxoviruses and flaviviruses, as well as influenza virus. The Mda5 detector seems to be essential for the antiviral response to encephalomyocarditis virus and measles virus.

In 2006, papers by Hornung et al. and Pichlmair et al. provided new insight

into how viral RNA can be distinguished from cellular RNA. It was widely held that discrimination was achieved by recognition of dsRNA produced during viral replication or by unique secondary structures. These two papers provided

evidence that RigI binds to RNAs with a 5′ phosphate group, in hindsight an obvious discriminator. Unlike cellular mRNAs that have 5′ cap structures, viral RNAs are often uncapped and carry a 5′ triphosphate group. While certainly an important

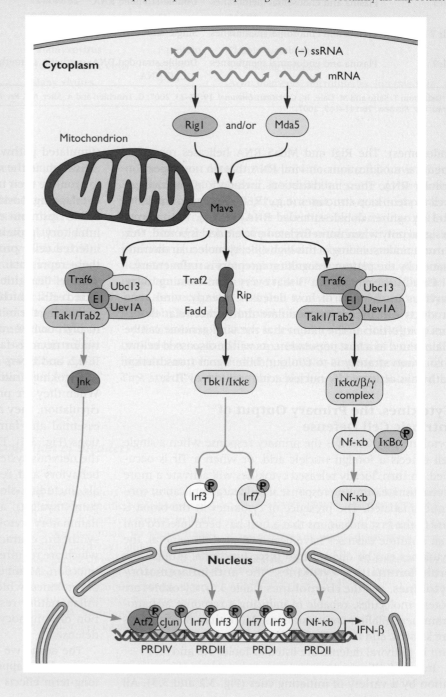

finding, the 5′ phosphate recognition must be just the tip of the iceberg for discrimination. Some anomalies are apparent. Picornaviral RNAs do not have cap structures or triphosphates on their 5′ ends, yet infection can be detected by Mda5 (but not RigI). Similarly, the abundant human 7SL RNA has a 5′ triphosphate group yet does not activate an interferon response. Other aspects of RNA structure are certain to be involved.

Hornung, V. J. Ellegast, S. Kim, K. Brzózka, A. Jung, H. Kato, H. Poeck, S. Akira, K. Conzelmann, M. Schlee, S. Endres, and G. Hartmann.** 2006. 5′-Triphosphate RNA is the ligand for RIG-I. *Science* **314:**994–997.

McWhirter, S., B. tenOever, and T. Maniatis.** 2005. Connecting mitochondria and innate immunity. *Cell* **122:**645–658.

Pichlmair, A., O. Schulz, C. Tan, T. Näslund, P. Liljeström, F. Weber, and C. Sousa.** 2006. RIG-I-mediated antiviral responses to single-stranded RNA bearing 5′-phosphates. *Science* **314:**997–1001.

Saito, T., and M. Gale, Jr.** 2007. Principles of intracellular viral recognition. *Curr. Opin. Immunol.* **19:**17–23.

Yoneyama, M., M. Kikuchi, T. Natsukawa, N. Shinobu, T. Imaizume, M. Miyagishi, K. Taira, S. Akira, and T. Fujita.** 2004. The RNA helicase RIG-I has an essential function in double-stranded RNA induced innate antiviral response. *Nat. Immunol* **5:**730–737.

constellation and concentration of each cytokine constitute a major means of communication between cells, and the innate and adaptive arms of the immune system. Indeed, cytokines participate in almost every phase of the host response to viral infection, including control of inflammation, induction of an antiviral state in uninfected cells, and regulation of immune responses.

Many viral gene products can mimic or modulate cytokine responses. These proteins have been called **virokines** if they mimic host cytokines, or **viroceptors** if they mimic host cytokine receptors. The arsenal includes remarkable proteins such as soluble IL-1 receptors, a variety of chemokine antagonists, and functional homologs of IL-10 and IL-17. Viral DNA genomes encode most of the well-known virokines and viroceptors, but viral RNA genomes contain some surprises. For example, the envelope protein of respiratory syncytial virus is a mimic of fractalkine, the only known chemokine that is a membrane protein. The viral envelope protein competes with fractalkine binding to its receptor, which also functions as a viral receptor.

Interferons, Cytokines of Early Warning and Action

The **interferons**, cytokines synthesized by mammals, birds, reptiles, and fish, are critical signaling proteins of the host frontline defense (Box 3.5). The discovery of interferon was first reported almost simultaneously in the 1950s by two groups of investigators. One group observed that chicken cells exposed to inactivated influenza virus contained a substance that interfered with the infection of other chicken cells by live influenza virus. The second group made their discovery using vaccinia virus infections. We now know that most cells synthesize new interferon when infected, and the released interferon inhibits replication of a wide spectrum of viruses. The broad spectrum of interferon action was a puzzle that was not resolved for more than 50 years.

There are three classes of interferons (Table 3.8). In following sections, we use the abbreviation IFN to mean both IFN-α and IFN-β (also called type I interferons). We will refer specifically to IFN-γ, which is induced only when certain lymphocytes are stimulated to replicate and divide after binding a foreign protein. In contrast, IFN-α and IFN-β are induced directly by viral infection of almost any cell type.

IFN Is Made by Infected Cells and by Immature Dendritic Cells

Virus-infected cells produce IFN when foreign nucleic acid is detected by RigI or Mda5 RNA helicase receptors, or by Tlr receptors (Fig. 3.2; Box 3.4). Other signals unique to viral infection can also lead to IFN production. For example, structural proteins of some viruses stimulate IFN synthesis upon binding of virions to cells. In other cases, virus-induced degradation of the inhibitor of Nf-κb (Iκbα) leads to activation of transcription of IFN genes (Fig. 3.4).

Virus-infected cells invariably produce IFN, but the uninfected macrophages and dendritic cells that patrol tissues, (sentinel cells; see Chapter 4) also make this cytokine when their Tlrs bind products released from infected cells. Such products include viral proteins, viral nucleic acids, and cellular stress proteins (e.g., heat shock proteins). The sentinel cell response, which is essential for amplification of the subsequent immune response, is discussed in the next chapter. If the infection is not contained and spreads to more cells, large quantities of IFN may be synthesized by specialized dendritic cell precursors in the blood called plasmacytoid cells. Such systemic production of IFN leads to many of the general symptoms of viral infection.

Production of IFN by infected cells and uninfected, immature dendritic cells at the site of infection is rapid, but transient; it occurs within hours of infection and declines in less than 10 h. Regulation of IFN synthesis

DISCUSSION
Pattern recognition receptors: the Toll-like receptors

Toll-like receptors (Tlrs) are the prototypical pattern recognition molecules. They are synthesized predominately by the sentinel cells (dendritic cells and macrophages), but can be found on other cells. Toll-like receptors are type I transmembrane proteins that are conserved from insects to humans. At least 12 members of this receptor family have been identified in mammals. Their ligands have been difficult to identify, because they are structurally diverse and vary among pathogens. However, some ligand-receptor pairs are known (Table 3.10). Ligands that might identify viral infections include CpG-containing DNA, dsRNA, and ssRNA. Unmethylated CpG tracts are present in bacterial and most viral DNA genomes, while dsRNA and ssRNA are commonly found in virus-infected cells.

After binding their unusual ligands, these Toll-like receptors, like many cytokine receptors, aggregate in the membrane, an event that stimulates binding of adapter proteins. Many downstream signaling steps in pattern recognition and inflammation are mediated by common components. For example, when ligands bind Tlrs, the IL-1 receptor-associated kinase (Irak) binds adapters such as Myd88, which engage the Tlr cytoplasmic domains to initiate signal transduction through Traf6 and then to protein kinase cascades to activate NF-κB. With the exception of Tlr3, all Tlrs engage Myd88. As indicated in the figure, the Tir domains of the Tklr and Myd88 adapter are important interfaces for complex formation. The DD domains of Myd88 and Irak act in a similar fashion. These complexes then can engage the Nf-κb, Erk, Jnk, and p38 mitogen-activated protein kinase signal transduction pathways to activate transcription of genes encoding inflammatory cytokines, IFN, and T-cell costimulatory molecules. The p38 pathway can lead to stabilization of short-lived mRNA and increased production of various cytokines. Short-lived mRNAs often have AU-rich pentameric elements (AREs) in their 3′ untranslated regions.

Tlrs can recognize extracellular as well as intracellular microbial ligands. Endocytosed proteins and virus particles end up in dendritic cell lysosomal compartments, where they can be digested. Some Tlrs, including Tlr3 and Tlr9, are located in endosomes and lysosomes, perfectly placed to bind these unusual viral products.

Tlrs are critical players in antiviral defense. Respiratory syncytial virus persists longer in the lungs of infected *tlr4*-null mice than in wild-type mice. NK cell responses and IL-12 synthesis are also reduced after challenge of these mice with this virus. One interpretation of these observations is that the Tlr4 protein is important for recognition of the infection and production of an antiviral response. An alternative idea is that virus propagation is dependent on signaling from this receptor in the dendritic cell. Other clues concerning the contribution of Tlrs to viral pathogenesis come from the study of viral proteins with the potential to disrupt their functions. Two vaccinia virus proteins, A46R and A52R, are similar in sequence to segments in the cytoplasmic domain of Tlrs and IL-1 receptors. These two viral proteins can inhibit IL-1- and Tlr4-mediated signal transduction, respectively. Vaccinia virus may modulate host immune responses by competing with this domain-dependent intracellular signaling.

Meylan, E., J. Tschopp, and M. Karin. 2006. Intracellular pattern recognition receptors in the host response. *Nature* **442**:39–44.

Saito, T., and M. Gale, Jr. 2007. Principles of intracellular viral recognition. *Curr. Opin. Immunol.* **19**:17–23.

Sioud, M. 2006. Innate sensing of self and nonself RNAs by Toll-like receptors. *Trends Mol. Med.* **12**:167–176.

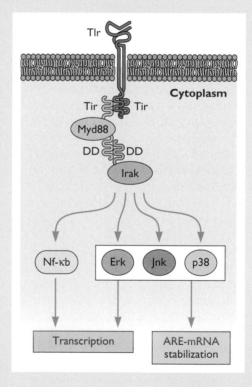

and secretion is complex, and much remains to be discovered. For example, we know that many *IFN-α* genes are differentially expressed after infection (Box 3.6). In addition, transcription of the human *IFN-β* gene is stimulated by infection, but only for a short period. The *IFN-β* enhancer possesses several remarkable properties that allow precise temporal control of transcription (Box 3.7). The biological significance of such a mix of IFN-α and IFN-β is unclear. Moreover, the quantity of IFN released from cells infected by different isolates of a given virus is astonishingly

Table 3.4 Toll-like receptors recognize microbial macromolecular patterns[a]

Toll-like receptor	Pattern recognized
Tlr1	Triacyl lipoproteins
Tlr2	Lipoproteins, viral glycoproteins; gram-positive peptidoglycan
Tlr3	Double-stranded RNA
Tlr4	Lipopolysaccharide, viral glycoproteins
Tlr5	Flagellin
Tlr6	Diacyl lipoproteins
Tlr7	Single-stranded RNA
Tlr8	Single-stranded RNA
Tlr9	CpG DNA; unmethylated CpG oligonucleotides
Tlr10	Unknown
Tlr11	Profilin

[a]Data from G. Barton and R. Medzhitov, *Curr. Opin. Biol.* **14**:380–383, 2002; S. Akira and K. Takeda, *Nat. Rev. Immunol.* **4**:499–511, 2004.

variable. In the case of vesicular stomatitis virus infection, the released IFN concentration can vary over a 10,000-fold range, depending on the serotype of the infecting virus. As discussed later, many viral proteins affect the quantity of IFN, as well as its action.

IFN Affects Only Cells with IFN Receptors

IFN functions only when it occupies its receptor on the surfaces of cells. A cell without IFN receptors may synthesize IFN, but cannot be affected by this cytokine. Binding of IFN to its receptor initiates a signal transduction cascade that culminates in increased transcription of many genes. A simplified outline of this signaling pathway is shown in Fig. 3.6.

The Jak/Stat pathway contains proteins that respond to binding to appropriate receptors of not only IFN, but also IL-6 and other cytokines. There are four known Jak kinases and seven structurally and functionally related *stat* genes. Their targeted disruption in mice has revealed much about their functions. For example, a mouse in which the *stat1* gene has been inactivated has no innate response to viral or bacterial infection, whereas deletion of *stat4* and *stat6* abrogates specific functions of the adaptive response. *stat* gene homologs are encoded in the genomes of *Drosophila melanogaster* and *Dictyostelium discoideum*, underscoring the ancient evolutionary origin of this pathway. Signaling via Jak/Stat activates transcription dependent on specific DNA sequences (Fig. 3.6). These sequences are found in the promoters of the more than 300 IFN-activated genes. Regulation is also exerted through suppressors of cytokine signaling proteins, phosphatases, and proteases.

IFN-γ Signaling

Unlike the common signaling cascades induced by IFN-α/β, IFN-γ signaling is mediated by several pathways (Box 3.8). The best-understood pathway leads to formation of the transcriptional activator called gamma-activated factor (Gaf), which binds to specific promoter sequences of IFN-γ-activated genes. While cross talk between pathways does occur, the results of such mixed signals are neither predictable nor understood (even in principle). In the milieu of an infected tissue, multiple cytokines are produced and many signaling pathways are operating in both infected and uninfected cells. In theory, the orchestration of these multiple signaling pathways enables the whole (control of infection and appropriate host response) to be greater than the sum of its parts. How the integrated output is achieved remains a mystery.

IFN Action Produces an Antiviral State

As the name so aptly indicates, IFN interferes with the replication of a wide variety of viruses in cultured cells and animals. Shortly after infection, newly made IFN released from infected cells and local immature dendritic cells can be found circulating in the body, but its concentration is highest at the site of infection, where it is bound by any cell with the appropriate receptor. Cells that bind and respond to IFN are unable to support propagation of many different viruses; they are said to be in an **antiviral state**.

What does IFN do to a cell to make it inhospitable for the replication of almost any virus? We now know that IFNs can induce the synthesis of more than 300 cellular proteins, but their mix and concentrations vary according to cell type and specific IFN. Which subset of the hundreds of IFN-inducible proteins establishes the antiviral state in any given cell remains an open question. It is possible that many antiviral constellations exist, depending on the cell type, virus, and cocktail of IFN and other cytokines sensed by that cell. Many of the products of IFN-inducible genes possess potent broad-spectrum antiviral activities, but the relevant molecular mechanisms of only a few are understood. IFN not only induces death of the infected cells, but also ensures that uninfected cells in the vicinity are induced to kill themselves should they become infected. Such a local cauterizing response has led some to characterize IFN action as a **firebreak** to infection.

The IFN-induced proteins are functionally diverse and participate in signal transduction, chemokine action, antigen presentation, regulation of transcription, the stress response, and control of apoptosis. Some of these proteins are induced by other stimuli, including dsRNA, bacterial lipopolysaccharides, Tnf-α, or IL-1.

Because IFN induces the synthesis of many deleterious gene products and is potentially lethal to any cell that has

Table 3.5   Three primary classes of cytokines[a]

Functional group	Selected members	Activity
Proinflammatory	IL-1, Tnf, IL-6, IL-12	Promote leukocyte activation
Anti-inflammatory	IL-10, IL-4, Tgf-β[b]	Suppress activity of proinflammatory cytokines; return system to basal "circulate and wait" state
Chemokines	IL-8	Recruit immune cells during early stages of immune response

[a]The terms "lymphokines" and "monokines" were originally used to denote secreted proteins produced by activated lymphocytes or monocytes, respectively. Similarly, the term "interleukin" identified proteins such as IL-2 that communicated signals between different populations of white blood cells and other nonhematopoietic cells.

[b]Tgf-β, transforming growth factor β.

specific receptors (most cells in our bodies), the production of large quantities of IFN in an infected individual has dramatic physiological effects. These responses include such common symptoms as fever, chills, nausea, and malaise. All infections lead to IFN production, one reason why these "flu-like" symptoms are so common.

Soon after its discovery, IFN was touted as a broad-spectrum antiviral drug. However, as was quickly discovered, the side effects often are worse than the infection. Nevertheless, IFN has been effective in the treatment of some persistent infections, particularly those caused by hepatitis B and C viruses.

Table 3.6   Some cytokines that function in the immune response to viral infection

Cytokine	Source	Target or action[a]
IFN-α/β	Immature dendritic cells; many types	Induces antiviral state; inhibits cell proliferation; stimulates growth and cytolytic function of NK cells; increases expression of MHC class I and decreases expression of MHC class II molecules
IFN-γ	T cell, NK cell	Activates macrophages; promotes adhesion of Th cells to vascular endothelium; inhibits IL-6 effects; induces antiviral state
Tnf-α	T cell, macrophage	Activates neutrophils; induces inflammatory response and fever, and initiates catabolism of muscle and fat (cachexia); induces adhesion molecules on vascular endothelial cells; potentiates lysis of some virus infected cells
Gm-Csf	T cell, macrophage, fibroblast, endothelial cell	Induces myelomonocytic cell growth and differentiation; important maturation protein for dendritic cells
Mip-1α	Macrophage, T cell, B cell, neutrophil, Langerhans' cell	Chemoattractant for T and B cells, monocytes, and eosinophils
Mip-1β	T cell, macrophage, B cell	Chemoattractant for T cells and monocytes
IL-1	Macrophage, T cell, B cell, epithelial cell	Costimulator of T cells; initiates inflammatory T-cell and B-cell response; affects brain to produce fever; causes metabolic wasting (cachexia); induces acute phase protein synthesis in the liver
IL-2	T cell	Induces proliferation of T cells; stimulates growth and cytolytic function of NK cells; induces antibody synthesis in B cells
IL-4	Mast cell, bone marrow	Induces B-cell proliferation and differentiation; required for production of IgE; inhibits macrophage function/blocks effects of IFN-γ on macrophages; stimulates growth and differentiation of Th2 cells; stimulates synthesis of adhesion molecules on vascular endothelium
IL-6	Macrophage, T cells, vascular endothelial cell, fibroblast	Made in response to IL-1 and Tnf; stimulates B-cell growth and costimulates T cells; stimulates hepatocytes to synthesize acute-phase proteins
IL-10	T cell, B cell	Promotes growth and differentiation of B cells; inhibits macrophage function
IL-12	Macrophage, monocyte	Potent stimulator of NK cell growth and killing activity; promotes differentiation of Th cells to Th1 subset; stimulates differentiation of immature CD8[+] T cells to functionally active CTLs; stimulates expansion and activation of autoreactive CD4[+] T cells

[a]MHC, major histocompatibility complex; IgE, immunoglobulin E; CTLs, cytotoxic T lymphocytes.

Table 3.7 Some chemokine receptors and their ligands[a]

Receptor[b]	Chemokine ligand
CCr1	Mip1α, Rantes
CCr2	Mcp1, Mcp3, Mcp4
CCr5	Mip1α, Mip1β
CCr6	Mip3α
CCr7	Mip3β
CXCr1	IL-8
CXCr2	IL-8
CXCr3	Ip-10

[a]Information in this table taken from C. R. Mackay, *Curr. Biol.* **7:**R384–R386, 1997.

[b]The four families of chemokine receptors are distinguished by the pattern of cysteine residues near the amino terminus and are abbreviated CXC, CC, C, and CX3C. Only two types are listed in this table. The CXC family has an amino acid between two cysteines; the CC family has none; the C family has only one cysteine; and the CX3C family has three amino acids between two cysteines. Subfamilies of these major four groups also exist.

Selected IFN-Induced Gene Products and Their Antiviral Actions

dsRNA-activated protein kinase. Often viral and cellular protein synthesis in infected cells stops abruptly. In many cases, this lethal defense is mediated by a cellular dsRNA-activated protein kinase (Pkr) (also described in Volume I, Chapter 11). Establishment of the Pkr-mediated antiviral state is a two-step process, in which IFN promotes the increased production and accumulation of an inactive protein that subsequently can become activated only when the cell is infected.

All mammalian cells contain low concentrations of inactive Pkr, a serine/threonine kinase with both antiviral and antiproliferative and antitumor activities. The signal

Figure 3.3 Processes inherent in viral infection that activate cellular defenses, signal transduction, and host gene expression.

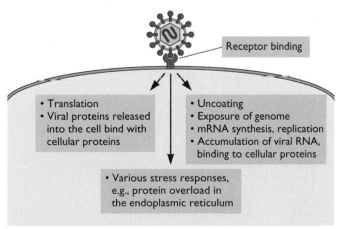

transduction cascade initiated by IFN binding to its receptor leads to a dramatic increase in the concentration of inactive Pkr. If the cell is infected, this enzyme can be activated by binding viral dsRNA. The enzyme then phosphorylates the alpha subunit of the eIF2 translation initiation protein (eIF2α), effectively rendering it incapable of supporting protein synthesis in the cell (see Volume I, Chapter 11). Phosphorylated eIF2α also can trigger autophagy, an intrinsic cell defense, discussed below.

Many viral genomes encode proteins that can block the lethal actions of Pkr (Table 3.9), but our understanding of the biological significance of this phenomenon is incomplete. An exception is the herpes simplex virus type 1 ICP34.5 protein, which redirects the cellular protein phosphatase 1 to dephosphorylate eIF2α, after it has been inactivated by Pkr (Box 3.9). While wild-type virus is fully virulent in mice, ICP34.5-null mutants are markedly attenuated, particularly in brain infections. Significantly, this mutant regains wild-type virulence in mice lacking the *pkr* gene. This experiment provides convincing evidence that Pkr mediates defense against herpes simplex virus infection in mice.

RNase L and 2′-5′-oligo(A) synthetase. Another well-studied antiviral system induced by IFN comprises two enzymes and dsRNA. Ribonuclease L (RNase L) is a nuclease that can degrade most cellular and viral RNA species. Its concentration increases 10- to 1,000-fold after IFN treatment, but the protein remains inactive unless a second enzyme is synthesized. This enzyme, 2′-5′-oligo(A) synthetase, makes oligomers of adenylic acid, but only when activated by dsRNA. These unusual nucleotide oligomers then activate RNase L, which in turn begins to degrade all host and viral mRNA. We now know from studies of mouse mutants defective in RNase L that this enzyme is important for the IFN-β response to viral infection. The products of RNase L cleavage amplify the production of IFN-β.

Mx proteins. Unlike the broad-spectrum antiviral effects of Pkr and RNase L, at least one IFN-induced mouse protein and two related human proteins appear to be directed against specific viruses. Mouse strains that have an IFN-inducible gene called *mx1* are completely resistant to influenza virus infection. The Mx1 protein is part of a small family of IFN-inducible guanosine triphosphatase (GTPases) with potent activities against various (−) strand RNA viruses. After IFN induction, this protein accumulates in the nucleus and inhibits the unusual influenza virus "cap-snatching" mechanism (Volume I, Chapter 6). It is likely that the Mx1 protein interferes with the function of the viral polymerase subunit PB2, for overproduction of this viral protein overcomes the antiviral effect of Mx1 protein. The significance of the *mx1* gene in the biology of

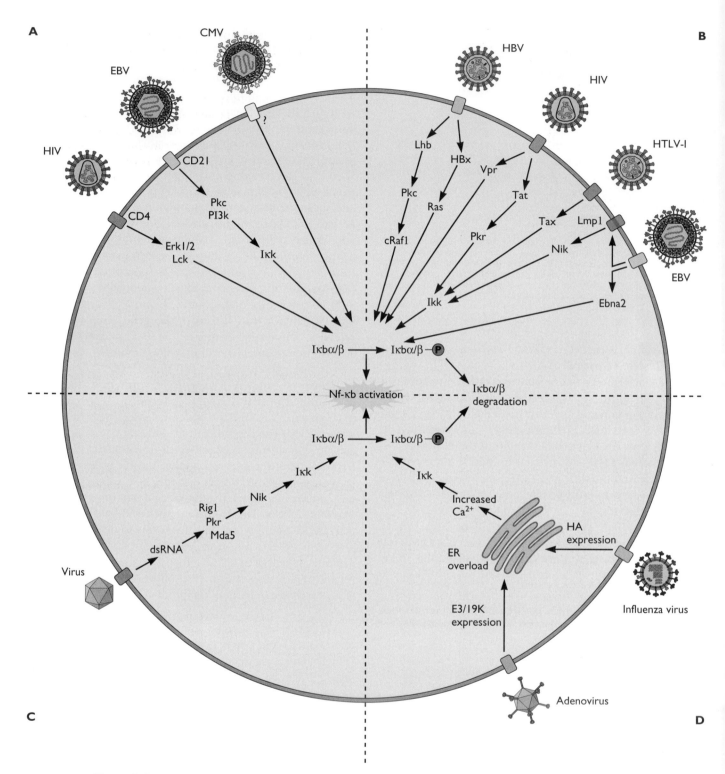

Figure 3.4 Activation of the transcription regulator Nf-κb by viral infection. Nf-κb is an important transcription control protein in the response to viral infection. Four diverse mechanisms that result in activation of Nf-κb, corresponding to the pathways described in Fig. 3.3, are illustrated. **(A)** Signal transduction pathways are activated on binding of a virus particle to its receptor; **(B)** viral proteins synthesized in the infected cell directly engage signal transduction pathways that culminate in Nf-κb activation; **(C)** Pkr binds double-stranded viral RNA, or RigI/Mda5 bind single-stranded viral RNA, leading to activation of Nf-κb; **(D)** overproduction of viral proteins in the endoplasmic reticulum (ER) leads to calcium release which, in turn, activates Nf-κb.

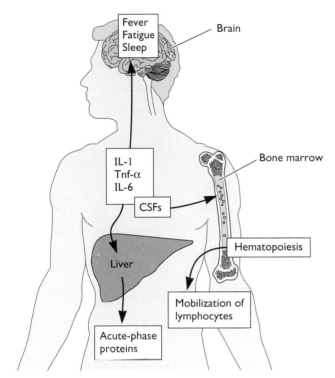

Figure 3.5 Systemic effects of cytokines in inflammation. A localized viral infection often produces global effects, including fever and lethargy, lymphocyte mobilization (swollen glands), and appearance of new proteins in the blood. The proinflammatory cytokines IL-1, IL-6, and Tnf all act on the brain (particularly the hypothalamus) to produce a variety of effects, including these typical responses to viral infection. These cytokines also act in the liver to cause the release of iron, zinc, and acute-phase proteins including mannose-binding protein, fibrinogen, C-reactive protein, and serum amyloid protein. Acute-phase proteins have innate immune defense capabilities; e.g., C-reactive protein binds phosphorylcholine on microbial surfaces and activates complement. The colony-stimulating factors (CSFs) activated by an inflammatory response have long-range effects in the bone marrow on hematopoiesis and lymphocyte mobilization. Adapted from A. S. Hamblin, *Cytokines and Cytokine Receptors* (IRL Press, Oxford, United Kingdom, 1993), with permission.

influenza virus or of mice is not at all clear, as influenza virus does not circulate among wild mice. Moreover, in these animals, about one-quarter of the population lacks a functional *mx1* gene with no clear consequences.

Two human genes related to the murine *mx1* gene are termed *mxA* and *mxB*. Expression of these genes is also induced by IFN, but unlike the murine protein, the human proteins reside in the cytoplasm. MxA, but not MxB, blocks replication of influenza virus. Interestingly, in contrast to murine Mx1, which inhibits only influenza virus, the human MxA protein also prevents replication of vesicular stomatitis virus, measles virus, and other (−) strand RNA viruses. The Mx proteins are related to members of the dynamin superfamily of GTPases, which regulate endocytosis and vesicle transport, but how this fact relates to their antiviral activities is unknown.

Interferon regulatory proteins. Members of the interferon regulatory (Irf) protein family are required for sustained IFN transcription after induction. Mice lacking the *irf1* gene are incapable of mounting an effective IFN response to viral infection. Other members of this gene family (*irf2* to *irf9*) were discovered because their protein products bound to the ISRE in promoters of IFN-regulated genes. The Irf2 protein is a repressor of transcription and cell growth. Irf4 is synthesized only in T and B cells, and Irf8 is made only in cells of the macrophage lineage. Mice defective for *irf8* gene expression are markedly more susceptible to infection and cannot synthesize proinflammatory cytokines. The protein Irf9 (also known as Isgf3 or p48) is the DNA-binding component of the transcriptional regulator Isgf3 (Fig. 3.6). Several viral Irf-like proteins that block IFN action have been identified (Table 3.9).

Nitric oxide synthase. Nitric oxide synthase is an IFN-γ-inducible protein with important antiviral activities. This enzyme exists as several isoforms, each of which has a distinctive tissue distribution. Nitric oxide synthase produces nitric oxide during the conversion of arginine to citrullene. Nitric oxide exerts a variety of antiviral effects, including inhibition of poxvirus and herpesvirus replication. Nitric oxide made by IFN-γ-activated NK cells accounts for much of their antiviral activity (see "NK cells" below).

Promyelocytic leukemia proteins. The promyelocytic leukemia (Pml) proteins are present in both the nucleoplasm and discrete multiprotein complexes known as nuclear bodies (Pml bodies, ND10 bodies, or PODs [discussed in Volume I, Chapter 9]). These structures are important in the intrinsic cellular response to infection because their components bind foreign DNA that enters the nucleus. Pml and other proteins present in the complexes are thought to exert their antiviral effects by transcriptional repression and nucleosome remodeling. Many viral infections promote disorganization of Pml bodies, in part as a measure to override global repression.

Ubiquitin-proteasome pathway components. The proteasome is a large multisubunit protease that destroys cytoplasmic and nuclear proteins targeted by polyubiquitination for proteolysis. This process is important for the destruction of abnormal or damaged proteins, in the turnover of short-lived regulatory proteins, and in the production of peptides for assembly of MHC-I complexes. All interferons induce transcription of a number of genes that

---

BOX 3.5

BACKGROUND
The interferon system is crucial for antiviral defense

Many steps in a viral life cycle can be inhibited by interferon, depending on the virus family and cell type. Binding of this cytokine to its receptor leads to increased transcription of more than 300 genes. The proteins so produced can inhibit viral penetration and uncoating, synthesis of viral mRNAs or viral proteins, replication of the viral genome, and assembly and release of progeny virions. More than one of these steps in a virus life cycle can be inhibited, providing a strong cumulative effect.

The contribution of interferon can be demonstrated in animal models in which the interferon response is reduced by treatment with anti-interferon antibodies, or in mice harboring mutations that delete

or inactivate interferon genes, interferon receptor genes, genes that regulate the interferon response, or genes that are induced by interferons.

Animals with a defective interferon response exhibit a reduced ability to contain viral infections, and often show an increased incidence of illness or death. When the IFN-α/β response is abrogated, there is a global increase in susceptibility to most viruses. When the IFN-γ response is blocked, viral pathogenesis is modestly affected, at best.

These observations suggest that IFN-α/β is crucial as a general antiviral defense, whereas the IFN-γ response has other roles.

Huang, S., W. Hendriks, A. Althage, S. Hemmi, H. Bluethmann, R. Kamijo, J. Vilcek, R. M. Zinkernagel, and M. Aguet. 1993. Immune response in mice that lack the interferon-gamma receptor. *Science* **259:**1742–1745.

Ryman, K., W. B. Klimstra, K. Nguyen, C. Biron, and R. Johnston. 2000. Alpha/beta interferon protects adult mice from fatal Sindbis virus infection and is an important determinant of cell and tissue tropism. *J. Virol.* **74:**3366–3378.

Stojdl, D., N. Abraham, S. Knowles, R. Marius, A. Brasey, B. Lichty, E. Brown, N. Sonenberg, and J. C. Bell. 2000. The murine double-stranded RNA-dependent protein kinase Pkr is required for resistance to vesicular stomatitis virus. *J. Virol.* **74:**9580–9585.

Zhou, A., J. Paranjape, S. Der, B. Williams, and R. Silverman. 1999. Interferon action in triply deficient mice reveals the existence of alternative anti-viral pathways. *Virology* **258:**435–440.

---

encode proteins of the ubiquitin-proteasome pathway. In fact, many interferon-stimulated genes encode ubiquitin ligases. Increased protein degradation may contribute to the antiviral response to some viruses. For example, proteasome inhibitors block the anti-hepatitis B virus action of IFN and IFN-γ. For this virus, activation of the proteasome may be **the** major antiviral effect, because the results of other experiments demonstrate that the Pkr and RNase L systems are completely ineffective. The hepatitis B virus X protein binds to various proteasomal subunits and may modulate protease activity.

Other IFN-induced proteins.   The most strongly induced IFN gene, *isg56*, is one member of a gene family encoding the structurally related proteins p54, p56, p58, and p60. The p58 protein interacts with the p48 subunit of the translation

initiation protein eIF3 and blocks the initiation of protein synthesis. Expression of intracellular nucleic acid detectors such as RigI and Mda5 is induced after IFN treatment. The expression of other intrinsic defense gene products, such as Trim5α, is increased as well. Other proteins with antiviral effects certainly remain to be discovered among the 300-plus IFN-induced proteins. For example, the IFN response is required to clear human cytomegalovirus infections, but Pkr, Mx, and RNase L proteins are not. Similarly, uncharacterized IFN-induced proteins block penetration and uncoating of simian virus 40 and some retroviruses. Others

---

BOX 3.6

BACKGROUND
Differential induction of IFN-α genes by viral infection

- Most cells produce IFN-α from more than 20 genes, depending on the species.
- *IFN-α* genes are expressed differentially. Different members of this gene family are induced by specific viruses. For example, transcription of *IFN-A4* but not *IFN-A1* is induced by infection of mice with Newcastle disease virus. The function of such mixtures of potent IFNs is not understood and not well studied.
- Differential expression occurs as a result of specific binding of the transcriptional activator proteins Irf3 and Irf7 to particular *IFN-α* promoters.

Pitha, P. M., and W. C. Au. 1995. Induction of interferon-alpha gene expression, p. 151. *In* P. M. Pitha (ed.), *Interferon and Interferon Inducers*. Academic Press, London, United Kingdom.

Table 3.8   The interferons: antiviral cytokines

Interferon[a]	Producer Cells	Inducers
IFN-α	Most if not all nucleated cells	Viral infection
IFN-β	Most if not all nucleated cells	Viral infection
IFN-γ	T cells, NK cells	T-cell receptor activation, IL-2, IL-12

[a]IFN-α and IFN-β are sometimes called type I interferons. IFN-γ is often called type II or immune interferon. The molecular mass of all IFNs is approximately 20 kDa. Human IFN-α is produced from more than 24 closely related, intronless genes. Human IFN-β is produced from a single intronless gene with about 30 to 45% homology to IFN-α genes. IFN-γ is produced from a single-copy gene with three introns. There is little homology to IFN-α and IFN-β.

Switching IFN-β transcription on and off

Viral infection activates transcription of the human *IFN-β* gene, but only for a short period. This on-off response is controlled by an enhancer located immediately upstream of the core promoter. Like other enhancers, this regulatory sequence contains binding sites for multiple transcriptional activators, including Nf-κb and members of the Ap-1 and Atf families. However, the *IFN-β* enhancer possesses several remarkable properties that allow precise temporal control of transcription.

- The enhancer also contains four binding sites for the architectural protein Hmg(y), which alters DNA conformation to direct the assembly of a precisely organized nucleoprotein complex on the enhancer.
- In contrast to typical modular enhancers, all binding sites **and** their natural arrangement are essential for activation of *IFN-β* transcription specifically in response to viral infection.
- Formation of the complex takes place in stages, and is not complete until several hours after infection.
- Activation of transcription then requires sequential recruitment of the histone acetylase Gcn5, the coactivator Cbp and RNA polymerase II, and the chromatin-remodeling complex Swi/Snf.
- In addition to modifying nucleosomes, Gcn5 acetylates Hmg(A1) at Lys71. This modification stabilizes the complex.
- The Hmg(A1) protein is also acetylated by Cbp at Lys65. However, **this**

modification impairs DNA-binding activity and results in disruption of the complex and cessation of *IFN-β* transcription.

- Remarkably, this inhibitory modification by Cbp is blocked for several hours by the prior Gcn5 acetylation of Hmg(A1). The "off" switch is delayed for a sufficient period to allow a burst of *IFN-β* transcription.

Munshi, N., T. Agalioti, S. Lomvardas, M. Merika, G. Chen, and D. Thanos. 2001. Coordination of a transcriptional switch by HMG1(Y) acetylation. *Science* **293:**1133–1136.

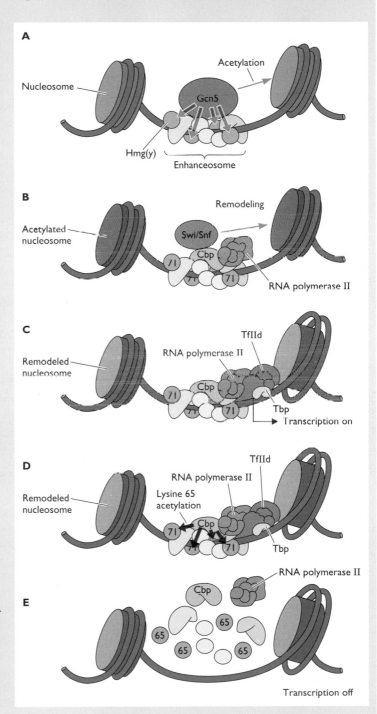

(A) Viral infection of human cells leads to assembly of a multiprotein complex on the *IFN-β* enhancer, which lies in a nucleosome-free region of the gene. The signals that direct binding of transcriptional activators (blue, yellow, and tan) and Hmg(A1) (green) are not fully understood. The precisely organized surface of the complex allows binding of Gcn5, which acetylates both histones in nearby nucleosomes and Lys71 of Hmg(A1) (green arrows). This modification stabilizes the enhanceosome. **(B)** A complex of Cbp and RNA polymerase II and the chromatin-remodeling protein Swi/Snf bind sequentially to the stabilized complex. The latter alters the adjacent nucleosome that contains the core promoter DNA (green arrow). **(C)** Such alteration allows binding of TfIId and activation of transcription. Because Lys71 of Hmg(A1) is acetylated, Cbp cannot acetylate Lys65. **(D)** Eventually, Cbp does acetylate Lys65 of Hmg(A1) (red arrows), but how the inhibition induced by Lys71 acetylation is overcome is not yet clear. **(E)** Regardless, Hmg(A1) modification by Cbp disrupts the complex and switches off transcription. Adapted from K. Struhl, *Science* **293:**1054, 2001, with permission.

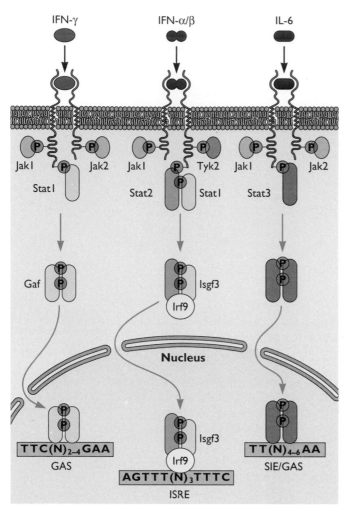

Figure 3.6 Overlapping signal transduction pathways for IFN-α/β, IFN-γ, and IL-6. IFN signals via the Jak/Stat pathway, characterized by a family of tyrosine kinases given the acronym Jak (Janus kinases; Janus, a Roman god, guardian of gates and doorways, is represented with two faces and therefore faces in two directions at once) and a set of transcription proteins named Stat (signal transduction and activators of transcription). The receptors for IFN-α/β, IFN-γ, and IL-6 are different, but all affect components of the Jak/Stat signal transduction pathway. All three interferon proteins and IL-6 bind to their receptors with high affinity (equilibrium dissociation constant [K_d] of about 10^{-10} M). Binding of IFN or IL-6 to the appropriate receptor leads to tyrosine phosphorylation of tyrosine kinases as well as of the receptor itself. These modifications are followed by tyrosine phosphorylation of the Stat proteins. In mammals there are seven Stat genes. The phosphorylated Stat proteins then form a variety of dimers that enter the nucleus. Within that organelle, Stat dimers bind, in some cases in conjunction with other proteins (e.g., Irf9), to specific transcriptional control sequences of IFN-α/β, IFN-γ, and IL-6-inducible genes called interferon-stimulated response elements (ISREs), gamma-activated site (GAS) elements, and Sis-inducible element (SIE), respectively. Later in the transcriptional response to IFN, a second transcriptional activator called Irf1 replaces Isgf3. For further information, see D. S. Aaronson and C. M. Horvath, *Science* **296:**1653–1655, 2002.

impair the maturation, assembly, and release of vesicular stomatitis virus, herpes simplex virus, and some retroviruses by unknown mechanisms.

Cellular micro-RNAs. Micro-RNAs (miRNAs) are single-stranded, noncoding host RNAs of 19 to 25 nucleotides that regulate gene function. While cellular miRNAs are active in the antiviral response in plants and invertebrates, their role in antiviral action in mammals is only now being analyzed. In one study, IFN-β treatment of the human hepatoma cell line Huh7 as well as freshly isolated primary murine hepatocytes, resulted in an induction of numerous cellular miRNAs. Moreover, eight of these miRNAs targeted hepatitis C virus genomic RNA: treatment of infected cells with synthetic miRNAs of the same sequence blocked virus replication. IFN-β treatment also reduces expression of liver-specific miR-122, an RNA known to be essential for hepatitis C virus replication. In another example, host miRNAs play central roles in shutting down human immunodeficiency virus type 1 transcription in blood mononuclear cells from infected donors. However, in other cell types, viral infection suppresses the host miRNAs that normally would repress proviral gene expression. Further work is in progress in many laboratories to determine if IFN induction of miRNAs is a general antiviral response in mammalian cells.

Viral Gene Products That Counter the IFN Response

The term "antiviral state" implies that the IFN response confers complete resistance to infection. However, it is misleading, because infections vary considerably in their sensitivity to the effects of this cytokine. The replication of some viruses, like vesicular stomatitis virus, is so sensitive to IFN that this property is used to titrate the cytokine. Other viruses can be more resistant to IFN. We now know that there are numerous viral mechanisms for confounding IFN production and action (Table 3.9). For example, viral soluble IFN receptors function as decoys, and viral regulatory proteins alter or block IFN-stimulated transcription.

Many viral genomes encode dsRNA-binding proteins that interfere with IFN induction. For example, the reovirus σ3 protein, the multifunctional influenza virus NS1 protein, and the hepatitis B virus core antigen are all well-characterized dsRNA-binding proteins with anti-IFN effects. The vaccinia virus E3L protein and the herpes simplex virus type 1 US11 protein also have dsRNA-binding properties that correlate with inhibition of IFN induction (Box 3.9). Interestingly, adenovirus VA-RNA I acts as a dsRNA decoy and blocks the activation of Pkr by directly binding to the enzyme.

BOX 3.8

BACKGROUND

IFN-γ: a powerful cytokine quite distinct from the type I interferons

- IFN-γ is produced primarily by NK and T cells, which are critical players in the innate and adaptive immune responses, respectively.
- Production of IFN-γ is stimulated by IL-1, IL-2, estrogen, and IFN-γ itself.
- IFN-γ synthesis is suppressed by glucocorticoids and the cytokines transforming growth factor β and IL-10.

- Transcription of *IFN-γ* is regulated by repressors and activators. A combination of positive and negative regulatory sequences in the *IFN-γ* promoter is essential for controlled expression, but the details remain to be elucidated.
- IFN-γ translation is also regulated in a remarkable fashion: a pseudoknot structure in the 5′ end of the mRNA

activates Pkr, which, in turn, blocks translation of IFN-γ mRNA.

Ben-Asouli, Y., Y. Banai, Y. Pel-Or, A. Shir, and R. Kaempfer. 2002. Human IFN-γ mRNA autoregulates its translation through a pseudoknot that activates the interferon-inducible protein kinase Pkr. *Cell* **108:**221–232.

Goodbourn, S., L. Didcock, and R. E Randall. 2000. Interferons: cell signaling, immune modulation, anti-viral responses and virus countermeasures. *J. Gen. Virol.* **81:**2341–2364.

Table 3.9   Some viral modulators of the interferon response[a]

Type of modulation	Representative viruses	Viral protein, if known	Mechanism of action
Inhibition of IFN synthesis	Epstein-Barr virus	Bcrf1	IL-10 homolog, inhibits production of IFN-γ
	Vaccinia virus	A18R	Regulates dsRNA production
	Foot-and-mouth disease virus	L	Host protein synthesis block
IFN receptor decoys	Vaccinia virus	B18R	Soluble IFN-α/β decoy receptor
Inhibition of IFN signaling	Adenovirus	E1A	Decreases quantity of Stat1 and P48; blocks Isgf3 formation; interferes with Stat1 and CBP/P300 interactions
	Vaccinia virus	VH1	Viral phosphatase reverses Stat1 activation
	Human papillomavirus 16	E7	Binds p48
	Hepatitis C virus	NS5a	Blocks formation of Isgf3 and Stat dimers
	Nipah virus	V protein	Prevents Stat1 and Stat2 activation and nuclear accumulation
Block function of IFN-induced proteins	Adenovirus	VA-RNA 1	Binds dsRNA, blocks Pkr
	Herpes simplex virus type 1	US11	Blocks Pkr activation
		γ34.5	Redirects protein phosphatase 1α to dephosphorylate eIF2α; reverses Pkr action
	Vaccinia virus	E3L	Binds dsRNA and blocks Pkr
		K3L	Pkr pseudosubstrate, decoy
	Human immunodeficiency virus type 1	TAR RNA	Blocks activation of Pkr
		Tat	Pkr decoy
	Hepatitis B virus	Capsid protein	Inhibits MxA
	Influenza virus	NS1	Binds dsRNA and Pkr; blocks action of ISG15
	Reovirus	σ3	Binds dsRNA, inhibits Pkr and 2′-5′ oligo (A) synthase

[a]For further examples and details, see B. B. Finlay and G. McFadden, *Cell* **124:**767–782, 2006.

BOX 3.9

EXPERIMENTS
Three herpes simplex virus proteins modulate the IFN response in the cytoplasm and the nucleus

Herpes simplex virus is only marginally sensitive to IFN in cultured cells, yet this cytokine plays a major role in limiting acute infection in animals. At least three viral proteins modulate the IFN response.

- ICP34.5 acts as a regulatory subunit of cellular protein phosphatase 1 and reverses Pkr-induced eIF2α phosphorylation. Consequently, this viral protein prevents the translational block and autophagy induction established by IFN.

- US11 is an RNA-binding protein that prevents Pkr activation.
- ICP0 functions in the nucleus to launch the replication cycle. This protein prevents the IFN-induced block to RNA polymerase II transcription.

Varicella-zoster virus, a closely related alphaherpesvirus, is similarly insensitive to IFN in cultured cells, but it has no counterparts to the ICP34.5 or US11 genes. It is unclear if the ICP0 homolog provides the only IFN defense for this common herpesvirus of humans.

Harle, P., B. Sainz, Jr., D. J. Carr, and W. P. Halford. 2002. The immediate-early protein, ICP0, is essential for the resistance of herpes simplex virus to interferon-alpha/beta. *Virology* **293:**295–304.

Mossman, K., and J. R. Smiley. 2002. Herpes simplex ICP0 and ICP34.5 counteract distinct interferon-induced barriers to virus replication. *J. Virol.* **76:**1995–1998.

An inescapable inference from the various countermeasures encoded by the genomes of diverse viruses is that IFN is an essential host defense component (Fig. 3.7). But numerous questions remain. For example, some infections (e.g., Newcastle disease virus) are inhibited only by IFN-α, while others (e.g., herpes simplex virus type 1) are inhibited primarily by IFN-β. IFN production is induced after infection by vaccine strains of measles virus, while little IFN is made after wild-type virus infection. When animals are infected, the IFN response varies depending on the route of infection (Box 3.10). It is clear that much remains to be learned about the antiviral effects of IFN and virus countermeasures.

Apoptosis (Programmed Cell Death)

Cell suicide is a potent intrinsic defense. The biochemical alterations initiated by infection can induce a process of controlled self-destruction called **apoptosis** (Fig. 3.8 and 3.9). Apoptosis normally functions to eliminate particular cells during development and differentiation. Cells in organs are in numerical equilibrium despite rather large changes in cell birth and death in response to physiological stimuli. Accordingly, apoptotic pathways are strictly controlled by a variety of mechanisms that monitor the processes of growth regulation, cell cycle progression, and metabolism. Death by apoptosis does not result in inflammation or activation of other defensive reactions. However, many diseases (including cancer) are associated with malfunctions in regulation of apoptosis. Survival signals from the cell's environment, and internal signals reporting on cell integrity, normally keep the apoptotic response in check. When these signals are perturbed by a variety of events such as viral infection, cell death invariably ensues. Cellular debris resulting from apoptosis is taken up by macrophages and dendritic cells, which are thereby stimulated to produce cytokines and migrate to local lymph nodes. Exogenous proteins are processed into peptides, and presented to T cells in the lymph nodes. If nonself peptides are detected, an immune response may be activated.

Controlled cell suicide, or apoptosis, can be activated by a large variety of both external and internal stimuli. Regardless of the nature of the initiation signal, all converge on common effectors, the **caspases.** Caspases are members of a family of *c*ysteine proteases that specifically cleave after *asp*artate residues. These proteases are first synthesized as precursors with little or no activity. A mature caspase with full activity is produced after cleavage by another protease (often another caspase). Alternatively, increasing the concentration of some caspase precursors results in cleavage-independent activation. These protease cascades are not unlike blood clotting or the complement cascade (see Chapter 4, "Complement"). The principle is similar: a modest initial signal can be amplified significantly, culminating in an all-or-none response.

Two convergent caspase activation cascades are known: the extrinsic and intrinsic pathways. The **extrinsic pathway** begins when a cell surface receptor binds a proapoptotic ligand (e.g., the cytokine Tnf-α). Binding changes the cytoplasmic domain of the receptor so that death-inducing signaling proteins are recruited (Figure 3.9B). This complex of proteins attracts pro-caspase 8, which is activated on binding. Caspase 8 cleaves and activates pro-caspase 3, the final effector for both extrinsic and intrinsic pathways.

The **intrinsic pathway**, often called the mitochondrial pathway, integrates stress responses, as well as internal developmental cues. Common intracellular initiators include DNA damage and ribonucleotide depletion. In

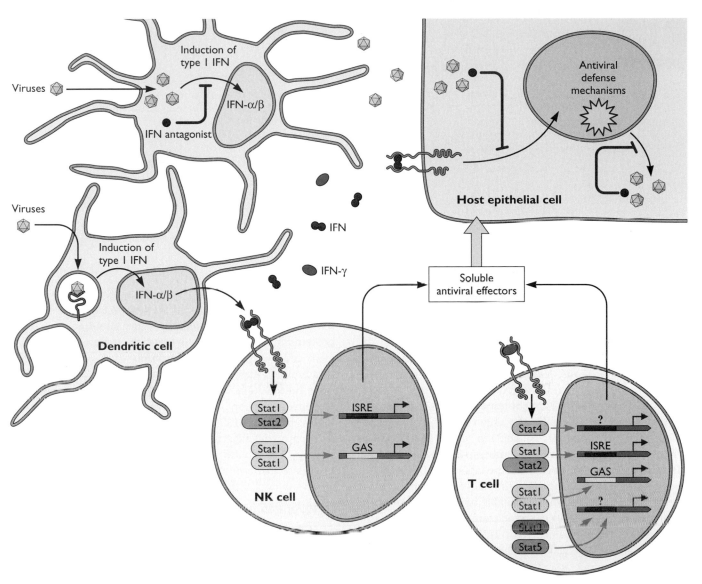

Figure 3.7 Virus-mediated modulation of IFN production and action. The pattern recognition receptors present in most cells, including dendritic cells, detect viral nucleic acid and viral proteins on the cell surface, in endosomes, and in the cytoplasm. IFN is produced as a consequence of these intrinsic defenses. Secreted IFN then binds to cells with IFN receptors and activates different but overlapping signal transduction cascades depending on the cell type. These cascades result in new gene expression including expression of IFN and other antiviral proteins. Viral gene products modulate most steps in the IFN response from the infected cell to the responding cell. This modulation changes the dynamics of cytokine production and action in ways that are not fully understood. For example, the dendritic cells (blue cytoplasm) detect viral infection or products of viral infection and produce IFN (purple ovals) and IFN-γ (blue circles). However, viral infection may lead to modulation of IFN production in these primary defense cells (red line, IFN antagonist). The IFN produced by dendritic cells can bind to receptors on innate immune cells (e.g., NK cells [yellow cytoplasm]) or T cells (yellow cytoplasm), leading to production of IFNs and other IFN inducible genes (indicated by the question mark). The combination of NK cell and T-cell action should produce soluble antiviral effectors leading to destruction of other infected host cells (e.g., epithelial cells, orange cytoplasm). However, viral gene products produced in these infected target cells can modulate IFN signaling or block recognition of the infected cell by NK cells or T cells. As a result, viral infected cells are exposed to a rapidly changing pallette of cytokines produced, not only by the infected cell, but also by innate and adaptive immune cells reacting to the infection. Adapted from A. Garcia-Sastre and C. Biron, *Science* **312**:879–882, 2006, with permission.

**BOX
3.10**

WARNING
Routes of infection make a difference

Most natural infections begin at mucosal surfaces. For ease of experimentation, many investigators resort to using rather unnatural routes, including injecting virus into the bloodstream or the peritoneal cavity. Can we assume that the innate defenses activated by unnatural infections are similar to those following infection by a natural route? Infections of transgenic animals lacking innate defense genes have provided some insight.

Mice are relatively resistant to infection by vesicular stomatitis virus no matter the route of infection. However, Pkr-null mutant mice become highly susceptible to this virus, **but only** after respiratory infection.

These findings show that the Pkr response in the respiratory tract is a primary defense against the virus. Defense after injection of virions into the bloodstream or the peritoneal cavity is mediated by some other mechanism.

The lesson is to be wary of generalizations about the innate immune response. Not all routes of infection are defended in the same manner.

Levy, D. E. 2002. Whence interferon? Variety in the production of interferon in response to viral infection. *J. Exp. Med.* **195:**F15–F18.

these situations, the cell cycle regulatory protein p53 is activated (see Volume I, Chapter 7) and apoptosis ensues. One single family of proteins, the Bcl-2 family, controls the process (Fig. 3.9A). These proteins are the master regulators that inhibit apoptosis. Their activity is regulated by proapoptotic proteins, which, curiously, are also Bcl-2 family members. The differential binding of antiapoptotic Bcl-2 proteins to BH3-only proteins enable tissue-specific regulation and stress-specific responses (Fig. 3.9A).

In general, apoptosis is held in check because the antiapoptotic Bcl-2 family members Bcl-xL and Mcl-1 directly block the translocation of proapoptotic Bax and Bak to the mitochondria. If these proteins reach mitochondrial membranes, they become permeable and internal stores of cytochrome *c* are released. Cytochrome *c* in the cytoplasm binds to a cellular protein (Apaf-1), which oligomerizes in the presence of deoxyadenosine 5' triphosphate (dATP) or ATP. The oligomeric assembly then binds and cleaves pro-caspase 9, which in turn activates pro-caspase 3. The extrinsic and intrinsic signaling pathways can converge in other ways. For example, if the extrinsic pathway is activated, mature caspase 8 may cleave a proapoptotic protein called Bid that then translocates to the mitochondria to trigger the intrinsic pathway. There is ample evidence to suggest that the intrinsic pathway can serve to amplify the extrinsic pathway.

Once caspase 3 is activated (no matter what the initial signal), the end results are always the same: cell and organelle dismantling, vesicle and membrane bleb formation, phosphatidylserine exposure on the cell surface, and DNA cleavage to nucleosome-sized fragments (Fig. 3.8A).

Apoptosis as a Defense against Viral Infection

Because virions engage cell receptors, and because viral replication engages all or part of the host's transcription, translation, and replication machines, a variety of signals may activate the extrinsic and intrinsic pathways

(Box 3.11). In many infections, the target cell is quiescent and hence unable to provide the enzymes and other proteins needed by the infecting virus. Consequently, viral proteins induce the cell to leave the resting state. However, cell cycle checkpoint proteins then respond to this unscheduled event by inducing apoptosis. Typically, viral early proteins activate the cell cycle (e.g., adenoviral E1A proteins or simian virus 40 large T protein). Not surprisingly, most viral infections trigger apoptosis. Accordingly, to ensure that infected cells survive long enough to produce progeny, viral genomes encode gene products that modulate this potentially lethal process.

Viral Gene Products That Inhibit Apoptosis

The discovery of viral proteins that modulate the apoptotic pathway proved exceedingly valuable in dissecting the complex pathways and regulatory circuits in normal cells. As noted above, apoptosis is normally held in check by regulatory proteins called inhibitors of apoptosis (IAPs). The prototype IAP gene was described in baculovirus genomes by the late Lois Miller and colleagues in 1993. This seminal work led to the discovery of cellular orthologs in yeasts, worms, flies, and humans. Mutant viruses unable to inhibit apoptosis were detected originally because the host DNA of infected cells was unstable, the cells lysed prematurely, and, as a consequence, viral yields were reduced, resulting in small plaques. Since then, we have discovered many viral proteins that regulate or block apoptosis (Table 3.10).

Remarkably, human cytomegalovirus encodes an abundant, 2.7-kb noncoding RNA (β2.7) that binds to and inhibits a mitochondrial protein complex that triggers apoptosis. Not only is apoptosis blocked, but mitochondrial function is maintained so that cells do not die quickly. The mitochondrial membrane potential is maintained, enabling

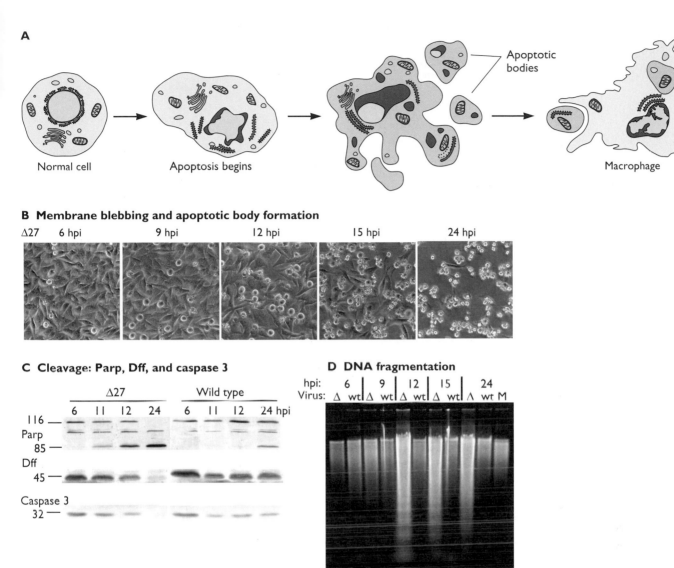

A

Normal cell

Apoptosis begins

Apoptotic bodies

Macrophage

B Membrane blebbing and apoptotic body formation

Δ27   6 hpi   9 hpi   12 hpi   15 hpi   24 hpi

C Cleavage: Parp, Dff, and caspase 3

Δ27   Wild type

6 11 12 24   6 11 12 24 hpi

116 —
Parp
85 —

Dff
45 —

Caspase 3
32 —

D DNA fragmentation

hpi: 6 | 9 | 12 | 15 | 24
Virus: Δ wt| Δ wt| Δ wt| Δ wt| Δ wt M

Figure 3.8 Apoptosis, the process of programmed cell death. (A) Apoptosis can be recognized by several distinct changes in cell structure. A normal cell is shown at the left. When programmed cell death is initiated, as indicated by the second cell, the first visible event is the compaction and segregation of chromatin into sharply delineated masses that accumulate at the nuclear envelope (dark blue shading around periphery of nucleus). The cytoplasm also condenses, and the outline of the cell and nuclear membranes changes, often dramatically. The process can be rapid: within minutes the nucleus fragments and the cell surface convolutes, giving rise to the characteristic "blebs" and stalked protuberances illustrated. These blebs then separate from the dying cell and are called apoptotic bodies. Macrophages (the cell at the right) engulf and destroy these apoptotic bodies. Adapted from J. A. Levy, *HIV and the Pathogenesis of AIDS,* 2nd ed. (ASM Press, Washington, DC, 1998), with permission. **(B)** Apoptosis in human epithelial cells infected with herpes simplex virus type 1. Infection by herpes simplex virus type 1 initiates apoptosis, but *de novo* synthesis of infected-cell proteins prevents cell death. The viral immediate-early ICP27 protein is one of several proteins involved in this inhibition: infection by a mutant virus with a deletion of the ICP27 gene (Δ27) induces apoptosis but does not block cell death. Phase-contrast microscopy of human epithelial cells infected with Δ27 is shown. Membrane blebbing and apoptotic-body formation typical of apoptosis increase with time after infection. hpi, hours postinfection. **(C)** Determination of apoptosis induction after infection by monitoring proteolytic processing of three proteins: DNA fragmentation factor (Dff), poly(ADP-ribose) polymerase (Parp), and caspase 3. Proteins are detected with specific antibodies. The processing of Parp, a 116-kDa protein, produces an 85-kDa product. Apoptosis-induced processing of Dff (45 kDa) and caspase 3 (32 kDa) results in the loss of reactivity with the specific antibodies. **(D)** DNA fragmentation after infection. Appearance of a ladder of short DNA fragments is indicative of apoptosis. The time course of DNA laddering in cells infected with Δ27 or wild-type (wt) herpes simplex virus type 1 is shown. Reprinted from M. Aubert and J. A. Blaho, *Microbes Infect.* **3:**859–866, 2001, with permission.

A Intrinsic death receptor pathway

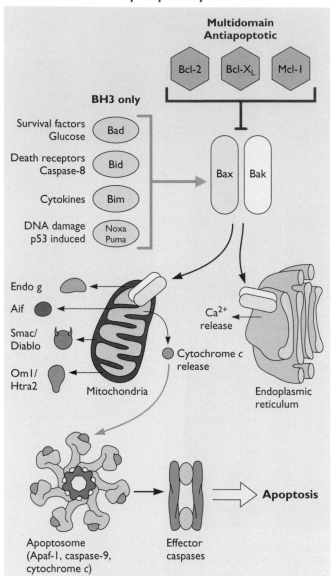

B Extrinsic death receptor pathway

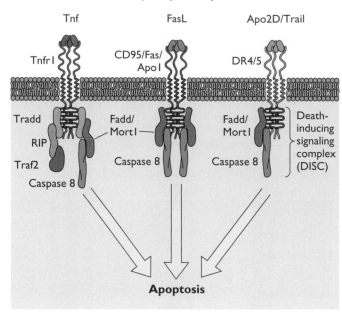

Figure 3.9 Pathways to apoptosis. (A) The process of apoptosis is controlled by the Bcl-2 family of proteins. The central antiapoptosis regulators are Bcl-2, Bcl-X$_L$, and Mcl-1 (pink hexagons). These proteins keep Bax and Bak (proapoptotic proteins [yellow]) from assembling on the mitochondrial or endoplasmic reticulum membranes and causing release of cytochrome c and calcium, respectively. A variety of other proteins are released from mitochondria after Bax/Bak action, as indicated. Four other classes of Bcl-2 regulatory proteins that also bind to different subsets of the Bcl-2 proteins are indicated at the top left (blue ellipses). These four classes act under conditions where survival is threatened (Bad), when the extrinsic pathway is stimulated (Bid), when certain cytokines are produced (Bim), and when DNA damage is detected and p53 is induced (Noxa and Puma). The oligomerization of Apaf-1, caspase 9, and cytochrome c forms a large structure called the apoptosome, which then activates effector caspases that produce the characteristic events of controlled cell suicide. **(B)** The extrinsic death receptors and their death-inducing signaling complexes. Three central receptors found on the surfaces of cells that can initiate the apoptosis pathway are illustrated. These are the Tnf receptor, the Fas ligand receptor (CD95), and the Apo2/Trail receptor (DR4/5). When these receptors engage their cognate ligand, the cytoplasmic domains of each protein complex forms a scaffold for assembly of the death-inducing signaling complex (DISC). Important cytoplasmic proteins in this complex are shown. Caspase 8 is activated when it binds these complexes, which initiates the apoptopic pathway. Adapted from N. Danial and S. Korsmeyer, *Cell* **116**:205–219, 2004, with permission.

BOX 3.11

BACKGROUND
The many ways by which virus infections perturb apoptotic pathways

At the Cell Surface

Production of apoptosis-inducing cytokines after virions bind their receptors

Alteration of membrane integrity or composition via membrane fusion or virion passage into the cytoplasm via receptor mediated endocytosis

In the Cytoplasm

Production of metabolic inhibitors (e.g., arrest of host translation)

Modification of cytoskeleton (e.g., disruption of actin microfilaments)

Disruption of signal transduction pathways (e.g., death domain proteins and kinase- and phosphatase-binding proteins)

In the Nucleus

Degradation of and damage to DNA

Alteration of gene expression (e.g., increased expression of heat shock genes)

Disruption of the cell cycle (e.g., inactivation of p53 or pRb)

Hay, S., and G. Kannourakis. 2002. A time to kill: viral manipulation of the cell death program. *J. Gen. Virol.* **83:**1547–1564.

Miller, L. K., and E. White (ed.). 1998. Apoptosis in viral infections. *Semin. Virol.* **8:**443–523.

continuing synthesis of ATP. It remains to be seen how many other viral RNA molecules are inhibitors of apoptosis.

Apoptosis Is Monitored by Sentinel Cells

Specialized phagocytes, called dendritic cells and macrophages, monitor most living tissues (see also Volume I, Chapter 4). These mobile phagocytes are critical players in early defense, as well as in activating a more global immune response. We call them **sentinel cells** because their function is to gather information (as packets of proteins) by taking up cellular debris and extracellular proteins released

from dying cells. Then, after migrating to local lymph nodes, the sentinel cells bind to and present their collected peptide fragments to lymphocytes of the adaptive immune system (the T cells in particular). This cell-cell communication informs T cells about the nature of the insult that is killing cells in peripheral tissues, and the T cells respond accordingly. The sentinel cells, as well as the damaged and dying cells, produce cytokines that can induce apoptosis in nearby infected cells. For example, Tnf-α is a cytokine that can induce apoptosis when it binds to the Tnf receptor (a so-called "death receptor") (Fig. 3.9B).

Table 3.10 Some viral regulators of apoptosis[a]

Cellular target	Virus	Gene	Function
Bcl-2	Adenovirus	E1B 19K	Bcl-2 homolog
	Epstein-Barr virus	LMP-1	Increases synthesis of Bcl-2; mimics CD40/Tnf receptor signaling
Caspases	Adenovirus	14.7K	Inactivates caspase 8
Cell cycle	Hepatitis B virus	pX	Blocks p53 mediated apoptosis
	Human papilloma virus	E6	Targets p53 degradation
	Simian virus 40	Large T	Binds and inactivates p53
Fas/TNF receptors	Adenovirus	E3 10.4/14.5K	Internalizes Fas
	Cowpox	CrmB	Neutralizes Tnf and LT-α
	Myxoma virus	MT-2	Secreted Tnf receptor homolog
vFLIPs; DED box-containing proteins	Human herpesvirus 8	K13	Blocks activation of caspases by death receptors
Oxidative stress	Molluscum contagiosum virus	MC066L	Inhibits UV- and peroxide-induced apoptosis; homologous to human glutathione peroxidase
Transcription	Human cytomegalovirus	IE1, IE2	Inhibits Tnf-α but not UV-induced apoptosis

[a]Data from D. Tortorella, B. Gewurz, M. Furman, D. Schust, and H. Ploegh, *Annu. Rev. Immunol.* **18:**861–926, 2000; S. Redpath, A. Angulo, N. Gascoigne, and P. Ghazal, *Annu. Rev. Microbiol.* **55:**531–560, 2001; S. Hay and G. Kannourakis, *J. Gen. Virol.* **83:**1547–1564, 2002.

In a curious twist of molecular biology, the vaccinia virus envelope is derived from membranes of a dying cell. These membranes have cellular markers of apoptosis including phosphatidylserine. This phospholipid normally is available to bind to receptors that are present on the surface of phagocytic cells and that initiate the destruction of the dying cell and endocytosis of debris. When vaccinia virions, with their envelopes marked by apoptotic phospholipids, bind to the cell surface of a susceptible cell, they trigger an endocytic engulfment response normally appropriate for apoptotic debris and enter the cell (a viral mimic of the Trojan horse).

The Hostile Cytoplasm: Other Intrinsic Defenses

At least five other widely conserved cellular processes function in cellular defense against viral infections. These processes are autophagy, epigenetic silencing, RNA interference, cytosine deamination, and Trim protein interference. The first four tend to be general defenses activated upon infection, whereas Trim proteins function only against retroviruses, In addition, this mechanism is constitutive (not induced by infection), and is found only in some cells.

Autophagy

Cells can be induced to degrade the bulk of their contents by formation of specialized membrane compartments related to lysosomes. This process is called **autophagy,** and is evoked by stress such as nutrient starvation or viral infection. Infection by many viruses induces a state of metabolic stress that normally triggers intrinsic defenses. Such stress-induced alterations in translation are modulated in part by eIF2α kinases. It has been proposed that phosphorylated eIF2α triggers autophagy, which in turn leads to engulfment and digestion of cytoplasmic virions or other viral components. Two lines of evidence are consistent with the proposal. In one, virus replication can be blocked when autophagy is induced. For example, Sindbis virus replication in neurons is inhibited after induction of autophagy. In the other, some viral genomes encode gene products that actively block this process. For example, herpes simplex virus infection transiently activates Pkr-mediated autophagy, but a viral protein (ICP34.5) blocks the process by reversing the phosphorylation of eIF2α. In addition, this protein binds to and inhibits the mammalian autophagy protein beclin 1. This interaction is essential for viral neurovirulence.

Autophagy also may play a role in the initial sensing of viral invaders by providing a mechanism for transfer of viral nucleic acid from the cytoplasm to intracellular compartments containing Toll-like receptors. The idea is that autophagy integrates viral induced stress responses with the molecular detectors of viral nucleic acid (e.g., cytoplasmic RigI/Mda5; and endosomal receptors Tlr3, Tlr7, Tlr8, and Tlr9).

Epigenetic Silencing

Epigenetic silencing is an intrinsic defense against DNA viruses that replicate in the nucleus. It is thought that upon entering the nucleus, foreign DNA molecules are quickly organized into transcriptionally silenced chromatin. Silencing is mediated by DNA methylation or chromatin modifications including histone deacetylation. These modifications can persist for long periods, often over many cell divisions. Organized collections of proteins in the nucleus called **Pml bodies** may mediate such repression. These structures are implicated in intrinsic antiviral defense for many reasons, including the fact that interferon stimulates synthesis of the proteins that comprise them (see "Promyelocytic leukemia proteins" above). As might be predicted, viral proteins to counter epigenetic silencing have been identified. For example, the human cytomegalovirus protein pp71 binds to a cellular protein called Daxx that interacts with histone deacetylases to relieve transcriptional repression. The global repression of Pml-bound DNA can be relieved by viral proteins such as the ICP0 protein of herpes simplex virus type 1. This protein accumulates at Pml bodies and induces the proteasome-mediated degradation of several of their protein components. The human cytomegalovirus IE1 proteins, the Epstein-Barr Ebna5 protein, and the adenovirus E4 Orf3 protein all affect Pml protein localization or synthesis. The ways in which the components of Pml bodies associate with DNA and repress transcription are active areas of study.

Epigenetic silencing manifests itself in many ways, but those studying gene transfer with retrovirus vectors often find that expression of their favorite gene is low or completely off. We now understand that integrated retrovirus DNA is subject to reversible epigenetic silencing, a prominent process in embryonic or adult stem cells. Histone deacetylases associate with newly integrated proviral DNA soon after infection and act to repress viral transcription (Volume I, Chapters 8 and 9).

RNA Silencing

RNA silencing is a mechanism of sequence-specific degradation of RNA observed among diverse plants and animals. It is likely to have arisen early in the evolution of eukaryotes to detect and destroy foreign nucleic acids. RNA silencing is related to a process called RNA interference (RNAi) that was first found to occur in *Caenorhabditis elegans* and subsequently detected in fungi, insects, and algae (Volume I, Chapter 10).

Table 3.11 Some viral gene products that suppress RNA interference

Virus	Gene product	Mechanism
Human adenovirus type 5	VA-RNA I and VA-RNA II[a]	Competition for binding to exportin-5 and Dicer
Ebola virus	VP35 protein	Binding to dsRNA
Influenza A virus	NS1 protein	Binding to dsRNA
Vaccinia virus	E3L protein	Binding to dsRNA
Human immunodeficiency virus type 1	Tat	Inhibition of Dicer?

[a]Both RNAs are cleaved by Dicer, and the products are incorporated into RNA-induced silencing complexes.

We are learning more about RNA interference as an antiviral defense, as many viral genomes, including those of plant, insect, fish, and human pathogens, encode suppressors. Many of these suppressors are RNA-binding proteins without a preference for small interfering RNAs (siRNAs) (Table 3.11). Others sequester these RNAs, inhibit their production, or affect amplification of the process. Other examples are certain to be discovered.

RNA interference is now used routinely as a tool in the laboratory to block cellular gene expression as well as viral replication. Some hold hope that this intrinsic defensive response may be harnessed for development of highly specific and effective antiviral drugs (see Chapter 9).

As previously discussed, cellular miRNAs are induced by IFN treatment and may play an important role in antiviral defense. Probably a more intriguing outcome of this research was the discovery of miRNAs in DNA viral genomes (discussed in Volume I, Chapter 10). In human cytomegalovirus-infected cells, a viral miRNA reduces expression of the major histocompatibility complex class I polypeptide sequence B (MicB) protein. As a result, NK cells have reduced capacity to kill the infected cell. Similarly, a human herpesvirus 8-encoded miRNA mimics host miR-155 required for B-cell development. This viral miRNA may contribute to viral lymphomagenesis. The interplay of viral gene products and host miRNA defenses is only now being described for a number of DNA and RNA viruses.

Cytosine Deamination (Apobec [Apolipoprotein B Editing Complex])

All mammalian genomes contain *apobec*3 genes. The Apobec family of proteins play a variety of roles, including RNA editing of host genes (Chapter 6). Several members of the Apobec3 family are induced by IFN and are intrinsic antiretroviral proteins packaged into virions. After infection, such cellular enzymes affect the process of reverse transcription such that newly replicated retroviral DNA is degraded. When the viral reverse transcriptase begins to copy viral RNA into DNA, Apobec deaminates single-stranded DNA, specifically the nascent minus strand, which is synthesized first. The enzyme converts C's to U's with the consequence that when the deaminated minus-DNA strand is copied, the U pairs with A, producing a G-to-A transition. The new proviral genome therefore is mutated in a very characteristic pattern (many GC pairs become AT pairs). However, the deamination event has two consequences: in one, uracil-containing DNA is quickly attacked by uracil DNA glycosidase, leaving an abasic site that is a target for endonucleases. If the uracil-containing DNA is copied and is integrated, viable progeny cannot be produced. For example, all tryptophan codons (TGG) could be converted into stop codons (TAA). One retrovirologist called Apobec a WMD—a weapon of mass deamination.

The action of Apobec should be lethal for retroviruses that incorporate this enzyme into their virions. However, human immunodeficiency virus counters this potential lethal defense by producing the Vif protein. Vif binds to the Apobecs as well as to a particular host ubiquitin ligase complex and promotes the ubiquitinylation and subsequent degradation of the enzymes by the proteasome.

Trim (Tripartite Interaction Motif) Proteins

A distinct class of intrinsic defense proteins prevent the cell from being infected by certain retroviruses (see Chapter 6). The hypothesis is that these genes evolved independently in various species to protect against endemic retroviruses. They are present in normal cells, and some have been recognized for years as mediators of processes called "restriction" or "exclusion." We know by the presence of large numbers of retroviral proviruses in vertebrate genomes that retroviruses have been around for millions of years. We also know that many cell types are quite resistant to infection by some retroviruses, despite carrying functioning receptors. Recently, human immunodeficiency virus infections were found to be restricted in some, but not all, cell types. The race was on to identify these constitutive inhibitors in hopes of finding new mechanisms to stop the AIDS pandemic.

The facts were simple: human immunodeficiency virus type 1 is unable to replicate in Old World monkeys, but virions can enter their cells. Infection is blocked soon after entry, but before reverse transcription. No proviral DNA is produced, and the infection is aborted. It was possible to introduce a rhesus macaque complementary DNA (cDNA) library into permissive cells and identify the dominant gene that blocked replication. The protein responsible for the inhibition of human immunodeficiency virus replication in rhesus macaques is called Trim5α. Trim stands for "tripartite interaction motif." A critical fact is that human Trim5α does not restrict human immunodeficiency virus replication. If humans had the rhesus macaque Trim protein gene, we might not have the devastating AIDS pandemic today.

Rhesus macaque Trim5α targets the human immunodeficiency virus capsid protein, but not other retroviral capsid proteins. When synthesis of this protein was reduced using small interfering RNA molecules, the block against human immunodeficiency virus infection in rhesus cells was relieved, but had no effect on murine leukemia virus infection. Trim5α appears to block infection by binding to some retroviral capsids to disrupt an ordered uncoating process or sequester particles to a nonproductive infection pathway. Trim5α is now known to be a ubiquitin ligase and promotes ubiquitinylation of capsids and their subsequent degradation by the proteasome. IFN treatment increases Trim5α mRNA production in both human and rhesus cells.

The idea that a retroviral capsid can be a target for intrinsic defense is not new, but took some time for support to develop. Host restriction (or exclusion) of mouse retrovirus infection has been known for more than 30 years. The prototypical host gene blocking early retroviral replication events was identified using the Friend strain of murine leukemia retrovirus. The locus is called FV1 (Friend virus susceptibility). The FV1 protein blocks replication of murine leukemia virus soon after reverse transcription. Importantly, host Fv1 action depends on the infecting virus coat protein. A single residue at position 110 of the capsid coat determines sensitivity to FV1. Remarkably, the mouse Fv1 gene is the capsid gene of an endogenous retrovirus resident in the mouse genome! How Fv1 acts remains unclear, but restriction depends on a specific interaction with the capsid protein of the incoming virus. This finding is a remarkable demonstration of selection events turning endogenous retroviral gene expression against potential infection by other retroviruses.

Perspectives

Every moment of our lives, we encounter a myriad of virus particles. Our physical, chemical, and biological defenses ensure that the vast majority of these encounters are of no consequence. Nevertheless, and invariably, we all experience many viral infections during our lifetimes. Obviously, our defenses, while powerful, can be surmounted. For every host defense, there is a viral countermeasure.

Infections, whether successful or not, reflect initial probabilistic events that are not well understood. For example, we have limited information on the probability of infection by a single virion, the probability that a cell will detect and respond to the infection, or the probability that sentinel cells will detect the products of a single-cell infection.

The difficulties of translating information derived from tissue culture models, to animal models and then to human infections are substantial. Nevertheless, from what we know today, we can articulate several principles of host-virus interactions that operate during the first minutes to hours of an infection.

First, all infections begin with events in a single cell. Many cells can be infected at one time, but the way in which each responds dictates the course of events from there on out. If the infection is curtailed by cell-autonomous defenses (the intrinsic defenses), secondary innate and tertiary adaptive immune responses will not be activated. If infection is not curtailed, the actions of these relatively few infected cells dictate subsequent events.

The second principle is that every cell has receptors on the surface and inside the cell that bind microbial proteins and nucleic acids. On binding their cognate ligand, these pattern recognition receptors initiate reactions leading to production of potent secreted proteins called cytokines. Cytokines are the primary response mediators that emanate from a single infected cell. The cytokine receptors and cells that carry them provide the integrating network that produces the appropriate effector action so characteristic of an individual host's response to infection (Fig. 3.10).

The third principle is that these early events are noisy. Individual cells do not produce the same amount of a given cytokine, or the same cytokine cocktail. One reason is that natural infections are not synchronous. In addition, some infected cells curtail the infection rapidly, while others are slower in responding. Some infected cells undergo apoptosis, while some do not. This rather large variation in single-cell responses to a single infectious agent is amplified or dampened, depending on the number of infected cells, the tissue containing the infected cells, and characteristics of the host (age and state of health, for example). Such potentially large variation at the earliest stages of infection influences the extent and duration of subsequent responses. Because many of the subsequent events in host defense are not reversible (such as killing of a cell, or activation of an antibody response instead of a cytotoxic T-cell response), extreme outcomes are possible for the host (life, death, uncomplicated acute infection that is cleared, or a persistent infection that lasts for the life of a host).

The remarkable fact is that while initially cell autonomous, the subsequent cytokine-initiated events escalate into a choreographed global response of ever-increasing complexity and intensity. The process recruits more and more effector cells, which in turn produce more cytokines and defensive proteins. If all works well, the infection is stopped, the host survives, and the host defenses stand down to fight another day.

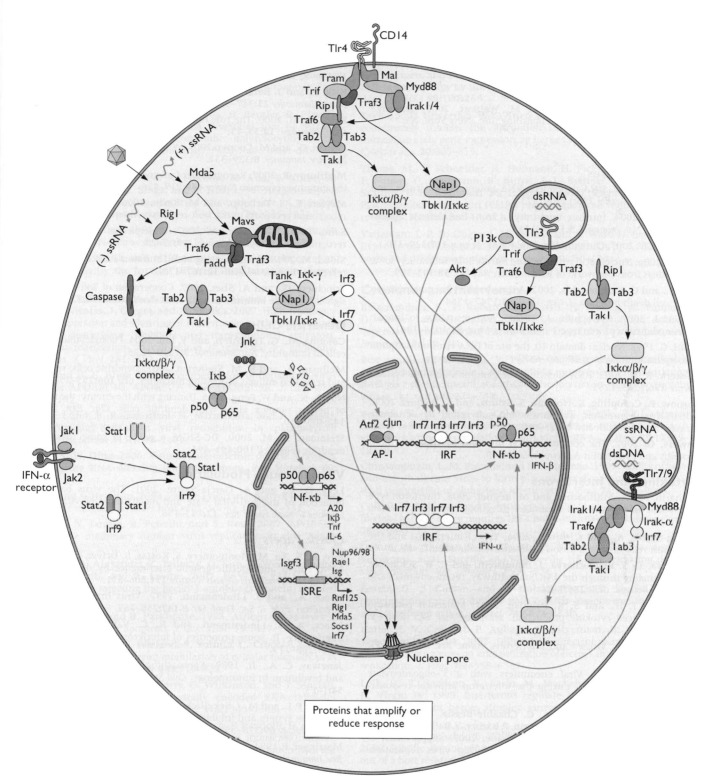

Figure 3.10 Summary of some established intrinsic defense responses. Every cell has receptors on the surface and inside that bind microbial proteins and nucleic acids. A generic cell is indicated here, but it should be clear that not all cells have the same constellation of responses. Upon binding their cognate ligand, these pattern recognition receptors initiate reactions leading to production of potent secreted proteins called cytokines. Cytokines, such as IFN, are the primary response mediators that emanate from a single infected cell. The cytokine receptors and cells that carry them provide the integrating network that produces the appropriate effector action so characteristic of an individual host's response to infection.

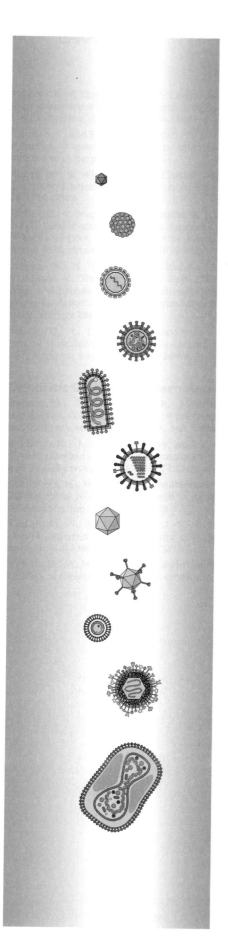

Introduction
  Innate and Adaptive Immune Defenses

The Innate Immune Response
  General Features
  Sentinel Cells
  Natural Killer Cells
  Complement
  The Inflammatory Response

The Adaptive Immune Response
  General Features
  Cells of the Adaptive Immune System
  Adaptive Immunity: the Action of
  Lymphocytes That Carry Distinct
  Antigen Receptors
  Antigen Presentation and Activation
  of Immune Cells
  The Cell-Mediated Adaptive Response
  The Antibody Response
  The Immune System and the Brain

**Immunopathology: Too Much
of a Good Thing**
  Immunopathological Lesions
  Viral Infection-Induced
  Immunosuppression
  Systemic Inflammatory Response
  Syndrome
  Autoimmune Diseases
  Heterologous T-Cell Immunity
  Superantigens "Short-Circuit" the
  Immune System
  Mechanisms Mediated by Free Radicals

Perspectives

References

Immune Defenses

Introduction

The cascade of antiviral defense starts by the reactions of a single infected cell. If, despite single-cell defenses, viral replication continues unabated, a threshold that signals the need for more aggressive and global defenses is crossed. Almost immediately thereafter, a remarkable defense, called the **immune response**, begins. This highly coordinated response depends on the interplay of secreted proteins, receptor-mediated signaling, and intimate cell-cell communication. The immune response is to be contrasted with intrinsic defense, which is cell autonomous and not dependent on white blood cells. White blood cells participate at every level of immune defense (Table 4.1). Several remarkable life-or-death decisions are made quickly: the nature of the invader is established, and an appropriate mixture of soluble proteins and white blood cells is mobilized to remove the invader. Amazingly, infections of just a few cells can be detected in the context of billions of uninfected cells.

Three critical steps in immune defense are **recognition**, **amplification**, and **control**. A viral infection must be recognized early, defenses must be activated quickly and amplified if needed, and then all responses must be turned off when the infection has ceased. Pathologies and immune dysfunction can result if any of these processes is defective, modulated, or bypassed. The paramount importance of the immune system in antiviral defense is amply documented by the devastating viral infections observed in children who lack normal immune function, as well as in patients with acquired immunodeficiency syndrome (AIDS).

Innate and Adaptive Immune Defenses

The immune response to viral infection consists of an innate (nonspecific) and an adaptive (specific) defense (see Fig. 3.1). The **innate response** is the **first line of immune defense**, because it functions continually in a normal host without any prior exposure to the invading virus (Box 4.1). Indeed, most viral invasions are repelled by intrinsic defenses and the innate immune system

Table 4.1   White blood cells that participate in the innate and adaptive defense systems

Cell	Source/function
Lymphocytes	Responsible for specificity of immune responses; recognize and bind to foreign antigens; derived from bone marrow
T cells	Differentiate into Th cells that secrete cytokines and CTLs; regulatory or suppressor T cells are a type of Th cell; all have T-cell receptors
B cells	Produce antibody
NK cells	Natural killer cells; large granular cytolytic cells; do not have T-cell receptors
Mononuclear phagocytes	Responsible for phagocytosis and antigen presentation; derived from bone marrow; monocyte lineage; includes macrophages, Kupffer cells in the liver, alveolar macrophages in the lungs
Dendritic cells	Responsible for induction of immune response; antigen presentation to Th cells; migratory cells found in every tissue except the brain
Interdigitating dendritic cells	Bone marrow derived; present in most organs, lymph nodes and spleen; includes Langerhans cells in the skin
Follicular dendritic cells	Not bone marrow derived; present in germinal layers of lymphoid follicles in spleen, lymph nodes, and mucosal lymphoid tissue; trap antigens and antigen-antibody complexes for display to B cells in lymphoid tissue
Plasmacytoid dendritic cells	Class of immature dendritic cells found in the blood and T-cell zones of lymph nodes; capable of synthesizing large quantities of IFN to protect immune cells from viral infection
Granulocytes	Contain abundant cytoplasmic granules; inflammatory cells
Neutrophils	Polymorpholeukocytes; respond to chemotactic signals, phagocytose foreign particles; major leukocyte in the inflammatory response
Eosinophils	Function in defense against certain pathogens (like worms) that induce IgE antibody; not critical in antiviral defense; involved in hypersensitivity (allergic) reactions
Basophils	Circulating counterparts to tissue mast cells; mediate hypersensitivity caused by IgE antibody response

before viral replication outpaces host defense. However, once this threshold is passed, second-line defenses (the **adaptive response**) must be mobilized if the host is to survive.

The **adaptive defense** consists of the antibody response and the lymphocyte-mediated response, often called the **humoral response** and the **cell-mediated response**, respectively. This system is called "adaptive" because it not only differentiates infected from noninfected self, but also is tailored individually to the particular foreign invader such that an appropriate combination of soluble molecules (antibodies and cytokines) and lymphocytes (Table 4.1) participate in the action. The tailoring of the specific adaptive response requires close communication with cells that participate in the innate response. Such communication occurs by binding of cytokines and intimate cell-cell interactions among dendritic cells and lymphocytes in the lymph nodes. Considerable evidence indicates that the cells and cytokines of the innate response provide essential information to the adaptive immune system about the nature of the potential hazard confronting the host. In fact, it is possible to demonstrate that the adaptive response cannot be established without the innate immune system.

A defining feature of the adaptive defense system is **memory**; subsequent infections by the same agent are met almost immediately with a robust and highly specific response that usually stops the infection as soon as it starts, with minimal reliance on the innate defenses.

The cellular agents of the innate and adaptive immune responses are the **myelomonocytes** (monocytes, macrophages, dendritic cells, and a variety of granulocytes) and the **lymphocytes** (natural killer, T, and B cells)

BOX 4.1

BACKGROUND
Innate immune defense stands alone

- Other than intrinsic defenses of cells, it is the **only** immune defense available for the first few **days** after viral infection.
- It can discern the general nature of the invader (viruses, bacteria, protozoans, fungi, or worms).
- It can inform the adaptive response when infection reaches a dangerous threshold.

(Table 4.1). One fascinating aspect of the immune system is that its effector cells are dispersed throughout the body, yet the response can be directed quickly to the focal point of infection. It is the powerful cytokines that coordinate the activity of this dispersed cellular defense system.

The Innate Immune Response

General Features

When intrinsic cell defenses are unable to stop the spread of infection, the combination of cell death, local increasing concentrations of cytokines, and release of other stress-related molecules around the area of infection leads to activation of the next phase of host defense, the innate immune response (see Fig. 3.1). The innate immune defense system comprises **cytokines** released from infected cells as part of the intrinsic defense response, local **sentinel cells** (dendritic cells and macrophages), a complex collection of serum proteins termed **complement**, and cytolytic lymphocytes called **natural killer cells (NK cells)**. Neutrophils and other granulocytic white blood cells also play important roles in innate defense in response to the initial burst of cytokines from dendritic cells, macrophages, and infected cells.

The innate immune response is crucial in antiviral defense, because it can be activated quickly if intrinsic defenses are overwhelmed and can begin functioning within minutes to hours of infection. Such rapid action contrasts with the activation of the adaptive response, which is orders of magnitude slower than the replication cycles of some viruses. It takes days to weeks to orchestrate the effective response of antibodies and activated lymphocytes specifically tailored for the infecting virus. While the rapidity of the innate response is important, this response must also be transient, because its continued activity is damaging to the host. Below, we discuss crucial components of the innate response. Despite their apparent diversity, every component initiates an immediate action that, in turn, is detected and amplified by cells of the adaptive immune system. Viral genomes encode a surprising variety of gene products that modulate every step of innate defense.

Sentinel Cells

Dendritic cells and macrophages are crucial sentinel cells present in peripheral compartments (e.g., the skin and various mucosal surfaces). These cells are active very early in infection and play central roles in classifying the infecting agent, mounting a strong intrinsic response, and then communicating with cells of the immune system. Dendritic cells bind cytokines produced by infected cells and take up viral proteins from dead and dying cells (Fig. 4.1). Sentinel cells are specially equipped not only to initiate immediate immune defense, but also to convey information of the attack to the adaptive immune system. This latter ability is emphasized by their common name, **professional antigen-presenting cells**. Even if a viral protein is introduced by injection into the skin or muscle, local dendritic cells and macrophages will most likely bind some of the molecules and stimulate an immune response to that protein. Indeed, most vaccines would not be effective without dendritic cells.

Dendritic cells play two major roles in antiviral responses: they directly inhibit viral replication at the onset of infection by producing large quantities of cytokines such as alpha/beta interferon (IFN), and subsequently they trigger adaptive, T-cell-mediated immunity appropriate to the infection. Dendritic cells exist in two functionally distinct states called **immature** and **mature** (Fig. 4.1). The immature cells are found in the periphery of the body, around body cavities, and under mucosal surfaces. Several dendritic cell types can be identified by cell surface markers, and probably differ in function. In general, dendritic cells are proficient at endocytosis and can synthesize copious quantities of cytokines when stimulated. Soluble proteins are taken up avidly and retained in endosomes until the dendritic cell matures, which may be hours, if not days, after the actual uptake of proteins. Immature dendritic cells also capture proteins from dead or dying cells by taking up complexes containing heat shock proteins and unfolded proteins, as well as cellular debris and vesicles that are produced by apoptosis. As the cell matures, internalized proteins are processed into peptides and moved to the cell surface complexed to receptors that can be recognized by lymphocytes of the adaptive immune system (see "Antigen Presentation and Activation of Immune Cells" below).

Immature dendritic cells carry Toll-like receptors (Tlrs), the RigI and Mda5 RNA detector proteins, and receptors for various proinflammatory cytokines. They act as a bridge between intrinsic defenses and immune defenses. When these receptors bind the appropriate ligands, most immature dendritic cells undergo dramatic morphological and functional changes, and differentiate. The mature cells no longer have the capacity for endocytosis, and they display a new repertoire of cell surface receptors. Some of these are chemokine receptors that direct migration of the mature dendritic cell (loaded with stored viral proteins) to the local lymph nodes (often called **homing**). Other new receptors are T-cell adhesion receptors and T-cell costimulatory molecules, essential for binding to and activation of naive T cells on arrival in the lymph node. A remarkable change in morphology also occurs as mature cells extend long eponymous dendritic processes that increase their

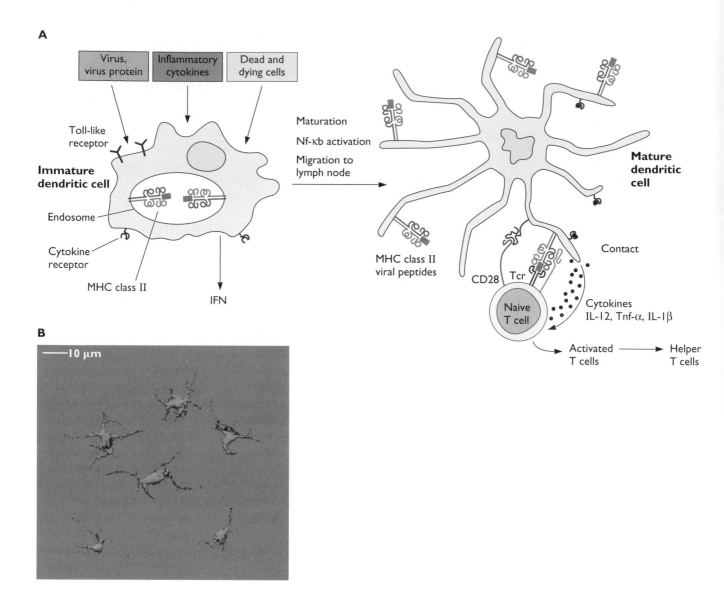

Figure 4.1 Dendritic cells provide cytokine signals and packets of protein information to naive T cells.
(A) Immature dendritic cells actively take up extracellular proteins by endocytosis and store the proteins internally. They do not express MHC class II complexes on their surfaces. Binding of ligands to the Toll-like receptors or cytokine receptors induces differentiation into mature dendritic cells. These cells no longer have the capacity for endocytosis of proteins and display a new repertoire of cell surface receptors. Some of these are chemokine receptors that enable the dendritic cell to migrate to the local lymph nodes. The proteins ingested by the immature cell are now processed into peptides and loaded on to MHC class II proteins for subsequent transport to the cell surface. Mature cells extend long dendritic processes to increase surface area for binding of naive T cells in the lymph node. Mature dendritic cells release proinflammatory cytokines as indicated to stimulate T-cell differentiation. Naive but antigen-specific T cells bind to the MHC class II-peptide complexes via their T-cell receptors. The interaction is strengthened by the presence of increased costimulatory ligands (e.g., CD28) on the mature dendritic cell. The T cell is activated, begins the maturation process into its final effector state, and moves out of the lymph node into the circulation. **(B)** Langerhans cells are abundant in mouse ear epithelium and are visualized here by confocal microscopy in a live tissue preparation by their production of an MHC class II-enhanced green fluorescent protein (EGFP) fusion protein. Figure provided by Marianne Boes, Jan Cerny, and Hidde Ploegh (Harvard Medical School, Boston, MA).

surface area. Mature dendritic cells become potent mobile signaling centers, releasing proinflammatory cytokines that act locally and at a distance.

Once within lymphoid tissue, mature dendritic cells instruct the adaptive immune response by directly engaging and stimulating naive, antigen-specific T lymphocytes. The density of mature dendritic cells and the combination of cytokines they secrete in the lymph node dictate the type of adaptive response that ensue (see "Th1 and Th2 Cells" below). Dendritic cell maturation provides an essential link between innate and adaptive immunity.

While dendritic cells function by sampling proteins and cytokines found in the area of infection, they may also be infected by the virus that they detect (Box 4.2). One obvious outcome is that the mobile, infected cells travel to the lymph node where the virus can be transmitted to T and B cells. Indeed, human immunodeficiency virus type 1 virions can bind to a lectin (DC-Sign [dendritic-cell-specific, Icam-3-grabbing nonintegrin]) on immature dendritic cells with grave consequences. This protein is essential for establishing contact of a mature dendritic cell with a naive T cell in lymphoid tissue. Virus particles so bound do not replicate in the dendritic cell, but rather are retained just below the cell surface during migration to the lymph node, where they subsequently infect CD4+ T cells (see Chapter 6). Two mosquito-borne viruses, Venezuelan equine encephalitis virus and dengue virus, replicate initially in immature dendritic cells at the site of inoculation. These infected cells then exhibit attributes of mature cells, in that they migrate to lymph nodes and propagate the infection with severe consequences to the host. Maturation appears to be triggered by viral RNA binding the RNA detectors, RigI and Mda5. This property, maturation of dendritic cells by infection, has inspired new approaches for vaccination in which attenuated versions of these viruses are used to deliver selected immunizing antigens directly to the dendritic cells, which then stimulate a strong immune response.

As might be expected, viral gene products modulate dendritic cell functions. For example, infection by either herpes simplex virus type 1 or vaccinia virus inhibits maturation of dendritic cells by blocking a signal transduction cascade, or by interfering with cytokine stimulation of maturation, respectively. In contrast, dendritic cells infected with murine cytomegalovirus or measles virus are fully capable of maturing, but the mature cells cannot stimulate T cells. Such interference with a critical component of the innate response is likely to contribute to the profound immunosuppression that follows infection by cytomegalovirus and measles virus. The contribution of dendritic cell infection to immunosuppression is a topic of considerable interest.

Natural Killer Cells

NK cells are in the immediate front line of innate defense: they are ready to recognize and kill virus-infected cells. Like dendritic cells, they recognize infected cells in the company of vast numbers of uninfected cells. However, the mechanism of recognition is completely different: NK cells recognize "missing self" or "altered self." NK cells are abundant lymphocytes (representing about 2% of circulating lymphocytes [Table 4.1]) that patrol the blood and lymphoid tissues. They are large, granular lymphocytes, distinguished from others by the absence of antigen receptors found on B and T cells (see below). When an NK cell binds an infected target cell, it releases a mix of cytokines (notably IFN-γ and tumor necrosis factor alpha [Tnf-α]) that contribute to a local inflammatory response and alert cells of the adaptive immune system. They also can produce prodigious quantities of interleukin-4 (IL-4) and IL-13, the major cytokines that stimulate antibody production. NK cells also participate later in adaptive defense by

---

BOX 4.2

BACKGROUND

Infection of the sentinels: dysfunctional immune modulation

When viruses infect the dendritic cells, the immune system's first command-and-control link is compromised. Some of the many possible consequences of sentinel cell infection, any one of which could suppress the immune response locally or systemically, include the following:

- interference with recruitment to peripheral sites of infection

- impairment of antigen uptake or processing
- infection and destruction of immature dendritic cells
- interference with maturation
- impairment of migration to lymphoid tissue
- interference with activation of T cells

López, C. B., J. S. Yount, and T. M. Moran. 2006. Toll-like receptor-independent triggering of dendritic cell maturation by viruses. *J. Virol.* **80:**3128–3134.

Mellman, I., and R. Steinman. 2001. Dendritic cells: specialized and regulated antigen processing machines. *Cell* **106:**255–258.

Raftery, M., M. Schwab, S. Eibert, Y. Samstag, H. Walczak, and G. Schonrich. 2001. Targeting the function of mature dendritic cells by human cytomegalovirus: a multilayered viral defense strategy. *Immunity* **15:**997–1009.

binding to infected cells coated with immunoglobulin G (IgG) antibody and killing them (see "Antibody-Dependent Cell-mediated Cytotoxicity" below).

The number of NK cells increases quickly after viral infection and then declines as the acquired immune response is established. These cells are stimulated to divide whenever infected cells and sentinel dendritic cells make IFN. NK cells kill after contact with the target by releasing perforins and granzymes that perforate membranes and trigger caspase-mediated cell death, respectively. In humans, NK cells are particularly important in controlling primary infection by many herpesviruses, as patients with NK cell deficiencies suffer from severe infections with varicella-zoster virus, human cytomegalovirus, and herpes simplex virus. While a role for direct NK cell-mediated killing in antiviral defense is difficult to establish experimentally, NK cell production of IFN-γ clearly provides significant antiviral action.

NK Cell Recognition of Infected Cells: Detection of "Missing Self" or "Altered Self" Signals

As discussed below, a collection of cell surface proteins called the major histocompatibility complex (MHC) proteins are important receptors in the adaptive immune response. MHC class I proteins are found on the surfaces of most cells of the body. The MHC class I molecules are the **self antigens** that, when missing, cause the NK cell to kill the target cell. A mechanism for detection of **missing self** is illustrated in Fig. 4.2. At least two receptor-binding interactions are required for such discrimination: one to activate the NK cell and the other to block this activation if the cell is not infected. The activating signal is delivered

when an NK cell receptor binds a pathogen-specific ligand (e.g., virus-infected cells may present new glycoproteins on their surface). As a consequence, a signal transduction cascade is initiated and the NK cell is stimulated to secrete a burst of cytokines and kill the cell. However, a negative regulatory signal is produced when an MHC class I-specific receptor on the NK cell engages MHC class I molecules on the surface of the same target cell. Because many infected cells carry fewer MHC class I molecules on their surfaces (discussed in Chapter 5), they are prime NK cell targets. The unusual two-receptor recognition system employed by NK cells ensures that normal cells that synthesize MHC class I proteins are not killed by NK cells, even if they have distinct ligands on their surfaces. Much work is in progress to test this idea.

NK MHC Class I Receptors Produce the Inhibitory Signals

Human NK cells synthesize two inhibitory MHC class I receptors of either the C-type lectin family or the immunoglobulin family (called killer cell immunoglobulin-like inhibitory receptors, or Kirs). NK cells also can recognize and spare target cells carrying HLA-E, an unusual MHC class I protein that binds peptides derived from the signal sequences of other MHC class I molecules. The presence of HLA-E protein complexed with signal peptide informs the NK cell that MHC class I synthesis is normal. An intriguing finding is that infection by human cytomegalovirus induces synthesis of HLA-E protein, thereby diverting potential NK cell recognition and lysis.

Viral Proteins Modulate NK Cell Actions

Many viral genomes encode proteins that block or confound NK cell recognition and killing (Fig. 4.3). At least five distinct categories of modulation can be described. These gene products are homologs of MHC class I, proteins that regulate MHC class I production, proteins that interfere with the NK activation receptor and cognate ligand interactions, and proteins that modulate cytokine pathways relevant to NK cell function. NK modulators have been identified in several virus families including *Flaviviridae, Papillomaviridae, Herpesviridae, Retroviridae,* and *Poxviridae.* Some viral genomes encode more than one distinct NK modulator. For example, human cytomegalovirus encodes at least seven gene products that modulate the NK cell response. One striking example of viral interference with NK cell activity is provided by the hepatitis C virus E2 envelope protein, which binds to CD81, a protein on the surface of NK cells, and blocks activation signals. As a result, the NK cell no longer recognizes infected cells. An example of variation of surface MHC class I expression on infected target cells features the Nef protein of human

Figure 4.2 NK cells distinguish normal, healthy target cells by a two-receptor mechanism. Both positive (stimulating) and negative (inhibiting) signals may be received when an NK cell contacts a target cell. The converging signal transduction cascades from the two classes of receptor regulate NK cell cytotoxicity and release of cytokines. The inhibitory receptors dominate all interactions with normal, healthy cells. Their ligands are the MHC class I proteins. When NK cells contact MHC class I molecules on the surface of the target cell, signal transduction blocks the response of activating receptors.

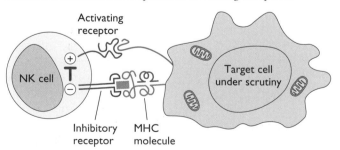

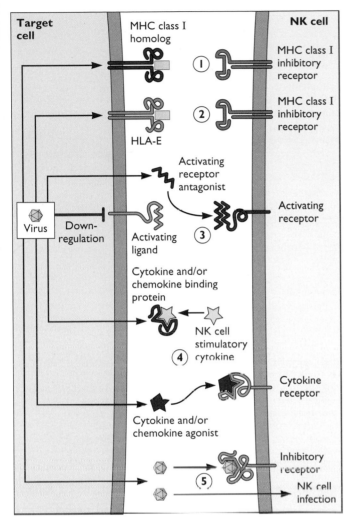

Figure 4.3 Virus-encoded mechanisms for modulation of NK cell activity. (Left) An infected target cell. **(Right)** An NK cell. The infected target cell should be lysed by an activated NK cell. However, five categories of NK-cell-modulating strategies are illustrated (circled numbers). Viral proteins produced in the infected cell are labeled in red. (1) Inhibition by a viral protein with homology to cellular MHC class I proteins. (2) Inhibition of expression or cell surface localization of host HLA-A or HLA-B (human MHC class I homologs) resulting in an increase in the amount of host HLA-E (or HLA-C) on the target cell surface. (3) Release of virus-encoded cytokine-binding proteins that block the action of NK cell-activating cytokines (also, viral proteins can reduce the amount of the activating ligand on the surface of the infected cell so that the NK cell is not activated). (4) Inhibition of action of NK cell-stimulating cytokines by binding these cytokines or by producing a chemokine antagonist. (5) Effect of newly produced virions, which can engage the NK cell, block an inhibitory NK cell receptor, or infect the NK cell itself to disrupt various effector functions or even kill the cell.

immunodeficiency virus type 1. Nef affects the cell surface expression of certain classes of MHC class I molecules, but not those involved in inhibition of NK cells (e.g., the HLA E proteins). Cells infected with ectromelia virus or molluscum contagiosum virus (both poxviruses), modulate the killing functions of NK cells by expressing proteins that bind IL-18, thereby inhibiting IFN-γ production by NK cells.

Complement

The **complement system** was identified in 1890 as a heat-labile serum component that lysed bacteria in the presence of antibody (it "complemented" antibody). The complement system in the blood is a major primary defense and a clearance component of **both** the innate and adaptive immune responses (Box 4.3). There are three distinct complement pathways: the **classical, alternative,** and **mannan-binding pathways**. As part of the innate defense system, complement action can be initiated by direct recognition of a microbial invader by C1q or C3b proteins in the alternative pathway (Fig. 4.4). The mannan-binding lectin pathway triggers complement action upon binding of a lectin similar to C1q with mannose-containing carbohydrates on bacteria or viruses. Importantly, complement can also function as an effector of the adaptive defense system by the binding of C1q to antibody-antigen complexes on the surface of an invader or infected cell (the classical pathway [Fig. 4.4]).

The Complement Cascade

Complement comprises at least 30 distinct serum proteins and cell surface membrane proteins that act sequentially to produce a wide range of activities from direct cell lysis to the augmentation of adaptive immune responses (Fig. 4.4). Unfortunately, the nomenclature of the complement proteins is baroque because they were named in order of their discovery, but not in terms of their function.

In all three pathways, a cascade of protease reactions activates two critical proteases called **C3 convertase** and **C5 convertase** (note that the three pathways yield the same enzyme activity, but the proteins comprising each convertase are different). A crucial property of C3 and C5 convertase enzymes is that they are bound covalently to the surface of the pathogen or the infected cell. The action of surface-bound C3 convertase on its substrate yields C3b, the primary effector of all three complement pathways, and C3a, a potent soluble mediator of inflammation. C3b remains on the pathogen's surface, where it binds more complement components to stimulate a protease cascade that produces many other bioactive proteins. The protease cleavage products stimulate inflammation, attract lymphocytes, potentiate the adaptive response, and kill infected

BOX 4.3

TERMINOLOGY
The complement cascade has four major biological functions

Lysis

Membrane disruption and lysis occur when specific activated complement components (C6, C7, C8, and C9) polymerize on a foreign cell or enveloped virus, forming pores or holes that disrupt the lipid bilayer and compromise its function. The cell or virus is disrupted by osmotic effects.

Activation of Inflammation

Inflammation is stimulated by several peptide products of complement proteins produced during the complement cascade. These peptides (C3a, C4a, and C5a) bind to vascular endothelial cells and various classes of lymphocytes to stimulate inflammation and to enhance responses to foreign antigens.

Opsonization

Complement proteins (typically C3b and C1q) can bind to virus particles so that phagocytic cells carrying appropriate receptors can then engulf the coated viruses and destroy them; this process is called opsonization. Complement receptors such as Cr1 present on phagocyte surfaces bind C3b-coated particles and stimulate their endocytosis.

Solubilization of Immune Complexes

Noncytopathic viral infections commonly result in pathological accumulations of antigen-antibody complexes in lymphoid organs and kidneys. Complement proteins can disrupt these complexes, by binding to both antigen and antibody, and facilitate their clearance from the circulatory system.

Figure 4.4 Activation and regulation of the complement system. The complement system can be activated through three pathways: classical, lectin, and alternative. Complement component 1 (C1) comprises C1q (a pattern recognition protein), C1r, and C1s. The complement cascade is activated when C1 binds an antigen-antibody complex on the surface of an infected cell or a virus particle; C1 also links the classical and lectin activation pathways by interacting with the mannose-binding lectin (MBL)-associated serine protease (MASPs). These complexes contain proteases that cleave complement proteins C2 and C4, which then form the C3 and C5 convertases for the classical and lectin pathways. The alternative pathway activates complement without going through the C1-C2-C4 complex. For the alternative pathway, factor B is the C2 equivalent. Factor B is cleaved by factor D. Factor P (properdin) stabilizes the alternative pathway convertases. All three pathways culminate in the formation of the C3 and C5 convertases (orange box) that produce the three primary actions of activated complement: inflammation, cell lysis, and coating of foreign antigens so that they can be taken up by phagocytes (opsinization). The C3a and C5a proteins are potent simulators of the inflammatory response (also called anaphylatoxins). The membrane attack complex is formed by the complement proteins C5b-C9 and forms a hole in membranes, leading to lysis of cells. The C3b (opsonin) coats bacteria and virions and also amplifies the alternative pathway. See C. Kemper and J. Atkinson, *Nat. Rev. Immunol.* **7:**9–18, 2007.

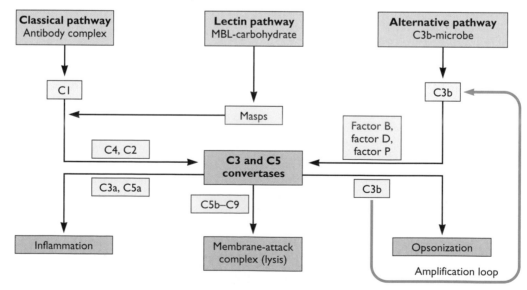

cells. C3b also stimulates phagocytic cells to take up the C3b-coated complex.

More than 90% of plasma complement components are made in the liver. However, these components also are produced elsewhere, including the major portals of pathogen entry. For example, the initiator complex C1 is synthesized mainly in the gut epithelium, and mannan-binding lectin is found in the respiratory tract. In addition, monocytes, macrophages, lymphocytes, fibroblasts, endothelial cells, and cells lining kidney glomeruli or synovial cavities all make most proteins of the complement system. Astrocytes in the brain can synthesize the full panoply of complement proteins when stimulated by inflammatory cytokines. As the brain is exquisitely sensitive to inflammation, it is not clear how the inflammatory action of peptides produced during the complement cascade is controlled.

One important consequence of complement cascade activation is the initiation of a local broad-spectrum defense (Box 4.3). Complement components released locally during the protease cascade aid in recruitment of monocytes and neutrophils (Table 4.2) to the site of infection, stimulate their activities, and also increase vascular permeability. The antiviral effects of complement are both direct and indirect. The membrane attack complex lyses infected cells and inactivates enveloped viruses, while phagocytes engulf and destroy virions coated with C3b protein. Complement components stimulate a local inflammatory response that can limit infection, and convey the nature of the invader to the adaptive immune system. The activated complement system "instructs" the humoral and T-cell responses much as activated dendritic cells communicate with T cells. Indeed, we now understand that the complement machinery plays a pivotal role in regulating both B- and T-cell-mediated immune responses. It represents a true bridge between frontline early defenses and adaptive immunity.

The molecular linkage between complement and adaptive defense is somewhat selective, as is evident by the antibodies made during the Th1 response (see "Th1 and Th2 Cells" below). These antibodies are predominantly of the IgG2a isotype, an isotype that can actively stimulate the complement cascade. In contrast, antibodies produced during the Th2 response typically are IgG1 (mice), IgG4 (humans), and IgE, which bind neither to complement nor to macrophage Fc receptors (Table 4.3). Curiously, many antibodies specific for viral envelope proteins cannot be recognized by C1q, a property likely to be a consequence of selective pressure to escape complement-mediated lysis.

"Natural Antibody" Protects against Infection

The classical complement pathway of humans and higher primates can be activated by a particular collection of antibodies present in serum prior to viral infection (historically called "natural antibody"). Synthesis of some of these antibodies is triggered by the antigen galactose $\alpha(1,3)$-galactose (α-Gal) found as a terminal sugar on glycosylated cell surface proteins. Lower primates, most other animals, and bacteria synthesize the enzyme galactosyltransferase, which attaches α-Gal to membrane proteins. Importantly, humans and higher primates do not make this antigen, as they lack the enzyme. Because of constant exposure to bacteria producing α-Gal in the gut, human serum contains high levels of antibodies specific for this antigen; indeed, more than 2% of the IgM and IgG populations is

Table 4.2 Biological activities of proteins and peptides released during the complement cascade

Substance	Biological activity
C5b, C6, C7, C8, and C9	Lytic membrane attack complex
C3a	Peptide mediator of inflammation, smooth-muscle contraction; vascular permeability increase; degranulation of mast cells, eosinophils, and basophils; histamine release; platelet aggregation
C3b	Opsonization of particles and solubilization of immune complexes; facilitation of phagocytosis
C3c	Neutrophil release from bone marrow; leukocyte lysis
C3dg	Molecular adjuvant; profound influence on adaptive response
C4a	Smooth-muscle contraction; vascular permeability increase
C4b	Opsonin for phagocytosis, processing, and clearance of antibody-antigen immune complexes
C5a	Peptide mediator of inflammation, smooth-muscle contraction; vascular permeability increase; degranulation of mast cells, basophils, and eosinophils; histamine release; platelet aggregation; chemotaxis of basophils, eosinophils, neutrophils, and monocytes; hydrolytic enzyme release from neutrophils
Bb	Inhibition of migration and induction of monocyte and macrophage spreading
C1q	Opsonin for phagocytosis, clearance of apoptotic cells, and processing and clearance of antibody-antigen immune complexes

Table 4.3   The major cell-mediated and humoral immune responses to viral infections

Response	Effector	Activity
Cell mediated	IFN-γ secreted by Th and CTLs	Induces antiviral state
	CTLs	Destroys virus-infected cells
	NK cells and macrophages	Destroys virus-infected cells directly or by antibody-dependent cell-mediated cytotoxicity
Humoral	Primarily secretory IgA	Inhibits virion-host attachment
	Primarily IgG	Inhibits fusion of enveloped viruses with host membrane
	IgG and IgM antibody	Enhances phagocytosis (opsonization) after binding to virions
	IgM antibody	Agglutinates virions
	Complement activated via IgG and IgM	Lyses enveloped viruses; opsonization by C3b/antibody complex

directed against this sugar. It is this antibody that triggers the complement cascade and subsequent lysis of foreign cells and enveloped viruses bearing α-Gal antigens. The anti-α-Gal antibody-complement reaction is probably the primary reason why humans and higher primates are not infected by enveloped viruses of other animals, despite the ability of many of these viruses to infect human cells efficiently in culture. In support of this assertion, when such viruses are grown in nonhuman cells they are sensitive to inactivation by fresh human serum, but when grown in human cells they are resistant. Anti-α-Gal antibodies provide a mechanism for cooperation of the adaptive immune system and the innate complement cascade to provide immediate, "uninstructed" action.

Regulation of the Complement Cascade

Any amplified antiviral defense system as lethal as the actions of the complement cascade must be fail-safe and regulated with precision. Spontaneous activation of any one of the three pathways must be blocked, and triggering by minor infections (which occur regularly) must be avoided. Some regulation is intrinsic to the complement proteins themselves. For example, many are large and therefore cannot leave blood vessels to attack infected tissues unless there is local tissue damage that exposes cells directly to blood. Consequently, minor infections do not activate a substantial complement response. Many cascade intermediates are short-lived, with millisecond half-lives, and therefore do not exist long enough to diffuse far from the site of infection. Further control is maintained by complement-inhibitory proteins present in the serum and on the surface of many cells (e.g., the complement receptor type 1 protein [Cr1], decay-accelerating protein [Daf, or CD55], protectin [CD59], and membrane cofactor protein [CD46]). These proteins are the only regulators that can limit the alternative-pathway cascade by binding complement components such as C3b and C4b. Enveloped viruses

that do not carry CD55 or Cr1 on their surfaces are susceptible to the action of complement, particularly via the alternative pathway. Others, such as human immunodeficiency virus type 1 and the extracellular form of vaccinia virus, incorporate CD46, CD55, and CD59 in their envelopes and are thereby protected from complement-mediated lysis.

Many viral genomes encode proteins that interfere with the complement cascade. For example, alphaherpesvirus glycoprotein C binds the C3b component, and several poxvirus proteins bind C3b and C4. The smallpox virus SPICE protein (smallpox inhibitor of complement enzymes) inactivates human C3b and C4b and is a major contributor to the high mortality of smallpox (see Box 2.11).

Several viral receptors, including those for measles virus and certain picornaviruses, are complement control proteins. Epstein-Barr virus particles bind to CD21 (the Cr2 complement receptor) with profound consequences for the host and virus. This interaction activates the Nf-κb pathway, which then allows transcription from an important viral promoter. Epstein-Barr virus binding to the complement receptor enables replication in resting B cells otherwise incapable of supporting viral transcription.

Pattern Recognition by C1q, the Collectins, and the Defensins

The action of the complement initiator protein C1q exemplifies a definitive property of intrinsic and innate defense: C1q can recognize molecular patterns characteristic of pathogens. It has properties of a pattern recognition receptor much like the Toll-like receptors. C1q is a calcium-dependent, sugar-binding protein (a **lectin**) in the collectin family of proteins. These proteins bind to polysaccharides on a wide variety of microbes and act as opsonins or activators of the complement cascade. Defensins represent another class of antimicrobial lectins. They are small (29- to 51-residue), cysteine-rich, cationic proteins produced by leukocytes and epithelial cells that are active against bacteria, fungi and enveloped viruses. Collectins and defensins bind the glycoproteins of a number of enveloped viruses, including human immunodeficiency virus type 1, herpes simplex viruses, Sindbis virus, and influenza virus. Some collectins and defensins have antiviral activity in cells in culture. The basis for their antiviral activity appears to be inhibition of membrane fusion. An attractive hypothesis is that they function by cross-linking surface glycoproteins and blocking displacement of other proteins from the fusion site. While these interesting lectins display antiviral activity in the laboratory, their physiological contributions have not been well studied. Some have been modified for testing as antiviral compounds to be delivered systemically or topically.

The Inflammatory Response

As we have seen, during the earliest stages of infection, cells produce cytokines as various intrinsic defenses are activated. The rapid release of cytokines and the appearance of soluble mediators of the complement cascade at the site of infection initiate new responses with far-reaching consequences (Fig. 4.5). The multifunctional cytokine Tnf-α, one of the cytokines of early warning, is produced by activated monocytes and macrophages. The action of Tnf-α induces marked changes in nearby capillaries that attract, and facilitate entry of, circulating white blood cells to the site of infection. Tnf-α also can induce an antiviral response when it binds to receptors on infected cells. Within seconds, the combination of infection and binding of Tnf-α to its receptor initiates a signal transduction cascade that activates caspases, resulting in cell death (see Fig. 3.9B). Viral proteins that modulate the function of Tnf-α are well known.

One very visible response to Tnf-α is **inflammation**. The four classical signs of inflammation are redness, heat, swelling, and pain. These symptoms result from increased blood flow, increased capillary permeability, influx of phagocytic cells, and tissue damage. Increased blood flow occurs when vessels that carry blood away constrict, resulting in engorgement of the capillary network in the area of infection. This response produces redness (erythema) and an increase in tissue temperature. Capillary permeability increases, facilitating an efflux of fluid and cells from the engorged capillaries into the surrounding tissue. The fluid that accumulates has a high protein content, in contrast to that of normal fluid found in tissues, and contributes to the swelling. The cells that migrate into the damaged area are largely mononuclear phagocytes (Table 4.1). They are attracted by molecules synthesized by virus-infected cells, by cytokines elaborated by local defensive systems, and by secondary reactions that facilitate adherence of phagocytic cells to capillary walls near sites of damage. Neutrophils are found in abundance in the blood, but normally are absent from tissues. They are the earliest phagocytic cells to be recruited to a site of infection, and are classic cellular markers of the inflammatory response. Neutrophils also secrete a variety of cytokines and toxic products that can determine subsequent events. Some cytokines made by infected cells, dendritic cells, and macrophages, as well as soluble complement components, are chemokines that direct the migration of monocytes and granular cells to regions of cell damage (Fig. 4.5). Monocytes are also important in the healing reactions that take place after the infection is cleared. Not only does the inflammatory response control viral infections and contribute to the overt pathogenic effects that follow, but also, like most innate responses, it is essential for the initiation of adaptive immune defenses.

Noncytopathic viruses do not induce a strong inflammatory response and, as a consequence, have dramatically different interactions with the host. It is now understood that the early reactions in inflammation determine the type of immune response that will predominate, which in turn can influence the outcome of a viral infection.

Despite an incomplete understanding of the process, for many years scientists have induced inflammation deliberately by the use of adjuvants, such as Freund's adjuvant (killed mycobacterial cells in oil emulsion) or aluminum hydroxide gels. The adjuvant-induced inflammation mimics an infection to provide the environment for the induction of a strong immune response to injected viral proteins in vaccines. In fact, most vaccines would not work without adjuvants to stimulate the adaptive immune response (see Chapter 8 for a discussion of adjuvants and vaccines).

The nature and extent of the inflammatory response to viral infection depend on the tissue that is infected, as well as on the cytopathic nature of the virus. Many of the cells and proteins that participate in an inflammatory response come from the bloodstream and are directed by proinflammatory cytokines and soluble complement effectors to the site of infection. As a result, tissues that have reduced access to the circulatory system (e.g., the brain and the interior of the eyeball) normally avoid the destructive effects of the inflammatory response. However, as a consequence of their so-called "privileged" state, the kinetics, extent, and final outcome of viral infections of these tissues can be markedly different from those of tissues with more intimate access to the circulatory system.

The inflammatory response is held under strict control by many regulatory proteins. A critical protein complex in the cytoplasm is called the "inflammasome"—a structure of more than 700 kDa. The inflammasome complex has properties of a cytoplasmic pattern recognition receptor, as well as a signaling initiator. When a microbial molecule is bound, the inflammasome activates the proinflammatory caspases 1 and 5. These two caspases then lead to the

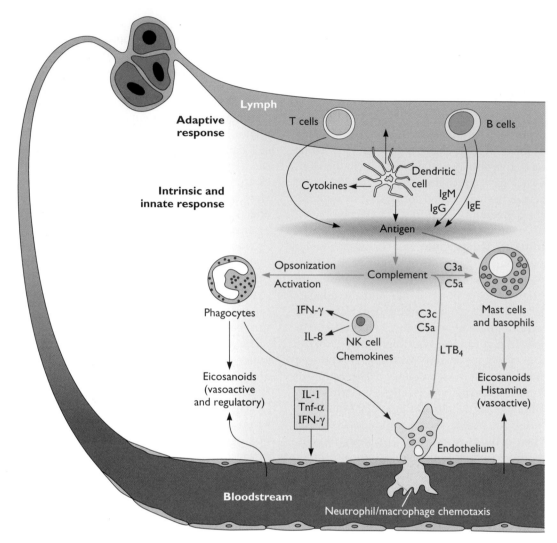

Figure 4.5 Inflammation provides integration and synergy with the main components of the immune system. Viral infections at entry sites in the body often trigger an inflammatory response. A stylized section of infected tissue served by the lympoid system (top, green) and the circulatory system (bottom, red) is shown. Inflammation reactions can be initiated in several ways, for example by cytokines such as IFN released by immature dendritic cells as they detect infection, by the classical or alternative pathway of complement activation, or by mast cells that migrate to sites of cell damage responding to cytokine release, where they can be activated by IgE antibody and antigen. C3a, C3c, and C5a are protease digestion products of the complement cascade that stimulate the inflammatory response. C3a increases vascular permeability and activates mast cells and basophils, C3c stimulates neutrophil release, and C5a increases vascular permeability and chemotaxis of basophils, eosinophils, neutrophils, and monocytes and stimulates neutrophils. The cytokines IL-1, IFN-γ, and Tnf-α act on the local capillary endothelium to enhance leukocyte adhesion and migration. IL-8 and other chemokines promote lymphocyte and monocyte chemotaxis. IL-1 and Tnf-α bind to receptors on epithelial and mesenchymal cells to cause division and collagen synthesis and stimulate prostaglandin and leukotriene synthesis (eicosanoid compounds). LTB$_4$ is a particularly active leukotriene that is vasoactive and chemotactic. The activities of cells that enter an infected site where inflammation reactions are occurring are controlled by locally produced cytokines, particularly Tnf-α, IL-1, and IFN-γ. Adapted from D. Male et al., *Advanced Immunology*, 3rd ed. (Mosby, St. Louis, MO, 1996).

processing and secretion of the proinflammatory cytokines IL-1β and IL-18. The role of the inflammasome in bacterial infections is well established, and it is likely to function similarly in viral infections.

Viral gene products known to modulate the inflammatory response include soluble cytokine receptors of poxviruses, the complement component-binding proteins of herpesviruses, and various cytokines and growth factors of beta- and gammaherpesviruses. Mutants that do not synthesize these proteins have markedly reduced virulence (virulence is discussed in Chapter 1). Such modulation can be critical for both host and virus. For example, if the normal inflammatory response induced by adenovirus infection is not suppressed by viral proteins encoded in the E3 gene complex, severe damage and even death of the host can result.

The Adaptive Immune Response

General Features

The adaptive response comprises two complex actions, the **humoral response** (antibody) and the **cell-mediated response** (helper and effector cells) (Fig. 4.6 and Table 4.3). As we discuss the cells and processes that characterize each response, it is important to understand that **both** are essential in antiviral defense and function in concert. However, the relative contribution of each in any given infection varies with the nature of both the infecting virus and the host. In general, antibodies bind to virus particles in the bloodstream and at mucosal surfaces, decreasing the number of cells that could have been infected, whereas T cells recognize and kill infected cells.

Like the innate response, the adaptive response to viral infection must distinguish infected from uninfected cells. This feat is accomplished in a markedly different fashion than occurs in the innate immune system. Highly specific molecular recognition is mediated by two antigen receptors: membrane-bound antibody on B cells or the T-cell receptor, and one of two membrane glycoprotein oligomers that display fragments of internal cellular proteins on the cell surface. These latter proteins are members of the MHC protein family. MHC class I proteins display protein fragments on the surface of almost all cells, whereas MHC class II proteins generally are found only on the surfaces of mature dendritic cells, macrophages, and B cells (the professional antigen-presenting cells).

While B- and T-cell receptors both bind foreign antigens, they do so in very different ways. The B-cell receptor is a membrane-bound antibody and binds discrete **epitopes** (contiguous sequences or unique conformations) in intact proteins. In contrast, the T-cell receptor binds short, linear **peptides** derived from proteolytically processed proteins. The act of binding to an epitope or peptide by either receptor has profound effects on the cell bearing that receptor: it responds by producing cytokines, by replicating and dividing, by killing the cell that bears the foreign protein or peptide, or by synthesizing antibodies. The entire sequence of events initiated by the binding of foreign peptides or epitopes comprises the adaptive immune response.

As in all defense mechanisms, an uncontrolled or inappropriate adaptive response can be damaging. Therefore, precise initiation and rapid cessation of the response are as important as its amplification. Regulation is achieved largely by multiple interactions among the surfaces of lymphocytes and infected cells, the short life spans of the activated cells, and the short half-lives of the chemical mediators. For example, complement proteins of the innate immune system play important roles in initiating and regulating the T-cell response (Fig. 4.7). When such regulation is disrupted actively or passively by viral infection, or when the cells of the immune system themselves are infected by viruses, severe consequences ensue. Many pathological effects associated with infection are manifestations of inappropriate reactions of the host immune defense system (see "Injury induced by viruses" below). On the other hand, all successful viral genomes encode proteins that modulate these host defenses. In some instances, viral infections can be long-lived, persistent, or latent in the face of an active adaptive immune system (these situations are discussed in Chapter 5).

Unlike the innate response, the adaptive response is tailored specifically to a particular invading organism or substance and requires more time (days to weeks) after exposure to become fully active. However, once a specific response has been established and the viral infection is subdued, the individual is immune to subsequent infection by the same invader (Fig. 4.8). Such memory of previous infections is one of the most powerful features of the adaptive immune response, and makes vaccines possible. While the primary response takes many days to reach its optimum, a subsequent encounter with an invader by an immune individual engenders a ferocious response that occurs within hours of the infection. The innate defenses also are stimulated during this secondary response to infection. The cellular and molecular mechanisms for maintenance of memory are the focus of considerable research and debate, but a subset of B and T lymphocytes called **memory cells** is maintained after each encounter with a foreign antigen. These cells survive for years in the body and are ready to respond immediately to any subsequent encounter by rapid proliferation and efficient production of their protective products. Because such a secondary response is usually stronger than the primary one, childhood infection

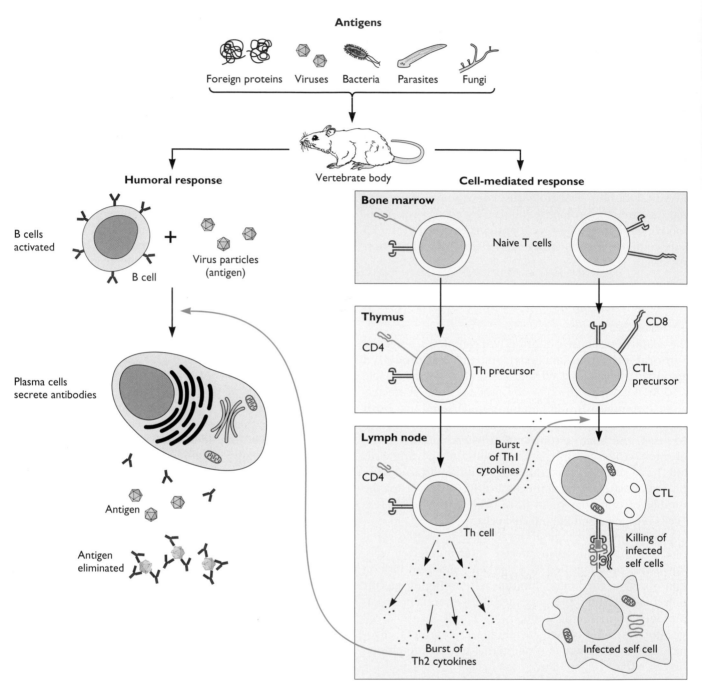

Figure 4.6 The humoral and cell-mediated branches of the adaptive immune system. A variety of foreign proteins and particles (antigens) may stimulate adapative immune responses after recognition by intrinsic and innate defense systems. **(Left)** The humoral branch comprises lymphocytes of the B-cell lineage. Antibodies are the important effector molecules produced by this response. The process begins with the interaction of a specific receptor on precursor B lymphocytes with antigens. Binding of antigen promotes differentiation into antibody-secreting cells (plasma cells). **(Right)** The cell-mediated branch comprises lymphocytes of the T-cell lineage that arise in the bone marrow and are selected in the thymus. The activation process is initiated in lymph nodes when the T-cell receptor on the surface of naive T lymphocytes bind viral peptides on dendritic cells complexed with the MHC class II protein. Two subpopulations of naive T cells are illustrated: the Th-cell precursor and the CTL precursor. The Th cell recognizes antigens bound to MHC class II molecules and produces powerful cytokines that "help" activated B cells to differentiate into antibody-producing plasma cells (Th2 cytokines) or CTL precursors (Th1 cytokines) to differentiate into CTLs capable of recognizing and killing virus-infected cells. The Th1 or Th2 cytokines are produced by different subsets of Th cells and promote or inhibit cell division and gene activity of B-cell or CTL precursors.

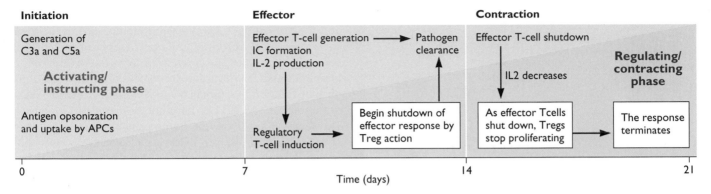

Figure 4.7 Regulation of the T-cell response by complement and regulatory T cells. The extent of the adaptive immune response is regulated in part by regulatory T cells (Treg cells). This model of an acute viral infection provides a view of how complement (part of the innate immune response) may regulate the three phases of a T-cell-mediated response. **(Initiation)** Soon after infection, antigen-presenting cells (APCs) take up viral proteins and make their way to local lymph nodes, where the T-cell response is initiated. The complement cascade stimulated at the site of infection produces a variety of effector proteins, including C3a, C5a, and C3b. The C3b opsonin facilitates the uptake of C3b-coated antigens by antigen-presenting cells, while C3a and C5a stimulate their maturation. **(Effector)** Mature antigen-presenting cells then engage potential effector T cells in lymph nodes, resulting in production of IL-2 and activation of CD46-stimulated Treg cells. Ligands for CD46 include C3b-opsonized immune complexes (IC formation). Effector CTLs and Th1 cells are stimulated by the antigen-presenting cells and leave the lymph node to attack the site of infection and clear the infected cells and virions. The balance between activated CTLs and CD46-stimulated Treg cells determines the extent of CTL action as well as the degree of immunopathology promoted by CTL action. Too many CTL cells can cause damage, but too few cannot clear the viral infection; conversely, too many Treg cells shut down the effector response prematurely, while too few Treg cells promote continued CTL action and potential immunopathology. **(Contraction)** The dynamics of CTL and Treg cell proliferation promote the controlled contraction of the CTL response. Both CTLs and Treg cells rapidly decline in numbers at this stage. The contraction occurs in part because CD46-stimulated Treg cells divide more quickly than CTLs, and through the action of Treg cytokines, the CTL response shuts down. Because the activated CTLs and Th cells produce IL-2 necessary for Treg-cell replication, the pool of Treg cells then diminishes as the system returns to its unstimulated state. During this phase, memory CTL and memory Treg cells also are produced. Adapted from C. Kemper and J. Atkins, *Nat. Rev. Immunol.* **7:**9–18, 2007.

protects adults, and immunity conferred by vaccination can last for years.

Cells of the Adaptive Immune System

The adaptive response depends on two important cell groups: lymphocytes and antigen-presenting cells (Table 4.1). Lymphocytes are white blood cells produced by hematopoiesis in the bone marrow. They are migratory cells, leaving the bone marrow to circulate in the blood and lymphatic system, settling in various lymphoid organs throughout the body (Fig. 4.9). The small organs called lymph nodes are essential for initiation of the adaptive immune response: when lymph nodes are removed, viral infections do not stimulate adaptive immunity. Lymphoid tissues are the collection centers and sites of communication for cells in the circulatory system.

T cells and **B cells** represent two major classes of lymphocytes. T cells (lymphocytes that mature in the thymus) and particular cytokines are the primary components of the cell-mediated response. Immature T cells differentiate

into two critical effector T cells, the T-helper (Th) cell and the cytotoxic T cell (cytotoxic T lymphocyte [CTL]). Plasma cells, whose immediate precursors are B cells (lymphocytes derived from the bone marrow, the mammalian equivalent of the avian bursa), make antibodies that bind to foreign molecules and define the antibody response (also known as the humoral response).

During the early maturation of T and B cells, cells that react against self tissues and proteins die. The remaining T and B cells have the capacity to recognize nonself molecules, but they remain in a dormant state in lymphoid tissue until they physically interact with particular lymphocytes. These lymphocytes (called professional antigen-presenting cells) move back and forth from peripheral tissues to lymphoid tissues. Their function is to expose on their surfaces peptides and proteins gathered from the peripheral tissues so that they can be bound by T- and B-cell receptors. When the peptides are recognized as nonself, the T or B cells are stimulated to divide and carry out their immune effector actions. Dendritic cells are crucial

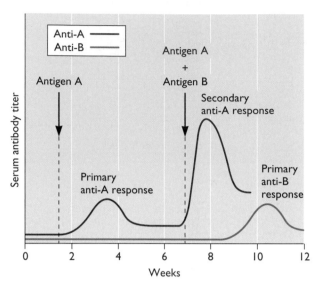

Figure 4.8 The specificity, self-limitation, and memory of the adaptive immune response. This general profile of a typical adaptive antibody response demonstrates the relative concentration of serum antibodies after time (weeks) of exposure to antigen A or a mixture of antigens A and B. The antibodies that recognize antigens A and B are indicated by the red and blue lines, respectively. The primary response to antigen A takes about 3 to 4 weeks to reach a maximum. When the animal is injected with a mixture of both antigens A and B at 7 weeks, the secondary response to antigen A is more rapid and more robust than the primary response. However, the primary response to antigen B again takes about 3 to 4 weeks. These properties demonstrate immunological memory. Antibody levels (also termed titers) decline with time after each immunization. This property is called self-limitation or resolution. From A. K. Abbas et al., *Cellular and Molecular Immunology* (The W. B. Saunders Co., Philadelphia, PA, 1994), with permission.

professional antigen-presenting cells, as discussed in Chapter 3. Immature B cells and cells of the monocyte lineage (e.g., macrophages) are also considered to be professional antigen-presenting cells.

The Mucosal and Cutaneous Arms of the Immune System

The adaptive immune system exhibits considerable decentralization of important cells and tissues. However, every major organ and body surface harbors components that are coordinated to mount a focused, adaptive immune response when signaled by the innate immune system. The **mucosal immune system** is usually the first adaptive defense to be engaged after infection. The lymphoid tissues below the mucosa of the gastrointestinal and respiratory tracts (often called mucosa-associated lymphoid tissue) (Fig. 4.9B) are vital in antiviral defense. These clusters of lymphoid cells include the collection called Peyer's patches in the lamina propria of the small intestine, the tonsils in

the pharynx, the submucosal follicles of the upper airways, and the appendix. A specialized epithelial cell of mucosal surfaces is the **M cell** (microfold or membranous epithelial cell), which samples and delivers antigens to the underlying lymphoid tissue by transcytosis. M cells have invaginations of their membranes (pockets) that harbor immature dendritic cells, B and CD4$^+$ T lymphocytes, and macrophages. The secreted antibody IgA (important in antiviral defense at mucosal surfaces [see below]) is made by B cells that accumulate at adhesion sites in these M cell membrane pockets. After viral proteins transit through M cells, they emerge to be in intimate contact with all the appropriate immune cells. This process represents an essential step for the development of mucosal immune responses.

The skin, the largest organ of the body, possesses its own complex community of organized immune cells. Lymphocytes and Langerhans cells comprise the **cutaneous immune system** (also called skin-associated lymphoid tissue) (Fig. 4.9C). These cells are important in the initial response and resolution of viral infections of the skin. In particular, Langerhans cells, the predominant scavenger antigen-presenting cells of the epidermis, function as the sentinels or outposts of early warning and reaction. These abundant, mobile cells sample antigens and migrate to regional lymph nodes to transfer information to T cells, and to activate B lymphocytes directly. Certain T cells in the circulation have tropism for the skin and, after binding to the vascular endothelium, can enter the epidermis to interact with Langerhans cells and keratinocytes. These skin-tropic T cells play important roles in production of the virus-specific skin rashes and pox characteristic of measles virus and varicella-zoster virus infections.

Virus particles can interact with lymphoid cells associated with mucosal and cutaneous immune systems at the primary site of infection. Such events can suppress immune responses by killing or misregulation of immune cells. These interactions can govern the outcome of the primary infection and often establish the pattern of infection characteristic of a given virus. The M cells in the mucosal epithelium have been implicated in the spread from the pharynx and the gut to the lymphoid system of a variety of viruses, including poliovirus, enteric adenoviruses, human immunodeficiency virus type 1, and reovirus. These cells also have been suggested to be sites of persistent or latent infection for a number of other viruses, including herpes simplex virus.

Adaptive Immunity: the Action of Lymphocytes That Carry Distinct Antigen Receptors

Unlike the germ line-encoded pattern recognition receptors of the intrinsic defenses and innate immune system,

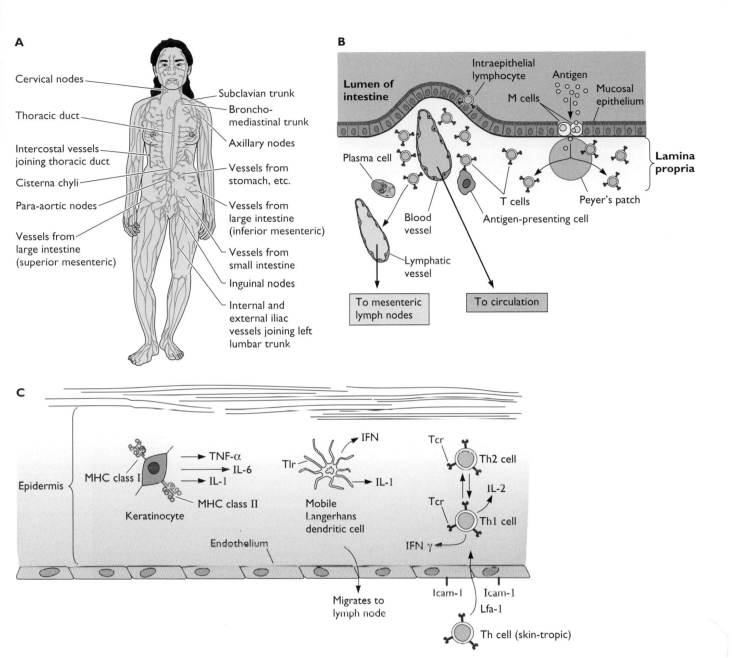

Figure 4.9 Components of the human lymphatic and mucosal immune systems. (A) The primary lymphatic system is illustrated in green. Many clusters of lymphatic tissue, called nodes, are found throughout the lymphatic system. Lymph nodes provide the interactive environment where mobile dendritic cells patrolling the local area exchange information with lymphocytes in the circulation. **(B)** Cellular components of the mucosal immune system in the gut (mucosa-associated lymphoid tissue). The lumen of the small intestine is at the top of the figure. The mucosal epithelial cells are shown with their basal surface oriented toward the lamina propria. Cross sections of a lymphatic vessel and a capillary are shown, illustrating their juxtaposition to cells of the mucosal immune system. M cells have large intraepithelial pockets filled with B and CD4$^+$ T lymphocytes, macrophages, and dendritic cells. M cells and intraepithelial lymphocytes are important in the transfer of antigen from the intestinal lumen to the lymphoid tissue in Peyer's patches, where an immune response can be initiated. **(C)** The cutaneous immune system (skin-associated lymphoid tissue) comprises three cell types: keratinocytes, Langerhans cells, and T cells. Keratinocytes actively secrete various cytokines, including Tnf-α, IL-1, and IL-6, and have phagocytic activity. They also synthesize both MHC class I and MHC class II proteins and can present antigens to T and B cells if stimulated by IFN-γ. Langerhans cells are migratory dendritic cells and are the major antigen-presenting cells in the epidermis. When products of viral infections in the skin are detected, Langerhans cells secrete IFN and undergo maturation. Mature dendritic cells migrate to the local draining lymph node, where they present viral peptides on both MHC class I and MHC class II proteins to antigen-specific T cells. Special skin-tropic T cells can cross the endothelium to enter the epidermis, where they can mature into Th1 or Th2 cells depending on the antigen and cytokine milieu. Activated T cells synthesize cytokines, including IFN-γ, that activate MHC expression from keratinocytes and Langerhans cells. Tcr, T-cell receptor. Adapted from A. K. Abbas et al., *Cellular and Molecular Immunology* (The W. B. Saunders Co., Philadelphia, PA, 1994), with permission.

the T- and B-cell antigen receptors are formed by somatic gene rearrangement during development of the organism. Their specificities are built without regard to any one target. Subsequently, cells that can bind a particular antigen are killed if they recognize self antigens, or stimulated to divide and prosper if they recognize a foreign invader or infected cell. Lymphocytes are responsible for specificity, memory, and self-nonself discrimination.

As noted above, each T or B cell carries on its surface a specific receptor that binds and responds to a particular peptide or epitope (Fig. 4.10). Antiviral defense is possible because T and B cells bearing these specific receptors have survived the normal selective process that eliminates those that respond to self peptides and self epitopes. The cells emerging from such selections enter the circulation and pass through the lymphatic system, or are retained in various tissues in the body (Fig. 4.6). They are said to be **naive,** because they are not completely differentiated and are not armed to produce their ultimate immune effector response. However, they are able to react to foreign signals with remarkable diversity and specificity. After such recognition, they rapidly acquire their specific immune effector activity.

The initial encounter with any foreign epitope, whether in lymphoid tissues or elsewhere in the body, involves only a few cells. For example, the frequency of B or T lymphocytes that recognize infected cells on first exposure is as few as 1 in 10,000 to 1 in 100,000. So few cells are certainly not sufficient for protection against an infection that is spreading in the host. What makes the adaptive response so powerful is that the initial response is amplified substantially during the ensuing 1 or 2 weeks: the number of virus-specific lymphocytes increases more than 1,000-fold. The original encounter stimulates these uncommitted, naive lymphocytes to differentiate, divide, and produce antibodies or cytokines. The antigen-stimulated cell is said to be **activated.** It then becomes fully differentiated and undergoes numerous rounds of cell division such that each daughter cell has the same specific immune reactivity as the original parent (often called a clonal response).

B Cells

B cells are produced in the bone marrow. As they mature, each synthesizes an antigen receptor, which is a membrane-bound antibody (Fig. 4.6 and 4.10). When an antigen binds specifically to a membrane-bound antibody, a signal transduction cascade is initiated. As a consequence, new gene products are made and the cell begins to divide rapidly. The daughter cells produced by each division differentiate into effector plasma cells and a small number of memory B cells. As their name implies, memory B cells, or their clonal progeny, are long-lived and continue to produce the parental, membrane-bound antibody receptor. In contrast, plasma cells live for only a few days and no longer make membrane-bound antibody, but instead synthesize the same antibody in secreted form. A single plasma cell can secrete more than 2,000 antibody molecules per second.

T Cells

T-cell precursors are also produced in the bone marrow, but in contrast to a B-cell precursor, a T-cell precursor must migrate to the thymus gland to mature (Fig. 4.6). The notation "T" in "T cell" reminds us that the thymus is required for their development. T cells can be distinguished from other lymphocytes by a special receptor on their surface, called the T-cell receptor. Subsets of T cells have distinct functions.

The maturation process comprises two types of selection: positive selection for T cells that can bind appropriate surface molecules via the T-cell receptor and negative selection that efficiently kills T cells that recognize target cells displaying self peptides on their surfaces. As a result, only 1 to 2% of all immature T cells entering the thymus emerge potentially able to defend the host against viral infections. These naive T cells are now able to respond to nonself antigens in lymphoid tissues. When they encounter such antigens presented to them by mature dendritic cells, they radically change their complement of cell surface proteins. Some differentiate into effector cells that can kill target cells when they leave the lymph node (CTLs),

Figure 4.10 The antigen receptors on the surfaces of B and T cells. (A) Each B cell has about 100,000 molecules of a unique membrane-bound receptor antibody. Every receptor antibody on a given B cell has identical bivalent specificity for one antigen epitope. **(B and C)** Each T lymphocyte has about 100,000 T-cell receptors (Tcr), each with identical specificity. **(B)** T cells bearing the surface membrane protein CD4 always recognize peptide antigens bound to MHC class II proteins and generally function as Th cells. **(C)** T cells bearing the surface membrane protein CD8 always recognize peptide antigens bound to MHC class I proteins and generally function as cytotoxic T (Tc) cells.

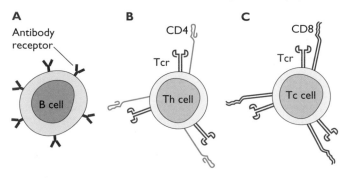

and some differentiate into helper cells that can stimulate B cells to make antibody (Th cells). Yet others become memory cells that retain the capacity to differentiate into effector cells when they reencounter the stimulating antigen, and some (Treg cells) function to shut down the T-cell-mediated response at the end of an immune response (Fig. 4.7).

As noted above, T cells synthesize the T-cell receptor, which is capable of binding peptides. The T-cell receptor is a disulfide-linked heterodimer composed of either alpha and beta or gamma and delta protein chains. The peptide-binding site of the T-cell receptor and the epitope-binding site of the B-cell receptor are very similar structures, formed by the folding of three regions in the amino-terminal domains of the proteins that participate in antigen recognition (the so-called hypervariable regions). However, unlike the B-cell receptor, which can recognize the epitope as part of an intact folded protein, the T-cell receptor can recognize **only** a peptide fragment produced by proteolysis. Furthermore, the peptide must be bound to MHC cell surface proteins (see below). When the T-cell receptor engages an MHC molecule carrying the appropriate antigenic peptide, a signal transduction cascade that leads to gene expression is initiated. As a result, the stimulated T cell is capable of differentiating to form memory and various effector T cells.

Th Cells and CTLs Are Distinguished by Unique Cell Surface Proteins

In general, lymphocytes can be distinguished by the presence on their surfaces of specific proteins called **cluster-of-differentiation (CD) markers** (e.g., CD3, CD4, and CD8). The presence of these proteins can be detected with antibodies raised against them in heterologous organisms; they are often referred to as "CD antigens." The over 247 individual CD markers known are invaluable in identifying lymphocytes of a particular lineage or differentiation stage. Two well-known subpopulations of T cells are defined by the presence of either the CD4 or the CD8 surface proteins (Fig. 4.11), which are coreceptors for MHC class II and MHC class I, respectively. When immature T cells leave the bone marrow, they do not synthesize CD4 or CD8 proteins (they are said to be "double-negative"). They differentiate sequentially in the thymus, initially producing both CD4 and CD8 proteins ("double-positive") and then either CD8 or CD4 ("single-positive"). These single-positive cells are the naive T cells that migrate to peripheral sites.

CD4$^+$ T cells are generally Th cells capable of interacting with B cells and antigen-presenting cells that have MHC class II proteins on their surfaces. After such interactions, CD4$^+$ Th cells mature into Th1 or Th2 cells (see below). Th cells synthesize cytokines and growth factors that stimulate (help) the specific classes of lymphocytes with which they interact. **CD8$^+$ T cells** differentiate into CTLs that can interact with almost all cells of the body expressing the more ubiquitous MHC class I proteins. Cytotoxic T cells recognize foreign peptides complexed with MHC class I proteins and, when productively engaged, actively destroy the cell presenting the peptides. Mature cytotoxic T cells play important roles in eliminating virus-infected cells from the body by cell lysis and by production of IFN-γ and Tnf-α.

Th1 and Th2 Cells

When naive Th cells engage mature dendritic cells in lymphoid tissue, cytokines and receptor ligand interactions

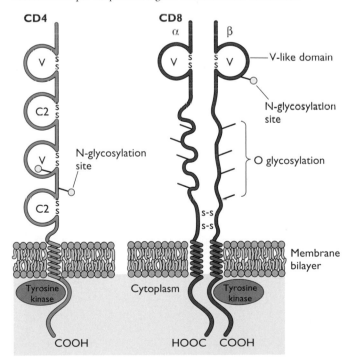

Figure 4.11 Simplified representations of CD4 and CD8 coreceptor molecules. These two molecules associate with the T-cell receptor on the surface of T cells. The CD4 molecule is a glycosylated type 1 membrane protein and exists as a monomer in membranes of T cells. It has four characteristic immunoglobulin-like domains labeled here as V and C2. The V domains are similar to the variable domain of immunoglobulin in the tertiary structure. The first two domains form a binding site for MHC class II proteins. The cytoplasmic domain interacts with specific tyrosine kinases, endowing CD4 with signal transduction properties. The CD8 molecule is a type 1 membrane protein with both N and O glycosylation. It is a heterodimer of an α chain and a β chain covalently linked by disulfide bonds that interacts with MHC class I proteins. The two polypeptides are quite similar in sequence, each having an immunoglobulin-like V domain thought to exist in an extended conformation. Tyrosine kinases also associate with the CD8 cytoplasmic domain and participate in signal transduction reactions.

stimulate the T cell to differentiate into one of two Th cell types called Th1 and Th2 (Fig. 4.6; Fig. 4.12). These two cell types can be distinguished by the cytokines they produce and the processes they invoke. Th1 cells are important for controlling most, but not all, viral infections. They promote the cell-mediated response by stimulating the maturation of cytotoxic T-cell precursors. They accomplish this, in part, by producing IL-2 and IFN-γ, cytokines that stimulate inflammation (the proinflammatory response). In addition, Th1 cells provide stimulating cytokines to the antigen-presenting dendritic cell so that it can communicate with naive CD8⁺ T cells. We know that if IL-12 is present at the time of antigen recognition, immature Th cells differentiate into Th1 cells. IL-12 also stimulates NK and Th1 cells to secrete IFN-γ, a cytokine important in increasing the activity of inflammatory cells such as macrophages.

In the presence of IL-4, immature Th cells differentiate into Th2 cells that stimulate the antibody response rather than the cell-mediated, proinflammatory response. The initial source of IL-4 may be the natural killer T cells (NKT cells) discussed below. Th2 cells promote the antibody response by inducing maturation of immature B cells and resting macrophages. They also reduce the inflammatory response by producing IL-4, IL-6, and IL-10, but not IL-2 or IFN-γ. Th2 cells are more active after invasion by extracellular bacteria or multicellular parasites. Nevertheless, the Th2 response is critical for controlling infections that produce large quantities of virus particles in the blood.

In general, Th1 and Th2 responses have a yin-yang relationship: as one increases, the other decreases (Fig. 4.12). While IFN-γ turns up the Th1 response, it also inhibits the synthesis of IL-4 and IL-5 by Th2 cells, effectively dampening the latter response. On the other hand, production of Th2 cytokines is an important mechanism to shut off the proinflammatory and potentially dangerous Th1 response.

Viral infection induces the production of proinflammatory Th1 cytokines such as IL-12. In contrast, bacterial infection induces synthesis of cytokines, such as IL-4, that promote the Th2 response. The mechanisms that provide such precise regulation are under study. One idea is that mature dendritic cells automatically produce proinflammatory cytokines as their default pathway, and always activate a Th1 response unless appropriate Th2 signals are provided. A similar idea gaining credence is that when dendritic cells detect CpG sequences, single-stranded RNA ssRNA, or double-stranded RNA dsRNA sequences via their Toll-like receptors, Nf-κb is activated and Th1 cytokine genes are transcribed.

We know that many viral proteins modulate the Th1-Th2 balance in interesting ways. For example, infection of B cells by Epstein-Barr virus and equine herpesvirus type 2 should stimulate an active Th1 response. However, both viral genomes encode proteins homologous to IL-10, a regulatory cytokine that represses the Th1 response. Viral IL-10 foils the Th1 antiviral defense that would kill infected B cells, while promoting differentiation into memory B cells that are important for long-term survival of the viral genome.

For most viral infections, a given Th response represents a spectrum of some Th1 and some Th2 cells, and consequently a mixture of cytokines. Establishment of the proper repertoire of Th cells therefore is an important early event in host defense; an inappropriate response has far-reaching consequences. For example, synthesis of the Th2 cytokine IL-4 by an attenuated mousepox virus recombinant resulted in lethal, uncontained spread of virus in an immune animal (see Box 2.8). How the fundamental Th1-Th2 decision is made is under much scrutiny in laboratories around the world. A practical reason for such interest

Figure 4.12 Th cells: the Th1 versus the Th2 response. Immature CD4⁺ Th cells differentiate into two general subtypes called Th1 and Th2, defined functionally according to the cytokines they secrete. Th1 cells produce cytokines that promote the inflammatory response and activity of cytotoxic T cells, and Th2 cells synthesize cytokines that stimulate the antibody response. The cytokines made by one class of Th cell tend to suppress production of those of the other class.

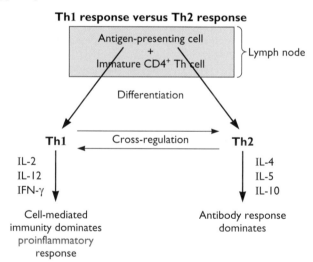

Th1 response versus Th2 response

| Antigen-presenting cell + Immature CD4⁺ Th cell | Lymph node |

Differentiation

Th1 ← Cross-regulation → Th2

Th1: IL-2, IL-12, IFN-γ → Cell-mediated immunity dominates proinflammatory response

Th2: IL-4, IL-5, IL-10 → Antibody response dominates

Immune cross-regulation by cytokines			
Th1 response		Th2 response	
Enhance	Suppress	Enhance	Suppress
IL-2	IL-4	IL-4	IFN-γ
IL-12	IL-10	IL-5	IL-12
IFN-γ		IL-6	
		IL-10	

is that the design of potent and effective vaccines depends on stimulating the appropriate spectrum of response.

Th17 Cells

This new class of CD4$^+$ helper cells plays central roles in control of the inflammatory response. These cells are found in the skin and the lining of the gastrointestinal tract and at other interfaces between the external and internal environments. When dendritic cells present antigens to them in the presence of Tgf-β and IL-6, Th17 cells secrete IL-17 and IL-21. In addition, the stimulated Th17 cells now express the receptor for IL-23, which leads to massive proliferation of the activated Th17 cells. The activated cells stimulate a strong inflammatory response, secrete defensins, and recruit neutrophils to the site of activation. Th17 cells have been implicated in autoimmune diseases involving chronic inflammation and in the control of bacterial infections. Their importance in controlling viral infections is only now being understood. For example, individuals with large numbers of Th17 cells in their gut mucosa appear to be able to control lentivirus infections much better than individuals with reduced numbers of these helper cells.

Memory T Cells

Memory T cells are long-lived, mature T cells. Each of their T-cell receptors binds to a specific nonself peptide. When the specific nonself peptide is bound, the cells divide rapidly, producing active effector T cells. Memory T cells can carry either CD8 or CD4 surface proteins. Whether memory cells are sequestered in various depots in the body for future use, or are constantly produced, is controversial. One idea of how memory is maintained is that follicular dendritic cells in lymph nodes bind antibody-antigen complexes and keep such complexes on their surfaces for long periods. When the circulating antibody concentration drops, as happens after an infection is cleared, antigen is released gradually from the immune complexes. Such slow release of antigen would provide continual stimulation of the immune response.

If memory cells are infected directly by viruses, as is the case for human immunodeficiency virus type 1, the host's secondary immune response to a wide variety of previous infectious agents can be compromised.

Regulatory T Cells

The regulatory T-cell (Treg) subset of T cells has been recognized for some time (initially they were called suppressor T cells). However, their function was controversial until recently. Now it is clear that Treg cells are pivotal players in the end-stage immune response to most if not all infectious agents. Their primary function is to terminate

the immune response and bring the immune system back to ground state (immune homeostasis; Fig. 4.7). These cells are also important for immune suppression, self-tolerance, and control of the inflammatory response. Treg cells serve to maintain a balance between protection and immune pathology. Ironically, Treg action may limit the effectiveness of vaccines because they shut down the immune response.

NKT Cells

NKT cells have a T-cell receptor, a property that distinguishes them from natural killer (NK) cells described previously. NKT cells do have some surface molecules in common with NK cells, hence their name. NKT cells play critical roles in early innate and adaptive responses. Unlike conventional T cells that recognize peptides bound to MHC class I molecules on target cells, NKT cells recognize glycolipid molecules bound to CD1d, a distant cousin of the MHC class I proteins. If a foreign glycolipid is recognized, the NKT cell can act as a Th cell or a cytotoxic cell. NKT cells account for 20 to 30% of lymphocytes present in the liver, and are capable of releasing IFN-γ after viral infection. Their role in antiviral defenses is only now being appreciated.

$\gamma\delta$ T Cells

$\gamma\delta$ T cells, a small subset of T cells, develop in the thymus as do all T cells, but after that, the similarity ends. These cells reside at the crucial interface between the outside world and tissues. Their name derives from their unique cell surface $\gamma\delta$ T-cell receptor. They are abundant in epithelial cell layers, including the gut mucosa, skin, and lining of the vagina. Curiously, the antigens recognized by $\gamma\delta$ T cells are not bound to classical cell surface MHC proteins and include not only peptides but also intact proteins and organic molecules that contain phosphorus. These cells do not interact with professional antigen-presenting cells (dendritic cells or macrophages), nor do they express CD8 or CD4 proteins on their cell surface. Long ignored because they were difficult to identify, purify, and study, the $\gamma\delta$ T cells now are seen as crucial players in front-line immune surveillance and antiviral action.

Antigen Presentation and Activation of Immune Cells

Naive T cells engage the professional antigen-presenting cells by binding to MHC class II proteins (Fig. 4.6). This encounter results in differentiation of T cells into effector cells, the **CTLs**. These cells enter the circulation and are able to distinguish infected cells from uninfected cells by specific interactions of the T-cell receptor with MHC class I proteins (Box 4.4). MHC class I proteins are found on the surfaces of nearly all nucleated cells. The MHC class I protein

BOX
4.4

EXPERIMENTS
Virology provides Nobel Prize-winning insight: MHC restriction

In 1974 at the John Curtin School of Medical Research in Canberra, Australia, Rolf Zinkernagel and Peter Doherty performed a classic experiment that provided insight into how CTLs can recognize virus-infected cells. Initially, they teamed up to determine the mechanism of the lethal brain destruction observed when mice are infected with lymphocytic choriomeningitis virus, an arenavirus that does not directly kill the cells it infects. They anticipated that the brain damage was due to CTLs responding to replication of the noncytopathic virus in the brain.

When they infected mice of a particular MHC type with the virus and then isolated T cells, these cells lysed virus-infected target cells in vitro **only** when the target cells and the T cells were of identical MHC haplotype. Uninfected target cells were not lysed, even when they shared identical MHC alleles. This requirement for MHC matching was called **MHC restriction**.

Their Nobel Prize-winning insight was that a CTL must recognize two determinants present on a virus-infected cell: one specific for the virus and one specific for the MHC of the host. We now know that CTLs recognize a short peptide derived from viral proteins and only engage the peptide when it is bound to MHC class I proteins present on the surface of target cells.

Zinkernagel, R. M., and P. C. Doherty. 1974. Restriction of in vitro T-cell mediated cytotoxicity in lymphocytic choriomeningitis within a syngeneic or semiallogenetic system. *Nature* **248**:701–702.

comprises two subunits called the α chain (often called the heavy chain) and β_2-microglobulin (the light chain). Lymphocytes possess the highest concentration of MHC class I protein, with about 5×10^5 molecules per cell. In contrast, fibroblasts, muscle cells, and liver hepatocytes carry much smaller quantities, sometimes 100 or fewer molecules per cell. There are three MHC class I loci in humans (*A, B,* and *C*) and two in mice (*K* and *D*). Because there are many allelic forms of these genes in outbred populations, MHC class I genes are said to be **polymorphic**. For example, at the human MHC class I locus *hla-B*, more than 149 alleles with pairwise differences ranging from 1 to 49 amino acids have been identified. When cells bind cytokines such as IFN and IFN-γ, transcription of the MHC class I α chains, β_2-microglobulin, and the linked proteasome and peptide transporter genes (see below), is markedly increased.

Synthesis of MHC class II proteins occurs primarily in the professional antigen-presenting cells (dendritic cells, macrophages, and B cells). Other cell types, including fibroblasts, pancreatic β cells, endothelial cells, and astrocytes, can produce MHC class II molecules, but only on exposure to IFN-γ. As with MHC class I, there are many alleles of MHC class II genes.

The defining property of both MHC class I and class II proteins is that they specialize in binding short peptides produced inside the cell and presenting them for recognition by lymphocytes with T-cell receptors on their surfaces. Both classes of MHC protein have a peptide-binding cleft that is sufficiently flexible to accommodate the binding of many peptides (Fig. 4.13 and 4.14). Even so, not all possible peptides are bound. The ability of MHC molecules to bind and display peptides on the cell surface varies from individual to individual as a result of the many MHC alleles. Such diversity of MHC alleles plays an important role in an individual's capacity to respond to various infections. The more diversity, the more robust the immune response. This fact has dramatic consequences for the spread of viral diseases in a given population. For example, individuals in inbred populations lose MHC diversity over time and also have a reduced capacity to respond to infections. Protective immunity is difficult to establish, and epidemics are likely.

T Cells Recognize Infected Cells by Engaging the MHC Class I Receptors

The immune system must destroy virus-infected cells while ignoring uninfected cells. As viruses can infect many cell types, the recognition/destruction system certainly operates on all cells with high fidelity. Virus-infected cells are identified, in part, because they display small viral peptides as well as cell peptides complexed to MHC class I proteins on their surfaces. The viral and cellular peptides are produced by **endogenous antigen presentation** (Fig. 4.13).

In uninfected and infected cells, a fraction of most newly synthesized proteins is broken down in a controlled manner by the proteasome. The targeted protein is marked for destruction by the covalent attachment of multiple copies of a small protein called ubiquitin and, following adenosine 5′-triphosphate (ATP)-dependent unfolding, the protein is degraded in the inner chamber of the proteasome. The peptide products are released and transported into the endoplasmic reticulum (ER) by a specific peptide transporter system. Within the ER, peptides bind to newly synthesized MHC class I proteins, an interaction that allows MHC class I molecules to adopt their native conformation for transport to the cell surface via the secretory pathway. Hence, the MHC class I pathway displays an "inside-out" picture of the cell to the T cell. Patrolling CTLs move over the surface of potential target cells, engaging

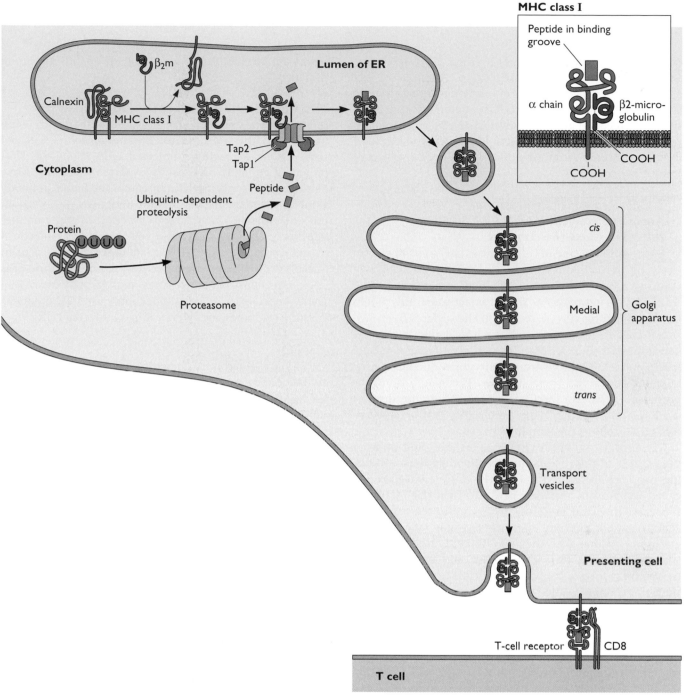

Figure 4.13 Endogenous antigen processing: the pathway for MHC class I peptide presentation. Intracellular proteins of both host cell and virus are degraded in the cytoplasm in a ubiquitin-dependent process. Proteins are marked for destruction by polyubiquitinylation. These modified proteins are then taken up and degraded by the proteasome. The resulting short peptides are transported into the ER lumen by the Tap1-Tap2 heterodimeric transporter in a reaction requiring ATP. Once in the ER lumen, the peptides associate with newly synthesized MHC class I molecules that bind weakly to the Tap complex. Assembly of the α chain and β$_2$-microglobulin of the MHC class I molecule is facilitated by the ER chaperone calnexin, but formation of the final native structure requires peptide loading. The MHC class I complex loaded with peptide is released from the ER to be transported via the Golgi compartments to the cell surface, where it is available for interaction with the T-cell receptor of a cytotoxic T cell carrying the CD8 coreceptor. **(Inset)** The MHC class I molecule is a heterodimer of the membrane-spanning type I glycoprotein α chain (43 kDa) and β$_2$-microglobulin (12 kDa) that does not span the membrane. The α chain folds into three domains, 1, 2, and 3. Domains 2 and 3 fold together to form the groove where peptide binds, and domain 1 folds into an immunoglobulinlike structure. Adapted from D. Male et al., *Advanced Immunology*, 3rd ed. (Mosby, St. Louis, MO, 1996).

MHC class I peptide complexes by their T-cell receptors. Binding of a viral peptide-MHC class I complex by the T-cell receptor triggers a series of reactions that activate the CTL for killing of the infected cell (see below). Surprisingly, the T-cell receptor has a low affinity, 1 μM or less, for its peptide-MHC class I ligand. How high-fidelity recognition is obtained from low-affinity binding has been a subject of intense research.

T Cells Recognize Professional Antigen-Presenting Cells by Engaging the MHC Class II Receptors

Both antibody and CTL responses are controlled precisely by the cytokines produced by Th cells. Such precision is achieved by specific antigen recognition by the MHC class II proteins. A Th cell is activated only when the peptide antigen is presented on the surfaces of professional antigen-presenting cells, such as dendritic cells. As dendritic cells mature, MHC class II glycoproteins loaded with peptides produced from their stores of endocytosed antigens appear on their surfaces. The mature antigen-presenting cells also carry high surface concentrations of costimulatory T-cell adhesion molecules that bind receptors on Th cells in the lymphoid tissue.

The process by which viral proteins are taken up from the outside of the cell and digested, and by which resulting peptides are loaded onto MHC class II molecules, is called **exogenous antigen presentation** (Fig. 4.14). In this case, the viral proteins are not produced inside the cell, as it is not infected, and their digestion takes place in endosomes rather than the proteasome. Furthermore, the peptides and MHC class II molecules are brought together by vesicular fusion. As with MHC class I, the complex is then transported to the surface of the antigen-presenting cell, where it is available to interact with appropriate T cells in the lymph node. Interaction of T-cell receptors on the naive CD4$^+$ Th cell with the MHC class II-peptide complex induces concerted changes in the Th cell, leading to its activation and differentiation (Fig. 4.6). Full activation requires the interactions of other surface proteins and costimulatory molecules (Fig. 4.15).

Th cells activated in this fashion produce IL-2, as well as a high-affinity receptor for this cytokine. The secreted IL-2 binds to the newly synthesized receptors to induce autostimulation and proliferation of the Th cell. Such clonal expansion of specific Th1 or Th2 cells then promotes the activation of CTLs and B lymphocytes (Fig. 4.6).

While MHC class II proteins are definitive components of the professional antigen-presenting cell, some MHC class I molecules also can be loaded with peptides produced via the exogenous route in dendritic cells (a process called **cross-presentation**). This pathway may be a significant mechanism for activating the adaptive immune response.

The Cell-Mediated Adaptive Response

In general, the cell-mediated response facilitates recovery from a viral infection, because it eliminates virus-infected cells without damaging uninfected cells. While the Th2-promoted antibody response is important for some infections in which virus particles spread in the blood, antibody alone is often unable to contain and clear an infection. Indeed, antibodies have little or no effect in many natural infections that spread by cell-to-cell contact (e.g., infections by neurotropic viruses such as alphaherpesviruses) or in infections by viruses that infect circulating immune cells (e.g., infections by lentiviruses and paramyxoviruses). These infections can be stopped only by CTL-produced antiviral cytokines and overt killing of infected cells.

CTLs

CTLs are superbly equipped to kill virus-infected cells, and once they complete one killing, they can detach and kill again. These lethal effector cells mature by a multi-step pathway that fully arms them for killing. At least two reactions are required for realization of their full killer potential. These reactions are the interaction of their T-cell receptor with foreign antigens presented by MHC class I molecules, and the binding of additional surface proteins (the coreceptors) on the precursor CTL to their ligands on the infected cell.

Signaling from the T-cell receptor when it engages the peptide antigen-MHC complex requires clustering (aggregation) of a number of T-cell receptors and reorganization of the T-cell cytoskeleton in a particular structure called the **immunological synapse** (Fig. 4.16). Only after these reactions have taken place can the CTL lyse an infected cell.

The term "immunological synapse" was coined because the proteins that mediate target and T-cell recognition show an unexpected degree of spatial organization at the site of T-cell–target cell contact. This focal collection of membrane proteins and their respective binding partners has functional analogy to the neuronal synapse, a site of informational transfer between neurons. The synapse structure contributes to stabilizing signal transduction by the T-cell receptor for the prolonged periods required for gene activation. In addition, membrane proteins in the structure engage the underlying cytoskeleton and polarize the secretion apparatus so that a high local concentration of effector molecules is attained at the site of contact. Small numbers of peptide ligands complexed to MHC class I molecules apparently can stimulate a T cell because they serially engage a large total number of T-cell receptors on the opposing cell surface in the immunological synapse. Unengaged T-cell receptors

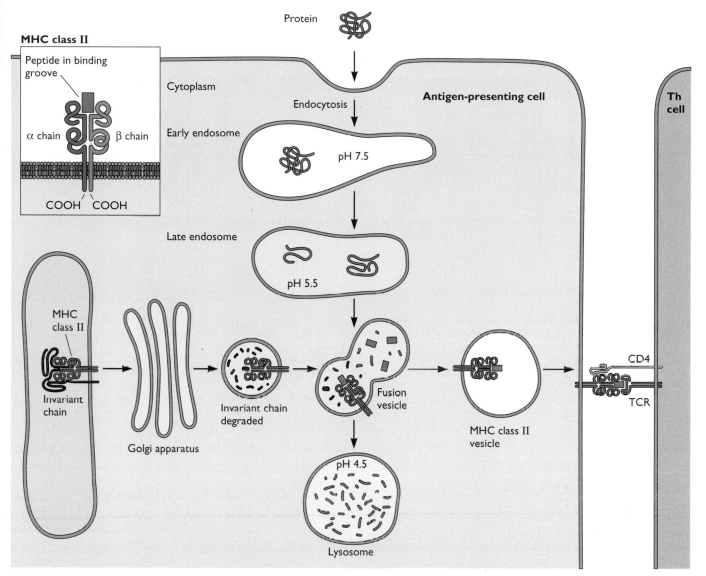

Figure 4.14 Exogenous antigen processing in the antigen-presenting cell: the pathway for MHC class II peptide presentation. Peptides in the ER lumen of the antigen-presenting cell are prevented from binding to the MHC class II peptide groove by association of a protein called the invariant chain with MHC class II molecules. The complex is transported through the Golgi compartments to a post-Golgi vesicle, where the invariant chain is removed by proteolysis. This reaction activates MIIC class II molecules to accept peptides. The peptides are derived, not from endogenous proteins, but from extracellular proteins that enter the antigen-presenting cell. In some antigen-presenting cells, the proteins enter by endocytosis (top) and are internalized to early endosomes with neutral luminal pH. Endocytotic vesicles traveling to the lysosome via this pathway are characterized by a decrease in pH as they "mature" into late endosomes. The lower pH activates vesicle proteases that degrade the exogenous protein into peptides. Internalized endosomes with their peptides fuse at some point with the vesicles containing activated MHC class II. The newly formed peptide-MIIC class II complex then becomes competent for transport to the cell surface, where it is available for interaction with the T-cell receptor (Tcr) of a Th cell carrying the CD4 coreceptor. **(Inset)** The MHC class II molecule is a heterodimer of the membrane-spanning type I α-chain (34-kDa) and β-chain (29-kDa) glycoproteins. Each chain folds into two domains, 1 and 2, and together the α and β chains fold into a structure similar to that of MHC class I. The two amino-terminal domains from α and β chains form the groove in which peptide binds. Unlike the closed MHC class I peptide groove, the MHC class II peptide-binding groove is open at both ends. The second domain of each chain folds into an immunoglobulinlike structure. Human genomes contain three MHC class II loci (*DR*, *DP*, and *DQ*), and mouse genomes have two (*IA* and *IE*). MHC class II molecules are dimers comprising α and β subunits. The three sets of human genes give rise to four types of MHC class II molecules. Adapted from D. Male et al., *Advanced Immunology*, 3rd ed. (Mosby, St. Louis, MO, 1996).

A

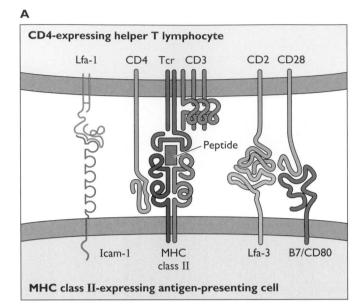

B

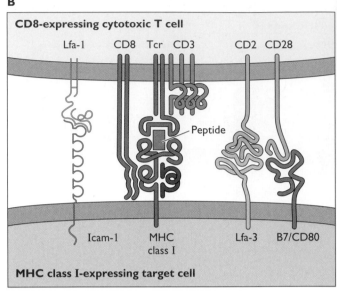

Figure 4.15 T-cell surface molecules and ligands. The interactions of these receptors and ligands are important for antigen recognition and initiation of signal transduction and other T-cell responses. **(A)** Interaction of a Th cell producing the CD4 coreceptor with an antigen-presenting cell. This cell exhibits an MHC class II-peptide complex in addition to Icam-1, Lfa-3, and CD80 (B7) membrane proteins. These complexes all are capable of binding cognate receptors on the Th cell as illustrated. **(B)** Interaction of a CTL producing the CD8 coreceptor with its target cell. The target cell exhibits an MHC class I-peptide complex in addition to Icam-1, Lfa-3, and CD80 (B7) membrane proteins. These complexes all can be recognized and bound by receptors on the CTL as illustrated.

subsequently entering this zone have an increased likelihood of binding specific ligand and signaling.

The primary interaction of the MHC-peptide complex with the T-cell receptor is often called **signal 1**. This interaction is not sufficient to activate the T cell. A second essential signal, often called **signal 2**, is produced by clustering and binding of accessory or costimulatory molecules on the CTL to appropriate ligands that form the immunological synapse with the infected cell (Fig. 4.16). One interaction required to produce signal 2 is the binding of CTL CD28 protein with the target cell CD80 protein. Other interactions include those that facilitate adhesion of T cell and target cell to one another, such as binding of CD2 or leukocyte function antigen 1 (Lfa-1) on the CTL with Lfa-3 or Icam-1, respectively, on the infected cell.

Given the central role of the T-cell receptor and formation of the immunological synapse in adaptive immune defense, it should come as no surprise that viral gene products can affect the structure, function, and localization of the T-cell receptor and the various coreceptors. Indeed, infection by several members of the *Retroviridae* and *Herpesviridae* families leads to reduction of T-cell receptor function. Viral infection also can affect the abundance of various accessory molecules on cell surfaces and

therefore alter CTL recognition and subsequent effector function.

CTLs kill by two primary mechanisms: transfer of cytoplasmic granules from the CTL to the target cell, and induction of apoptosis. These killing systems are formed during the differentiation process. The maturing CTL fills with cytoplasmic granules that contain macromolecules required for lysis of target cells, such as **perforin**, a membrane pore-forming protein, and **granzymes**, members of a family of serine proteases. Granules are released by CTLs in a directed fashion when in direct membrane contact with the target cell, and are taken up by that cell via receptor-mediated endocytosis. Perforin, as its name implies, makes holes in the endosomal membrane, allowing the release of granzymes that induce apoptosis of the infected cell. CTL killing by the perforin pathway is rapid, occurring within minutes after contact and recognition. Activated CTLs also can induce apoptotic cell death via binding of the Fas ligand on their surfaces to the Fas receptor on target cells. Fas pathway killing is much slower than perforin-mediated killing. Many activated CTLs also secrete IFN-γ, which, as we have seen, is a potent inducer of both the antiviral state in neighboring cells, and synthesis of MHC class I and II proteins. Activated CTLs also secrete powerful

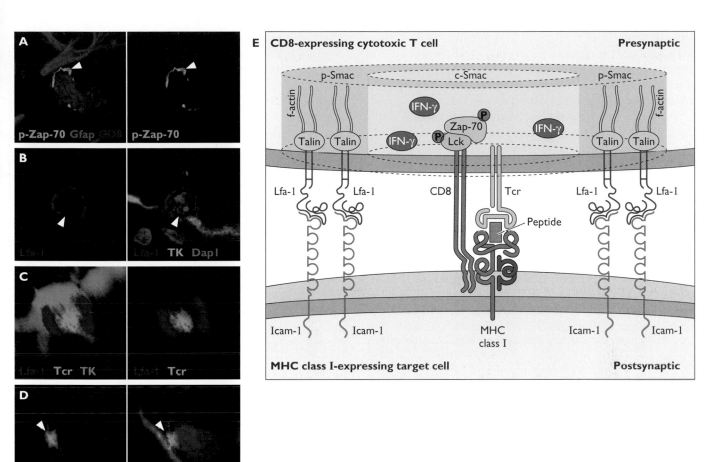

Figure 4.16 The immunological synapse. The morphological characteristics of an *in vivo* immunological synapse between CD8⁺ CTLs and adenovirus-infected astrocytes is illustrated. Colors are specific antibody reactions to identify proteins. The striatum of rats was injected with a recombinant adenovirus vector expressing the herpes simplex virus thymidine kinase (TK) gene. **(A)** Interaction between a CD8⁺ CTL (red) and an infected astrocyte (Gfap, magenta [marks astrocytes]) stimulates T-cell receptor (Tcr) signaling, resulting in phosphorylation and polarization of tyrosine kinases such as Zap70 (green) toward the interface with the infected cell. The white arrow indicates polarized pZap70. **(B)** Adhesion molecules such as Lfa1(red) aggregate to form a **peripheral** ring (p-SMAC: peripheral supramolecular activation cluster) at the junction formed by the immunological synapse. The postsynaptic astrocyte process can be identified by staining with antibody to TK, a marker of adenovirus infection (green). Note the characteristic absence of Lfa1 at the central portion of the immunological synapse between the T cell and the infected astrocye (white arrow). **(C)** A rotated image from a three-dimensional (3-D) reconstruction demonstrates the typical **central** polarization (c-SMAC) of Tcr molecules (green) toward the infected astrocyte (TK, white), and the peripheral distribution of Lfa1 in the p-SMAC (red). **(D)** The effector molecule IFN-γ (green) within a Tcr⁺ (red) CTL is directed toward the site of close contact with an infected target cell (TK, white); the white arrow indicates the T-target cell contact zone. The diameter of a CTL is ~10 μm. **(E)** Schematic cross section of an immunological synapse showing the characteristic polarized arrangement of the cytoskeleton (actin and talin proteins indicated) and organization of the adhesion molecule Lfa1 toward the peripheral supramolecular activation cluster (p-SMAC). The Tcr molecules are directed toward the central-supramolecular activation cluster (c-SMAC), The phosphorylated TKs (Zap70 and Lck) and effector IFN-γ molecules are in the center of the immunological synapse. Figure kindly provided by Pedro Lowenstein, Kurt Kroeger, and Maria Castro. See C. Barcia et al., *J. Exp. Med.* **203**:2095–2107, 2006.

cytokines, such as IL-16, and chemokines such as Rantes (regulated on activation, normal T-cell expressed and secreted) protein. Their release by virus-specific CTLs following recognition of an infected target cell may assist in coordination of the antiviral response.

Typically, for a cytopathic virus infection, CTL activity appears within 3 to 5 days after infection, peaks in about a week, and declines thereafter. The magnitude of the CTL response depends on such variables as titer of infecting virions, route of infection, and age of the host. The critical contribution of CTLs to antiviral defense is demonstrated by **adoptive-transfer** experiments in which virus-specific CTLs from an infected animal can be shown to confer protection to nonimmunized recipients (see also Box 4.5). However, CTLs can also cause direct harm by large-scale cell killing. Such immunopathology often follows infection by noncytopathic viruses, when cells can be infected yet still function. For example, the liver damage caused by hepatitis viruses is actually due to CTL killing of persistently infected liver cells. Other examples of immunopathology are discussed later in this chapter.

Control of CTL Proliferation

By using assays described in Box 4.6, several investigators discovered a hitherto unknown, massive CTL precursor expansion after acute primary infections by viruses such as lymphocytic choriomeningitis virus and Epstein-Barr virus. For example, more than 50% of CTLs from the spleen of a lymphocytic choriomeningitis virus-infected mouse were specific for a **single** viral peptide. The response reached a maximum 8 days after infection, but up to 10% of virus-specific T cells were still detectable after a year. Such results are in contrast to those for hepatitis B virus or human immunodeficiency virus infection: less than 1% of the CTLs from spleens of infected patients are specific for a single viral peptide.

These findings are stimulating a variety of new avenues of study. We must understand why there is a large primary expansion of particular CTL precursors for some infections and not others. It may be that this expansion is necessary to thwart the infection, or it may simply reflect an overreaction or lack of control in these controlled laboratory infections. The mechanisms that control this rapid expansion remain to be discovered. Other studies to determine the relationship of T-cell function (e.g., cell killing) to peptide specificity are under way.

As discussed in Chapter 5, viral proteins can blunt the deadly CTL response with far-reaching effects ranging from rapid death of the host to long-lived, persistent infections. Many such proteins confound CTL recognition by disguising or reducing antigen presentation by MHC class I molecules (see Fig. 5.5). In the case of human immunodeficiency virus type 1, the viral genome encodes three proteins that interfere with CTL action: Nef and Tat induce Fas ligand production and subsequent Fas-mediated apoptosis of CTLs, while Env engages the CXCr4 chemokine receptor, triggering the death of the CTL. Human cytomegalovirus-infected cells contain at least six viral proteins that interfere with the MHC class I pathway and also evoke apoptosis of virus-specific CTLs by increasing synthesis of Fas ligand (see Table 5.3).

Noncytolytic Control of Infection by T Cells

Complete clearance of intracellular viruses by the adaptive immune system does not depend solely on the destruction of infected cells by CTLs. The production of cytokines, such as IFN-γ and Tnf-α, by CTLs can lead to purging of viruses from infected cells without cell lysis. Such a mechanism requires that the infected cell retain the ability to activate antiviral pathways induced by binding of these cytokines to their receptors, and that viral replication be sensitive to the resulting antiviral response.

In certain circumstances, such as infection of the liver by hepatitis B and C viruses, there are orders of magnitude more infected cells than there are virus-specific CTLs. Furthermore, if vital organs such as the brain or liver are

---

BOX 4.5   DISCUSSION

An adoptive-transfer assay for T-cell-mediated delayed-type hypersensitivity

The delayed-type hypersensitivity response in a previously infected mouse typically is measured by an ear-swelling assay. Viral antigen is injected intradermally on the dorsal pinna of a mouse ear, while the other ear is injected with control antigen.

Twenty-four hours later, the antigen-injected ear becomes inflamed, and swells as a result of delayed-type hypersensitivity, but the control-injected ear does not.

The same swelling reaction is observed in an animal that has not been infected when it is injected with purified, activated T cells from an infected animal. This adoptive-transfer experiment shows that delayed-type hypersensitivity is due to T cells.

DISCUSSION
Measuring the antiviral cellular immune response

The Classic Assay: Limiting Dilution and Chromium Release

For the past 45 years, virologists have determined the presence of CTLs in blood, spleen, or lymphoid tissues of immune animals by using the limiting-dilution assay and chromium release from lysed target cells. This assay measures CD8$^+$ CTL precursors, or memory CTLs, based on two attributes: (i) the action of foreign proteins and peptides to stimulate CTL proliferation and (ii) the ability of activated CTLs to lyse target cells.

Lymphocytes are obtained from an animal that has survived virus infection and are cultured for 1 to 2 weeks in the presence of whole inactivated virus, its proteins, or synthetic viral peptides. Under these conditions, virus-specific CTL precursors begin to replicate and divide (clonal expansion).

The expanded CTL population is then tested for its ability to destroy target cells that display viral peptides on MHC class I molecules. The target cells loaded with the viral antigen in question are then exposed to chromium-51, a radioactive isotope that binds to most intracellular proteins. After being washed to remove external isotope, cultured CTLs are incubated with the target cells, and lysis is measured by the release of chromium-51 into the supernatant.

Serial dilution before assay provides an estimate of the number of CD8+ CTL precursors in the original cell suspension (providing the name of the assay: "limiting dilution"). This assay does provide a quantitative measure of cellular immunity, but it is time-consuming, technically demanding, and expensive.

Identifying and Counting Virus-Specific T Cells

The limiting-dilution assay defines T cells by function, but not by their peptide specificity. Until recently, scientists have tried with little success to identify individual T cells based on their peptide recognition properties. This failure has been attributed to the low affinity and high "off" rates of the MHC-peptide complex and T-cell receptor. Without this information, it was difficult, if not impossible, to measure antigen-specific T-cell responses.

An important advance is the use of an **artificial MHC tetramer** as an antigen-specific, T-cell-staining reagent. The extracellular domains of MHC class I proteins are produced in *Escherichia coli*. These engineered MHC class I molecules have an unusual C-terminal 13-amino-acid sequence that enables them to be biotinylated. The truncated MHC class I proteins are folded in vitro with a synthetic peptide that will be recognized by a specific T cell. Biotinylated tetrameric complexes are purified and mixed with isolated T-cell pools from virus-infected animals. Individual T cells that bind the biotinylated MHC class I-peptide complex are detected by a variety of immunohistochemical techniques. Cell-sorting techniques can be used, and stained cells are viable.

Another assay for counting single T cells is the **enzyme-linked immunospot assay (ELISPot)**. In this assay, the cytokines are used as a surrogate for the T cell of interest. Fresh lymphoid cells are put into culture medium on plates that have been coated with antibody specific for cytokines under study. These cells are stimulated with virus-specific peptides. CTLs that recognize the peptide secrete cytokines locally where they are bound by the antibody on the plate. After the plate is washed, the foci of bound cytokine are stained and enumerated.

The **intracellular cytokine assay** is a relatively rapid method to count specific CTLs. Fresh lymphoid cells are treated with brefeldin A. This fungal metabolite blocks the secretory pathway and prevents the secretion of cytokines. The cells are then fixed with a mild cross-linking chemical that preserves protein, such as glutaraldehyde. Treated cells are permeabilized so that a specific antibody for a given cytokine can react with cytokines retained. Cells that react with the antibody can be quantified in a fluorescence-activated cell sorter. With appropriate software and calibration, the staining intensity corresponds to the level of cytokine expression, and the number of cells responding to a particular epitope and MHC class I molecule can be determined.

Measuring the Antiviral Antibody Response

Antibodies are the primary effector molecules of the humoral response. There are many methods to detect antibodies. However, a standard method in virology is the **neutralization assay**. Here, viral infectivity is determined in the presence and absence of antibody. Two variations on this general theme include the **plaque reduction assay** and the **neutralization index**. In the plaque assay, a known number of plaque-forming units (PFU) are exposed to serial dilutions of the antibody or serum in question. The highest dilution that will reduce the plaque count by 50% is taken as the plaque reduction titer of the serum or antibody. To compute a neutralization index, the titer of a virus stock is compared in the presence and absence of test antibody or serum. The index is calculated as the difference in viral titers. Obvious requirements for these assays is that the virus in question can be propagated in cultured cells, and that a measure of virus growth such as plaque formation is available.

Other important assays to determine binding of antibody to viral proteins are immunoprecipitation, Western blot, enzyme-linked immunosorbent assay, and hemagglutination inhibition.

---

infected, CTL killing can do more harm than good. When hepatitis B virus-specific CTLs are transferred to another animal (adoptive transfer), the IFN-γ and Tnf-α produced appear to clear the infection from thousands of cells without their destruction.

Noncytolytic clearing of infection by IFN-γ and Tnf-α produced by CTLs has now been documented for many viral infections, including those caused by primate lentiviruses, arenaviruses, adenoviruses, coronaviruses, hepadnaviruses, and picornaviruses. Additional cytokines produced by a variety of immune system cells are likely to participate in viral clearance. In addition, the defensins (described previously) secreted by CTLs are known to promote the noncytolytic clearing of human immunodeficiency virus. CD4⁺ T cells can also accomplish noncytolytic clearing of some infections with little involvement of CTLs. Such cases include infections by vaccinia virus, vesicular stomatitis virus, and Semliki Forest virus. CTL production of powerful, secreted, antiviral cytokines provides a simple explanation for how CTLs are able to control massive numbers of infected cells. However, how cytokines effect nonlethal purging of infection is not well understood.

Rashes and Poxes: Examples of T-Cell-Mediated, Delayed-Type Hypersensitivity

Many infections, including those of measles virus, smallpox virus, and varicella-zoster virus, produce a characteristic rash or lesion over extensive areas of the body, even though the primary infection began at a distant mucosal surface. This phenomenon results when the primary infection escapes the local defenses and virions or infected cells spread in the circulation to initiate many foci of infected cells in the skin. Th1 cells and macrophages that were activated by the initial infection home in on these secondary sites and respond by aggressive synthesis of cytokines, including IL-2 and IFN-γ. Such cytokines then act locally to increase capillary permeability, a reaction partially responsible for a characteristic local reaction referred to as **delayed-type hypersensitivity** (Box 4.5). This reaction is responsible for many virus-promoted rashes and lesions with fluid-filled vesicles.

The Antibody Response

Specific Antibodies Are Made by Activated B Cells Called Plasma Cells

When the B cell emerges from the bone marrow into the circulation and travels to lymph and lymphoid organs, it differentiates and synthesizes antibody **only** when its surface antibody receptor is bound to the cognate antigen. The activating signal requires clustering of receptors complexed with antigen. Such receptor clustering activates signaling via Src family tyrosine kinases that associate with the cytoplasmic domains of the closely opposed receptors. B-cell coreceptors, such as CD19, CD21, and CD8, enhance signaling by recruiting tyrosine kinases to clustered antigen receptors and coreceptors. Like dendritic cells, the B cell is an antigen-presenting cell that uses the MHC class II system and exogenous antigen processing (Fig. 4.14).

Binding of antigen to the B-cell receptor is only part of the activation process. Cytokines from Th cells also are required. When the T-cell receptor of Th2 cells recognizes MHC class II-peptide complexes present on the B-cell surface, these Th2 cells produce a locally high concentration of stimulatory cytokines, as well as CD40 ligand (a protein homologous to Tnf). The engagement of CD40 ligand with its B-cell receptor facilitates a local exchange of cytokines that further stimulates proliferation of the activated B cell and promotes its differentiation. Fully differentiated plasma cells produce prodigious amounts of specific antibodies: the rate of synthesis of IgG can be as high as 30 mg/kg of body weight/day.

Antibodies

Antibodies (immunoglobulins) have the common structural features illustrated in Fig. 4.17. Five classes of immunoglobulin—IgA, IgD, IgE, IgG, and IgM—are defined by their distinctive heavy chains—α, δ, ε, γ, and μ, respectively. Their properties are summarized in Table 4.4. IgG, IgA, and IgM are commonly produced after viral infection. During B-cell differentiation, "switching" of the constant region of heavy-chain genes occurs by somatic recombination and is regulated in part by specific cytokines. Consequently, each cell produces a specific type of antibody after such switching.

During the **primary antibody response**, which follows initial contact with antigen or viral infection, the production of antibodies follows a characteristic course. The IgM antibody appears first, followed by IgA on mucosal surfaces or IgG in the serum. The IgG antibody is the major antibody of the response and is remarkably stable, with a half-life of 7 to 21 days. Specific IgG molecules remain detectable for years, because of the presence of memory B cells. A subsequent challenge with the same antigen or viral infection promotes a rapid antibody response, the **secondary antibody response** (Fig. 4.8).

Virus Neutralization by Antibodies

While the cell-mediated response clearly plays an important role in eliminating virus-infected cells, the antibody response is crucial for preventing many viral infections and may also contribute to resolution of infection. Virions that infect mucosal surfaces will be exposed to secretory IgA antibodies. Similarly, virions that spread in the blood will be exposed to circulating IgG and IgM antibody molecules. Antibodies in the blood and lymph circulation are an important defense against infections caused by rabies virus, vesicular stomatitis virus, and enteroviruses. We know that immunodeficient animals can be protected from some lethal viral infections by injection with virus-specific antiserum or purified monoclonal antibodies (also called passive immunization). In studies of humans infected with eastern and western equine encephalitis viruses (alphaviruses in the family *Togaviridae*), patients with high antibody titers recovered, while those with low antibody titers were more likely to die of viral encephalitis.

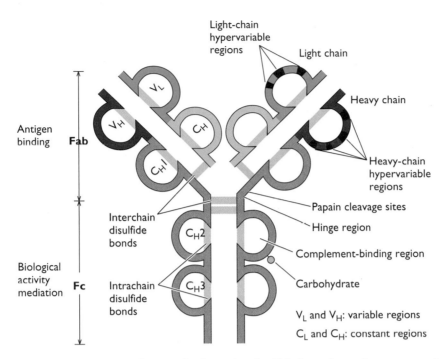

Figure 4.17 The structure of an antibody molecule. This is a schematic representation of an IgG molecule delineating the subunit and domain structures. The light and heavy chains are held together by disulfide bonds (yellow bars). The variable regions of the heavy (V_H) and light (V_L) chains, as well as the constant regions of the heavy (C_H) and light (C_L) chains, are indicated on the left part of the molecule. The hypervariable regions and invariable regions of the antigen-binding domain (Fab) are emphasized. The constant region (Fc) performs many important functions, including complement binding (activation of the classical pathway) and binding to Fc receptors found on macrophages and other cells. Clusters of papain protease cleavage sites are indicated, as this enzyme is used to define the Fab and Fc domains.

Perhaps the best example of the importance of antibodies in antiviral defense is the success of the poliovirus vaccine in preventing polio. We have learned that the type of antibody produced can influence the outcome of a viral infection significantly. Poliovirus infection stimulates strong IgM and IgG responses in the blood, but it is mucosal IgA that is vital in defense. This isotype can neutralize poliovirus directly in the gut, the site of primary infection. The live attenuated Sabin poliovirus vaccine is effective because it elicits a strong mucosal IgA response. IgA is synthesized by plasma cells that underlie the mucosal epithelium. This antibody is secreted from these cells as dimers of two conventional immunoglobulin subunits. The dimers then bind the polymeric immunoglobulin receptor on the basolateral surface of epithelial cells (Fig. 4.18). This complex is then internalized by endocytosis,

Table 4.4   The five classes of immunoglobulins

Property	IgA	IgD	IgE	IgG	IgM
Function	Mucosal; secretory	Surface of B cell	Allergy; anaphylaxis; epithelial surfaces	Major systemic immunity; memory responses	Major systemic immunity; primary response; agglutination
Subclasses	2	1	1	4	1
Light chain	κ, λ	κ, λ	κ, λ	κ, λ	κ, λ
Heavy chain	α	δ	ε	γ	μ
Concn in serum (mg/ml)	3.5	0.03	0.00005	13	1.5
Half-life (days)	6	2.8	2	25	5
Complement activation					
Classical	–	–	–	+	++
Alternative	–/+	–	+	–	–

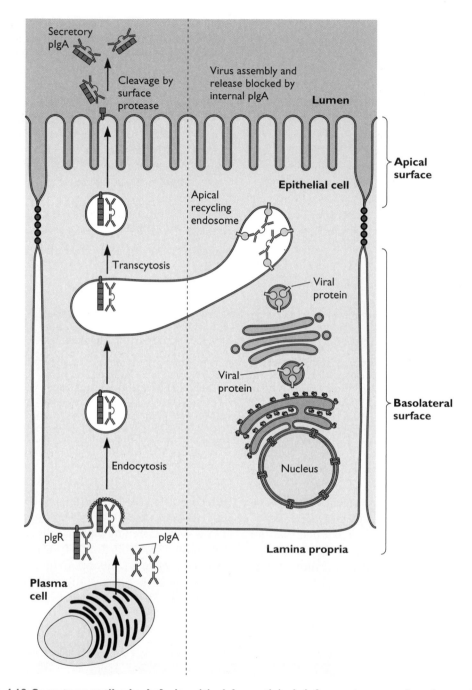

Figure 4.18 Secretory antibody, IgA, is critical for antiviral defense at mucosal surfaces. A single polarized epithelial cell is illustrated. The apical surface is shown at the top, and the basal surface is shown at the bottom. **(Left)** Antibody-producing B cells (plasma cells) in the lamina propria of a mucous membrane secrete the IgA antibody (also called polymeric IgA [pIgA]). pIgA is a dimer, joined at its Fc ends. For IgA to be effective in defense, it must be moved to the surface of the epithelial cells that line body cavities. This process is called transcytosis. **(Right)** Interestingly, a virus particle infecting an epithelial cell potentially can be bound by internal IgA if virus components intersect with the IgA in the lumen of vesicles during transcytosis. This process is likely to occur for enveloped viruses, as their membrane proteins are processed in many of the same compartments as those mediating transcytosis. Adapted from M. E. Lamm, *Annu. Rev. Microbiol.* **51:**311–340, 1997, with permission.

and moved across the cell (**transcytosis**) to the apical surface. Protease cleavage of the receptor releases dimeric IgA into mucosal secretions, where it can interact with incoming virions.

IgA may also block viral replication inside infected mucosal epithelial cells (Fig. 4.18). Because IgA must pass through such a cell en route to secretion from the apical surface, it is available during transit for interaction with viral proteins produced within the cell. The antigen-binding domain of intracellular IgA lies in the lumen of the ER, the Golgi compartment, and any transport vesicles of the secretory pathway. It is therefore available to bind to the external domain of any type I viral membrane protein that has the cognate epitope of that IgA molecule. Such interactions have been demonstrated with Sendai virus and influenza virus proteins during infection of cultured cells. In these experiments, antibodies colocalized with viral antigen only when the IgA could bind to the particular viral envelope protein. These studies suggest that clearing viral infection of mucosal surfaces need not be limited to the lymphoid cells of the adaptive immune system.

It is widely assumed that the primary mechanism of antibody-mediated viral neutralization is via steric blocking of virion-receptor interaction (Fig. 4.19). While some antibodies do prevent virions from attaching to cell receptors, the vast majority of virus-specific antibodies are likely to interfere with the concerted structural changes that are required for entry. Antibodies also can promote aggregation of virions, and thereby reduce the effective concentration of infectious particles. Many enveloped viruses can be destroyed in a test tube when antiviral antibodies and serum complement disrupt membranes (the classical

Figure 4.19 Interactions of neutralizing antibodies with human rhinovirus 14. (A) The normal route of infection. The virus attaches to the Icam-1 receptor and enters by endocytosis. As the internal pH of the endosome decreases, the particle uncoats and releases its RNA genome into the cytoplasm. **(B)** Possible mechanisms of neutralization of human rhinovirus 14 by antibodies. With well-characterized monoclonal antibodies, at least five modes of neutralization have been proposed and are illustrated: (1) blocked attachment—binding of antibody molecules to virus results in steric interference with virus-receptor binding; (2) blocked endocytosis—antibody molecules binding to the capsid can alter the capsid structure, affecting the process of endocytosis; (3) blocked uncoating—antibodies bound to the particle fix the capsid in a stable conformation so that pH-dependent uncoating is not possible; (4) blocked uncoating, inside cell—antibodies themselves may be taken up by endocytosis and interact with virions inside the cell after infection starts; (5) aggregation—because all antibodies are divalent, they can aggregate virus particles affecting concentration of virus particles and facilitating their destruction by phagocytes. (A) Adapted from T. J. Smith et al., *Semin. Virol.* **6:**233–242, 1995, with permission.

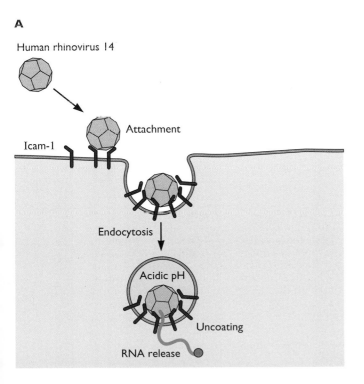

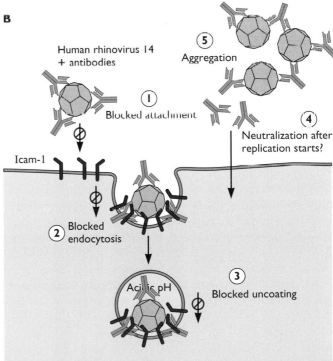

complement activation pathway). Nonneutralizing antibodies are also prevalent after infection; they bind specifically to virus particles, but do not interfere with infectivity. In some cases, such antibodies can even enhance infectivity: antibody bound to virions is recognized by Fc receptors on macrophages, and the entire complex is brought into the cell by endocytosis.

Much of what we know about antibody neutralization comes from the isolation and characterization of "antibody escape" mutants or **monoclonal antibody-resistant mutants**. These mutants are selected after propagating virus in the presence of neutralizing antibody. The analysis of the mutant viruses allows a precise molecular definition, not only of antibody-binding sites but also of parts of viral proteins important for entry. Antigenic drift (see Chapters 5 and 10) is a consequence of selection and establishment of antibody escape mutants in viral populations.

Antibodies can provoke other remarkable responses in virus-infected cells. For example, in a process analogous to CTL purging, antibodies that bind to the surface proteins of many enveloped viruses (e.g., alphaviruses, paramyxoviruses, and arenaviruses) can clear these viruses from persistently infected cells. This process is noncytolytic and complement independent. In this case, antibodies act synergistically with IFN and other cytokines. Virus-specific antibodies bound to surfaces of infected cells can inhibit virion budding at the plasma membrane and can also reduce surface expression of viral membrane proteins by inducing endocytosis. Antiviral antibodies injected into the peritoneal cavities of mice can block the neurotropic transmission of reovirus, poliovirus, and herpesvirus to the brain, even when the virions are injected at sites far removed from the peritoneum. The molecular bases for these intriguing effects remain elusive.

Antibody-Dependent Cell-Mediated Cytotoxicity: Specific Killing by Nonspecific Cells

The Th1 response results in production of a particular isotype of IgG that can bind to antibody receptors on macrophages and some NK cells. These receptors are specific for the carboxy-terminal, more conserved region of an antibody molecule, the Fc region (Fig. 4.16). If an antiviral antibody is bound in this manner, the amino-terminal antigen-binding site is still free to bind viral antigen on the surface of the infected cell. In this way, the antiviral antibody targets the infected cell for elimination by macrophages or NK cells. This process is called **antibody-dependent cell-mediated cytotoxicity** (often referred to as ADCC). The antibody provides the specificity for killing by a less discriminating NK cell. While this mechanism is well documented in cultured cells, its importance in controlling viral infections in animals is unknown.

The Immune System and the Brain

Cells of the central nervous system (brain and spinal cord) can initiate a robust and transient innate defense, but, surprisingly, they are unable to mount an adaptive response. A primary reason for this deficit is that the central nervous system is devoid of lymphoid tissue and dendritic cells. In addition, the central nervous system of vertebrates is separated from many cells and proteins of the bloodstream by tight endothelial cell junctions that comprise the so-called **blood-brain barrier**. As a consequence of these features, viral infections of the central nervous system can have unexpected outcomes. For example, if virus particles are injected directly into the ventricles or membranes covering the brain that are in contact with the bloodstream, the innate immune system is activated, a strong inflammatory response occurs, and the adaptive response ensues. In contrast, if virus particles are injected directly into brain tissue, avoiding the blood vessels and ventricles, only a transient inflammatory response is found. The adaptive response is not activated; antibodies and antigen-activated T cells are not made.

Although the central nervous system is unable to initiate an adaptive immune response, it is not isolated from the immune system. Indeed, the blood-brain barrier is open to entry of activated immune cells circulating in the periphery. Antigen-specific T cells regularly enter and travel through the brain, performing immune surveillance. Moreover, at least two glial cell types, astrocytes and microglia, have a variety of cell surface receptors that can engage these T cells. Astrocytes, the most numerous

Table 4.5 Cells and mechanisms associated with immunopathology

Proposed mechanism	Virus
CD8+ T-cell mediated	Coxsackievirus B
	Lymphocytic choriomeningitis virus
	Sin Nombre virus
	Human immunodeficiency virus type 1
	Hepatitis B virus
CD4+ T-cell mediated	Theiler's virus
Th1	Mouse coronavirus
	Semliki Forest virus
	Measles virus
	Visna virus
	Herpes simplex virus
Th2	Respiratory syncytial virus
B-cell mediated (antibody)	Dengue virus
	Feline infectious peritonitis virus

cell type in the central nervous system, respond to a variety of cytokines made by cells of the immune system. All natural brain infections begin in peripheral tissue, and any infection that begins outside the central nervous system activates the adaptive immune response. However, if the infection spreads to the brain, the resulting immune attack on this organ can be devastating. A careful scientist can inject virus particles experimentally into the brain without infecting peripheral tissue, thereby avoiding an adaptive response. On the other hand, if the animal is first immunized by injecting virus particles into a peripheral tissue, the adaptive response is activated as expected. Subsequent injection of an identical virion preparation into the brain of the immunized animal elicits massive immune attack on that organ: any virus-infected target in the brain is recognized and destroyed by the peripherally activated T cells. In both natural and experimental infections, the inflammatory response is not transient but sustained, resulting in capillary leakage, swelling, and cell death. Swelling of the brain in the closed confinement of the skull has many deleterious consequences, and, when this is coupled with bleeding and cell death, the results are disastrous. It is clear that although the brain is "immunoprivileged" in some sense, it is not completely isolated and, when infected, is vulnerable to attack by T cells produced and stimulated by the immune system.

Immunopathology: Too Much of a Good Thing

The clinical symptoms of viral disease in the host (e.g., fever, tissue damage, aches, pains, and nausea) result primarily from the host response to infection (Table 4.5). Damage caused by the immune system is called **immunopathology**, and it may be the price paid by the host to eliminate a viral infection. For noncytolytic viruses it is likely that the immune response is the sole cause of disease. In fact, most viral infections with an immunopathological component are noncytolytic and persistent (see Chapter 5). Most immunopathology is induced by activated T cells, but there are examples of disease caused by antibodies or an excessive innate response (Box 4.7).

Immunopathological Lesions

Lesions Caused by CTLs

The best-characterized example of CTL-mediated immunopathology occurs during the infection of mice with lymphocytic choriomeningitis virus. Infection is not cytopathic and induces tissue damage only in immunocompetent animals. Experiments using adoptive transfer of T-cell subtypes, depletion of cells, and gene knockout and transgenic mice clearly show that tissue damage requires CTLs (Box 4.8). The mechanism by which these cells cause damage is not clear, but may be a result of

---

BOX 4.7

EXPERIMENTS
Defective viral vectors and lethal immunopathology

In September 1999, an 18-year-old man participated in a clinical trial to test the safety of a defective adenovirus designed as a gene delivery vector. It seemed like a routine procedure: normally, even replication-competent adenoviruses cause only mild respiratory disease. Most humans harbor adenoviruses as persistent colonizers of the respiratory tract and, indeed, produce antibodies against the virus. The young man was injected with a large dose of the viral vector. Four days after the injection, he died of multiple-organ failure. What caused this devastating response to such an apparently benign virus?

Some relevant facts:

- Natural adenovirus infection never occurs by direct introduction of virions into the circulation. Most infections occur at mucosal surfaces with rather small numbers of infecting virions.

- Most humans have antibodies to the adenoviral vectors used for gene therapy.
- A large dose of virus was injected directly into his bloodstream.

One compelling idea is that most of the infecting defective virions were bound by antibody present in the young man's blood. As a consequence, the innate immune system, primarily complement proteins, responded to the antibody-virus complex, resulting in massive activation of the complement cascade. The amplified complement cascade caused widespread inflammation in the vessel walls of the liver, lungs, and kidneys, resulting in multiple-organ failure.

Given the complexities of the immune response to infections, this idea is most certainly an oversimplification. Nevertheless, this fatal trial resulted in a large-scale

reassessment of gene delivery methods and clinical protocols. Research is ongoing to find methods and tests to identify, reduce, or eliminate immunopathology so that the promise of gene therapy can be realized.

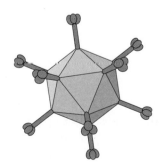

BOX
4.8

BACKGROUND
Transgenic mouse models prove useful in defining the antiviral contribution of CTLs

Studies of mice carrying mutations in genes that encode proteins required for CTL killing, B-cell function, and IFN responses have provided a wealth of information on the contributions of these gene products in immune defense. The only requirement is that the virus of interest be able to infect the transgenic mouse.

Genetic manipulation of the CTL response has been particularly informative. In the case of some noncytopathic viruses, perforin-mediated, but not granzyme-mediated, or Fas-mediated, killing is an important component of host CTL defense. When virus was not cleared in perforin-defective mice, the resulting persistent infection led to a lethal overproduction of cytokines, including Tnf-α and IFN-γ. In contrast, granzyme-deficient mice recover completely from viral infection. How the various CTL killing proteins are coordinated to control other viral infections is an important area of research.

Transgenic models also demonstrate an unexpectedly limited contribution of the CTL killing functions in defense against certain cytopathic viruses. Vaccinia virus and vesicular stomatitis virus pathogenesis was not affected by lack of CTL killing systems: rather, recovery from primary infection depended largely on production of neutralizing IgG antibodies. In fact, cytokines secreted by CD4+ Th cells and CTLs are required for **early** immune defense, but not for recovery from infection by many different cytopathic viruses. One way to think about these observations is that recognition and lysis of infected cells by CTLs may be too slow to be effective as a primary defense against acute cytopathic infections: the rapid innate defenses and early cytokine responses of lymphocytes are the critical frontline systems.

Rall, G., D. Lawrence, and C. Patterson. 2000. The application of transgenic and knockout mouse technology for the study of viral pathogenesis. *Virology* **271:**220–226.

Yeung, R., J. Penninger, and T. Mak. 1994. T-cell development and function in gene knockout mice. *Curr. Opin. Immunol.* **6:**298–307.

their cytotoxicity. For example, knockout mice lacking perforin (the major cytolytic protein of CTLs) develop less severe disease after infection. CTLs may also release proteins that recruit inflammatory cells to the site of infection, which in turn elaborate proinflammatory cytokines.

Liver damage caused by hepatitis B virus also appears to depend on the action of CTLs. Production of the hepatitis B virus envelope proteins in transgenic mice has no effect on the animals. When the mice are injected with hepatitis B virus-specific CTLs, liver lesions that resemble those observed in acute human viral hepatitis develop. First, CTLs attach to hepatocytes and induce apoptotic cell death. Next, cytokines released by these lymphocytes recruit neutrophils and monocytes, which cause even more extensive cell damage. Death from hepatitis can be prevented by the administration of antibody to IFN-γ or by depletion of macrophages. CTLs are required for immunopathology, but tissue damage is mediated largely by nonspecific cytokines and cells recruited to the site of infection.

Myocarditis (inflammation of the heart muscle) caused by coxsackievirus B infection of mice also requires the presence of CTLs. In particular, perforin is a major determinant of myocarditis. Mice lacking the perforin gene develop a mild form of heart disease yet are still able to clear the infection. Chemokines contribute to disease by controlling directional migration of lymphocytes into infected tissues. For example, mice lacking the chemokine macrophage inflammatory protein 1α do not develop myocarditis. These observations indicate that inflammation of heart muscle following infection is a result of the combined action of immune-mediated tissue damage and virus-induced cytopathology.

Lesions Caused by CD4+ T Cells

CD4+ T lymphocytes elaborate far more cytokines than do CTLs and recruit and activate many nonspecific effector cells. Such inflammatory reactions are usually called delayed-type hypersensitivity responses. Most of the recruited cells are neutrophils and mononuclear cells, which are protective but can cause tissue damage. Immunopathology is the result of release of proteolytic enzymes, reactive free radicals such as peroxide and nitric oxide (see below), and cytokines such as Tnf-α. For noncytopathic persisting viruses, the CD4+-mediated inflammatory reaction is largely immunopathological. This response is protective against cytopathic viruses, although there may be cases in which immunopathology occurs.

CD4+ Th1 cells. The cytokines produced by CD4+ Th1 cells cells facilitate the inflammatory response but not the antibody response. These cells are necessary for demyelination caused by viral infection of the nervous system and provoke central nervous system disease of rodents infected with several different viruses. When mice are infected with Theiler's murine encephalomyelitis virus (a picornavirus), proinflammatory cytokines produced by CD4+ Th1 cells activate macrophages and microglial cells that mediate demyelination of neurons. It is not known how demyelination occurs, but it has been proposed that the activated phagocytic cells release superoxide and nitric oxide free radicals in addition to Th1 cell proinflammatory cytokines, and the combination destroys oligodendrocytes, which are the source of myelin. That a similar demyelinating pathology is caused by so many different viral infections is consistent with the hypothesis that an underlying immunopathology is at work.

Herpes stromal keratitis is one of the most common causes of vision impairment in developed countries of the world. The eye damage is caused almost entirely by immunopathology. In humans, herpes simplex virus infection of the eye induces lesions on the corneal epithelium, and repeated infections result in opacity and reduced vision. Studies of a mouse model for this disease have demonstrated the importance of CD4+ Th1 cells in an interesting example of immunopathology. The surprise was that while viral replication occurs in the corneal epithelium, CD4+ T-cell-mediated inflammation was restricted to the underlying uninfected stromal cells. An important observation was that viral replication in the cornea had ceased by the time that CD4+ T cells attacked the stromal cells. Herpes stromal keratitis is now thought to result from a damaging inflammatory reaction directed to uninfected cells in the stroma that is stimulated by secreted cytokines produced by infected cells in the corneal epithelium (bystander cell activation).

CD4+ Th2 cells. The cytokines produced by CD4+ Th2 cells facilitate the humoral response. Respiratory syncytial virus disease is an important cause of lower respiratory tract disease in infants and the elderly. Models for this particular disease have been difficult to produce, but some success has come with immunosuppressed mice. When these animals are infected, lesions of the respiratory tract are minor, but they become severe after adoptive transfer of viral antigen-specific, CD4+ Th2 cells. The respiratory tract lesions contain many eosinophils, which may be responsible for pathology. One possibility is that the cytokines produced by CD4+ Th2 cells recruit and stimulate proliferation of these eosinophils.

The Balance of Th1 and Th2 Cells

An inappropriate Th1 or Th2 response can have pathogenic effects. For example, respiratory syncytial virus causes respiratory infections which can be prevented by vaccination that induces a Th1 response in young children. These children were protected and suffered no ill effects. However, when children were vaccinated with a formalin-inactivated whole-virus vaccine that elicited a Th2 response, they not only remained susceptible to infection but also developed an atypically severe disease characterized by increased infiltration of eosinophils into the lungs. This particular pathology had been predicted by adoptive transfer of CD4+ Th2 cells in mice (discussed above). Current vaccination efforts are based on preparations that induce only the Th1 response, which does not cause immunopathological disease.

Immunopathological Lesions Caused by B Cells

Virus-antibody complexes accumulate to high concentrations when extensive viral replication occurs at sites inaccessible to the immune system or continues in the presence of an inadequate immune response. Such complexes are not cleared efficiently by the reticuloendothelial system and continue to circulate in the blood. They become deposited in the smallest capillaries and cause lesions that are exacerbated when the complement system is activated (Fig. 4.20). Deposition of such immune complexes in blood vessels, kidneys, and brain may result in vasculitis, glomerulonephritis, and mental confusion, respectively. This type of immunopathology was first described in mice infected with lymphocytic choriomeningitis virus. Although immune complexes have been demonstrated in humans, viral antigens have been found in the complexes only in hepatitis B virus infections.

Antibodies may also cause an enhancement of viral infection. This mechanism probably accounts for the pathogenesis of dengue hemorrhagic fever. This disease is transmitted by mosquitoes and is endemic in the Caribbean, Central and South America, Africa, and Southeast Asia, where billions of people are at risk. The primary infection is usually asymptomatic, but may result in an acute febrile illness with severe headache, back and limb pain, and a rash. It is normally self-limiting, and patients recover in 7 to 10 days. There are four viral serotypes, and antibodies to any one serotype do not protect against infection by another. After infection by another serotype of dengue virus, nonprotective antibodies bind virus particles and facilitate their uptake into normally nonsusceptible peripheral blood monocytes carrying Fc receptors. Consequently, the infected monocytes produce proinflammatory cytokines, which in turn stimulate T cells to produce more cytokines. This vicious cycle results in high concentrations of cytokines and other chemical mediators that are thought to trigger the plasma leakage and hemorrhage characteristic of dengue hemorrhagic fever. There may be so much internal bleeding that the often fatal dengue shock syndrome results. Dengue hemorrhagic fever occurs in approximately 1 in 14,000 primary infections. However, after infection with a dengue virus of another serotype,

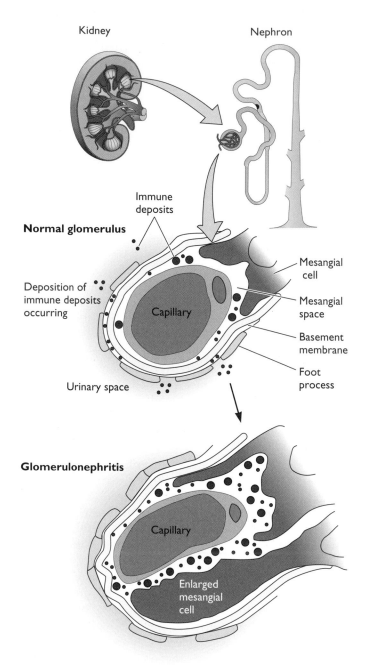

Figure 4.20 Deposition of immune complexes in the kidneys, leading to glomerulonephritis. (Top) Normal glomerulus and its location in the nephron and kidney. **(Middle)** Normal glomerulus. Red dots are immune complexes. The smaller complexes pass to the urine, and the larger ones are retained at the basement membrane. **(Bottom)** Glomerulonephritis. Complexes have been deposited in the mesangial space and around the endothelial cell. The function of the mesangial cell is to remove complexes from the kidney. In glomerulonephritis, the mesangial cells enlarge into the subepithelial space. This results in constriction of the glomerular capillary, and foot processes of the endothelial cells fuse. The basement membrane becomes leaky, filtering is blocked, and glomerular function becomes impaired, resulting in failure to produce urine. Adapted from C. A. Mims et al., *Mims' Pathogenesis of Infectious Disease* (Academic Press, Orlando, FL, 1995), with permission.

the incidence of hemorrhagic fever increases dramatically to 1 in 90, and the shock syndrome is seen in as many as 1 in 50.

Viral Infection-Induced Immunosuppression

Modulation of the immune defenses by viral gene products can range from a mild and rather specific attenuation to a marked global inhibition of the response (Table 4.6). The molecular mechanisms of immune modulation are discussed above. Immunosuppression by viral infection first was observed over 100 years ago because patients were unable to respond to a skin test for tuberculosis during and after measles infection. However, progress in understanding the phenomenon was slow until the human immunodeficiency virus epidemic was under way and the well-known, devastating immunodeficiency syndrome stimulated unprecedented efforts. Immunosuppression by common human viruses such as rubella and measles viruses is also a serious public health concern. For example, the vast majority of the million children who die from measles virus each year in Third World countries succumb to other infections that arise during this transient immunosuppression.

Systemic Inflammatory Response Syndrome

An important tenet of immune defense is that when viral replication exceeds a certain threshold, immune defenses are mobilized and amplified, resulting in a global response. How this threshold is determined is not known. Normally, the global response is well tolerated, and it quickly contains the infection. However, if the threshold is breached too rapidly, or if the immune response is not proportional to the infection, the large-scale production and systemic release of inflammatory cytokines and stress mediators can overwhelm and kill an infected host. Such a disastrous outcome often results if the host is naive and has not coevolved with the invading virus (zoonotic infections; see Chapter 10), or if the host is very young, malnourished, or otherwise compromised. This type of pathogenesis is called the **systemic inflammatory response syndrome** and is sometimes referred to as a "cytokine storm." The lethal effects of the 1918 influenza virus have been attributed to this response. Similar syndromes (e.g., toxic shock syndrome and toxic sepsis) are triggered by other microbial pathogens.

Autoimmune Diseases

Autoimmune disease is caused by an immune response directed against host tissues (often described as "breaking immune tolerance"). In laboratory animals, viral infection

Table 4.6 Some mechanisms of immunosuppression by viruses

Mechanism	Representative virus	Immunosuppression	Specific for infecting virus only
Infection of immune cells	Human immunodeficiency virus, canine distemper virus, lymphocytic choriomeningitis virus	Marked	No
Tolerance after infection of fetus	Rubella virus	Moderate	Yes
Disruption of cytokine defense pathways	Measles virus	Moderate	No
Virokines and viroceptors	Poxviruses, herpesviruses	Mild	Yes

can trigger potent autoimmune responses, but it has been difficult to find evidence for any role for viral infections in human autoimmune disease. How can viral infections promote autoimmune disease in animals? One model posits that replication at an anatomically sequestered site leads to the release and subsequent recognition of self antigens. A modified version of this hypothesis proposes that infection leads to exposure of cellular self antigens normally hidden from the immune system. Cytokines, or even virus-antibody complexes that modulate the activity of proteases in antigen-presenting cells, might cause the unmasking of self antigens. Cytokines produced during infection may stimulate inappropriate surface expression of host membrane proteins that are recognized by host defenses. Another possibility is that during virion assembly, host proteins normally not exposed to the immune system are packaged in virions (Volume I, Chapter 4); these host proteins are delivered to cells during infection and are recognized by the immune system.

A popular hypothesis for virus-induced autoimmunity states that viral and host proteins share antigenic determinants. This idea is often called **molecular mimicry**. Infection leads to the elaboration of immune responses to such shared determinants, resulting in an autoreactive response. Although many peptide sequences are shared among viral and host proteins, direct evidence for this hypothesis has been difficult to obtain. One reason is the long lag period between events that trigger human autoimmune diseases and the onset of clinical symptoms. To circumvent this problem, transgenic mouse models in which the products of foreign genes are expressed as self antigens have been established. This approach permits a precise study of how the autoreactive immune response is initiated and what cell types and soluble mediators are involved (Box 4.9).

Heterologous T-Cell Immunity

The immune memory cells of the adaptive immune system enable a subsequent rapid and specific response to previously encountered infections. Surprisingly, these memory cells are not always as specific as once thought. This realization came from many studies, but one fact was particularly informative. It is well known that common infections can run surprisingly different courses in different individuals. Why this should be so is complicated, but from experiments with genetically identical mice, it become clear that the history of previous infections can dictate the outcome of a new infection. The phenomenon is called **heterologous T-cell immunity**: memory T cells specific for a particular virus epitope can be activated during infection with a completely unrelated virus. The presence of a cross-reactive memory T cell has the potential to influence the new immune response to be protective or pathological or even to change the balance between Th1 and Th2 responses. Cross-reacting T-cell epitopes are another type of molecular mimicry that plays a significant role in the outcome of an infection.

When the T-cell receptor on a memory T cell engages its cognate epitope, the cell begins to proliferate and to secrete cytokines. These cytokines almost instantly affect the subsequent immune response. If a heterologous memory T cell responds to a similar epitope, it can be activated by an unrelated infection. Consequently, the immediate immune response is not tailored to the new pathogen and instead produces a response appropriate for the unrelated previous infection. When mice are immunized against one of several viruses and then challenged with a panel of other viruses, the animals show partial, but not necessarily reciprocal protection to the heterologous infection. For example, infection with lymphocytic choriomeningitis virus, an areanavirus, provided substantial protection against vaccinia virus, a poxvirus, but not vice versa. For other virus pairs such as murine cytomegalovirus and vaccinia virus, the protection was partially reciprocal. The significance of these findings to human infections is only now emerging. For example, patients experiencing Epstein-Barr virus-induced mononucleosis may have a strong T-cell response to a particular influenza virus epitope rather than the typical response to an immunodominant

BOX
4.9

EXPERIMENTS
Viral infections promote or protect against autoimmune disease

Promotion

Transgenic mice that synthesize proteins of lymphocytic choriomeningitis virus in β cells of the pancreas or oligodendrocytes have been developed. Expression of these viral proteins has no consequence; the mice are healthy. The viral transgene products are present in the mouse throughout development, and therefore are self antigen.

Infection of these transgenic animals with lymphocytic choriomeningitis virus stimulates an immune response in which the self antigen is recognized, leading to insulin-dependent diabetes mellitus or central nervous system demyelinating disease, respectively.

The action of virus-specific CTLs leads to inflammation, insulitis, and diabetes. Curiously, in uninfected animals, virus-specific lymphocytes are present in the peripheries as determined from in vitro studies, but they do not respond to the viral transgene product. The mechanisms that maintain such unresponsiveness to self antigens and prevent autoimmunity can be elucidated with such transgenic models. It was determined that tolerance to self antigens is circumvented depending on the number of autoreactive T cells and their affinity for peptides presented by MHC class II proteins, the number of memory cells induced by infection, and the cytokine milieu.

Hypothesis: If an individual becomes infected in the first few months after birth with an organ-tropic virus that establishes a persistent infection, tolerance to the virus is established. Later in life, an infection with the same virus leads to an immune-mediated attack on virus-infected cells in the organ.

Protection

Both the nonobese diabetic mouse strain and a particular strain of rat spontaneously develop type 1 diabetes. Infection of either animal with lymphocytic choriomeningitis virus reduces the incidence of diabetes. IL-12 is known to exacerbate autoimmune disease in these experimental models by promoting expansion and activation of autoreactive CD4+ T cells. Moreover, production of IFN results in reduction in IL-12 production by monocytes and dendritic cells.

Explanation: Lymphocytic choriomeningitis virus infection induces synthesis of IFN that, in turn, protects against autoimmune disease by decreasing IL-12 expression.

Fujinami, R. 2001. Viruses and autoimmune disease—two sides of the same coin? *Trends Microbiol.* **9:**377–381.

von Herrath, M., and M. Oldstone. 1996. Virus-induced autoimmune disease. *Curr. Opin. Immunol.* **8:**878–885.

Epstein-Barr virus epitope. It appears that the Epstein-Barr virus infection activated memory T cells established by previous exposure to influenza virus. These individuals had a different course of mononucleosis, often more severe, than did individuals with no previous exposure to influenza virus.

T-cell cross-reactivities among heterologous viruses are more frequent than commonly expected. These reactions modulate the course of disease in animal models and are likely to exert similar influences in human infections. The important principle is that prior infection affects the defense against pathogens that have not yet been encountered.

Superantigens "Short-Circuit" the Immune System

Some viral proteins are extremely powerful T-cell mitogens known as **superantigens**. These proteins bind to MHC class II molecules on antigen-presenting cells, and interact with the Vβ chain of the T-cell receptor. As approximately 2 to 20% of **all** T cells produce the particular Vβ chain that binds the superantigen, superantigens short-circuit the interaction of MHC class II-peptide complex and the T cell. Rather than activation of a small, specific subset of T cells (only 0.001 to 0.01% of T cells usually respond to a given antigen), **all subsets** of T cells producing the Vβ chain to which the superantigen binds are activated and proliferate. Superantigens clearly interfere with a coordinated immune response and divert the host's defenses.

All known superantigens are microbial products, and many are produced after infection by viruses. The best-understood viral superantigen is encoded in the U3 region of the mouse mammary tumor virus long terminal repeat. This retrovirus is transmitted efficiently from mother to offspring via milk. However, the virus replicates poorly in most somatic tissues, so the question is, how do virions get into the milk? When B cells in the neonatal small intestine epithelium are infected, the viral superantigen is produced and recognized by T cells carrying the appropriate T-cell receptor Vβ chain. Consequently, extraordinarily large numbers of T cells are activated, and produce growth factors and other molecules that stimulate the infected B cells. These cells then carry the virus to the mammary gland. Infection of mice with mutants harboring a deletion of the superantigen gene results in limited viral replication and minimal transmission to offspring via milk. Expression of the superantigen stimulates the immune system to facilitate the spread of infection from the gut to the mammary gland.

Mechanisms Mediated by Free Radicals

Two free radicals, superoxide (O_2^-) and nitric oxide (NO), are produced during the inflammatory response and may play important roles in virus-induced pathology. Superoxide is produced by the enzyme xanthine oxidase, present in phagocytes. The production of O_2^- is significantly increased in the lungs of mice infected with influenza virus or cytomegalovirus. Inhibition of xanthine oxidase protects mice from virus-induced death. However, O_2^- is not toxic for some cells and viruses, and its effects might be the result of formation of peroxynitrite (ONOO$^-$) through interaction with nitric oxide.

Nitric oxide is produced in abundance in virus-infected tissues during inflammation as part of the innate immune response (Fig. 4.21). This gas has been shown to inhibit the replication of many viruses in cultured cells and in animal models. It probably acts intracellularly to inhibit viral replication, but the molecular sites of action are not well understood. Nitric oxide is produced by three different IFN-inducible isoforms of nitric oxide synthase. While low concentrations of NO have a protective effect, high concentrations or prolonged production have the potential to contribute to tissue damage. For example, treating infected animals with inhibitors of nitric oxide synthase prevents tissue damage. Although NO is relatively inert, it reacts rapidly with O_2^- to form peroxynitrite (ONOO$^-$), which is much more reactive than either molecule and may be responsible for cytotoxic effects on cells (Fig. 4.21). In the central nervous system, NO is produced by activated astrocytes and microglia, and may be directly toxic to oligodendrocytes and neurons.

Perspectives

The breadth and complexity of the host response to infection can be initially confusing, if not overwhelming. To illuminate the important principles, it is useful to consider a hypothetical acute viral infection that is cleared by the host response (Fig. 4.22). Remember also this central fact: all successful viruses encode gene products that modulate their host's defenses (Table 4.7; Figure 4.23).

To initiate the primary infection, physical barriers are breached and virus particles enter permissive cells. Almost immediately, viral proteins and viral nucleic acids are bound by pathogen recognition receptors. Signal transduction cascades release latent transcription activation proteins from the cytoplasm. The infected cell

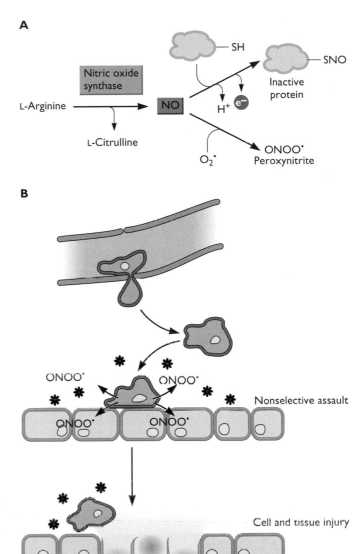

Figure 4.21 Consequences of nitric oxide production. **(A)** Formation of NO and ONOO$^-$. Nitric oxide, which is produced from L-arginine by nitric oxide synthase, reacts with proteins, inactivating them, or with O_2^-, forming peroxynitrite, ONOO$^-$. Adapted from C. S. Reiss and T. Komatsu, *J. Virol.* **72:**4547–4551, 1998, with permission. **(B)** Biological effects of NO/O_2^-/ONOO$^-$. Phagocytes move into the area of active virus replication and produce NO/O_2^-/ONOO$^-$, which destroys cells, causing tissue injury but perhaps limiting virus yields. Adapted from T. Akaike et al., *Proc. Soc. Exp. Biol. Med.* **217:**64–73, 1998, with permission.

now synthesizes cytokines, such as IFN. As new viral proteins are produced, the cell initiates other intrinsic defenses, such as apoptosis or autophagy. Local sentinel cells, the immature dendritic cells and macrophages, engage the locally released cytokines and internalize

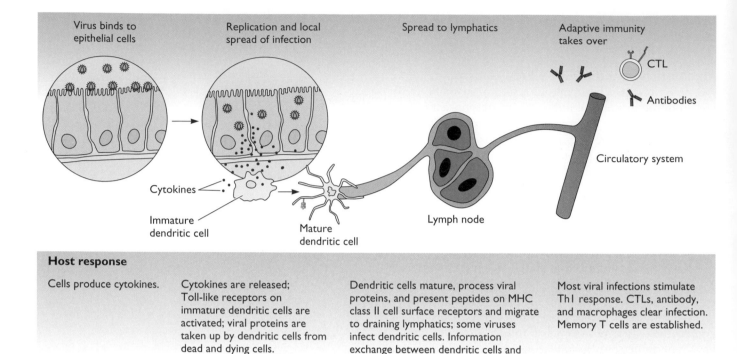

Virus binds to epithelial cells

Replication and local spread of infection

Spread to lymphatics

Adaptive immunity takes over

CTL

Antibodies

Circulatory system

Cytokines

Immature dendritic cell

Mature dendritic cell

Lymph node

Host response

Cells produce cytokines.	Cytokines are released; Toll-like receptors on immature dendritic cells are activated; viral proteins are taken up by dendritic cells from dead and dying cells.	Dendritic cells mature, process viral proteins, and present peptides on MHC class II cell surface receptors and migrate to draining lymphatics; some viruses infect dendritic cells. Information exchange between dendritic cells and naive T cells takes place in lymphatics.	Most viral infections stimulate Th1 response. CTLs, antibody, and macrophages clear infection. Memory T cells are established.

Figure 4.22 The stages of viral infection and responses of the host. Four stages are illustrated for a typical viral infection at an epithelial surface, a typical site of viral entry. The virus must attach to epithelial cells and then initiate replication. Infected cells produce cytokines that signal local dendritic cells, which respond accordingly. As a result, most often, local defenses contain the infection. However, if virus replication is not stopped, the adaptive immune system is activated by contact with mature dendritic cells and cytokines in local lymph nodes. The adaptive response leads to production of activated effector T and B cells that are released into the circulation to clear the infection and provide precise memory of the particular invader.

viral proteins produced by infected cells. The first response of the immature dendritic cell is to produce massive quantities of IFN and other cytokines. If viral anti-IFN or antiapoptotic gene products are made, progeny virions are released. If the newly infected cells have already bound IFN, protein synthesis is inhibited when viral nucleic acid is produced. Soon thereafter, NK cells can recognize the infected cells because of new surface antigens and a low or aberrant display of MHC class I proteins. The IFN produced by infected cells stimulates the NK cells to intensify their activities, which include target cell destruction and synthesis of IFN-γ (Box 4.10). In some cases, serum complement can be activated to destroy enveloped viruses and infected cells. In general, the intrinsic and innate defenses bring most viral infections to an uneventful close before the adaptive response is required.

If viral replication outpaces the innate defense, a critical threshold is reached: increased IFN production by circulating immature dendritic cells elicits a more global host response, and flulike symptoms are experienced by the infected individual. As viral replication continues, viral antigens are delivered by mature dendritic cells to the local lymph nodes or spleen to establish sites of information exchange with T cells. T-cell recirculation is shut down because of the massive recruitment of lymphocytes into lymphoid tissue. The swelling of lymph nodes so often characteristic of infection is a sign of this stage of immune action.

Within days, Th cells and CTLs appear; these cells are the first signs of activation of the adaptive immune response. Th cells produce cytokines that begin to direct the amplification of this response. The synthesis of antibodies, first of IgM and then of other isotypes, quickly

Table 4.7   Immune modulation strategies deduced from virus-infected cells[a]

Strategy	Example(s)
Secreted modulators	Virokines (ligand mimics)
	Viroceptors (receptor mimics)
Modulators on the infected cell surface	Complement inhibitors
	Coagulation regulators
	Immune receptors
	Adhesion molecules
Stealth	Latency
	Infection of immunoprivileged tissue
Antigenic hypervariability	Error-prone replicase
	Antigenic drift by antibody selection
	Epitope drift by CTL selection
Bypassing or killing of lymphocytes	Direct cytopathic infection
	Blockage of NK cell recognition
	Interference with signal transduction, maturation, or effector functions
Blockage of adaptive immune response	Alteration of MHC-I or MHC-II production
	Blockage of antigen processing
	Blockage of transcription of immune response genes
Inhibition of complement	Soluble receptor mimics
	Viral Fc receptors
Inhibition of cytokine action	Blockage of signal transduction cascade
Modulation of apoptosis	Blockage of cell death signaling pathways
	Scavenging of free radicals
	Blockage of death receptors or ligands
Interference with pattern recognition receptors	Alteration of ligands
	Production of decoy ligands
	Blockage of downstream signaling

[a]Adapted from G. Finlay and G. McFadden, *Cell* **124**:767–782, 2006.

follows. The relative concentrations of these isotypes are governed by the route of infection and the pattern of cytokines produced by the Th cells. As the immune response is amplified, CTLs kill or purge infected cells, and antibodies bind to virus particles and infected cells. Specific antibody-virus complexes can be recognized by macrophages and NK cells to induce antibody-dependent cell-mediated cytotoxicity, and can also activate the classical complement pathway. Both of these processes lead to the directed killing of infected cells and enveloped viruses by macrophages and NK cells.

An inflammatory response often occurs as infected cells die, and innate and adaptive responses develop. Cytokines, chemotactic proteins, and vasodilators are released at the site of infection. These proteins, invading white blood cells, and various complement components all contribute to the swelling, redness, heat, and pain characteristic of the inflammatory response. Many viral proteins modulate this response and the subsequent activation of immune cells.

If infection spreads from the primary site, second and third rounds of replication can occur in other organs. T cells that were activated at the initial site of infection can cause delayed-type hypersensitivity (usually evident as a characteristic rash or lesion) at the sites of later replication. Immunopathology, particularly after infection by noncytopathic viruses, is caused by the cytokines, antibodies, and cells in response to the infection and can contribute to the severity of the disease. Finally, the combination of innate and adaptive responses clears the infection, and the host is immune because of the presence of memory T and B cells and long-lived antibodies. The high concentrations of immune lymphocytes drop dramatically as these cells

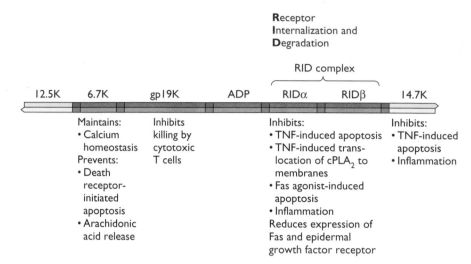

Figure 4.23 The adenovirus type 2 E3 region, a cluster of five genes encoding proteins that mediate host defense modulation. The proteins were named initially according to their apparent molecular masses (e.g., 14.7K for 14.7 kDa). Recently, some of the proteins have been given names that reflect their known functions. RID is an acronym for "receptor internalization and degradation." The RID protein complex is composed of RIDα and RIDβ. RID proteins have multiple functions, as indicated. These functions may or may not represent the same molecular mechanism. Integral membrane proteins are indicated by reddish shading. gp19K is a glycoprotein that reduces MHC class I protein synthesis, and inhibits killing of infected cells by CTLs. The 14.7K protein inhibits Tnf-induced apoptosis. The 6.7K protein maintains calcium homeostasis and prevents death-receptor-initiated apoptosis and arachidonic acid release. cPLA$_2$, cytoplasmic phospholipase A2. Adapted from W. S. M. Wold and A. E. Tollefson, *Semin. Virol.* **8:**515–523, 1998, with permission.

die by apoptosis, and the system returns to its normal, preinfection state. The adaptive response can be avoided completely, or in part, when organs or tissues that have poor immune responses or none are infected, when new viral variants are produced rapidly because of high mutation rates, or when progeny virions spread directly from cell to cell. Consequently, there can be dramatic differences in patterns of infection ranging from short-lived (acute) to lifelong (persistent), the topic of Chapter 5.

BOX 4.10

EXPERIMENTS
Setting the threshold for activation of innate immunity

The threshold that will trigger an innate immune response when crossed is not the same for every individual. However, the molecular mechanisms that set this threshold are poorly understood. Some insight into the problem came recently from an unexpected source: studies of γHV68 latent infections in mice. γHV68 is a gammaherpesvirus that establishes a latent infection in mouse B cells. The conventional view had been that latent infections are quiescent to the point of being inapparent to the host. Unexpectedly, latent γHV68 herpesvirus infections confer prolonged cross-protection against a variety of bacterial pathogens, including *Listeria* (listeriosis) and *Yersinia* (plague). The latent infection activates macrophages because immune recognition of the latently infected B cells leads to production of IFN-γ. The authors conclude that the latent virus infection sets the level of innate immunity. Moreover, not only is the latent infection an active immunological state, but also infected animals enjoy a benefit from this curious symbiotic relationship.

Barton, E. S., D. W. White, J. S. Cathelyn, K. A. Brett-McClellan, M. Engle, M. S. Diamond, V. L. Miller, and H. W. Virgin IV. 2007. Herpesvirus latency confers symbiotic protection from bacterial infection. *Nature* **447:**326–329.

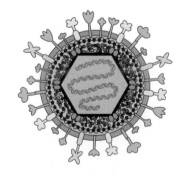

References

Textbooks

Murphy, K. M., P. Travers, and M. Walport. 2007. *Janeway's Immunobiology*, 7th ed. Garland Publishing, New York, NY.

Nathanson, N (ed.). 2007. *Viral Pathogenesis and Immunity*, 2nd ed. Academic Press, London, United Kingdom.

Reviews

Dendritic Cells

Alsharifi, M., A. Mullbacher, and M. Regner. 2008. Interferon type I responses in Primary and secondary infections. *Immunol. Cell Biol.* **86:**239–245.

Colonna, M., G. Trinchieri, and Y. Liu. 2004. Plasmacytoid dendritic cells in immunity. *Nat Immunol.* **5:**1219–1226.

López, C. B., J. S. Yount, and T. M. Moran. 2006. Toll-like receptor-independent triggering of dendritic cell maturation by viruses. *J. Virol.* **80:**3128–3134.

Novak, N., and W. Peng. 2005. Dancing with the enemy: the interplay of herpes simplex virus with dendritic cells. *Clin. Exp. Immunol.* **142:**405–410.

Steinman, R. M. 2000. DC-SIGN: a guide to some mysteries of dendritic cells. *Cell* **100:**491–494.

Zuniga, E., B. Hahm, K. Edelmann, and M. Oldstone. 2005. Imunosuppressive viruses and dendritic cells: a multifront war. *ASM News* **71:**285–290.

NK Cells

Biron, C. A., and L. Brossay. 2001. NK cells and NK T-cells in innate defense against viral infections. *Curr. Opin. Immunol.* **13:**458–464.

Fauci, A., D. Mavilio, and S. Kottilil. 2005. NK cells in HIV infection: paradigm for protection or targets for ambush. *Nat. Rev. Immunol.* **5:**835–844.

Lodoen, M., and L. Lanier. 2005. Viral modulation of NK cell immunity. *Nat. Rev. Microbiol.* **3:**59–69.

Moretta, A., C. Bottino, M. C. Mingari, R. Biassoni, and L. Moretta. 2002. What is a natural killer cell? *Nat. Immunol.* **3:**6–8.

Orange, J. S., M. S. Fassett, L. A. Koopman, J. E. Boyson, and J. L. Strominger. 2002. Viral evasion of natural killer cells. *Nat. Immunol.* **3:**1006–1012.

Park, S.-H., and A. Bendelac. 2000. CD1-restricted T-cell responses and microbial infection. *Nature* **406:**788–792.

Complement and Inflammation

Carroll, M. 2004. The complement system in regulation of adaptive immunity. *Nat. Immunol.* **5:**981–986.

Inohara, N., M. Chamaillard, C. McDonald, and G. Nunez. 2005. NOD-LRR proteins: role in host-microbial interactions and inflammatory disease. *Annu. Rev. Biochem.* **74:**355–383.

Lachman, P. J., and A. Davies. 1997. Complement and immunity to viruses. *Immunol. Rev.* **159:**69–77.

Weiss, R. A. 1998. Transgenic pigs and virus adaptation. *Nature* **391:**327–328.

Virus Immune Modulation

Coscoy, L. 2007. Immune evasion by Kaposi's sarcoma-associated herpesvirus. *Nat. Rev. Immunol.* **7:**391–401.

Finlay, G., and G. McFadden. 2006. Anti-immunology: evasion of the host immune system by bacterial and viral pathogens. *Cell* **124:**767–782.

Lorenzo, M. E., H. L. Ploegh, and R. S. Tirabassi. 2001. Viral immune evasion strategies and the underlying cell biology. *Semin. Immunol.* **13:**1–9.

Martin, M., and M. Carrington. 2005. Immunogenetics of viral infections. *Curr. Opin. Immunol.* **17:**510–516.

Mossman, K., and A. Ashkar. 2005. Herpesviruses and the innate immune response. *Virol. Immunol.* **18:**267–281.

Pasieka, T., T. Baas, V. Carter, S. Proll, M. Katze, and D. Leib. 2006. Functional genomic analysis of herpes simplex virus type 1 counteraction of the host immune response. *J. Virol.* **80:**7600–7612.

Redpath, S., A. Angulo, N. R. J. Gascoigne, and P. Ghazal. 2001. Immune checkpoints in viral latency. *Annu. Rev. Microbiol.* **55:**531–560.

Tortorella, D., B. E. Gewurz, M. H. Furman, D. J. Schust, and H. L. Ploegh. 2000. Viral subversion of the immune system. *Annu. Rev. Immunol.* **18:**861–926.

Adaptive Immunity

Belkaid, and Y. B. Rouse. 2005. Natural regulatory T cells in infectious disease. *Nat. Immunol.* **6:**353–360.

Delon, J., and R. N. Germain. 2000. Information transfer at the immunological synapse. *Curr. Biol.* **10:**R923–R933.

Dimmock, J. J. 1993. Neutralization of animal viruses. *Curr. Top. Microbiol. Immunol.* **183:**1–149.

DiVico, A., and R. Gallo. 2004. Control of HIV-1 infection by soluble factors of the immune response. *Nat. Rev. Microbiol.* **2:**401–413.

Doherty, P. C., and J. P. Christensen. 2000. Accessing complexity: the dynamics of virus-specific T-cell responses. *Annu. Rev. Immunol.* **18:**561–592.

Guidotti, L. G., and F. V. Chisari. 2001. Non-cytolytic control of viral infections by the innate and adaptive immune response. *Annu. Rev. Immunol.* **19:**65–91.

Lamm, M. E. 1997. Interaction of antigens and antibodies at mucosal surfaces. *Annu. Rev. Microbiol.* **51:**311–340.

Martin, M., and M. Carrington. 2005. Immunogenetics of viral infections. *Curr. Opin. Immunol.* **17:**510–516.

Medzhitov, R. 2007. Recognition of microorganisms and activation of the immune response. *Nature* **449:**819–826.

Central Nervous System Defense

Hickey, W. F. 2001. Basic principles of immunological surveillance of the normal central nervous system. *Glia* **36:**118–124.

Lowenstein, P. R. 2002. Immunology of viral-vector-mediated gene transfer into the brain: an evolutionary and developmental perspective. *Trends Immunol.* **23:**23–30.

Immunopathology

Davies, M., and P. Hagen. 1997. Systemic inflammatory response syndrome. *Br. J. Surg.* **84:**920–935.

Iannacone, M., G. Sitia, and L. Guidotti. 2006. Pathogenetic and antiviral immune responses against hepatitis B virus. *Fut. Virol.* **1:**189–196.

Lane, T. E., and M. J. Buchmeier. 1997. Murine coronavirus infection: a paradigm for virus-induced demyelinating disease. *Trends Microbiol.* **5:**9–14.

Reiss, C. S., and T. Komatsu. 1998. Does nitric oxide play a critical role in viral infections? *J. Virol.* **72:**4547–4551.

Rouse, B. T. 1996. Virus-induced immunopathology. *Adv. Virus Res.* **47:**353–376.

Sansonetti, P. 2006. The innate signaling of danger and the dangers of innate signaling. *Nat. Immunol.* **7:**1237–1242.

Simas, J. P., and S. Efstathiou. 1998. Murine gammaherpesvirus 68: a model for the study of gammaherpesvirus pathogenesis. *Trends Microbiol.* **7**:276–282.

Streilein, J. W., M. R. Dana, and B. R. Ksander. 1997. Immunity causing blindness: five different paths to herpes stromal keratitis. *Immunol. Today* **18**:443–449.

Classic Papers

Janeway, C. A., Jr. 2001. How the immune system works to protect the host from infection: a personal view. *Proc. Natl. Acad. Sci. USA* **98**:7461–7468.

Zinkernagel, R. M. 1996. Immunology taught by viruses. *Science* **271**:173–178.

Zinkernagel, R. M., and P. C. Doherty. 1974. Restriction of *in vitro* T-cell-mediated cytotoxicity in lymphocytic choriomeningitis within a syngeneic or semiallogeneic system. *Nature* **248**:701–702.

Selected Papers

Dendritic Cells

Kadowaki, N., S. Antonenko, J. Y.-N. Lau, and Y.-J. Liu. 2000. Natural interferon α/β producing cells link innate and adaptive immunity. *J. Exp. Med.* **192**:219–225.

Ludewig, B., K. Maloy, C. Lopez-Macias, B. Odermatt, H. Hengartner, and R. M. Zinkernagel. 2000. Induction of optimal anti-viral neutralizing B-cell responses by dendritic cells requires transport and release of virus particles in secondary lymphoid organs. *Eur. J. Immunol.* **30**:185–196.

Trevejo, J., M. Marino, N. Philpott, R. Josien, E. Richards, K. Elkon, and E. Falck-Pedersen. 2001. TNF-α-dependent maturation of local dendritic cells is critical for activating the adaptive immune response to virus infection. *Proc. Natl. Acad. Sci. USA* **98**:12162–12167.

NK Cells

Brown, M., A. Dokun, J. Heusel, H. Smith, D. Beckman, E. Blattenberger, C. Dubbelde, L. Stone, A. Scalzo, and W. Yokoyama. 2001. Vital involvement of a natural killer cell activation receptor in resistance to viral infection. *Science* **292**:934–937.

Complement and Inflammation

Cecchinato, V., C. J. Trindade, A. Laurence, J. M. Heraud, J. M. Brenchley, M. Ferrari, L. Zaffiri, E. Tryniszewska, W. P. Tsai, M. Vaccari, R. Washington Parks, D. Venzon, D. C. Douek, J. J. O'Shea, and G. Franchini. 2008. Altered balance between Th17 and Th1 cells at mucosal sites predicts AIDS progression in simian immunodeficiency virus-infected macaques. *Mucosal Immunol.* **1**:279–288.

Lubinski, J., L. Wang, D. Mastellos, A. Sahu, J. D. Lambris, and H. Friedman. 1999. *In vivo* role of complement interacting domains of herpes simplex virus type 1 glycoprotein gC. *J. Exp. Med.* **90**:1637–1646.

Rosengard, A., Y. Liu, Z. Nie, and R. Jimenez. 2002. Variola virus immune evasion design: expression of a highly efficient inhibitor of human complement. *Proc. Natl. Acad. Sci. USA* **99**:8808–8813.

Rother, R. P., and S. P. Squinto. 1996. The α-galactosyl epitope: a sugar coating that makes viruses and cells unpalatable. *Cell* **86**:185–188.

Vanderplasschen, A., E. Mathew, M. Hollinshead, R. Sim, and G. Smith. 1998. Extracellular enveloped vaccinia virus is resistant to complement because of incorporation of host complement control proteins into its envelope. *Proc. Natl. Acad. Sci. USA* **95**:7544–7549.

Collectins and Defensins

Leikina, E., H. Delanoe-Ayari, K. Melikov, M. Cho, A. Chen, A. Waring, W. Wang, Y. Xie, J. Loo, R. Lehrer, L. Chernomordik.

2005. Carbohydrate-binding molecules inhibit viral fusion and entry by crosslinking membrane glycoproteins. *Nat. Immunol.* **6**:995–1001.

Selsted, M., and A. Ouellette. 2005. Mammalian defensins in the antimicrobial immune response. *Nat. Immunol.* **6**:551–557.

Adaptive Immunity

Callan, M., L. Tan, N. Annels, G. Ogg, J. Wilson, C. O'Callaghan, N. Steven, A. McMichael, and A. Rickinson. 1998. Direct visualization of antigen-specific CD8[+] T-cells during the primary immune response to Epstein-Barr virus *in vivo*. *J. Exp. Med.* **187**:1395–1402.

Nagashunmugam, T., J. Lubinski, L. Wang, L. T. Goldstein, B. Weeks, P. Sundaresan, E. Kang, G. Dubin, and H. Friedman. 1998. *In vivo* immune evasion mediated by herpes simplex virus type 1 IgG Fc receptor. *J. Virol.* **72**:5351–5359.

Selin, L. and R. Welsh. 2004. Plasticity of T cell memory responses to viruses. *Immunity* **20**:5–16.

Sigal, L., S. Crotty, R. Andino, and K. Rock. 1999. Cytotoxic T-cell immunity to virus-infected non-haematopoietic cells requires presentation of exogenous antigen. *Nature* **398**:77–80.

Central Nervous System Defense

Stevenson, P., J. Austyn, and S. Hawke. 2002 Uncoupling of virus-induced inflammation and anti-viral immunity in the brain parenchyma. *J. Gen. Virol.* **83**:1735–1743.

Stevenson, P., S. Hawke, D. Sloan, and C. Bangham. 1997. The immunogenicity of intracerebral virus infection depends on the anatomical site. *J. Virol.* **71**:145–151.

Stevenson, P., S. Freeman, C. Bangham, and S. Hawke. 1997. Virus dissemination through the brain parenchyma without immunologic control. *J. Immunol.* **159**:1876–1884.

Thomas, C., G. Schiedner, S. Kochaneck, M. Castro, and P. Lowenstein. 2000. Peripheral infection with adenovirus induces unexpected long term brain inflammation in animals injected intracranially with first generation, but not with high capacity adenovirus vectors: toward realistic long term neurological gene therapy for chronic diseases. *Proc. Natl. Acad. Sci. USA* **97**:7482–7487.

Immunopathology

Deshpande, S., S. Lee, M. Zheng, B. Song, D. Knipe, J. Kapp, and B. Rouse. 2001. Herpes simplex virus-induced keratitis: evaluation of the role of molecular mimicry in lesion pathogenesis. *J. Virol.* **75**:3077–3088.

Evans, C., M. Horwitz, M. Hobbs, and M. Oldstone. 1996. Viral infection of transgenic mice expressing a viral protein in oligodendrocytes leads to chronic central nervous system autoimmune disease. *J. Exp. Med.* **184**:2371–2384.

Gebhard, J., C. Perry, S. Harkins, T. Lane, I. Mena, V. Asensio, I. Campbell, and J. Whitton. 1998. Coxsackievirus B3-induced myocarditis: perforin exacerbates disease, but plays no detectable role in virus clearance. *Am. J. Pathol.* **153**:417–428.

Klimstra, W., K. Ryman, K. Bernard, K. Nguyen, C. Biron, and R. Johnston. 1999. Infection of neonatal mice with Sindbis virus results in a systemic inflammatory response syndrome. *J. Virol.* **73**:10387–10398.

Posavad, C., D. Koelle, M. Shaughnessy, and L. Corey. 1997. Severe genital herpes infections in human immunodeficiency virus-infected individuals with impaired herpes simplex virus-specific CD8[+] cytotoxic T lymphocyte responses. *Proc. Natl. Acad. Sci. USA* **94**:10289–10294.

Tripp, R., A. Hamilton-Easton, R. Cardin, P. Nguyen, F. Behm, D. Woodland, P. Doherty, and M. Blackman. 1997. Pathogenesis of an infectious mononucleosis-like disease induced by a murine β-herpesvirus: role for a viral superantigen? *J. Exp. Med.* **185**:1641–1650.

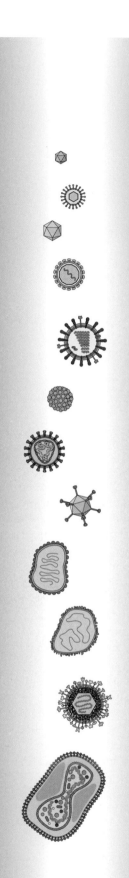

5

Introduction

Life Cycles and Host Defenses

Mathematics of Growth Correlate with Patterns of Infection

Acute Infections

Definition and Requirements

Acute Infections Tend To Be Efficiently Contained and Cleared

Antigenic Variation Provides a Selective Advantage in Acute Infections

Acute Infections Present Common Public Health Problems

Persistent Infections

Definition and Requirements

An Ineffective Intrinsic or Innate Immune Response Can Promote a Persistent Infection

Modulation of the Adaptive Immune Response Perpetuates a Persistent Infection

Persistent Infections May Be Established in Tissues with Reduced Immune Surveillance

Persistent Infections May Occur When Cells of the Immune System Are Infected

Two Viruses That Cause Persistent Infections

Measles Virus

Lymphocytic Choriomeningitis Virus

Latent Infections

General Properties

Herpes Simplex Virus

Epstein-Barr Virus

Slow Infections: Sigurdsson's Legacy

Abortive Infections

Transforming Infections

Perspectives

References

Patterns of Infection

Introduction

Viral infections of individuals in populations differ from infections of cultured cells in the laboratory. In the former, initiation of the infection and its eventual outcome rest upon complex variables such as host defenses, composition of the host population, and the environment. Despite such complexity and the plethora of viruses and hosts, common patterns of infection do appear. In general, natural infections can be rapid and self-limiting (**acute infections**) or long-term (**persistent infections**). These patterns can be surprisingly stable over time and characteristic for many virus families. Variations and combinations of these two modes abound (Fig. 5.1). It can be argued that all patterns begin with an acute infection, and differences in the subsequent management of that infection engender the many variations. For example, most **latent infections** begin as an acute infection of one cell type, but then when a different cell type is infected, no infectious particles are produced at all. Nevertheless, the genome persists to be reactivated in the future. Intermediate patterns that lie between rapid viral growth and latent infection can be thought of as "**smoldering infections**" in which low-level viral replication occurs in the face of a strong immune response. Similarly, **slow infections**, **abortive infections**, and **transforming infections** are more complicated variants of persistent infections.

While we can provide detailed descriptions of individual patterns of infection, we are in the early days of understanding the molecular mechanisms required to initiate or maintain any specific one. In this chapter, we discuss the principles that are the foundations of the observed patterns of infection.

Life Cycles and Host Defenses

A cursory examination of the animal viruses that grow in cultured cells identifies many distinctive life cycles with common features. Some infections rapidly kill the cell while producing a burst of new particles (**cytopathic viruses**). Others yield virions without causing immediate host cell death (**noncytopathic viruses**). Alternatively, some infections neither kill the cell nor produce any

135

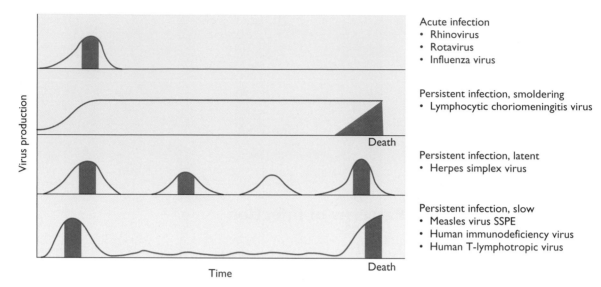

Figure 5.1 General patterns of infection. Relative virion production is plotted as a function of time after infection (blue line). The time when symptoms appear is indicated by the red shaded area. The top panel is the typical profile of an acute infection, in which virions are produced, symptoms appear, and the infection is cleared within 7 to 10 days after infection. The second panel is the typical profile of a persistent smoldering infection, in which virion production continues for the life of the host. Symptoms may or may not appear just before death, depending upon the virus. Virions are usually produced throughout the infection. The bottom two panels are variations of the profile of a persistent infection. The third panel depicts a latent infection in which an initial acute infection is followed by a quiescent phase and repeated bouts of reactivation. Reactivation may or may not be accompanied by symptoms, but generally results in the production of virions. The fourth panel depicts a slow virus infection, in which a period of years intervenes between a typical primary acute infection and the usually fatal appearance of symptoms. The production of infectious particles during the long period between primary infection and fatal outcome may be continuous (e.g., human immunodeficiency virus) or absent (e.g., measles virus SSPE). Adapted from F. J. Fenner et al., *Veterinary Virology* (Academic Press, Inc., Orlando, FL, 1993), with permission.

progeny. These life cycle characteristics are only part of the processes that produce stable patterns. Host responses also play central roles in the evolution of patterns of infection. As we discuss in Chapters 3 and 4, the interplay between the molecular biology of viral life cycles and host defense systems is of paramount importance to the outcome of any infection. These interactions are dynamic and, despite appearing rather chaotic at the molecular and cellular level, can be remarkably stable and predictable when averaged over many individuals infected over long periods.

Mathematics of Growth Correlate with Patterns of Infection

The changes in size of a viral population can be described by one simple concept: the rate of increase is the difference between the rate of replication and the rate of elimination. We can write this statement as

$$dN/dt = (b - d)N$$

where dN/dt is the rate of change of the population (N) with respect to the change in time. The terms b and d are the average rates of birth and death, respectively. The term $(b - d)$ is usually written as a constant r, the intrinsic rate of increase. Therefore, we obtain equation 5.1:

$$dN/dt = rN$$

$$\text{and } \ln N = rt$$

(5.1)

This is the equation for exponential population growth. Plotting $\ln N$ versus t yields a straight line with slope r (Fig. 5.2A).

If b far exceeds d (as is the case for infections in cultured cells), progeny accumulate. When b equals d, the population maintains a stable size. If we assume a linear relationship for increase and decrease of the population, then the slope of the increase of replication rate is equal to k_b and the slope of death or removal rate is equal to k_d. The stability of the population N then can be written as follows:

$$b_0 - k_b N = d_0 + k_d N$$

$$\text{or}$$

$$N = (b_0 - d_0)/(k_d + k_b)$$

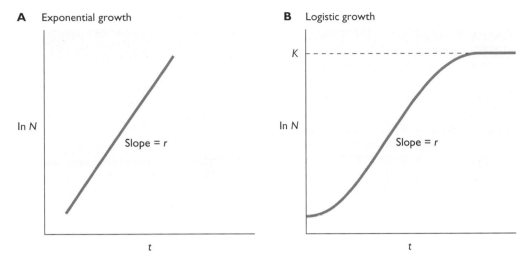

Figure 5.2 Two plots of standard growth equations. (A) A graph of simple exponential growth. **(B)** A graph of the pattern termed logistic growth illustrating K, the limit to growth. r is the slope in both types of plot.

This description of N is called the **carrying capacity (K)** of the environment. The term "environment" can define a single cell, an individual, or the entire host population. For any value of N greater than K, the viral population will decrease, and for any value of N less than K, the population will increase. The carrying capacity K is of particular interest in virology, as it defines the upper boundary of the growing population and, as we will note later, influences patterns of infection.

Therefore, by knowing that $r = (b_0 - d_0)$ and $K = (b_0 - d_0)/(k_b + k_d)$, we can substitute these values in equation 5.1 to obtain the basic equation for growth and regulation of a population, sometimes called the logistic growth equation (equation 5.2).

$$dN/dt = rN(K - N/K) \qquad (5.2)$$

Plotting ln N versus t yields the curve illustrated in Fig 5.2B. Here, K is easily seen to be the limit to growth, and the rate of increase is r.

Two fundamental viral growth strategies are apparent in nature, and these are strongly correlated with distinctive patterns of viral infection. The first is the **r-replication strategy**, in which large numbers of progeny are produced and growth is maintained by a steady, unbroken lineage of serial infections and never reaches a limit as long as there are susceptible hosts (equation 5.1; Fig. 5.2A). The second is the **K-replication strategy**, in which the host population is at or close to its saturation density (e.g., new susceptible hosts are rare or nonexistent [equation 5.2; Fig. 5.2B]). In addition, rates of viral propagation may be slow or vanishingly small.

r-replication strategies often manifest as **acute infections**: virulent, short reproductive cycles with production of many progeny. Pathogenesis may or may not be obvious, but high rates of shedding and efficient transmission are the rule. Acute infections following an r-replication strategy will "burn out" in the absence of susceptible hosts. One can mimic an r-selection environment in cell culture by low-multiplicity-of-infection (MOI) infections: susceptible cells abound to sustain multiple rounds of replication, but transmission stops abruptly when all the cells are infected.

K-replication strategies often appear as persistent or **latent infections**. In this pattern, infected hosts survive for extended times (resources for both host and viral reproduction are maximized). In cell culture, this selection environment can be mimicked by high-MOI infections: most cells are infected and faster replication confers no selective advantage. For some bacteriophages, this environment leads to efficient integration of viral DNA and formation of lysogens (formation of prophage).

Other examples of persistent infection appear as a mixture of r and K strategies: virion production occurs continuously over the life of most of the individuals in a population, but is balanced by immune elimination (a smoldering infection). In other situations (e.g., latent infection), the viral genome survives as long as the host prospers. Often infectious virions are produced only sporadically during the host's lifetime, when the host is in danger or stressed. Under these conditions, the mechanisms for viral shedding and transmission are so efficient that even a small number of virions have a high probability of surviving and infecting others in the population.

Viruses and their hosts exist along a continuum of values for r and K. Hosts can have generation times of minutes to years, which influence the selection of viruses that infect them. These host growth properties also play pivotal roles in evolution of viral growth patterns. Hosts that grow rapidly with generation times of minutes or hours (e.g., bacteria), or have rapid growth as part of their life cycles (cycling cells), tend to be infected by rapidly replicating viruses. Hosts with generation times of years (e.g., mammals, fish, and plants), tend to be infected by viruses that can establish persistent infections of some type. Even in these slow-growing hosts, rapidly growing viruses can be selected. However, in these cases, rapid growth is quickly met by host defenses that limit pathogenesis and spread.

The growth equations, as written in their simplest form above, can be used to model replication in single, identical cells in culture. However, to describe how a viral infection is propagated and maintained in a large population of hosts, more terms must be included. These terms would define the rate of shedding from infected individuals, the rate of transmission to other individuals, the probability that one infected individual will infect more than one person, and the number and density of susceptible individuals (Box 5.1). Some of these parameters are discussed in Chapter 10.

Acute Infections

Definition and Requirements

Acute infections are common and well studied. The term "acute" refers to rapid onset of disease with a short but occasionally severe course. Hallmarks of an acute viral infection are rapid production of infectious virions, followed by resolution and elimination of the infection by the host ("clearing" the infection). These infections occur only when intrinsic and innate defenses are transiently bypassed. Acute infections are the typical, expected course for agents such as influenza virus and rhinovirus (Fig. 5.3). The disease symptoms tend to be relatively brief and resolve over a period of days. In a healthy host, virions and virus-infected cells are destroyed (cleared) by the adaptive immune system within days. Nevertheless, during the rapid replication phase, some progeny invariably are shed and spread to other hosts **before** the infection is resolved (Boxes 5.2 and 5.3). If the initial infection escapes local defenses and spreads via hematogenous or neural routes to other parts of the body, several distinct rounds of replication may occur in the same animal, with new and distinctive symptoms. The pattern of multiple infections is characteristic of varicella-zoster virus, an alphaherpesvirus that causes the childhood disease chickenpox (Fig. 5.4).

Occasionally, an acute infection results in limited or even no obvious symptoms. Indeed, **inapparent** (asymptomatic) acute infections are quite common, and can be major sources of infections in populations (Box 5.3). Such infections are recognized by the presence of virus-specific antibodies with no reported history of disease. For example, over 95% of the unvaccinated population of the United States has antibody to varicella-zoster virus, but less than half report that they have had chickenpox.

Acute Infections Tend To Be Efficiently Contained and Cleared

Once immediate host defenses have been breached and an infection is established, a cascade of new defensive reactions occurs in the host (see Chapters 3 and 4). Symptoms and disease may or may not be obvious, depending upon the virus, the infected tissue, and the host defenses. The initial period before the characteristic symptoms of a disease are obvious is called the **incubation period**. During this time,

---

BOX 5.1

BACKGROUND
Viral spread in space and time: modeling epidemic spread

In the simple growth equations discussed in the text, susceptible cells or animals were assumed to be well mixed. In addition, viral growth was considered only with respect to time. In reality, of course, populations are not well mixed, population centers are separated geographically, and individuals are free to move around the country and the world. When spatial dimensions that describe these conditions are added to the basic r and K expressions, extremely rich patterns of infection emerge. In epidemics, the patterns often appear as waves of infection propagating through the nonhomogeneous population. For example, more people live in cities than in the country so cities form nodes of high propagation. As populations become immune and infected individuals move to different locations where susceptible hosts abound, more complex patterns result. For example, the spread of measles across the United Kingdom and sub-Saharan Africa exhibits complicated waveforms. These studies provide important data for epidemiologists and those who wish to implement effective control strategies.

Anderson, R., and R. May. 1979. Population biology of infectious diseases: part I. *Nature* **280**:361–367.

Ferrari, M., R. Grais, N. Bharti, A. J. Conlan, O. N. Bjørnstad, L. J. Wolfson, P. J. Guerin, A. Djibo, and B. T. Grenfell. 2008. The dynamics of measles in sub-Saharan Africa. *Nature* **451**: 679–684.

Grenfell, B. T., O. N. Bjørnstad, and J. Kappey. 2001. Traveling waves and spatial hierarchies in measles epidemics. *Nature* **414**:716–723.

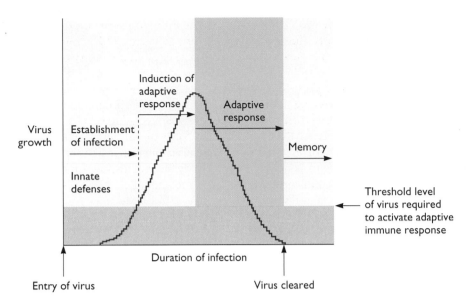

Figure 5.3 The course of a typical acute infection. Relative virus growth plotted as a function of time after infection. The concentration of virions increases with time, as indicated by the jagged red line. During the establishment of infection, only the innate defenses are at work. If the infection reaches a certain threshold level characteristic of the virus and host (purple), the adaptive responses initiate. After 4 to 5 days, effector cells and molecules of the adaptive response begin to clear infected tissues and virions (green). After this action, memory cells are produced, and the adaptive response ceases. Antibodies, residual effector cells, and memory cells provide lasting protection should the host be reinfected at a later date. Redrawn from C. A. Janeway, Jr., and P. Travers, *Immunobiology: the Immune System in Health and Disease* (Current Biology Ltd. and Garland Publishing, New York, NY, 1996), with permission.

viral genomes are replicating and the host is responding, producing cytokines such as interferon (IFN) that can have global effects that manifest as the classical symptoms of an acute infection (e.g., fever, malaise, aches, pains, and nausea). Remarkably, incubation periods can vary from 1 or 2 days to years (Table 5.1). Short incubation times usually indicate that actions at the primary site produce the characteristic symptoms of the disease. Long incubation times indicate that the host response, or the tissue damage required to reveal the symptoms of infection, result from actions other than those at the primary site of infection. The events that occur during the incubation period certainly influence the observed pattern of infection, but identification and characterization of these processes remain major challenges for those studying pathogenesis. Meeting these challenges is of paramount importance if we are to design

---

BOX 5.2

BACKGROUND
Uncomplicated acute infection by influenza virus

An influenza virus infection begins in the upper respiratory tract by inhalation of droplets produced when an infected individual sneezes or coughs. Viral replication occurs in ciliated columnar epithelial cells of the respiratory epithelium, and progeny virions spread to nearby cells. Infectious particles can be isolated from nasal secretions or throat swabs for 1 to 7 days, with the peak occurring on the fourth or fifth day after infection. About 48 h after the initial infection, symptoms appear

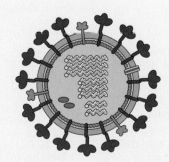

abruptly, such that infected individuals can almost pinpoint the hour that they noticed they had the flu. Symptoms last for about 3 days and then begin to abate.

The infection typically resolves within a week through action of the innate and acquired immune systems, but it may take several weeks before the individual feels completely well because of the lingering effects of the host responses.

BOX 5.3   DISCUSSION
Inapparent acute infections

It is important to distinguish an inapparent acute infection from an unsuccessful one. Inapparent infections are successful acute infections that produce no symptoms or disease. Sufficient virions are made to maintain the infection in the host population, but the quantity is below the threshold required to induce symptoms in infected individuals. The usual way an inapparent infection is detected is by a rise in antiviral antibody concentrations in an otherwise healthy individual. Well-adapted pathogens often follow this pattern, as demonstrated by poliovirus, in which more than 90% of infections are inapparent.

and implement useful diagnostic tools and treatments. In Chapter 3, we present more discussion of the adage that "whatever happens early in a viral infection dictates what happens later on."

The intrinsic and innate responses limit and contain most acute infections. When these defenses are lacking or compromised, acute infections can be disastrous, primarily because the infection becomes systemic and multiple organs can be damaged. In a naive host, the adaptive immune response (antibody and activated cytotoxic T lymphocytes [CTLs]) does not influence viral replication for several days, but is essential for final clearance of virions and infected cells. The adaptive response also provides **memory T and B cells** for defense against subsequent exposure to the same infectious agent.

Antigenic Variation Provides a Selective Advantage in Acute Infections

If an individual survives a typical acute infection, he or she often is immune to infections by the same virus. Nevertheless, some acute infections (e.g., infections with rhinovirus [the common-cold virus], influenza virus, and human immunodeficiency virus) occur repeatedly despite a robust immune response to them. These recurring infections are possible because selection pressures during the initial acute infections lead to shedding of virions that are resistant to immune clearance. In many cases, the structural properties of virions and the capacity of neutralizing antibodies to block infectivity are critical parameters.

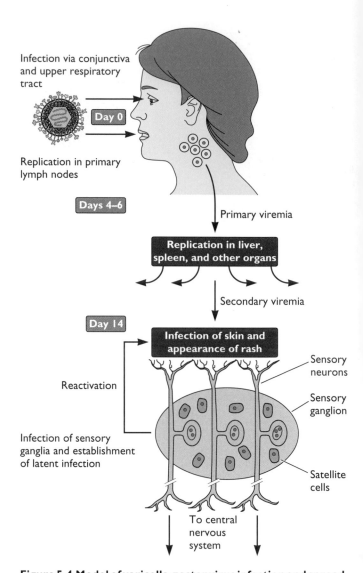

Figure 5.4 Model of varicella-zoster virus infection and spread. Infection initiated on the conjunctiva or mucosa of the upper respiratory tract spreads to regional lymph nodes. After 4 to 6 days, infected T cells enter the bloodstream, causing a **primary viremia**. These infected cells subsequently invade the liver, spleen, and other organs, initiating a second round of infection. Virions and infected cells are then released into the bloodstream in a **secondary viremia**. Infected skin-homing T cells efficiently invade the skin and initiate this third round of infection about 2 weeks from the initial infection. The characteristic vesicular rash of chicken pox appears as a result of immune defensive action. Next, virions produced in the skin infect sensory nerve terminals and spread to dorsal root ganglia of the peripheral nervous system, where a latent infection is established. The latent infection is maintained by active immune surveillance. Later in life, perhaps as the immune system wanes, reactivations are not contained and another infectious cycle is initiated. Virions leave the peripheral neurons to infect the skin. The characteristic recurrent disease called shingles is often accompanied by a long-lasting painful condition called postherpetic neuralgia. Normally, an infected individual experiences only one visible reactivation event, probably because reactivation stimulates the immune system. Such restimulation of the immune system is the rationale for administering the varicella-zoster live vaccine to adults to prevent reactivation and shingles.

Table 5.1   Incubation periods of some common viral infections

Disease	Incubation period (days)[a]
Influenza	1–2
Common cold	1–3
Bronchiolitis, croup	3–5
Acute respiratory disease (adenoviruses)	5–7
Dengue	5–8
Herpes simplex	5–8
Enterovirus disease	6–12
Poliomyelitis	5–20
Measles	9–12
Smallpox	12–14
Chickenpox	13–17
Mumps	16–20
Rubella	17–20
Mononucleosis	30–50
Hepatitis A	15–40
Hepatitis B and C	50–150
Rabies	30–100
Papilloma (warts)	50–150
AIDS	1–10 yr

[a]Until first appearance of prodromal symptoms.

Virions that can tolerate many amino acid substitutions in their structural proteins and remain infectious are said to have **structural plasticity** (e.g., influenza virus and human immunodeficiency virus). If viral replication occurs in the presence of antibody that neutralizes virions, antibody-resistant mutants are selected and shed. These mutant virions can reinfect individuals who were immune to the initiating virus. In contrast, there are other virions that cannot tolerate many amino acid changes in their structural proteins (e.g., those of poliovirus, measles virus, and yellow fever virus). For these viruses, even if the mutation rate is high, antibody-resistant virions have a low probability of being selected.

The principles underlying the selection and maintenance of antibody-resistant virions in natural infections are not well developed. Exceptions to rules such as structural plasticity are common. For example, the virions of rhinoviruses have remarkable structural plasticity, while those of poliovirus, a related picornavirus, do not. Over 100 different serotypes of rhinovirus are maintained in humans at all times, a property that accounts for the fact that individuals may contract more than one common cold each year. This fact also explains why it is difficult to produce a vaccine against the common cold. Why just three serotypes of poliovirus are circulating around the world is a mystery. Fortunately, this property ensures that the poliovirus vaccines that were effective in the 1950s are just as potent in the 21st century. Similarly, enveloped influenza

virus particles more resistant to antibodies are readily selected, while the enveloped particles of measles virus and yellow fever virus exhibit little variation in membrane protein amino acid sequence, and antibody-resistant variants are rarely observed. Consequently, an influenza vaccine is required every year, while a single measles virus vaccination lasts a lifetime.

Antigenic variation is the change of virion proteins in response to antibody selection. In an immunocompetent host, antigenic variation comes about by two distinct processes. **Antigenic drift** is the appearance of virions with slightly altered surface proteins (antigen) following passage in the natural host (Fig. 5.5). In contrast, **antigenic shift** is a major change in the surface protein(s) of a virion as genes encoding completely new surface proteins are acquired. This dramatic change in virion composition results when a host is coinfected with two viral serotypes. Viruses with segmented genomes can exchange segments, or coreplicating genomes can produce recombinant genomes. The new reassortant and recombinant viruses have exchanged blocks of genetic information, and the resulting hybrid virions may temporarily avoid immune defenses (Box 5.4; see also Chapter 10).

Acute Infections Present Common Public Health Problems

An acute infection is most frequently associated with serious epidemics of disease affecting millions of individuals every year (e.g., influenza and measles). The nature of an

Figure 5.5 Antigenic drift: distribution of amino acid residue changes in hemagglutinins (HA) of influenza viruses isolated during the Hong Kong pandemic era (1968 to 1995). The space-filling models represent the virus-receptor binding site (yellow) and the substituted amino acids (green). **(Left)** All substitutions in HAs of virions isolated between 1968 and 1995; **(middle)** amino acid substitutions that were retained in subsequent years; **(right)** amino acid substitutions detected in monoclonal antibody-selected variants of A/Hong Kong/68 HA. The α-carbon tracings of the HA1 and HA2 chains are shown in blue and red, respectively. Adapted from T. Bizebard et al., *Curr. Top. Microbiol. Immunol.* **260:**55–64, 2001, with permission.

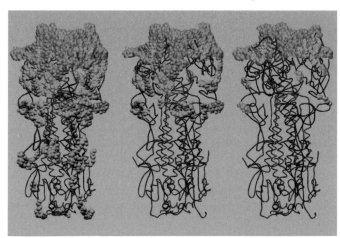

BOX 5.4	DISCUSSION

DISCUSSION
Recombination and antigenic shift during human immunodeficiency virus infections

While the contribution of antigenic shift to the dramatic influenza virus pandemics is well documented, the role of this process in human immunodeficiency virus pathogenesis is only now becoming appreciated. In this case, antigenic shift can occur only by recombination. The process requires packaging of two distinct genomes into a single particle for subsequent reverse transcription and copy-choice replication. At first glance, recombination appears to be an unlikely event. However, results of recent experiments indicate that recombination may be more frequent than expected. Splenocytes from human immunodeficiency virus-infected patients harbor as many as three or four distinct

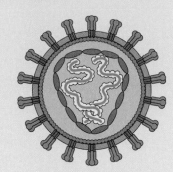

proviral genomes per cell, and give rise to huge numbers of recombinants. Antigenic shift arising from recombination can be thought of as a double-edged sword

for pathogen and host: virions can escape elimination by immune defenses with serious consequences to the host, but if the immune system is functional, CTLs and antibodies will recognize the new combinations, and broaden the immune repertoire against the infected cell.

Jung, A., R. Maier, J.-P. Vartanian, G. Bocharov, V. Jung, U. Fischer, E. Meese, S. Wain-Hobson, and A. Meyerhans. 2002. Multiply infected spleen cells in HIV patients. *Nature* **418**:144.

acute infection presents difficult problems for physicians, epidemiologists, drug companies, and public health officials. The main problem is that by the time people feel ill or mount a detectable immune response, most acute infections are essentially complete, and the infection has spread to the next host. Such infections can be difficult to diagnose retrospectively, or to control in large populations or crowded environments (e.g., day care centers, military camps, college dormitories, nursing homes, schools, and offices). Effective antiviral drug therapy requires treatment early in the infection, often before symptoms are manifested, because by the time the patient feels ill, the infection has been resolved. Antiviral drugs can be given in anticipation of an infection, but this strategy demands that the drugs be affordable, safe, and free of side effects. Moreover, as we discuss in Chapter 8, our arsenal of antiviral drugs is very small, and drugs effective for most common acute viral diseases simply do not exist.

Persistent Infections

Definition and Requirements

Persistent infections occur when the primary infection is not cleared efficiently by the adaptive immune response. Instead, virions, proteins, and genomes continue to be produced or persist for long periods, often for the life of the animal. Virions may be produced continuously or intermittently for months or years, even in the face of an active immune response (Fig. 5.1). In some instances, viral

genomes remain after viral proteins are no longer detected. Distinctions are often made between persistent infections that are eventually cleared (**chronic infections**) and those that last the life of the host (**latent infections** or **slow infections**).

The persistent pattern is surprisingly common (Table 5.2), particularly for noncytopathic viruses. For example, some arenaviruses, such as lymphocytic choriomeningitis virus, are inherently noncytopathic in their natural hosts and maintain a persistent infection if the host cannot clear the infected cells. Other viral life cycles include a distinct noncytopathic phase in addition to a cytolytic phase. Epstein-Barr virus infections include alternative transcription and replication programs that maintain the viral genome in some cell types with no production of viral particles. In other cases, ubiquitous infections, such as those produced by adenoviruses, circoviruses, and human herpesvirus 7, persist uneventfully in most human populations. Adenoviruses can be isolated from lymphoid tissue, including adenoids and tonsils, in most respiratory infections, but cultured lymphoid cells do not support efficient viral replication. It is possible that delayed kinetics of infection and replication observed in these cells contribute to the long-term maintenance of adenovirus in a fraction of lymphoid cells. What is clear from these examples is that no single mechanism is responsible for establishing a persistent infection. However, one common theme does emerge: when viral cytopathic effects and host defenses are reduced, a persistent infection is likely.

Table 5.2 Some persistent viral infections of humans

Virus	Site(s) of persistence	Consequence(s)
Adenovirus	Adenoids, tonsils, lymphocytes	None known
Epstein-Barr virus	B cells, nasopharyngeal epithelia	Lymphoma, carcinoma
Human cytomegalovirus	Kidneys, salivary gland, lymphocytes,[a] macrophages,[a] stem cells,[a] stromal cells[a]	Pneumonia, retinitis
Hepatitis B virus	Liver, lymphocytes	Cirrhosis, hepatocellular carcinoma
Hepatitis C virus	Liver	Cirrhosis, hepatocellular carcinoma
Human immunodeficiency virus	CD4+ T cells, macrophages, microglia	AIDS
Herpes simplex virus types 1 and 2	Sensory and autonomic ganglia	Cold sore, genital herpes
Human T-lymphotropic virus types 1 and 2	T cells	Leukemia, brain infections
Papillomavirus	Skin, epithelial cells	Papillomas, carcinomas
Polyomavirus BK	Kidneys	Hemorrhagic cystitis
Polyomavirus JC	Kidneys, central nervous system	Progressive multifocal leukoencephalopathy
Measles virus	Central nervous system	Subacute sclerosing panencephalitis, measles inclusion body encephalitis
Rubella virus	Central nervous system	Progressive rubella panencephalitis
Varicella-zoster virus	Sensory ganglia	Zoster (shingles), postherpetic neuralgia

[a] Proposed, but not certain.

An Ineffective Intrinsic or Innate Immune Response Can Promote a Persistent Infection

Primary infections are cleared by the adaptive immune response, but this action can occur only through close integration with intrinsic and innate defenses. If such early action and communication among defense systems does not occur, the host may die as the infection spreads out of control. Alternatively, the infection may persist. For example, apoptosis is a common intrinsic cellular defense that can limit or expand viral replication and spread. In some vertebrate cell lines, Sindbis virus infection is acute and cytopathic because apoptosis is induced. However, if the host Bcl2 protein blocks apoptosis, a persistent infection is established. Similarly, Sindbis virus causes a persistent infection of cultured postmitotic neurons because these cells are intrinsically resistant to virus-induced apoptosis. The *in vitro* studies are recapitulated in host animals: when Sindbis virions are injected into an adult mouse brain, a persistent noncytopathic infection is established. In contrast, when the same preparation is injected into neonatal mouse brains, the infection is cytopathic and lethal, because neonatal neurons lack the gene products to block virus-induced apoptosis.

The host IFN response plays unexpected roles in establishing patterns of infection. For example, bovine viral diarrhea virus, a pestivirus in the *Flaviviridae* family, establishes a lifelong persistent infection in the vast majority of cattle around the world. Remarkably, persistently infected animals have no detectable antibody or T-cell responses to viral antigens. Cytopathic and noncytopathic strains have been useful in understanding how a persistent infection occurs. Infection of pregnant cattle with the noncytopathic biotype during the first 120 days of pregnancy results in birth of persistently infected calves. In contrast, infection of the fetus by cytopathic virus is contained quickly and eliminated. We now understand that the phenotype depends on a rapid IFN response by the fetus that clears the infection. In contrast, noncytopathic infection of fetal tissue does not stimulate production of IFN. Consequently, the adaptive immune system is not activated, and because infection does not kill cells, a persistent infection is established.

Modulation of the Adaptive Immune Response Perpetuates a Persistent Infection

Interference with Production and Function of MHC Proteins

The CTL response is one of the most powerful adaptive host defenses against viral infection. This response depends, in part, upon the ability of host T cells to detect viral antigens present on the surfaces of infected cells, and to kill them. Recognition requires the presentation of viral peptides by major histocompatibility complex (MHC) class I proteins. The pathway by which endogenous peptide antigens are produced and transported to the cell surface is discussed in Chapter 4 (see Fig. 4.13). Obviously, any mechanism that prevents viral peptides from binding to MHC class I molecules, even transiently, provides a potential selective advantage. The production of MHC class I

proteins, as well as costimulatory molecules present on the surfaces of infected cells, by adding ubiquitin to the cytoplasmic domains. This modification stimulates endocytosis of the marked proteins. The related MK3 protein is also an E3 ubiquitin ligase, but in this case, ubiquitinylation promotes retrotranslocation and proteosomal destruction of MHC class I proteins soon after they appear in the endoplasmic reticulum. The genome of myxoma virus encodes a similar RING finger E3 ligase called MV-LAP that also directs proteasomal destruction by a mechanism analogous to that used by the K5 protein. Importantly, while the effects of K5 protein upon human infections cannot be assessed, myxoma virus mutants that lack the MV-LAP gene are markedly attenuated in rabbits (the natural host).

Early observations indicated that Epstein-Barr virus-infected individuals do not produce CTLs capable of recognizing the viral protein EBNA-1. This phosphoprotein is found in the nuclei of latently infected cells, and is regularly detected in malignancies associated with the virus (see Tables 5.4 and 5.5). T cells specific for other Epstein-Barr virus proteins are made in abundance, indicating that EBNA-1 must possess some special features. Indeed, this protein contains an amino acid sequence with a remarkable activity that renders it invisible to the host proteasome so that relevant EBNA-1 peptides are not produced at all. This inhibitory sequence can be fused to other proteins to inhibit their processing and the subsequent presentation of peptide antigens normally produced from them. The biological relevance of this mechanism is evident after acute infection of B cells. The adaptive immune system kills all productively infected cells, sparing only rare cells that produce EBNA-1. These cells harbor a latent viral genome.

MHC Class II Modulation after Infection

In the exogenous pathway of antigen presentation, proteins are internalized and degraded to peptides that can bind to MHC class II molecules (Fig. 4.14). These complexes are transported to the cell surface, where they can be recognized by the CD4+ T-cell receptor. Activated CD4+ T-helper (Th) cells stimulate the development of CTLs and help coordinate an antiviral response to the pathogen. Any viral protein that modulates the MHC class II antigen presentation pathway would therefore interfere with Th-cell activation.

Many viral gene products modulate the MHC class II pathways. For example, the human cytomegalovirus US2 protein has been reported to promote proteasomal destruction of the class II DR-alpha and DM-alpha molecules. The Epstein-Barr virus BZLF2 protein interacts with intracellular and cell surface MHC class II molecules to block T-cell activation. Herpes simplex virus strain KOS

infection results in removal of the MHC class II complex from the endocytic compartment. The human immunodeficiency virus Nef protein blocks the appearance of CD4 and MHC molecules on the cell surface. In endosomal compartments, Nef may interfere with acidification, affecting the loading of antigenic peptides onto MHC class II proteins.

Bypassing Deadly CTLs by Mutation of Immunodominant Epitopes

The Th and CTL populations found after some infections are surprisingly limited: the T cells respond to very few viral peptides. These peptides are said to be **immunodominant**. An extreme example of a limited CTL response is observed after infection of C57BL/6 mice with herpes simplex virus type 1. The virus-specific CTLs respond **almost entirely to a single peptide** in the viral envelope protein gB (the amino acid sequence of this peptide is SSIEFARL). Given that there are more than 85 open reading frames in the viral genome, it is remarkable that CTLs recognize only one peptide in this particular animal model of infection.

Focus of the T-cell response upon a small repertoire of viral peptides provides a ready opportunity to bypass T-cell recognition. A limited number of mutations in the coding sequence for these immunodominant peptides will render the infected cell invisible to the T-cell response produced early in infection. Viruses with these mutations are called **CTL escape mutants** and are thought to contribute to progressive accumulation of virus particles because of decreased clearing of infected cells. For example, CTL escape mutants, which are of central importance in human immunodeficiency virus pathogenesis, arise because of error-prone replication and the constant exposure to an activated immune response. In some well-documented cases, the T-cell peptide sequence is completely deleted from the viral protein synthesized by the CTL escape mutant. Understanding how immunodominant peptides are selected, maintained, and bypassed is essential if effective vaccines against human immunodeficiency virus are to be developed. For example, a vaccine directed toward a dominant T-cell peptide that is part of a critical structural motif in a viral protein may have value because CTL escape mutants will be less likely to survive and participate in subsequent spread of infection.

Immunodominant epitopes and CTL escape mutants play central roles in the increasingly common and dangerous infection caused by hepatitis C virus (more than 70 million people worldwide are infected [Appendix A]). The CTL response stimulated by acute infection is effective in less than 20 to 30% of individuals. An insidious persistent infection remains in the vast majority of patients.

After several years, this persistent infection can lead to serious liver damage and even fatal hepatocellular carcinoma. Persistently infected chimpanzees harbor viruses with CTL escape mutations in their genomes. In contrast, the viral population isolated from animals that resolved their acute infections rapidly included no such mutants. The conclusions from this work are clear: if CTL escape mutations occur early, a persistent infection is likely. If CTLs resolve the infection before escape mutants appear, no persistent infection is possible.

The CTL epitope need not be deleted or radically altered to escape CTL recognition. Indeed, when a T cell specific for a given viral peptide engages a similar, **but not identical**, peptide complexed to MHC class I, the T cell may respond partially or not at all. In this case, both mutant and parent viruses are likely to be maintained in the population. Altered viral peptides of this kind have been identified in viral isolates from persistent hepatitis B infections.

Killing Activated T Cells

Sometimes, when the CTL engages an infected cell, the CTL dies instead of its target. This unexpected turn of events is a remarkable example of viral defense. Activated T cells produce a membrane receptor called Fas on their surfaces. Fas is related to the Tnf family of membrane-associated cytokine receptors, and binds a membrane protein called Fas ligand (FasL). When Fas on activated T cells binds FasL on target cells, the receptor trimerizes, triggering a signal transduction cascade that results in apoptosis of the T cell. If viral proteins increase the quantity of FasL on the cell surface, any T cell (Th cell or CTL) that binds will be killed. Such a mechanism has been suggested to explain the relatively high frequency of "spontaneous" T-cell apoptosis in human immunodeficiency virus-infected patients. The viral Nef, Tat, and SU proteins, human T-lymphotropic virus Tax protein, and the human cytomegalovirus IE2 protein have all been implicated in increased production of FasL and resulting T-cell apoptosis. This seemingly unusual mechanism to kill T cells is not unique to viral infections. If it were, evolution would have removed the Fas system long ago. Indeed, Fas-mediated CTL killing is a normal activity in most complex organisms: its function is to remove T cells when they are no longer needed after infection, or when their presence in a tissue is detrimental. For example, certain delicate and irreplaceable tissues, such as the eye, remain free of potentially destructive T cells by maintaining a high concentration of FasL on cell surfaces.

Persistent Infections May Be Established in Tissues with Reduced Immune Surveillance

Cells and organs of the body differ in the degree of their immune defense. When virus particles infect such tissues, a persistent infection may be established if the organ is not otherwise compromised. Tissues with surfaces exposed to the environment (e.g., skin, glands, bile ducts, and kidney tubules) are exposed routinely to foreign matter, and therefore have a higher threshold for activating immune defenses. Persistent infections by members of the *Papillomaviridae* and the *Betaherpesvirinae* are common in these tissues. By replicating in cells present on lumenal surfaces of glands and ducts with poor immune surveillance (kidneys, salivary glands, and mammary glands), human cytomegalovirus particles are shed almost continually in secretions. Possibly the most extreme example of immune avoidance is represented by the papillomaviruses that cause skin warts. Productive replication of these infectious particles occurs only in the outer, terminally differentiated skin layer, where an immune response is impossible. Dry skin is continually flaking off, ensuring efficient spread of infection. This assertion can be verified by running a finger along a clean surface in the most hygienic hospital and noticing a white film, which is 70 to 80% keratin from human skin. Molecular biologists often discover this abundance of dried skin in the laboratory when examining silver-stained protein gels; the major band is often contaminating human keratin.

Certain compartments of the body, such as the central nervous system, vitreous humor of the eye, and areas of lymphoid drainage, are devoid of initiators and effectors of the inflammatory response, simply because these tissues can be damaged by the fluid accumulation, swelling, and ionic imbalances that characterize inflammation. In addition, because most neurons do not regenerate, immune defense by cell death is obviously detrimental. Accordingly, persistent infections of these tissues are common (Table 5.2).

Persistent Infections May Occur When Cells of the Immune System Are Infected

Many viruses infect cells of the immune system. Such infected lymphocytes and monocytes migrate to the extremes of the body, providing easy transport of virions to new sites of replication and shed. If these cells die or become impaired in an acute infection, the immune response could be ineffective, and a persistent infection may ensue. In the simplest case, the degree of immune response deficit would be directly proportional to the type and number of immune cells that are infected. We discuss the far-reaching consequences of systemic immunosuppression by viral infection in Chapter 4.

Human immunodeficiency virus provides a powerful reminder of how complicated and dynamic an infection of immune system cells can be (see Chapter 6). The virus infects not only CD4+ Th cells, but also monocytes, dendritic cells, and macrophages, all of which can transport virions to lymph nodes, the brain, and other organs. At first glance, one might expect the immune system to crash within a few days of the initial infection. However, this catastrophe does not happen, primarily because immune cells are continuously replenished. The new cells subsequently are infected and die, but on average, the immune system remains functional and does so for years. As a result, an untreated infected individual continues to produce prodigious quantities of virus particles in the face of a highly engaged immune system. The early stages of disease are characterized by continuous immune activation, and not by immunosuppression. It is only at the end stage of disease, when viral replication finally outpaces replenishment of immune cells, that massive immunosuppression occurs. Only then does the virus, as well as other unrelated secondary infections, spread in uncontrolled fashion, and the patient succumbs.

Two Viruses That Cause Persistent Infections

Measles Virus

Measles virus, a member of the family *Paramyxoviridae,* is a common human pathogen with no known animal reservoir (Appendix A). The genome organization and replication strategy are similar to those of the rhabdovirus vesicular stomatitis virus (Volume I, Appendix). Measles is one of the most contagious human viruses, with about 40 million infections occurring worldwide each year, resulting in more than 250,000 deaths (predominantly of children). This number is likely to be a substantial underestimate of the global burden, given the difficulties of record keeping in some countries. Normally, a single infection protects the individual for life. Consequently, the virus is maintained only in populations sufficient to produce a large number of new susceptible hosts (children). Population geneticists calculate that the critical community size required to maintain measles virus in the population is between 300,000 and 500,000.

Cellular receptors for measles virus include the CD46 and CD150 proteins. While CD150 is the receptor for all tested strains, CD46 is the predominant receptor for vaccine- and laboratory-adapted viruses. Many measles virologists think that other viral receptors must exist to account for the broad cellular tropism. After primary replication, measles virus infects local monocytes and lymphoid cells that migrate to draining lymph nodes. After replication in these tissues, a small proportion of monocytes, B cells, and T cells are infected and enter the circulation. Secondary infections of lymph tissues result in a secondary viremia, and replication continues in the epithelial cells of the lungs and the mouth. The characteristic mouth lesions, called

Koplik's spots, are caused by a delayed-type hypersensitivity reaction, analogous to that responsible for the typical measles skin rash. The course of acute infection, so-called uncomplicated measles, runs for about 2 weeks (Fig. 5.7). An acute infection causes cough, fever, and conjunctivitis and confers lifelong immunity.

The vast majority of measles victims have an uneventful recovery, but a characteristic systemic immunosuppressive effect lasts for a week or two after the infection is resolved. Consequently, secondary infections by unrelated pathogens during this period may be uncontested by host defenses; the results may be serious or fatal if immediate intervention and care are not provided. The large number of children in the Third World who die after measles infection succumb to complications of secondary infections. Several mechanisms for immunosuppression have been proposed. Measles virus infects cells of the immune system, and this action may deregulate the immune response. Interleukin-12 is produced when Toll-like receptors on antigen-presenting cells are stimulated by infection. Infected T and B cells, as well as macrophages, are arrested in the late G_1 phase of the cell cycle and cannot perform their normal functions. Uninfected lymphocytes can also be affected by direct contact with viral membrane proteins present on the surface of infected cells.

On rare occasions, measles virus genomes and structural proteins are not cleared by the adaptive immune system, and may persist for years in an infected individual. The mechanisms responsible are only now being characterized. Some studies of humans infected with measles virus reported a positive correlation between antibody concentration and persistent infection. The significance of this correlation has been debated, but some insight has come from studying the effect of antibodies during infection of cultured cells. Measles virus-specific antibodies bind to viral membrane proteins present on cell surfaces to induce endocytosis and proteolysis of the antibody-protein complexes. Because one of these proteins, the viral fusion protein, is responsible for cell-cell fusion that causes cell death, exposure to antibodies effectively blocks this mechanism of cell killing, thereby allowing viral persistence in cultured cells.

An important finding is that measles virus can enter the brain in infected lymphocytes that traverse the body during the viremia following primary infection. Such a secondary infection of a tissue with reduced immune surveillance has a number of consequences. One is **acute postinfectious encephalitis**, which occurs in about 1 in 3,000 infections. The other is a rare, but delayed and often lethal, brain infection called **subacute sclerosing panencephalitis (SSPE)**. This disease is a manifestation of a slow infection, an unusual variation of a persistent infection (see "Slow

A

Pleomorphic particles
100–300 nm

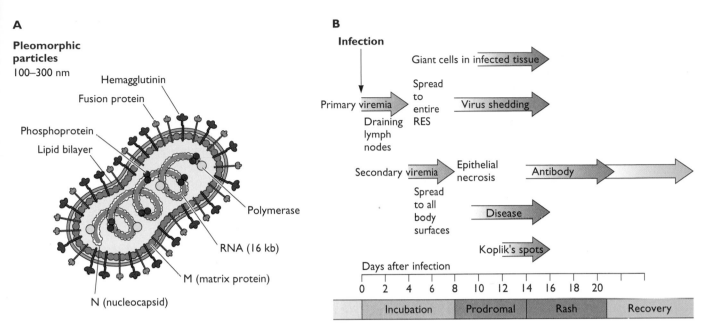

Figure 5.7 Infection by measles virus. (A) Diagrammatic representation of the structure of the pleomorphic measles virion. **(B)** Course of clinical measles infection and events occurring in the spread of the infection within the body. Four clinically defined temporal stages occur as infection proceeds (illustrated at the bottom). Characteristic symptoms appear as infection spreads by primary and secondary viremia from the lymph node to the entire reticuloendothelial system (RES) and finally to all body surfaces. The timing of typical reactions that correspond to the clinical stages is shown by the colored arrows. The telltale spots on the inside of the cheek (Koplik's spots) and the skin lesions of measles consist of pinhead-sized papules upon a reddened, raised area. They are typical of immunopathology produced in response to measles virus proteins. Redrawn from A. J. Zuckerman et al., *Principles and Practice of Clinical Virology*, 3rd ed. (John Wiley & Sons, Inc., New York, NY, 1994), with permission.

Infections: Sigurdsson's Legacy" below and Fig. 5.1). After young adults and children contract measles, about one in a million develop SSPE, with a 6- to 8-year incubation period. SSPE is more likely to occur if children are infected in their first year or two of life than if they are infected later. This disease begins in rare infected cells of the brain. In these cells, viral gene expression, especially synthesis of envelope proteins, is reduced. In addition, fully assembled particles cannot be detected in brains of afflicted patients. Alterations in the matrix (M) protein may lead to ineffective virion assembly. Even though particle assembly is not observed, nucleoprotein complexes are produced, and infectious genomes spread between synaptically connected neurons. The mechanism of such spread in the absence of assembled virions is of some interest because it does require the viral fusion protein, but not viral receptors.

We lack testable hypotheses that relate these provocative findings to the mechanism of persistent infection. More critical perhaps is that it is difficult, if not impossible, to test these ideas in human infections. Nevertheless, with appropriate animal models, we may be able to answer some fundamental questions. Does transient immunosuppression during an acute infection facilitate infection of the brain? Are defects in M protein synthesis and particle assembly necessary and sufficient to cause disease, or are they effects of other selection processes in the brain? Are the defects in viral gene expression the cumulative results of selection after years of exposure to host defenses? Transgenic mice that produce the human measles virus receptors are available, and should allow these important questions to be addressed in a rigorous and controlled fashion.

Lymphocytic Choriomeningitis Virus

Lymphocytic choriomeningitis virus, a member of the family *Arenaviridae*, was the first virus associated with aseptic meningitis in humans. Perhaps more importantly, in recent years its study has illuminated fundamental principles of immunology and viral pathogenesis, particularly those that underlie persistent infection and CTL recognition and killing. It was noted early on that the infection spreads from rodents (the natural host) to humans, in whom it can cause severe neurological and developmental damage (an example of a zoonotic infection [Chapter 10]). Infected rodents normally excrete large quantities of virions in feces and urine throughout their lives without any apparent detrimental effect. These mice are called "carriers" because of such lifelong production of virions. The carrier state is established for two reasons: infection is not cytopathic, and, if mice are infected congenitally or immediately after birth, viral proteins are not recognized as "foreign" and so the infection is not cleared by the immune system. However, if virions are injected into the brains of healthy adult mice, the mice die of acute immunopathological encephalitis. This disease is similar to that contracted by humans.

In the mouse model, CTLs are required both for clearing virus and for the lethal response to intracerebral infections. If adult mice are depleted of CTLs, injection of virus particles into the brain is no longer fatal. Instead, the mice produce virions throughout their lifetimes, precisely as seen in persistent infections of neonates. When virus-responsive CTLs are added back to persistently infected neonates, the infection is cleared after several weeks. In the case of neonatal persistent infection, the brain is not infected, so no encephalitis is promoted by the activated T cells. How the effector system is prevented from clearing the infection in these circumstances is currently under investigation. Recent experiments have implicated an active process of clonal deletion of T lymphocytes that are capable of recognizing the dominant lymphocytic choriomeningitis virus peptides.

Latent Infections

General Properties

Latent infections are characterized by three general properties: viral gene products that promote productive replication are not made, or are found only in low concentrations; cells harboring the latent genome are poorly recognized by the immune system; and the viral genome persists intact so that at some later time a productive infection can be initiated to ensure the spread of viral progeny to new hosts (Fig. 5.1). The latent genome can be maintained as a nonreplicating chromosome in a nondividing cell like a neuron (e.g., herpes simplex virus, varicella-zoster virus), or as an autonomous, self-replicating chromosome in a dividing cell (e.g., Epstein-Barr virus or cytomegalovirus), or be integrated into a host chromosome, where it is replicated in concert with the host genome (e.g., adeno-associated virus).

Such "long-term parking" of a viral genome in the latent infection is noteworthy for its stability: a balance among the regulators of viral and cellular gene expression must be maintained. There is no one mechanism to establish and maintain a latent infection, but one principle emerging is that epigenetic alterations of viral genomes play central roles. Generally, only a restricted set of viral gene products are found in latently infected cells. For neurons harboring

latent herpes simplex virus (an alphaherpesvirus), a unique RNA transcript, but no viral proteins are synthesized. In other cases, latently infected cells synthesize a small subset of viral proteins required for productive replication, a pattern exemplified by varicella-zoster virus, an alphaherpesvirus of humans (Fig. 5.4). Epstein-Barr virus (a gammaherpesvirus) latent infection of B cells is more complicated. At least nine viral proteins and small viral RNAs are required to support replication of the latent viral genome and modulate the host immune response (Table 5.4). For betaherpesviruses such as cytomegalovirus, viral micro-RNAs may function to establish a latent infection.

If latency is to have any value as a survival strategy, a mechanism for reactivation must exist so that infectious virions can spread to other hosts. Reactivation may be spontaneous or may follow trauma, stress, or other insults, conditions that may mark the host as unsuitable for continued latent infection.

Herpes Simplex Virus

Over three-quarters of all adults in the United States have antibodies to herpes simplex virus type 1 or 2 and harbor latent viral genomes in their peripheral nervous systems. Approximately 40 million infected individuals will experience recurrent herpes disease due to reactivation of their own personal viruses sometime in their lifetimes. Many millions more carry latent viral genomes in their nervous systems, but never report reactivated infections. Herpes simplex virus is an example of a well-adapted pathogen, as demonstrated by its widespread prevalence in humans, its only known natural hosts. However, why some people are more likely than others to be infected by this ubiquitous virus is poorly understood (Box 5.5). No animal reservoirs are known, although several laboratory animals, including rats, mice, guinea pigs, and rabbits, can be infected. The alphaherpesviruses, of which herpes simplex virus type 1 is the type species, are unique in establishing latent infections predominantly in terminally differentiated, nondividing neurons of the peripheral nervous system. Indeed, most other neurotropic viruses (e.g., rabies virus) initiate infections of peripheral tissue and spread to the central nervous system to cause devastating disease or death.

The Primary Infection

Herpes simplex virus infections usually begin in epithelial cells at mucosal surfaces (Fig. 5.8). Virions are released from the basal surface in close proximity to sensory nerve endings. Because sensory terminals are abundant, they are easily infected, but other autonomic nerve terminals also may be infected if deeper layers of the epithelium are involved. For example, endothelial cells of capillaries or cells surrounding hair follicles are in contact with sympathetic

Table 5.4   Epstein-Barr virus gene products synthesized in the latent infection

Gene product	Function
EBNA-1	Maintains replication of the latent Epstein-Barr virus genome during S phase of the cell cycle. It is a sequence-specific DNA-binding protein and binds to a unique origin of replication called *oriP* that is distinct from the origin used in the productive replication cycle.
EBNA-2	A transcriptional regulator that coordinates Epstein-Barr virus and cell gene expression in the latent infection by activating the promoters for the LMP-1 gene and cellular genes like CD23 (low-affinity immunoglobulin E Fc receptor) and CD21 (the Epstein-Barr virus receptor, CD23 ligand, and receptor for complement protein C3d)
EBNA-LP	Required for cyclin D2 induction in primary B cells in cooperation with EBNA-2
EBNA-3A and EBNA-3C	Play important roles early in the establishment of the latent infection
LMP-1	An integral membrane protein required to protect the latent infected B cell from the immune response. LMP-1 stimulates the synthesis of several surface adhesion molecules in B cells, a calcium-dependent protein kinase, and the apoptosis inhibitor Bcl2.
LMP-2	An integral membrane protein required to block the activation of the *src* family signal transduction cascade; an inhibitor of reactivation from latency. Two spliced forms exist: LMP-2A and LMP-2B; LMP-2B lacks receptor-binding domain and may act to modulate LMP-2A.
EBER-1 and EBER-2	Nonpolyadenylated, small RNA molecules that do not encode proteins; transcribed by RNA polymerase III; 166 and 172 nucleotides in length, respectively
miRNAs	At least 20 different miRNAs are processed from two viral transcripts, one set in the BART gene and one set near the BHRF1 cluster

BOX 5.5

DISCUSSION

The hygiene hypothesis: why people vary in susceptibility to herpes simplex virus infection

More than 80% of the adult population in the developed world harbor latent herpesviral genomes in their peripheral nervous system. Some individuals suffer from lesions after reactivation while some never report symptoms or lesions. What accounts for the high infectivity yet marked diversity in host response to infection?

Hypothesis: The effectiveness of innate immunity to stimulate appropriate adaptive immune responses (Th1 versus Th2) is conditioned by the individual's exposure to microbes early in life. In the extreme case, early life in a highly sanitized environment leads to reduced stimulation of innate immunity. One consequence is inadequate development of the adaptive immune system and reduced capacity to control infections later in life.

According to this hypothesis, the rising incidence of allergy and asthma, as well as of herpes simplex virus infections, in Western societies results from

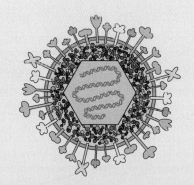

"hypersanitized" living conditions. Such conditions include use of sterilized baby food, excessive application of germicidal soaps and cleaners, and limited exposure of newborns to relatives and friends of the family. Individuals who had limited exposure to microbes in early life will experience more reactivations of latent herpesvirus with severe symptoms because of their inability to mount an effective Th1-dominated response. Instead, with

inadequate early stimulation of innate immunity by microbial infections, subsequent exposure to foreign antigens may stimulate an inappropriate Th2 response. Testing the hygiene hypothesis is not an easy matter; many observations that apparently support or refute the hypothesis are anecdotal or poorly controlled. Nevertheless, the idea has stimulated considerable research and debate.

Camateros, P., J. Moisan, J. Henault, S. De, E. Skamene, and D. Radzioch. 2006. Toll-like receptors, cytokines and the immunotherapeutics of asthma. *Curr. Pharm. Des.* **12:**2365–2374.

Rouse, B. T., and M. Gicrynska. 2001. Immunity to herpes simplex virus: a hypothesis. *Herpes* **8**(Suppl. 1):2A–5A.

Strachan, D. 1989. Hay fever, hygiene, and household size. *Br. Med J.* **299:**1259–1260.

Zock J., E. Plana, D. Jarvis, et al. 2007. The use of household cleaning sprays and adult asthma: an international longitudinal study. *Am. J. Respir. Crit. Care Med.* **176:**735–741.

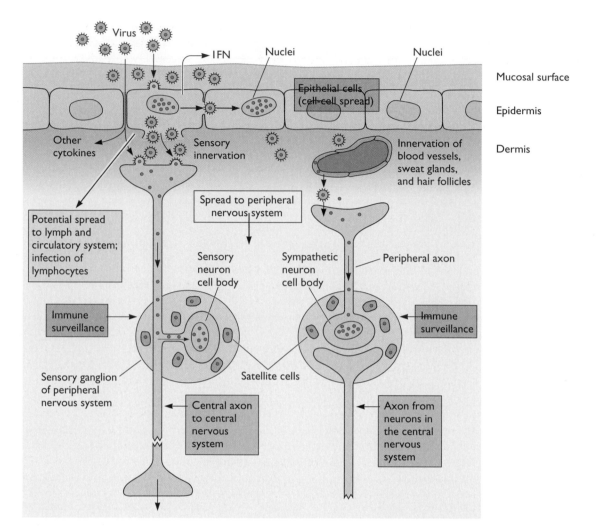

Figure 5.8 Herpes simplex virus primary infection of sensory and sympathetic ganglia. Viral replication occurs at the site of infection, usually in the mucosal epithelium; the infection may or may not manifest as a lesion. Host intrinsic and innate defenses, including IFN and other cytokines, normally limit the spread of infection at this stage. Virions may infect local immune effector cells, including dendritic cells and infiltrating natural killer cells. The infection also spreads locally between epithelial cells and may spread to deeper layers to engage fibroblasts, capillary endothelial cells, sweat glands, and other dermal cells such as those present in piloerector muscles around hair follicles. Particles that are released from basal surfaces infect nerve terminals in close contact. These axon terminals can derive from sensory neurons in dorsal root ganglia (left) or from autonomic neurons in sympathetic ganglia (right). Viron envelopes fuse with neuron axonal membranes, and the nucleocapsid with outer tegument proteins is transported within the axon to the neuronal cell body by microtubule-based systems (dynein motors), where it delivers the viral DNA to the nucleus. Spread of productive infection to the central nervous system from these peripheral nervous system ganglia is rare. Unlike the brain and spinal cord, peripheral nervous system ganglia are in close contact with the bloodstream and are exposed to lymphocytes and humoral effectors of the immune system (immune surveillance). Consequently, infected ganglia become inflamed and populated with lymphocytes and macrophages. Infection of the ganglion is usually resolved within 7 to 14 days after primary infection, virus particles are cleared, and a latent infection of some neurons in the ganglion is established (see also Fig. 5.9).

nerve endings and also may be exposed to virus. If infection occurs in the eye or other facial tissues, parasympathetic and cranial nerve endings may be invaded. Fusion of the virion envelope with any of these nerve endings releases the nucleocapsid with inner tegument proteins into the

axoplasm. Dynein motors then move the internalized nucleocapsid on microtubules over long distances to the particular neuronal cell bodies that innervate the infected peripheral tissue. A productive infection may be initiated in these neurons when the viral DNA enters the nucleus.

Establishment and Maintenance of the Latent Infection

Soon after this acute infection begins in neurons, the viral genome is silenced and coated with nucleosomes (Volume I, Chapter 8). The nucleosome-covered viral genome is tethered in some fashion in the nucleus to cellular chromatin. Only limited transcription occurs, and a quiescent, latent infection is established. As we will see, the establishment of this latent state is likely to depend on both viral regulatory proteins and RNA, as well as the intrinsic and innate immune defenses that protect these tissues.

In general, most neurons neither replicate their DNA nor divide, and so once a silenced viral genome is established in the nucleus, no further viral replication is required for it to persist. Standard antiviral drugs and vaccines are not able to cure a latent infection. Consequently, latency is absolute persistence, or, as one herpesvirologist put it, "herpes is forever."

Despite many years of study, many details of the molecular aspects of herpes simplex virus latency await discovery and explanation, but the general pathway is well established. The outline of possible regulatory steps necessary for the establishment, maintenance, and reactivation of a viral infection is shown in Fig. 5.9. A typical primary infection of a mouse, showing the time course of production of infectious virus and establishment of a latent infection, is illustrated in Fig. 5.10.

An often unappreciated fact is that in several animal models, peripheral ganglia undergo a rather robust acute infection with substantial production of virions followed by a strong inflammatory response. Nevertheless, after 1 or 2 weeks, infectious particles can no longer be isolated from the ganglia, the operational definition of an established latent infection. If the animal survives the primary infection, establishment of the latent infection is inevitable. The time frame for this process varies depending on the animal species, the concentration and genotype of the infecting virus, and the site of primary infection. Inflammatory cells may persist in the latently infected ganglia for months or years, perhaps as a result of continuous or frequent low-level reactivation and production of viral proteins in latently infected tissue.

Many questions remain. We do not understand why neurons are the favored site for a latent infection. Under particular laboratory conditions, it is possible to establish a quiescent infection in nonneuronal cells, but these conditions apparently are not available in natural infections. It is difficult to understand how neurons in ganglia survive the primary infection by this markedly cytolytic virus. Evidence suggests that the productive-cycle gene expression pathway is turned off **after** it has started and cells are subsequently purged of infection by local innate defenses. In addition, we do not understand why the infection stops in the first-order neurons of the peripheral nervous system and rarely spreads to the central nervous system, which is in direct synaptic contact with peripheral neurons.

The Latency-Associated Transcripts

Many latently infected neurons synthesize RNA molecules termed **latency-associated transcripts** (LATs) (discussed in Volume I, Chapter 8). Some researchers argue that all latently infected neurons synthesize LATs, while others report that only 5 to 30% do so. As in many studies of the herpes simplex latent state, the results depend on the animal model.

After infection of rabbits, herpes simplex virus type 1 mutants that do not synthesize LATs establish a latent infection, but spontaneous reactivation is markedly reduced. Despite this important finding, identifying functions for the LATs continues to be a challenge. The major LAT contains two prominent open reading frames with potential to encode two proteins, but there is little evidence that these proteins are produced during latency. Moreover, disruption of these open reading frames induces no latency phenotypes, and the sequences are not conserved in the closely related herpes simplex type 2 genome.

Remarkably, and in contrast, human ganglia latently infected with varicella-zoster virus, a distantly related alphaherpesvirus, do not synthesize a single LAT. Rather, at least five distinct viral transcripts are found. Despite considerable effort, scientists cannot ascribe functions of these viral transcripts or the corresponding proteins in either the establishment or the maintenance of the varicella-zoster latent infection. Suffice it to say that while alphaherpesvirus latency in neurons is a common feature of these viruses, it is probably not achieved by a single mechanism.

If the herpes simplex virus LATs are not translated, then the RNA molecules themselves may have biological activity. One idea is that they are micro-RNA precursors leading to degradation or reduced translation of host messenger RNAs (mRNAs). Micro-RNAs may be one common feature of herpesvirus latency systems, as they now are suspected to be important for latent infections by the betaherpesviruses and gammaherpesviruses. Another idea is that the herpes simplex virus type 1 LATs block apoptosis upon primary infection of neurons (or upon reactivation). Evidence exists indicating that they maintain the latent state through antisense inhibition of translation of ICP0 (a crucial transcription activator). A hypothesis with some support is that herpes simplex virus type 1 LATs mediate the transition to latency by altering chromatin structure, perhaps by a process similar to mammalian X-chromosome inactivation by the Xist RNA.

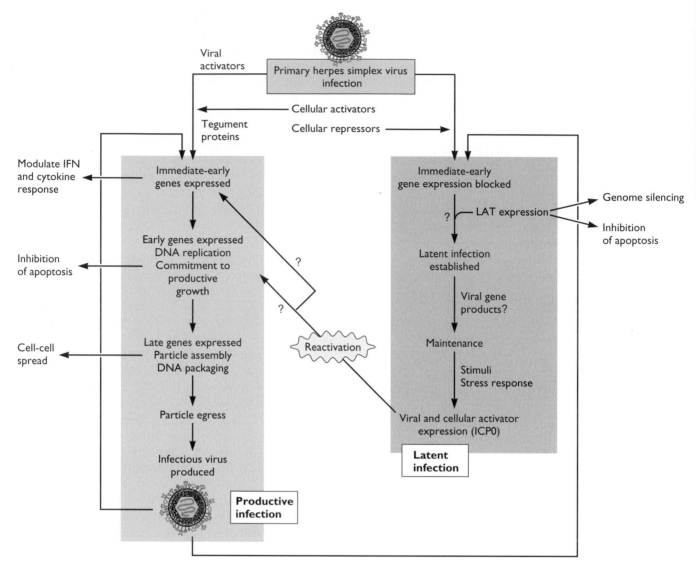

Figure 5.9 General flowchart for establishment, maintenance, and reactivation of a latent infection by herpes simplex virus. The green box at the top indicates the primary infection by virus particles at mucosal surfaces. The productive infection is shown by the pathway on the left, and the latent infection is indicated by the pathway on the right. The question marks indicate our lack of knowledge concerning synthesis and function of viral proteins at the indicated steps. Infectious particles produced by the productive pathway may infect other cells and enter either the productive or latent pathway as indicated. Infection can also spread from cell to cell without release of particles. Apoptosis induced by infection is inhibited by viral gene products. In addition, antiviral effects of IFN and other cytokines are modulated by viral gene products. The contribution of these processes in establishing the latent infection is not well understood. Reactivation is indicated by the diagonal arrow from the latent state to the start of the productive infection. The question marks note the current controversy as to whether reactivation requires "return to go" (immediate-early gene expression) or "start in the middle" (expression of early genes required for DNA replication). Experimental data indicate that synthesis of the immediate-early protein ICP0 is sufficient to activate latent infection. Adapted from M. A. Garcia-Blanco and B. R. Cullen, *Science* **254:**815–820, 1991, with permission.

Larger Numbers of Nonneuronal Cells than Neurons in Peripheral Ganglia

While it is commonplace to focus on neurons in this pattern of infection by herpes simplex virus, only 10% of the cells in a typical sensory ganglion are neurons; the remaining 90% are nonneuronal satellite cells and Schwann cells associated with a fibrocollagenous matrix. These nonneuronal cells are in intimate contact with ganglionic neurons. Some of the nonneuronal cells are infected during initial invasion of the ganglion, and may

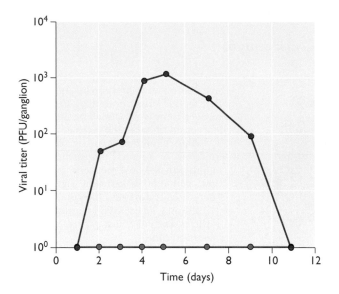

Figure 5.10 Replication of infectious herpes simplex virus type 1 in mouse trigeminal ganglia during acute infection. Mice were anesthetized and infected by a standard strain of herpes simplex virus by dropping approximately 10^5 plaque-forming units (PFU) onto the cornea of one eye that had been lightly scratched with a sterile needle. After a few minutes, the liquid was blotted and the animal was allowed to recover. At selected time points, animals were euthanized and the trigeminal ganglia were removed quickly and frozen. Each point on the graph (red line) represents the geometric mean titer in PFU from eight individual ganglia from two different experiments tested at the indicated time after infection. Uninfected animal controls are indicated by the blue line.

be the major source of infectious particles isolated from infected ganglia. These infected support cells also produce prodigious amounts of cytokines that can promote an antiviral response in the entire ganglia. In addition, in contrast to the brain and spinal cord, the peripheral nervous system is accessible to antibodies, complement, cytokines, and lymphocytes of the innate and adaptive immune system.

The intimate contact of peripheral neurons with epithelial cells (sites of primary infection) enables movement of virus particles in and out of the nervous system without exposure to circulating antibodies. Murine models demonstrate efficient establishment of latency in neurons even in the face of a robust antibody response in vaccinated animals or in animals that receive passive immunization with virus-specific antibodies prior to infection. This rather curious twist of immune avoidance presents extreme difficulties to those who strive to produce alphaherpesvirus vaccines for humans.

Reactivation

After reactivation of a latent infection in sensory ganglia, virions appear in the mucosal tissues innervated by that particular ganglion. This outcome is an effective means of ensuring transmission of virions after reactivation, because mucosal contact is widespread among affectionate humans. However, two apparently contradictory facts should be obvious. First, reactivation takes place **in an immune individual**, and second, an individual must be actively producing sufficient virions to infect another person. Both facts are true, and the contradiction is explained by the simple fact that the immune response reacts more slowly than shedding of infectious virions. The spread of infection among epithelial cells after reactivation may be facilitated by action of the viral protein ICP47. This protein blocks MHC class I presentation of viral antigens to T cells. Such activity may provide sufficient time for a few rounds of replication before elimination of the infected cell by activated CTLs.

The immune response after reactivation is usually robust and clears the infected epithelial cells in a few days, but not before virions are shed. The typical "cold sore" lesion of herpes labialis is the result of the inflammatory immune response attacking the infected epithelial cells that were in contact with axon terminals of reactivating neurons. Some individuals with latent herpes simplex virus experience reactivation every 2 to 3 weeks, while others report only rare or no episodes of reactivation. Importantly, reactivation may result in shedding infectious virions in the absence of obvious lesions or symptoms (Fig. 5.1).

Reactivation: Not "All or None"

The signaling mechanisms that reactivate the latent infection are under active study. Sunburn, stress, nerve damage, depletion of nerve growth factor, steroids, heavy metals, and trauma (including dental surgery) all promote reactivation. Despite the apparent systemic nature of most reactivation stimuli, when reactivation does occur in animal models, only about 0.1% of neurons in a ganglion containing the viral genome synthesize viral proteins and virions. Multiple levels of regulation must be operating, and the overwhelming thrust must be to maintain the latent state. The regulatory network employed is not an "on or off" circuit affecting all latently infected neurons. Its nonlinear response may be the result of some nonuniformity within the latent population, or of the signal transduction process. Not only are different types of neurons infected in peripheral ganglia, but also the number of viral genomes in a given neuron varies dramatically (Box 5.6). It is likely that one facet of competency for reactivation is the number of viral genomes within a given neuron: more genomes, more likely to reactivate.

Signaling Pathways in Reactivation

At first glance, the diversity of potential reactivation signals may be surprising. However, it is likely that they all converge to stimulate production or action of specific

BOX 5.6

EXPERIMENTS
Neurons harboring latent herpes simplex virus often contain hundreds of viral genomes

The number of neurons in a ganglion that will ultimately harbor latent genomes following primary infection depends upon the host, the strain of virus, the concentration of infecting virions, and the conditions at the time of infection. A mouse trigeminal ganglion contains about 20,000 neurons. It is possible to infect as few as 1% to as many as 50% of the neurons in a ganglion. In controlled experiments with mice, the number of latently infected neurons increases as the titer of infecting virions increases.

Many infected neurons contain multiple copies of the latent viral genome, varying from fewer than 10 to more than 1,000; a small number have more than 10,000

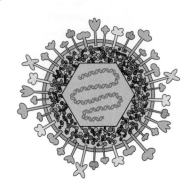

copies. This variation in copy number has been enigmatic. Does it reflect multiple infections of a single neuron, or is it the result of replication in a stimulated permis-

sive neuron after infection by one particle? If it is the latter, how does the neuron recover from what should be an irreversible commitment to the productive cycle?

When viral replication is blocked by mutation or antiviral drugs, the number of latently infected neurons with multiple genomes is reduced significantly. Therefore, in a natural infection, a single neuron may be infected by multiple virus particles, each of which participates in the latent infection.

Sawtell, N. M. 1997. Comprehensive quantification of herpes simplex virus latency at the single-cell level. *J. Virol.* **71:**5423–5431.

---

cellular proteins needed for transcription of the herpes simplex virus immediate-early genes and consequently activate the productive transcriptional program. Indeed, all of these exogenous signals have the capacity to induce the synthesis of cell cycle and transcription regulatory proteins that may render neurons permissive for viral replication. It is known that synthesis of the viral immediate-early protein ICP0 is sufficient to reactivate a latent infection in model systems (Fig. 5.9). The ICP0 protein and LATs appear to have opposing functions in modulating chromatin structure, leading to active transcription or gene silencing, respectively. In a single latently infected neuron, reactivation may be an all-or-none process requiring but a single reaction such as chromatin structural changes to "flip the switch" that triggers the cascade of gene expression of the productive pathway. Glucocorticoids are excellent examples of such activators, as they stimulate transcription rapidly and efficiently while inducing an immunosuppressive response. These properties explain the observation that clinical administration of glucocorticoids frequently results in reactivation of latent herpesvirus.

Latent infections may reactivate in the absence of obvious stress or activation signals. Such spontaneous reactivation appears to result from random, low-concentration signals that impinge on the neurons from the circulation (cytokines) or from the innervated tissue (local trauma and consequent nerve firing). These stress signals are sensed by individual neurons and lead to a small burst of viral transcription in only one or a few neurons. When the stimulus is strong enough, such sporadic transcrip-

tion ultimately passes a threshold, resulting in replication and reactivation. This idea is consistent with the very low levels of immediate-early and early transcripts that can be detected in latently infected ganglia. In turn, the immune system would be stimulated constantly by these low-level, nonproductive events. One thought is that the continuously activated immune system would be able to provide rapid control of massive reactivation, should it occur.

One question remains unanswered: what prevents an apparently systemic reactivation signal from turning on **all** latently infected neurons? Why are so few neurons activated at any given time even when glucocorticoids or trauma must affect all ganglia? One idea is that massive reactivation of latency would be met instantly with a lethal immune response. Killing all your sensory or sympathetic neurons is unlikely to be a selective force in evolution! The selective advantage may be that by ensuring reactivation in only a small fraction of latently infected neurons, the functional life of these important peripheral nervous system tissues is preserved so that the host can survive. This hypothesis begs the fundamental question of how reactivation of all latently infected neurons in an individual is avoided.

Epstein-Barr Virus

Epstein-Barr virus is named after Michael Epstein and Yvonne Barr, who, along with Bert Achong, discovered it in 1964. Epstein-Barr virus, also called human herpesvirus type 4, is the type species of the gamma subfamily of herpesviruses. It is one of the most common viral infections of humans (its only host). Indeed, in the United States, up to 95% of adults between the ages of 35 and 40 are seropositive and carry the viral genome in latently infected B cells. Two strains of Epstein-Barr virus are recognized that differ in their terminal internal repeats, as well as in production of nuclear antigens and small RNAs during the latent infection. Epstein-Barr virus 1 is about 10 times more prevalent in the United States and Europe than is Epstein-Barr virus 2, while both strains are equally represented in Africa. Most people are infected with the virus

Figure 5.11 Epstein-Barr virus primary and persistent infection. (Left) Primary infection. Epstein-Barr virus infects epithelial cells in the oropharynx (e.g., the tonsils). Virions produced can infect resting B cells in the lymphoid tissue. Virus-infected B cells express the full complement of latent viral proteins and RNAs and are stimulated to enter mitosis and replicate. They produce antibody and function as B cell blasts. The latently infected B cells are attacked by natural killer cells and CTLs. **(Right)** Persistent infection. Most infected B cells are killed as a result of innate and immune defenses, but a few (approximately 1 in 100,000) persist in the blood as small, nonproliferating memory B cells that synthesize only LMP-2A mRNA. These memory B cells are presumably the long-term reservoir of Epstein-Barr virus *in vivo* and the source of infectious virus when peripheral blood cells are removed and cultured. A limited immune response to these infected B cells leads to self-limiting proliferation, infectious mononucleosis, or unlimited proliferation (polyclonal B-cell lymphoma). When stimulated or propagated in culture, viral proteins needed to replicate and maintain the viral genome are again produced. Some latently infected B cells traffic to lymphoid tissues in close proximity to epithelial cells in the oropharynx. Here the B cells are stimulated to produce virions capable of infecting and replicating in epithelial cells. Infectious virions are produced and shed into the saliva for transmission to another host.

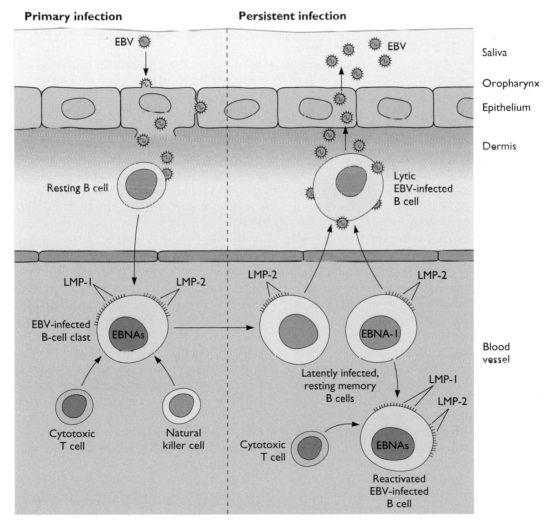

Table 5.5   Diseases associated with Epstein-Barr virus infections[a]

Disease	Characteristics
Acute infection	
Infectious mononucleosis	The best-known clinical presentation of infection; resolves in 1–2 wk, but fatigue symptoms may last longer
Oral hairy leukoplakia	Primary infection leading to a wartlike lesion of epithelial cells of the tongue seen in AIDS patients and transplant recipients
Abnormalities of latent infection, lymphoproliferative disorders, and malignancies	
B-lymphoproliferative disease	Frequently observed in individuals experiencing a primary viral infection following tissue transplantation; initially benign; if untreated, can lead to B-cell lymphoma
X-linked lymphoproliferative syndrome	Certain males have X-linked mutations that lead to a severe immunodeficiency after primary viral infection
Burkitt's lymphoma	The most common childhood cancer in equatorial Africa; cells from Burkitt's B-cell lymphomas exhibit a reciprocal translocation involving the *c-myc* locus on the long arm of chromosome 8 and one of the immunoglobulin loci on chromosome 2, 14, or 22; newly explanted B cells from tumors produce only EBNA-1
Hodgkin's disease	Mixed cells in tumor; 1–2% are malignant, and the remaining cells are infiltrating lymphocytes; association of virus with Hodgkin's lymphomas varies with geography; newly explanted B cells from tumors produce EBNA-1, LMP-1, and LMP-2A
Nasopharyngeal carcinoma	A cancer of epithelial cells and one of the most common cancers in China; tumor cells produce EBNA-1, LMP-1, and LMP-2, but not EBNA-2

[a]Data from G. C. Faulkner et al., *Trends Microbiol.* **8:**185–189, 2000.

at an early age and have no symptoms, but some develop **infectious mononucleosis** ("mono").

Epstein-Barr virus establishes latent infections in B lymphocytes (Fig. 5.11). It is one of the human herpesviruses consistently associated with human cancers (Table 5.5; Appendix A). As we will learn in Chapter 7, such viral oncogenesis is a by-product of the mechanisms by which a latent infection is established. In contrast to the nonpathogenic latent state of herpes simplex virus, the latent state of Epstein-Barr virus is implicated in several important diseases.

The Primary Infection

Epstein-Barr virions have the capacity to shuttle between epithelial cells and B cells. One line of research suggests that a different viral ligand complex on virions engages different entry receptors on either cell type. Other data are more consistent with a common receptor. In any case, infection initiates in epithelial cells, usually those of the mucosal epithelia in the oropharygeal cavity. The replication in epithelia is reminiscent of papillomavirus replication in that the infection is not completed in basal cells, but finishes in the more superficial differentiated spinous and

granular layers. These tissues are the sites for shedding of infectious virions. Lingual epithelium and tonsil tissue are rich in lymphoid cells and provide the perfect milieu for the next stage of infection. After productive infection of epithelial cells, released virions infect B lymphocytes in closely associated lymphoid tissue via a pathway different from that used to infect epithelial cells. These infected B cells do not produce infectious virions, because an entirely different transcriptional program ensues that leads to establishment of a unique latent infection. The viral genome exists as a circular, self-replicating episome in the B-cell nucleus (Volume I, Chapter 9). The viral episome becomes associated with nucleosomes and undergoes progressive methylation at CpG residues. When latently infected B cells come in close contact with epithelial cells, the latent infection may be reactivated, resulting in production of more infectious virions capable of infecting epithelial cells. Infectious particles are shed predominantly in the saliva, but shedding from lung and cervical epithelia has also been reported.

The Persistent Infection Is a Dynamic State

A dynamic state of latent and productive infection exists in infected individuals. Despite the presence of latently infected B cells, infectious virions are still produced, virus-specific CTLs circulate in the blood, and antibodies specific for viral proteins are produced in relatively large quantities. How latency is maintained in the face of an active immune response is an important question.

Children and teenagers are commonly afflicted, usually after oral contact (hence the name "kissing disease"). The acute infection requires expression of most viral genes and rapidly stimulates a strong immune response. Spread of infection to B cells in an individual with a normal immune system induces the infected B cells to divide, leading to substantial immune and cytokine responses. The resulting disease is called infectious mononucleosis. The ensuing immune response destroys most infected cells, but approximately 1 in 100,000 infected B cells survive. They persist as small, nonproliferating memory B cells that make only latent membrane protein 2A (LMP-2A) mRNA. They home to lymphoid organs and bone marrow, where they are maintained. These cells do not produce the B7 coactivator receptor, and therefore are not killed by CTLs (see Chapter 4). They proliferate indefinitely when stimulated or when propagated as cultured cells. They are the progenitors of the B-cell lines that grow out of peripheral blood of an infected patient.

When peripheral blood of an infected individual is cultured, growth factors in the media stimulate replication of the rare latently infected B cells, while the uninfected B cells die. It is important to understand that these cultured immortal lymphoblasts are most assuredly **not** the same as latently infected cells that circulate *in vivo*. Nevertheless, this class of virus-infected B cells often yields immortalized progeny capable of being propagated indefinitely in the laboratory. Consequently, these laboratory-produced cells are the best-understood models of Epstein-Barr virus latent infection. These cells synthesize a set of at least 10 gene products, including six nuclear proteins (termed EBNAs), three viral membrane proteins (LMPs) that are important in altering the properties of the cells, small RNA molecules called EBER-1 and EBER-2, and at least 20 micro-RNAs (Table 5.4). While considerable effort has focused on understanding the mechanism of transformation of B cells in culture, the viral genes so identified often are not expressed in human cancers associated with viral infection.

Three Programs of Viral Gene Expression Produce Different Phenotypes of Latent Infection

At least three distinct phenotypes or programs can be distinguished by the viral gene products made in an infected B cell. These are called latency 1 (EBNA-1, Bam A RNAs, EBER RNAs 1 and 2), latency 2 (EBNA-1, LMP-1, LMP-2, Bam A RNAs, EBER RNAs, micro-RNAs), and latency 3 (all the latent-cycle proteins and the RNAs). Viral infection stimulates B cells to divide rapidly and continuously. These B-cell blasts express all of the latency-associated genes (latency 3) (Fig. 5.11; Table 5.4). The viral proteins are required to establish the latent infection and to promote growth of the infected cell. This phenotype is also characteristic of B-cell lymphomas of immunodeficient patients. The latency 3 phenotype is similarly characteristic of nasopharyngeal carcinoma, Hodgkin's disease, and T-cell lymphomas. B cells expressing the latency 1 program (EBNA-1 only) are found in Burkitt's lymphoma, and have been difficult to detect in virus-infected individuals.

The Complicated Collection of Different B-Cell Phenotypes Is Best Understood in the Context of Normal B-Cell Biology

To enter the resting state and become a memory cell, an uninfected B cell must have bound its cognate antigen and received appropriate signals from helper T cells in germinal centers of lymphoid tissue. Remarkably, during latent infection, the viral LMP-1 and LMP-2a proteins mimic **all** of these steps such that the infected B cell is able to differentiate into a memory cell in the absence of other cues.

The Equilibrium Established between Active Immune Elimination of Infected Cells and Viral Persistence Is Noteworthy

Although immunocompetent individuals maintain CTLs directed against many of the viral proteins synthesized in latently infected B cells, these cells are not eliminated. Some viral proteins, such as LMP-1, inhibit apoptosis or immune recognition of latently infected cells. Moreover, EBNA-1 peptides are not presented to T cells, as discussed above. When the equilibrium between proliferation of latently infected B cells and the immune response that kills them is altered (e.g., after immunosuppression), the immortalized B cells can form lymphomas (Fig. 5.11; see also Chapter 7). It is a matter of debate if any viral protein is synthesized in the infected, nondividing B cell. Certainly, virus-infected proliferating cells produce viral proteins and are superb targets for the host's immune system. It is likely that the normal immune response selects nonproliferating B cells as the survivors of infection, and ensures that the latent infection is benign in the majority of cases.

Reactivation

The signals that reactivate latent Epstein-Barr virus infection in humans are not well understood, but considerable information has been obtained from studies of

6

Introduction
  Worldwide Scope of the Problem

HIV Is a Lentivirus
  Discovery and Characterization
  Distinctive Features of the HIV
  Replication Cycle and the Roles of
  Auxiliary Proteins

Cellular Targets

Routes of Transmission
  Sources of Virus Infection
  Modes of Transmission
  Mechanics of Spread

The Course of Infection
  Patterns of Virus Appearance and
  Immune Cell Indicators of Infection
  Variability of Response to Infection

Origins of Cellular Immune Dysfunction
  CD4⁺ T Lymphocytes
  Cytotoxic T Lymphocytes
  Monocytes and Macrophages
  B Cells
  Natural Killer Cells
  Autoimmunity

Immune Responses to HIV
  Humoral Responses
  The Cellular Immune Response
  Summary: the Critical Balance

Dynamics of HIV-1 Replication in AIDS Patients

Effects of HIV on Different Tissues and Organ Systems
  Lymphoid Organs
  The Nervous System
  The Gastrointestinal System
  Other Organ Systems

HIV and Cancer
  Kaposi's Sarcoma
  B-Cell Lymphomas
  Anogenital Carcinomas

Prospects for Treatment and Prevention
  Antiviral Drugs and Therapies
  Highly Active Antiretroviral Therapy
  Prophylactic Vaccine Development
  To Prevent Infection

Perspectives

References

Human Immunodeficiency Virus Pathogenesis

Introduction

Worldwide Scope of the Problem

Acquired immunodeficiency syndrome (AIDS) is the name given to end-stage disease caused by infection with human immunodeficiency virus (HIV). By almost any criteria, HIV qualifies as one of the world's deadliest scourges. First recognized as a clinical entity in 1981, by 1992 AIDS had become the major cause of death in individuals 25 to 44 years of age in the United States. Although the rate of increase has been reduced since 2000, the current worldwide statistics are still staggering, with the developing countries of Africa and parts of Asia being especially hard-hit (Fig. 6.1). An end-of-year report from the United Nations' AIDS program estimated the number of new HIV infections in 2007 to be 2.5 million, bringing the total number of infected people worldwide to approximately 33.2 million. This number corresponds to almost 1 in every 100 adults aged 15 to 49 in the world's population. Although the recent availability of drugs to treat HIV infection has decreased the annual death toll in wealthy countries, HIV/AIDS is still the leading cause of death in sub-Saharan Africa, with 1.6 million fatalities in 2007 alone. In certain parts of this region, 25 to 30% of the adult population has become infected. It is estimated that one-third of the children under 15 years of age in these areas have lost one or both parents to AIDS. In these places a whole generation of human beings has succumbed to this fatal disease. AIDS kills more people than any other infectious disease, and HIV continues to spread faster than any known persistent infectious agent in the last half century. The clinical emergence of this virus is likely to be the consequence of a number of political, economic, and societal changes, including the breakdown of national borders, economic distress with the migration of large populations, and the ease and frequency of travel throughout the world. International efforts have focused on bringing funds and expertise to bear on the HIV/AIDS pandemic in Africa and elsewhere, but the task is enormous.

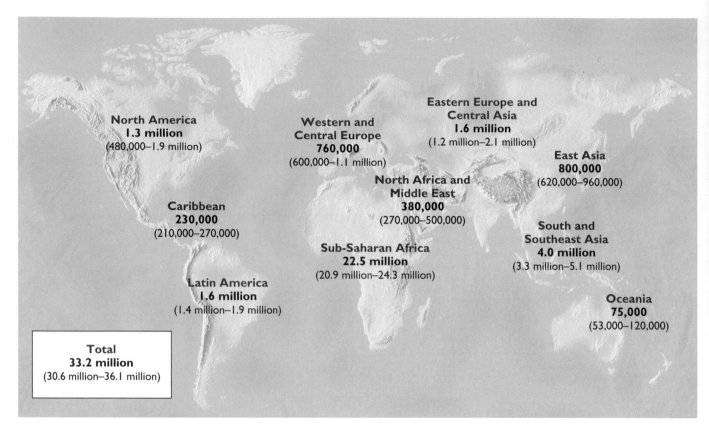

Figure 6.1 Estimated number of people living with HIV infection worldwide, 2007. Data from the Joint United Nations Programme on HIV/AIDS, November 2007.

Because of its medical importance, HIV has also become the most intensely studied infectious agent. Research on the virus has contributed to our understanding of AIDS and related veterinary diseases; provided new insights into virology, cellular biology, and immunology; and allowed researchers to develop strategies for its prevention and control of HIV infection. This chapter describes the many facets of HIV-induced pathogenesis and what has been learned through its analysis. The complexities illustrate the enormous scope of the challenges faced by biomedical researchers and physicians in their efforts to control this virus, which strikes at the very heart of the body's defense systems.

HIV Is a Lentivirus

Discovery and Characterization

The first clue to the etiology of AIDS came in 1983 with the isolation of a retrovirus from the lymph node of a patient with lymphadenopathy at the Pasteur Institute in Paris. Although not fully appreciated initially, the significance of this finding became apparent in the following year with the isolation of a cytopathic, T-cell-tropic retrovirus from combined blood cells of AIDS patients by researchers at the U.S. National Institutes of Health and of a similar retrovirus from blood cells of an AIDS patient at the University of California, San Francisco. Although the National Institutes of Health isolate was later shown to originate from a sample received from the Pasteur Institute (Box 6.1), the virus isolated at the University of California, San Francisco, and subsequent isolates at the National Institutes of Health laboratory were unique. As commonly happens, each laboratory gave its isolate a different name: LAV (lymphadenopathy-associated virus), HTLV-III (human T-cell lymphotropic virus type III), and ARV (AIDS-associated retrovirus). Electron microscopic examination revealed that these viruses were morphologically similar to a known group of retroviruses, the lentiviruses, and further characterization confirmed this relationship. In 1986, the International Committee on Taxonomy of Viruses recommended the current name, human immunodeficiency virus.

Lentiviruses comprise a separate genus of the family *Retroviridae* (Table 6.1). The equine infectious anemia lentivirus was one of the first viruses to be identified. Discovered

DISCUSSION
Lessons from discovery of the AIDS virus(es)

The first AIDS virus was obtained from a patient with lymphadenopathy by Françoise Barré-Sinoussi in collaboration with Jean-Claude Chermann and Luc Montagnier at the Pasteur Institute (1983). The isolate, named Bru, grew only in primary cell cultures. We now know that Bru belonged to a class of slow-growing, low-titer viruses that are common in early-stage infection.

Between 20 July and 3 August 1983, Bru-infected cultures at the Pasteur Institute became contaminated with a second AIDS virus, called Lai, which had been isolated from a patient with full-blown AIDS and which belonged to a class of viruses that grow well in cell culture. HIV-1 Lai rapidly overtook the cultures.

Unaware of this contamination, Pasteur scientists subsequently sent out virus samples from these cultures as "Bru" to several laboratories, including those of Robin Weiss in Britain and Malcolm Martin and Robert Gallo in the United States. Unlike earlier samples of Bru, this virus grew robustly in the laboratories to which it was distributed. Indeed, Lai was later discovered to have contaminated some AIDS patient "isolates" obtained by Weiss. In retrospect, such contamination is not surprising, as biological containment facilities were limited at the time, with the same incubators and hoods being used for maintaining HIV stocks and making new isolates.

Lai also contaminated cultures of blood cells combined from several AIDS patients in the Gallo laboratory at the National Institutes of Health. Because the properties of this virus were found to be different from those described for Bru, Gallo and coworkers reported the discovery of a second type of AIDS virus, which they believed to have originated from one of their AIDS patients.

This second claim, a race to develop blood screening tests, and the later revelation from DNA sequence analyses that the French and the Gallo viruses were one and the same (Lai) led to a much publicized scientific controversy with significant political overtones. Simon Wain-Hobson and colleagues at the Pasteur Institute eventually sorted out the chain of events in 1991 by comparing nucleotide sequences of stored samples of the original stocks of Bru and Lai. The controversy has since subsided—what remains are important lessons in virology: that contamination can be a real problem, that passage in the laboratory tends to select for viruses that replicate rapidly, and that rigorous characterization (nowadays by genome sequencing) is a prudent safeguard against costly mistakes.

Goudsmit, J. 2002. Lots of peanut shells but no elephant. A controversial account of the discovery of HIV. *Nature* **416:**125–126.

Wain-Hobson, S. J. P. Vartanian, M. Henry, N. Chenciner, R. Cheynier, S. Delassus, L. P. Martins, M. Sala, M. T. Nugeyre, D. Guetard, et al. 1991. LAV revisited: origins of the early HIV-1 isolates from Institut Pasteur. *Science* **252:**961–965.

Weiss, R., and M. Martin. Personal communication.

in 1904, this virus causes episodic autoimmune hemolytic anemia in horses. Lentiviruses of sheep (visna/maedi virus) and goats (caprine arthritis-encephalitis virus) have also been known for many years. All these viruses are associated with long incubation periods and are therefore called **slow viruses** (Chapter 5). The discovery of HIV led to a search for additional lentiviruses and their subsequent isolation from cats (feline immunodeficiency virus) and a variety of nonhuman primates (simian immunodeficiency virus [SIV]). In 1986, a distinct type of HIV that is prevalent in certain regions of West Africa was discovered. It was called

HIV-2. Individuals infected with HIV-2 also develop AIDS, but with a longer incubation period and lower morbidity.

Many independent isolates of both HIV-1 and HIV-2 have been characterized over the last decade. Nucleotide sequence comparisons allow us to distinguish two major groups among HIV-1 isolates: group M includes most HIV-1 isolates, and group O represents what appear to be relatively rare "outliers" (Box 6.2). Nine distinct subtypes are currently recognized in group M (called **clades** A to K, except E, which was found to be a recombinant), each of which is prevalent in a different geographic area. For example,

Table 6.1 Lentiviruses[a]

Virus	Host infected	Primary cell type infected	Clinical disorder(s)
Equine infectious anemia virus	Horse	Macrophages	Cyclical infection in the first year, autoimmune hemolytic anemia, sometimes encephalopathy
Visna/maedi virus	Sheep	Macrophages	Encephalopathy/pneumonitis
Caprine arthritis-encephalitis virus	Goat	Macrophages	Immune deficiency, arthritis, encephalopathy
Bovine immunodeficiency virus	Cow	Macrophages	Lymphadenopathy, lymphocytosis
Feline immunodeficiency virus	Cat	T lymphocytes	Immune deficiency
Simian immunodeficiency virus	Primate	T lymphocytes	Immune deficiency and encephalopathy
Human immunodeficiency virus	Human	T lymphocytes	Immune deficiency and encephalopathy

[a]Adapted from Table 1.1 (p. 2) of J. A., Levy, *HIV and the Pathogenesis of AIDS*, 3rd ed. (ASM Press, Washington, DC, 2007), with permission.

BOX	BACKGROUND
6.2	*The earliest record of HIV-1 infection*

The earliest record of HIV-1 infection comes from a serum sample obtained in 1959 from a Bantu male in the city now known as Kinshasa, in the Democratic Republic of Congo. Phylogenetic analyses place the viral sequence (ZR59) near the ancestral node of clades B and D. As this is not at the base of the M group, this group must have originated earlier (red arrowhead near top of figure), and back calculations suggest that the M group of viruses arose via transspecies transmission from a chimpanzee into the African population around 1930. Its rapid evolution, giving rise to at least 10 subtypes (clades A to K), seems to have occurred near the end of or just after World War II. Separate transspecies transmissions (red arrowheads in bottom half of figure) account for the origin of the N and O groups.

Sharp, P. M. 2002. Origins of human virus diversity. *Cell* **108**:305–312.

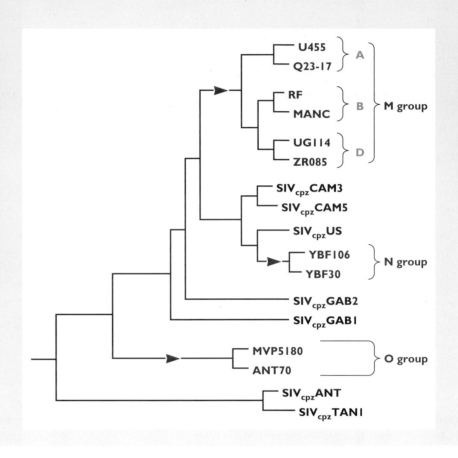

clade B is the most common subtype in North America and Europe. A new group, N, was proposed for HIV-1 in 1998 based on a virus isolate, YBF30, obtained from an AIDS patient in Cameroon. The nucleotide sequence of this virus is more closely related to group M than to group O. Identification of related strains in Cameroon supports a three-pronged radiation of HIV-1 groups. Eight distinct groups of HIV-2 have also been identified. Of these, groups A and B (found in different parts of West Africa) account for most infections worldwide.

Figure 6.2 shows the phylogenetic relationships among the lentiviruses, based on sequences of their *pol* genes. The African monkey and ape isolates are endemic to each of the species from which they were obtained and do not appear to cause disease in their native hosts (Box 6.3). However, a fatal AIDS-like disease is caused by infection of Asian macaques with virus originating from the African sooty mangabey (SIV$_{smm}$). Close contact between sooty mangabeys and humans is common, as these animals are

hunted for food and kept as pets. Such interaction and the observation that several isolates of HIV-2 are nearly indistinguishable in nucleotide sequence from SIV$_{smm}$ support the hypothesis that HIVs emerged via interspecies transmission from nonhuman primates to humans. This hypothesis is supported further by recent studies indicating that the known HIV-1 groups arose via at least three independent transmissions from chimpanzees. The strains of SIV$_{cpz}$ from the chimpanzee *Pan troglodytes troglodytes* are closest in sequence to HIV-1, implicating this subspecies as the origin of the human virus. In this chapter we use the abbreviation HIV to describe properties shared by HIV-1 and HIV-2, and specify the type when referring to one or the other.

As summarized in Table 6.1, lentiviruses cause immune deficiencies and disorders of the hematopoietic and central nervous systems and, sometimes, arthritis and autoimmunity. Lentiviral genomes are relatively large, with more genes than those of simpler retroviruses (Fig. 6.3).

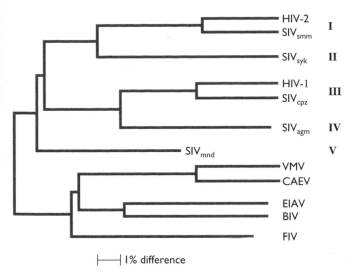

Figure 6.2 Phylogenetic relationships among lentiviruses. Representative lentiviruses were compared by using *pol* gene nucleotide sequences for establishing phylogenetic relationships. Five groups of primate lentiviruses (labeled I through V) are shown: HIV-1, HIV-2, SIV from the sooty mangabey monkey (SIV$_{smm}$), SIV from Sykes' monkey (SIV$_{syk}$), SIV from the chimpanzee (SIV$_{cpz}$), SIV from the African green monkey (SIV$_{agm}$), and SIV from the mandrill (SIV$_{mnd}$). Nonprimate lentiviruses are visna/maedi virus (VMV), caprine arthritis-encephalitis virus (CAEV), equine infectious anemia virus (EIAV), bovine immunodeficiency virus (BIV), and feline immunodeficiency virus (FIV). The scale indicates the percentage difference in nucleotide sequences in the *pol* gene. The branching order of the primate lentiviruses is controversial. Adapted from Fig. 2 of P. A. Luciw, p. 1881–1952, *in* B. N. Fields et al. (ed.), *Fields Virology*, 3rd ed. (Lippincott-Raven, Philadelphia, PA, 1996), with permission.

In addition to the three structural polyproteins Gag, Pol, and Env, common to all retroviruses, lentiviral genomes encode a number of additional **auxiliary proteins**. Two HIV auxiliary proteins (Table 6.2), Tat and Rev, perform **regulatory** functions that are essential for viral replication. The remaining four, Nef, Vif, Vpr, and Vpu, are not essential for viral reproduction in most immortalized T-cell lines and hence are known as **accessory** proteins. However, these proteins do modulate virus replication, and they are essential for efficient virus production *in vivo* and the ensuing pathogenesis.

Distinctive Features of the HIV Replication Cycle and the Roles of Auxiliary Proteins

Much of what we know about the function of the auxilliary proteins of HIV comes from studies of their effects on cells in culture, often produced transiently from plasmid expression vectors in the absence of other viral components (Volume I, Box 8.8). Although these methods are simple and sensitive, they do not necessarily reproduce conditions similar to those that occur upon viral infection. Preparation and analysis of viral mutants have also been used to investigate the functions of these proteins in cell culture. However, as the hosts for this virus are humans, it is difficult to evaluate the significance of many of the functions deduced from cell culture to pathology in the whole organism.

cis-*Acting Regulatory Sequences and Tat and Rev*

Tat interacts with TAR sequences in the long terminal repeat. As in all retroviruses, expression of integrated HIV DNA is regulated by sequences in the transcriptional control region of the viral long terminal repeat (LTR), which are recognized by the host cell's transcriptional machinery. The HIV-1 LTR functions as a promoter in a variety of cell types, but its basal level is very low. As described in Volume I, Chapter 8 (Fig. 8.10), the LTR of HIV includes an enhancer sequence that binds a number of cell-type-specific transcriptional activators, for example, Nf-κb (Volume I, Fig. 8.11). The release of Nf-κb from its cytoplasmic inhibitor in activated T-cells may explain why HIV replication requires T-cell stimulation.

Just downstream of the site of initiation of transcription in the HIV LTR is a unique viral regulatory sequence, TAR (Fig. 6.4). As described in Volume I, Chapter 8 (Fig. 8.13), TAR RNA forms a stable, bulged stem-loop structure that binds the regulatory protein Tat (Table 6.2), together with a number of host proteins, to stimulate transcription. In the absence of Tat, viral transcription usually terminates prematurely. The principal role of Tat is to enhance the processivity of transcription and thereby facilitate the elongation of viral RNA.

The Tat protein is released by infected cells and can be taken up by other cells and influence their function. Tat can act as a chemoattractant for monocytes, basophils, and mast cells. It also induces expression of a variety of important proteins in the cells that it enters, and some of these proteins can have a profound effect on virus spread and immune cell function. For example, in transient-expression assays Tat can up-regulate the expression of genes encoding the CXCr4 and CCr5 coreceptors in target cells and can enhance the expression of a number of chemokines. The Tat protein is reported be cytotoxic to some cultured cells and is neurotoxic when inoculated intracerebrally into mice. It has also been reported that transgenic expression of Tat in mice causes a disease that resembles Kaposi's sarcoma. Although the human disease is almost certainly caused by a herpesvirus, Tat contributes to the aggressive nature of this malignancy in AIDS patients by promoting the growth of spindle cells in the Kaposi's sarcoma lesions.

BOX
6.3

DISCUSSION
TRIM5 restriction, an example of coevolution of viral and host genes?

The tripartite motif (Trim) 5α protein was identified in a screen for "factors" that might be responsible for the inability of HIV-1 to replicate in the cells of Old World monkeys, including those of rhesus macaques. Trim5α is translated from a spliced macaque TRIM5 mRNA. The macaque protein has no apparent effect on the replication of SIV$_{mac}$, but blocks HIV-1 replication shortly after entry by binding to the capsid protein and mediating the degradation or premature disassembly of the infecting particle.

TRIM5 is a member of a large, multigene family that has proliferated during evolution and has been under positive selection for about 35 million years. There are hundreds of copies of TRIM genes in the primate genomes; 80 members of this family have been identified in humans. However, each primate species encodes a TRIM5 gene with different antiviral specificity, consistent with differences in the viral capsids with which the proteins interact.

One current theory is that TRIM5 genes evolved independently in each species to protect against particular endemic viruses. This idea is supported by the finding

that human Trim5α blocks replication of a 4-million-year-old endogenous virus (PtERV1) that was resurrected from the genome of the chimpanzee *Pan troglodytes*.

Susceptibility to retroviral infection is dependent on a number of host-virus

The rhesus macaque Trim5α protein contains a RING domain (RF), a B box, and a coiled-coil domain. Many Zn-binding RING domains have E3 ubiquitin ligase activity, and Trim5α can mediate RING-dependent auto-ubiquitinylation *in vitro*. This property supports the hypothesis that the antiviral activity of Trim5α is mediated by ubiquitinylation of itself and associated viral proteins, leading to degradation in the proteosome. The B box domain also has a Zn-binding motif, is likely to be involved in protein-protein interactions, and may contribute to ubiquitin ligation specificity. The SPRY domain interacts with the HIV-1 capsid protein and is responsible for antiviral activity. Trim5α exists as a trimer with the coiled-coil facilitating homo- and heteromultimerization with related Trim proteins.

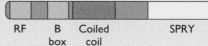

RF   B   Coiled   SPRY
      box   coil

interactions, and it is not yet clear how the antiviral activities of TRIM5 orthologs may affect the resistance or sensitivity of each primate species. However, it has been suggested that selective changes in the TRIM5 gene that occurred in the human lineage in response to PtERV1, or a related ancient virus, may have left our species more susceptible to HIV-1.

There is currently great interest in elucidating the antiviral mechanism of TRIM5 and related proteins. Such knowledge should uncover new vulnerabilities of HIV and may suggest additional targets for antiviral drug development.

Kaiser, S. M., H. S. Malik, and M. Emerman. 2007. Restriction of an extinct retrovirus by the human TRIM5α antiviral protein. *Science* **316**:1756–1758.

Luban, J. 2007. Cyclophilin A, TRIM5, and resistance to human immunodeficiency virus type 1 infection. *J. Virol.* **81**:1054–1061.

Stremlau, M., C. M. Owens, M. J. Perron, P. Kiessling, P. Autissier, and J. Sodroski. 2004. The cytoplasmic body component TRIM5α restricts HIV-1 infection in Old World monkeys. *Nature* **427**:848–853.

Towers, G. 2007. The control of viral infection by tripartite motif proteins and cyclophilin A. *Retrovirology* **4**:40.

Multiple splice sites and the role of Rev.   Unlike those of the simpler oncogenic retroviruses, the full-length HIV transcript contains numerous 5′ and 3′ splice sites. The regulatory proteins Tat and Rev and the accessory protein Nef are synthesized early in infection from multiply spliced messenger RNAs (mRNAs) (Fig. 21, appendix in Volume I). As Tat then stimulates transcription, these mRNAs are found in abundance at this early time. However, the accumulation of Rev protein brings about a change in the pattern of mRNAs, leading to a temporal shift in viral gene expression.

Figure 6.3 Organization of HIV-I and HIV-2 proviral DNA. Vertical positions of the colored bars denote each of the three different reading frames that encode viral proteins. The LTRs contain sequences necessary for transcriptional initiation and termination, reverse transcription, and integration. Adapted from Fig. 1 of M. Emerman, *Curr. Biol.* **6**:1096–1103, 1996, with permission.

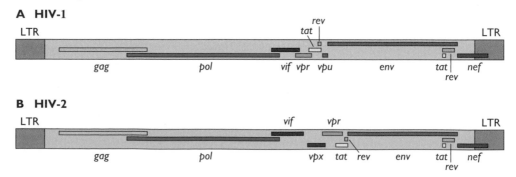

Table 6.2   HIV auxiliary proteins[a]

Protein[b]	Size (kDa)	Function	Location
Regulatory			
Tat	14	Transactivation; binds TAR to facilitate initiation and elongation of viral transcription	Primarily in cell nucleus
Rev	19	Regulation of viral mRNA expression; binds RRE and facilitates nuclear export of unspliced or singly spliced RNAs	Primarily in cell nucleus
Accessory			
Nef	27	Pleiotropic, can increase or decrease virus replication; down-regulates MHC-I and the CD4 receptor; influences T-cell activation; enhances virion infectivity	Cell cytoplasm, plasma membrane
Vif	23	Increases virus infectivity; helps in virion assembly and in viral DNA synthesis	Cell cytoplasm
Vpr	15	Helps in virus replication; causes G_2 arrest; facilitates nuclear entry of preintegration complex	Virion
Vpu[c]	16	Helps in virus release; disrupts Env-CD4 complexes; causes CD4 degradation	Integral cell membrane protein
Vpx[d]	15	Nuclear entry of preintegration complexes	Virion

[a]Adapted from Table 1.5 (p. 10) of J. A. Levy, *HIV and the Pathogenesis of AIDS*, 3rd ed. (ASM Press, Washington, DC, 2007), with permission.

[b]See Figure 6.3 for location of the viral genes on the HIV genome.

[c]Present only with HIV-1. Expression appears regulated by Vpr.

[d]Encoded only by HIV-2. May have originated via a duplication of Vpr.

Figure 6.4 Mechanisms of Tat activation. Some regulatory sequences in the HIV LTR are depicted in the expanded section at the top. The numbers refer to positions relative to the site of initiation of transcription. The opposing arrows in R represent a palindromic sequence that folds into a stem-loop structure (TAR) in the transcribed mRNA to which Tat binds (center). Tat is required for efficient elongation during HIV-1 RNA synthesis. The position of the RRE in the *env* transcript and the presence of *cis*-acting repressive sequences, also known as instability elements (INS), are also illustrated.

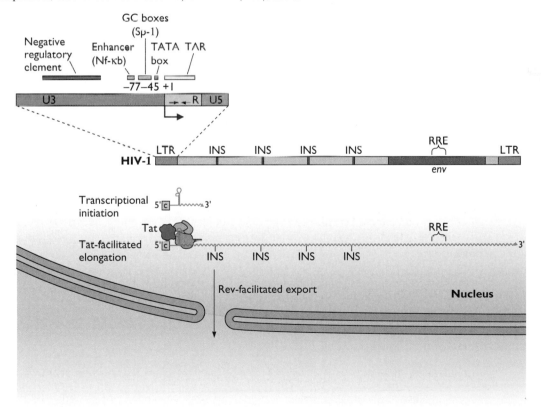

The Rev protein (Table 6.2) is an RNA-binding protein that recognizes a specific sequence within a structural element in the *env* region of the elongated transcript called the **Rev-responsive element (RRE)** (Fig. 6.4). As discussed in Volume I, Chapter 10 (Fig. 10.14 to 10.16), Rev mediates the nuclear export of any RRE-containing RNA. As the concentration of Rev increases, unspliced or singly spliced transcripts containing the RRE are exported from the nucleus. In this way, Rev promotes synthesis of the viral structural proteins and enzymes and ensures the availability of full-length genomic RNA for incorporation into new virus particles. The accessory proteins Vif, Vpr, and Vpu (for HIV-1) or Vpx (for HIV-2) are also expressed later in infection from singly spliced mRNAs that are dependent on Rev for export to the cytoplasm (Fig. 22, appendix in Volume I).

The dependence of HIV gene expression on Rev is due in part to *cis*-acting repressive sequences, also called **instability elements,** present in the unspliced or singly spliced transcripts. These sequences, some of which are characterized by a high A+U content, lie within regions in *gag* and *pol* mRNAs. Mutations in these sequences increase the stability, nuclear export, and translatability of the transcripts in the absence of Rev. The response to their presence appears to be cell dependent, but the mechanism(s) by which these sequences act, and exactly how Rev counteracts their effects, are not understood. This phenomenon does, however, provide an explanation for the puzzling failure of early attempts to express individual HIV-1 structural proteins and enzymes in primate or human cells from mRNAs that did not also encode Rev.

The Accessory Proteins

A very large number of seemingly disparate functions have been attributed to the accessory proteins of HIV. Because in many cases activities have been observed under conditions in which the proteins are overproduced from plasmid vectors in cultured cells, the biological significance of some of the proposed functions is not always clear. However, recent studies have uncovered a common mechanism for many of the activities of the accessory proteins: all seem to act as **adapter proteins,** partnering with multicomponent cellular complexes that target proteins for degradation.

Vif protein.   Vif stands for viral infectivity factor. This protein (Table 6.2 and Fig. 6.5) accumulates in the cytoplasm and at the plasma membrane of infected cells. Early studies showed that mutant viruses lacking the *vif* gene were approximately 1,000 times less infectious than the wild type in certain CD4+ T-cell lines and peripheral blood lymphocytes and macrophages. However, direct cell-to-cell transfer was only slightly lower than normal. Virions produced in the absence of Vif are therefore defective.

Figure 6.5 Organization of known or presumed functional regions in HIV-I accessory proteins. Locations of conserved and critical cysteine residues are indicated in Nef and Vif. Serines that require phosphorylation for protein function are noted in Vpu. Magenta regions in Vif are conserved, but their functions are unknown. Numbering corresponds to amino acid positions in proteins of the laboratory strain of HIV-1, NL43. Where known, functions are indicated. G^{myr} is the myristoylated glycine residue at position 2 in Nef, which facilitates membrane binding.

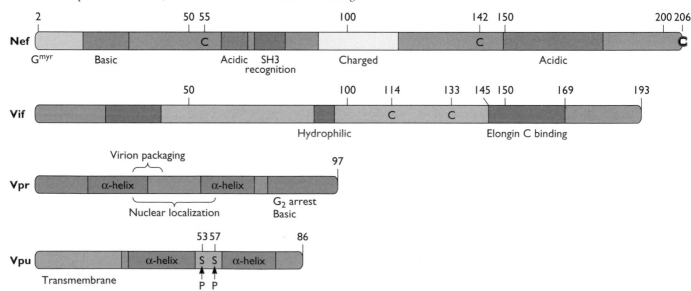

Production of Vif from a plasmid vector in susceptible host cells does not compensate for its absence in the cell that produces virions. Rather, Vif is needed at the time of virus assembly in the producing cells.

Vif is an RNA-binding protein. Small amounts can be detected in HIV particles and also in heterologous retroviral particles produced by cells that contain a Vif-expressing plasmid. Virions produced from *vif*-defective HIV genomes contain the normal complement of progeny RNA, and they are able to enter susceptible cells and to initiate reverse transcription, but full-length double-stranded viral DNA is not detected. These observations indicate that Vif is required in a step following virus entry that is essential for completion of reverse transcription. The requirement for Vif is strikingly cell type dependent. Experiments in which cells that are permissive for *vif* mutants were fused with cells that are nonpermissive established that the nonpermissive phenotype is dominant; the infectivity of virions produced in such heterokaryons was enhanced by Vif production. This observation suggested that Vif protein may suppress a host cell function that otherwise inhibits progeny virus infectivity.

All of these seemingly unusual properties were demystified with the discovery that Vif plays a critical role for the virus by blocking the antiviral action of members of an RNA-binding family of cellular cytidine deaminases, called apoplipoprotein B mRNA editing enzyme catalytic peptides 3 (Apobec3). These enzymes are synthesized in nonpermissive cells and incorporated into virus progeny via interactions with the viral RNA and possibly NC protein. Apobec3G (A3G) was the first family member to be identified as a Vif target. It was subsequently shown that Vif prevents its incorporation into virions by binding to A3G and mediating its depletion. In this role, Vif partners with cellular proteins (Cul5, elongins B ad C, and Rbx1) in an E3 ubiquitin ligase complex that polyubiquitinates A3G, leading to its degradation in proteosomes (Fig. 6.6). Vif expression also blocks the antiviral activities of human Apobec3F and 3C, which like A3G are produced in abundance in lymphoid cells, presumably by a similar mechanism.

A3G appears to exert a number of antiviral activities. It has been proposed that its binding to viral RNA may account, in part, for its inhibition of reverse transcription in newly infected cells. In addition, the enzyme catalyzes the deamination of deoxycytidine to form deoxyuridine (dU) in the first (−) strand of viral DNA to be synthesized by reverse transcriptase. The dU is a substrate for the cellular uracil-DNA glycosylase, and the abasic sites produced by its action are likely targets for endonucleolytic digestion in the newly formed (−) DNA strands. If dU is not removed, the (+) strand complement of the deaminated (−) strand would contain deoxyadenosine in place of the normal

deoxyguanosine at such sites (Fig.6.6). The frequency of G→A transitions is abnormally high in the genomes of *vif*-defective virions produced in nonpermissive cells, and incomplete protection from Apobec3 proteins by Vif may explain why such transitions are the most frequent point mutations in HIV genomes. It has been suggested that the Apobec3 proteins represent an ancestral mode of intrinsic cellular defense against retroviruses (see Chapter 3).

Vpr protein. The viral protein R, or Vpr (Table 6.2 and Fig. 6.5), derives its name from the early observation that it affects the rapidity with which the virus replicates in, and destroys, T cells. Most T-cell-adapted strains of HIV-1 carry mutations in *vpr*. The Vpr protein is encoded in an open reading frame lying between *vif* and *tat* in the genomes of primate lentiviruses (Fig. 6.2). The SIV and HIV-2 genomes include a second, related gene, *vpx*, which appears to have arisen as a duplication of *vpr*. The other lentiviruses do not contain sequences related to *vpr* but do include small open reading frames that might encode proteins with similar functions.

Vpr and Vpx are incorporated into virions. Vpr incorporation is dependent on specific interactions with a proline-rich domain at the C terminus of the Gag polyprotein. Vpr protein, in turn, mediates virion incorporation of the host's uracil DNA glycosylase, Ung2. About 100 to 200 molecules of Vpr are present in nucleocapsids. Its presence in virions is consistent with the observation that Vpr function is required at some early stage in the virus replication cycle.

Two principal functions have been recognized for HIV-1 Vpr. The protein causes a G_2 cell cycle arrest and may promote entry of viral nucleic acids into the nucleus. In HIV-2 these functions are segregated into Vpr and Vpx, respectively. Although Vpr itself does not damage host DNA, expression of the isolated gene in cultured cells elicits a response similar to the response to DNA damage, resulting in G_2 arrest and apoptosis. It has recently been discovered that Vpr, like Vif, is an adapter protein, which hijacks another ubiquitin ligase complex. In the case of Vpr, the complex includes the scaffold protein Cul4A, Rbx1 E2 ligase, and damaged DNA-binding protein 1 (Ddb1); targets for polyubiquitinylation and subsequent degradation include the Ung2 protein mobilized by Vpr binding. It has been proposed that binding of Vpr to the Ddb1-containing complex could prevent DNA repair, leading to the accumulation of damaged DNA, subsequent activation of the DNA damage response, and G_2 arrest. The biological advantage of preventing infected cells from entering mitosis is not clear, but the increased activity of the LTR promoters in the G_2 phase of the cell cycle may lead to enhanced virus production. The fact that other viruses, including paramyxovirus 5 and hepatitis B virus, encode proteins that

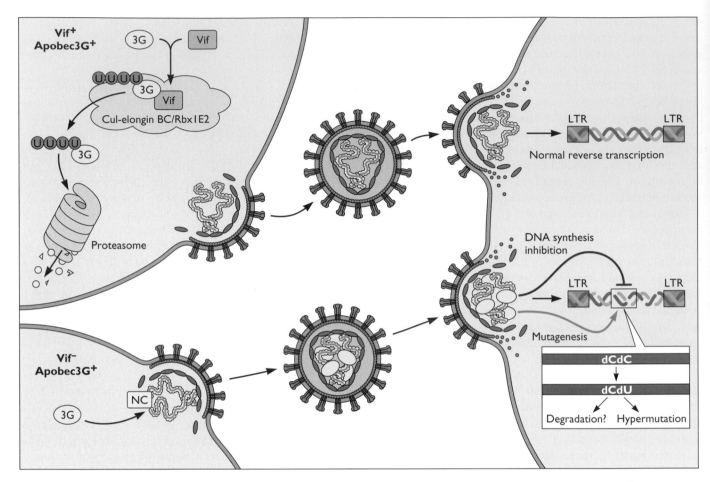

Figure 6.6 Mechanisms of action of Vif and Apobec3G. (Top) Vif counteracts the antiviral affects of Apobec3G (3G) by mediating its polyubiquitinylation, which leads to proteosomal degradation. **(Bottom)** In the absence of Vif, 3G is incorporated into newly formed virions through interaction between viral RNA and NC protein. In the newly infected cell, viral RNA reverse transcription is inhibited by 3G and cytosines in the newly synthesized DNA are converted to uracil, causing cDNA degradation or hypermutation through eventual C to A transversions. Adapted from B. Cullen, *J. Virol.* **80:**1067–1076, 2006, with permission.

also target Ddb1 suggests that this response may provide some physiological advantage to viral replication.

Vpr has been shown to bind to nuclear pore proteins. As noted in Volume I, Chapter 5, although not essential, these interactions may facilitate docking of the HIV-1 preintegration complex at the nuclear pore in preparation for import. Studies with SIV-infected macaques indicate that deletion of *vpr* attenuates viral pathogenicity. However, deletion of both *vpr* and *vpx* does reduce virus replication in these animals. It seems likely, therefore, that HIV-1 Vpr, which combines the functions of the two SIV gene products, is crucial to HIV pathogenesis.

Vpu protein. The small Vpu protein is unique to HIV-1 and the related SIV$_{cpz}$ (Fig. 6.2), hence the name viral protein U (Vpu). The predicted sequence of Vpu includes an N-terminal stretch of 27 hydrophobic amino acids that comprises a membrane-spanning domain (Fig. 6.4). Biochemical studies show that Vpu is an integral membrane protein that self-associates to form oligomeric complexes. In infected cells, the protein accumulates in the perinuclear region.

Synthesis of Vpu is required for the proper maturation and targeting of progeny virions and for their efficient release (Table 6.2). In its absence, virions containing multiple cores are produced and budding is targeted to multivesicular bodies rather than to the plasma membrane. Vpu also reduces the syncytium-mediated cytopathogenicity of HIV-1, perhaps because the efficient release of virions prevents the accumulation of sufficient Env protein at the cell surface to promote cell fusion.

The Vpu protein has structural and biochemical features similar to those of the influenza virus M2 protein. As noted in Volume I, Chapter 12, M2 is an ion channel protein that modulates the pH in the Golgi compartment, thereby protecting the newly formed hemagglutinin protein from changing conformation prematurely in the secretory pathway. Vpu appears to oligomerize in lipid bilayers, forming channel-like structures which are thought to play a role in virion release. In this connection, it is interesting that synthesis of Vpu also enhances the release of virions produced by other retroviruses, such as Moloney murine leukemia virus, and the lentiviruses visna/maedi virus and HIV-2. Recent studies suggest that Vpu enhances virus particle release by counteracting the activity of one or more cellular proteins that trap assembled virus particles at the cell surface, leading to their sequestration in endosomes.

A second function of Vpu is the degradation of CD4. Vpu traps newly formed CD4 receptor in the endoplasmic reticulum via specific interactions with its cytoplasmic domain. In this role Vpu, like Vif and Vpr, acts as an adapter protein that links the receptor to the Scf ubiquitin ligase complex (which includes Cul1, Skp1, and Roc1) mediating the entry of CD4 into the endoplasmic reticulum-associated proteasome degradation pathway. These two activities of Vpu are distinct, as the stimulation of virion release is independent of Env or CD4 expression.

Nef protein.   Most laboratory strains of HIV-1 that have been adapted to grow well in T-cell lines contain deletions or other mutations in the *nef* gene. Restoration of *nef* reduces the efficiency of virus replication in these cells, hence the name "negative factor." Multiple functions have been attributed to Nef (Table 6.2 and Fig. 6.7), and it is now clear that Nef does exert pleiotropic effects on infected cells. The functions reported for the Nef protein vary with different strains of the virus and with different cell types.

As noted above, Nef is translated from multiply spliced early transcripts. The 5′ end of Nef mRNA includes two initiation codons, and, as both are utilized, two forms of Nef are produced in infected cells. The apparent size of these proteins can vary because of differences in posttranslational modification. Nef is incorporated into the virion, as with Vpr, via interaction with the p6 domain at the C terminus of the Gag polyprotein. Virion-incorporated Nef appears to contribute to virion disassembly and may also enhance reverse transcription. Nef is synthesized in large quantities after proviral DNA integration. The protein is myristoylated posttranslationally at its N terminus (Fig. 6.5 and 6.6) and thereby anchored to the inner surface of the plasma membrane, probably in a complex with a cellular serine kinase. There are numerous potential threonine and serine phosphorylation sites in Nef, and the protein is phosphorylated, but the significance of such modification is as yet unknown.

Nef includes a protein-protein interaction domain (SH3), which mediates binding to components of intracellular signaling pathways, eliciting a program of gene

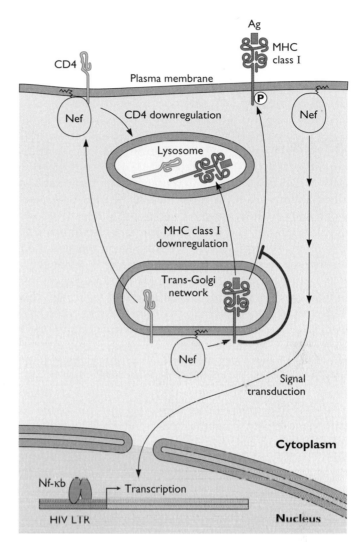

Figure 6.7 Intracellular functions attributed to Nef. Nef is myristoylated posttranslationally; the jagged protrusion represents myristic acid covalently linked to the glycine residue at position 2. Myristoylation enables Nef to attach to cell membranes, where it can interact with membrane-bound cellular proteins. Nef reduces the cell surface expression of CD4 by binding to sequences in the cytoplasmic domain of this receptor and enhancing clathrin-dependent endocytosis and the subsequent degradation of CD4 within lysosomes (left). In contrast, MHC class I expression on the cell surface is reduced by Nef binding in the membrane of the trans-Golgi network. This interaction interferes with the normal vesicular sorting required for passage of the receptor to the cell surface, and MHC class I is directed to the lysosome for degradation (right). Nef also affects signal transduction by increasing the activity of the cellular transcriptional activator Nf-κb and perhaps other cellular transcription proteins.

expression similar to that observed after T-cell activation. Such expression may provide an optimal environment for viral replication. Among the best-studied and possibly most physiologically relevant activities of Nef are its downregulation of surface concentrations of CD4 and major histocompatibility complex (MHC) class I molecules. As noted above, the former activity is shared with Vpu.

Nef binds to the cytoplasmic tail of CD4 and links this receptor to components of a clathrin-dependent trafficking pathway at the plasma membrane, leading to its internalization and delivery to lysosomes for degradation (Fig. 6.7, left). Reducing the amount of CD4 at the cell surface limits superinfection by HIV-1. It also limits the loss of Env protein via CD4 binding, thereby enhancing infectious-particle production. Nef decreases cell surface expression of MHC class I molecules by a different pathway that involves engagement with these molecules in the *trans*-Golgi network prior to their transport to the cell surface (Fig. 6.7, right). These MHC class 1 molecules are also directed to lysosomes for degradation, reducing their concentration on the cell surface. As a strong cytotoxic T-lymphocyte (CTL) response against viral infection requires recognition of viral epitopes presented by MHC class I molecules, this inhibitory activity of Nef allows infected cells to evade lysis by CTLs and could be a major factor contributing to HIV-1 pathogenesis. Nef-mediated down-regulation of a number of other cell surface molecules such as CD28, the costimulatory molecule for T-cell activation, MHC class II, and CCr5, might also affect the outcome of infection. Many other activities reported for Nef could also contribute to pathogenesis. For example, Nef is reported to enhance interaction and virus transmission between dendritic cells and T cells, and Nef-up-regulation of Fas ligand signaling could protect infected cells by promoting apoptosis in attacking CTLs.

Although the initial cell culture experiments suggested a negative effect on virus production, subsequent experiments with animals showed that Nef augments HIV pathogenesis quite significantly. Rhesus macaques inoculated with a Nef-defective mutant of SIV had low virus titers in their blood during early stages of infection, and the later appearance of high titers was associated with reversion of the mutation. More importantly, adult macaques inoculated with a virus strain containing a deletion of *nef* did not progress to clinical disease and were, in fact, immune to subsequent challenge with wild-type virus. The observation that *nef* had been deleted in HIV-1 isolates from some individuals who remained asymptomatic for long periods and from transfusion recipients who did not develop AIDS also suggests that this viral protein can contribute significantly to pathogenesis. Initial hopes that intentional deletion of *nef* might facilitate the development of a vaccine strain for humans were dashed when it was discovered that the humans infected with *nef* deletion mutants eventually developed AIDS and that newborn offspring of the female macaques that had been immunized with the *nef*-minus strain of SIV developed an AIDS-like disease.

Cellular Targets

As discussed in Volume I, Chapter 5, virus attachment and entry into host cells are dependent on the interaction between viral proteins and cellular receptors. The major receptor for the HIV envelope protein, SU, is the cell surface CD4 molecule. The viral envelope protein must also interact with a coreceptor to trigger fusion of the viral and cellular membranes and gain entry into the cytoplasm. The ability to bind to specific coreceptors is a critical determinant of the cell tropism of different HIV-1 strains. For example, binding to the α-chemokine receptor CXCr4 is a definitive feature of strains that infect T-cell lines. Infection with these strains also causes T cells to fuse, forming syncytia. Binding to the β-chemokine receptor CCr5 is characteristic of non-syncytium-inducing monocyte/macrophage-tropic strains. However, some of these strains can also infect T cells. Strains of HIV that bind to CXCr4 or CCr5 coreceptors are commonly referred to as X4 and R5 strains, respectively. The importance of these two chemokine receptors to HIV pathogenesis is demonstrated by two findings. People who carry a mutation in the gene encoding CCr5, and produce a defective receptor protein, are resistant to HIV-1 infection. So too are individuals who carry a mutation in the gene for the ligand of CXCr4 (Table 3.7). The latter mutation may lead to increased availability of the ligand, which then blocks virus entry by competing for coreceptor binding. This idea is consistent with earlier studies showing that chemokine binding to the receptors inhibits the infectivity of specific strains of HIV in cell culture (Fig. 6.8). Cells of the hematopoietic lineage that bear CD4 and one or more of these chemokine receptors are the main targets of HIV infection, and they produce the highest titers of progeny virions.

Several additional coreceptors for HIV and SIV have been identified in cell culture experiments in various laboratories, but their roles in natural infection remain to be determined. These additional coreceptors may allow the virus to enter a broader range of cells than first appreciated. Some of them are found on cells of the thymus gland and the brain, and they could play a role in infection in infancy or of cells in the central nervous system. It has also been proposed that binding to these additional coreceptors may trigger signals that affect virus replication in target cells, or that harm nonpermissive cells, producing a "bystander" effect.

Cell culture studies have identified additional mechanisms by which HIV may enter cells. For example, the virus can be transmitted very efficiently through direct cell contact. In addition, cells may be infected by virus particles that are endocytosed after binding to cell surface galactosyl ceramide or to Fc receptors (as antibody-virus complexes). HIV can infect many different types of human

T cell-tropic strain of HIV-1

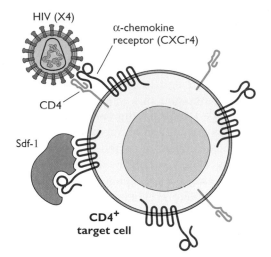

Macrophage-tropic strain of HIV-1

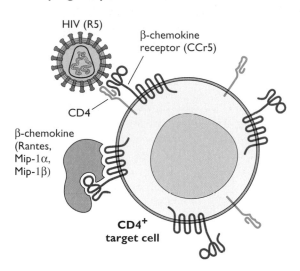

Figure 6.8 Coreceptors for T-cell- and macrophage/monocyte-tropic strains of HIV-1. CXCr4 is the major coreceptor for T-cell-tropic strains; entry of such strains (denoted X4) is inhibited by the receptor's natural ligand, Sdf-1. CCr5 is the major coreceptor for macrophage/monocyte-tropic strains (denoted R5), and their entry is inhibited by the receptor's natural ligands, Rantes, and the macrophage inflammatory proteins Mip-1α and Mip-1β. Primary T cells and monocytes produce both coreceptors; primary T cells are susceptible to both strains, but monocytes can be infected only by M-tropic strains for reasons that are not yet clear. Adapted from Fig. 3 of A. S. Fauci, *Nature* **384:**529 533, 1996, with permission.

cells in culture and has been found in small quantities in several tissues of the body. As discussed below, infection of these cells and tissues is likely to be relevant to HIV-1 pathogenesis in humans.

Routes of Transmission

Sources of Virus Infection
Even before HIV-1 was identified, epidemiologists had established the most likely routes of its transmission to be via sexual contact, via blood, and from mother to child. As might be anticipated, the efficiency of transmission is influenced greatly by the concentration of the virus in the body fluid to which an individual is exposed. Table 6.3 provides estimates of the percentage of infected cells and the concentration of HIV-1 in different body fluids. The highest values are observed in peripheral blood monocytes, in blood plasma, and in cerebrospinal fluid, but semen and female genital secretions also appear to be important sources of the virus.

Other routes of transmission are relatively unimportant or nonexistent, at least for HIV-1; among these are casual nonsexual contact, exposure to saliva or urine from infected individuals, and exposure to blood-sucking insects. Fortunately, HIV-1 infectivity is reduced upon air drying (by 90 to 99% within 24 h), by heating (56 to 60°C for 30 min), by exposure to standard germicides (such as 10% bleach or 70% alcohol), or by exposure to pH extremes (e.g., <6 or >10 for 10 min). This information and results from epidemiology studies have been used to establish safety regulations to prevent transmission in the public sector and in the health care setting.

Modes of Transmission
Modes of HIV-1 transmission vary in different geographic locations. In the United States, the major mode is via homosexual contact, although the number and relative proportion of heterosexual transmissions has increased since the mid-1990s; heterosexual transmission remains the most common world wide (Figure 6.9). A single contact can be sufficient for transmission of the virus if the infected partner is highly viremic. The presence of other sexually transmitted diseases also increases the probability of HIV-1 transmission, presumably because infected inflammatory cells may be present in both seminal and vaginal fluids. Genital ulceration and consequent direct exposure to infected blood cells also increase the likelihood of transmission. In both heterosexual and male homosexual contact, the recipient partner is the one most at risk.

Intravenous drug use is the next most common route of transmission, owing to the common practice of sharing contaminated needles and other drug paraphernalia. Here again, the probability of transmission is a function of the frequency of exposure and the degree of viremia among

Table 6.3 Isolation of infectious HIV-1 from body fluids[a]

Fluid	Virus isolation[b]	Estimated quantity of virus[c]
Cell-free fluid		
Cerebrospinal fluid	21/40	10–10,000
Ear secretions	1/8	5–10
Feces	0/2	None detected
Milk	1/5	<1
Plasma	33/33	1–5,000[d]
Saliva	3/55	<1
Semen	5/15	10–50
Sweat	0/2	None detected
Tears	2/5	<1
Urine	1/5	<1
Vaginal-cervical	5/16	<1
Infected cells		
Bronchial fluid	3/24	Not determined
PBMC	89/92	0.001–1%[d]
Saliva	4/11	<0.01%
Semen	11/28	0.01–5%
Vaginal-cervical fluid	7/16	Not determined

[a]Adapted from Table 2.1 (p. 28) of J. A. Levy, *HIV and the Pathogenesis of AIDS*, 3rd ed. (ASM Press, Washington, DC, 2007), with permission.

[b]Number of samples positive/number analyzed.

[c]For cell-free fluid, units are infectious particles per milliliter; for infected cells, units are percentages of total cells capable of releasing virus. Results from studies in the laboratory of J. A. Levy are presented.

[d]High levels associated with acute infection and advanced disease (~5 × 10⁶ PBMCs/ml/of blood).

a drug user's contacts. Of course, sexual partners of drug users are also at increased risk.

Until 1985, when routine HIV antibody testing of donated blood was established in the United States and other industrialized countries, individuals who received blood transfusions or certain blood products, such as clotting factors VIII and IX, were at high risk of becoming infected. Transfusion of a single unit (500 ml) of blood from an HIV-1-infected individual nearly always led to infection of the recipient. Appropriate heat treatment of clotting factor preparations and, more recently, their production by biotechnology have eliminated transmission from this source. This safeguard was small comfort to the many hemophiliacs who contracted the disease before this route was understood. Fortunately, other blood products, such as pooled immunoglobulin, albumin, and hepatitis B vaccine, were not implicated in HIV-1 transmission, presumably because their production methods include steps that destroy the virus.

Transmission of HIV from mother to child can occur across the placenta (5 to 10%) or at the time of delivery as a consequence of exposure to a contaminated genital tract (ca. 15%). The virus can also be transmitted via infected cells in the mother's milk during breast-feeding. Rates of transmission from an infected mother to a child range from as low as 11% to as high as 60%, depending on the severity of infection (i.e., the concentration of virus present) in the mother and the prevalence of breast-feeding (the frequency

Figure 6.9 Modes of transmission of HIV in the United States and worldwide. (A) Comparison of the number of AIDS cases in the United States by major transmission category and year of diagnosis. Adapted from Fig. 1 (p. 591) of Centers for Disease Control and Prevention, *Morb. Mortal. Wkly. Rep.* **55:**589–592, 2006. **(B)** Transmission of HIV worldwide, shown as percentage of cases in each major transmission category. From January 1998 data published by the Joint United Nations Programme on HIV/AIDS.

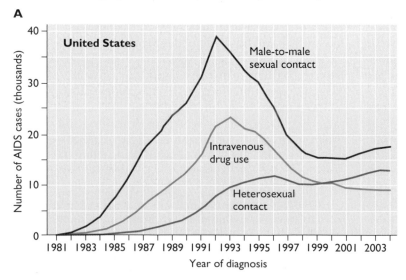

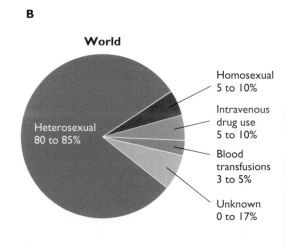

of the infant's exposure). Administration of antiviral drug therapy during pregnancy is an effective measure to reduce the amount of virus to which the newborn is exposed and, therefore, to reduce the frequency of transmission. Even a single treatment with an antiviral drug early in labor can reduce the incidence significantly. Unfortunately, because of the cost, this is not often an option in underdeveloped countries, where the risk is greatest. Campaigns focused on discouraging breast-feeding by infected mothers, although logical, have actually led to decreased infant survival, primarily because breast-feeding protects against a variety of other infections that are prevalent in these parts of the world. Each year an estimated 700,000 children are newly infected worldwide.

Mechanics of Spread

Except in cases of direct needle sticks or blood transfusion, HIV enters the body through mucosal surfaces, as do most viruses (see Chapter 1). In the case of sexual transmission, the initial target cells in the rectum or genital tract have not yet been identified. The most likely sources of transmission are virus-infected cells, as they can be present in much larger numbers than free infectious virus particles in vaginal or seminal fluids. Results of cell culture studies show that HIV-1 can be transferred directly to CD4⁻ epithelial cells via cell-cell contact. Whether such transfer is relevant to natural transmission is unknown. Results from analyses of tissue biopsy specimens of bowel mucosae and cervical and uterine epithelia suggest that cells in these layers can be infected in the absence of any injury. As noted in the preceding section, this infection might occur by interaction of the virus with galactosyl ceramide or Fc receptors on the mucosal cells.

The main routes of initial infection are likely to include the acquisition of virus by cells of the mucosal and cutaneous immune system described in Chapter 4; these include M cells, present in the bowel epithelium; dendritic, antigen-presenting $CD4^+$ Langerhans' cells in the vaginal and cervical epithelia; and $CD4^+$ T cells present in the intestinal and genital mucosa. In addition, dendritic cells express a glycoprotein on their surface, called DC-Sign (dendritic-cell-specific, Icam-3-grabbing nonintegrin), that binds the HIV-1 envelope protein with high affinity and can stabilize the virus for several days until it encounters a susceptible T cell (Fig. 6.10). Activated T cells, whose numbers are usually elevated at sites of genital infections that cause lesions, are also likely targets. Although the insertive partner is at relatively low risk for infection, transmission to the male

A

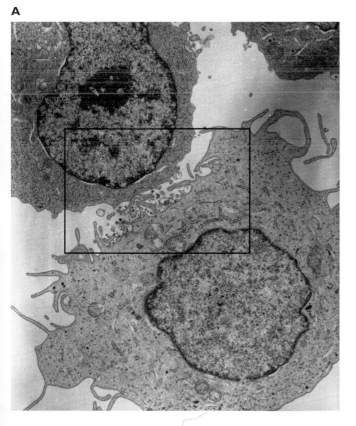

B

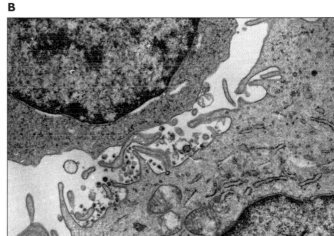

Figure 6.10 HIV particles in a virological synapse between a mature dendritic cell and a susceptible T cell. Large numbers of HIV particles are concentrated at the mature dendritic–T-cell junction (dendritic cell in lower section). The inset shows a higher-magnification image of the boxed area. Electron micrographic images were produced by Clive Wells and provided through the courtesy of Li Wu, Medical College of Wisconsin. Reprinted from J.-H. Wang, A. M. Janas, W. J. Olson, and L. Wu, *J. Virol.* **81:**8933–8943, 2007, with permission.

can occur through cells in the lining of the urethral canal of the penis, presumably from infected macrophages or Langerhans' cells in the cervix or the intestinal mucosa of the infected partner. Uncircumcised males have a twofold-increased risk of infection, suggesting that the mucosal lining of the foreskin may be susceptible to HIV infection. Both male and female hormones appear to facilitate HIV transmission by stimulating cell-cell contact (prostaglandins) or erosion of the vaginal lining (progestin).

Free virus, virus attached to dendritic cells, or virus-infected cells enter draining lymph nodes or the circulatory system, where they encounter the next major targets, namely, susceptible cells that bear the CD4 receptor. Non-activated peripheral blood mononuclear cells are not very permissive for infection, and few activated CD4$^+$ lymphocytes are circulating in the blood at any given time. The first CD4$^+$ cells in the blood to be infected are therefore probably macrophages. The macrophages are in a differentiated state, permissive for viral replication, and can pass progeny virions to activated lymphocytes in the lymph nodes. From this point the infection runs its protracted but usually inevitable course.

The Course of Infection

Patterns of Virus Appearance and Immune Cell Indicators of Infection

Pathological conditions associated with different phases in HIV-1 infection are summarized in Table 6.4.

The Acute Phase

In the first few days after infection, the virus is produced in large quantities by the activated lymphocytes in lymph nodes, sometimes causing the nodes to swell (lymphadenopathy) or producing flu-like symptoms. Virus released into the blood can be detected by infectivity with appropriate cell cultures or by screening directly for viral RNA or proteins (Fig. 6.11). As many as 5×10^3 infectious virions or 1×10^7 viral RNA molecules (i.e., ~5 $\times 10^6$ particles) per ml of plasma can be found during this stage. During this time, some 30 to 60% of CD4$^+$ T cells in the gut are lost. The memory T cells in this location are most susceptible to infection, and a percentage of the surviving, quiescent memory cells harbor replication-competent proviruses that cannot be transcribed in these cells. They form a long-lived latent viral reservoir. Interaction of such memory cells with their cognate antigens, sometimes many years after the initial HIV infection, will lead to their activation and subsequent transcription of their latent provirus. If antiviral treatments have not been continued, these progeny viruses can initiate a new round of infection.

The initial peak of viremia is greatly curtailed within a few weeks after initial infection, as the susceptible T-cell population is depleted and a cell-mediated immune response is mounted. The number of CTLs increases before neutralizing antibodies can be detected. The inflammatory response that occurs upon primary infection stimulates the production of additional CD4$^+$ T cells, which stems the depletion of this population. The CD4$^+$ T-cell count returns to near normal levels, but these cells represent a

Table 6.4   Pathological conditions associated with HIV-1 infection[a]

Acute phase

Mononucleosis-like syndrome: fever, malaise, pharyngitis, lymphadenopathy, headache, arthralgias, diarrhea, maculopapular rash, meningoencephalitis

Asymptomatic phase

Often none, but patients may present sporadically with one or more of the following symptoms: fatigue, mild weight loss, generalized lymphadenopathy, thrush, oral hairy leukoplakia, shingles

Symptomatic phase and AIDS

200–500 CD4$^+$ T cells/ml

  Generalized lymphadenopathy

  Oral lesions (thrush, hairy leukoplakia, aphthous ulcers)

  Shingles

  Thrombocytopenia

  Molluscum contagiosum

  Basal cell carcinomas of the skin

  Headache

  Condyloma acuminatum

  Reactivation of latent *Mycobacterium tuberculosis*

Fewer than 200 CD4$^+$ T cells/ml

  Protozoal infections: *Pneumocystis jiroveci, Toxoplasma gondii, Isospora belli, Cryptosporidium*, microsporidia

  Bacterial infections: *Mycobacterium avium-M. intracellulare, Treponema pallidum*

  Fungal infections: *Candida albicans, Cryptococcus neoformans, Histoplasma capsulatum*

  Viral infections and malignancies: cytomegalovirus, recurrent bouts of oral or genital herpes simplex virus infection, lymphoma (mostly Epstein-Barr virus some human herpesvirus 8), Kaposi's sarcoma (human herpesvirus 8), anogenital carcinoma (human papillomavirus)

  Neurological symptoms: aseptic meningitis; myelopathies such as vacuolar myelopathy; pure sensory ataxia, paresthesia/dysesthesia; peripheral neuropathies such as acute demyelinating polyneuropathy, mononeuritis multiplex, and distal symmetric polyneuropathy; myopathy; AIDS dementia complex

[a] Adapted from Table 1 (p. 597) of A. S. Fauci and R. C. Desrosiers, *in* J. M. Coffin et al. (ed.), *The Retroviruses* (Cold Spring Harbor Laboratory Press, Plainview, NY) with permission.

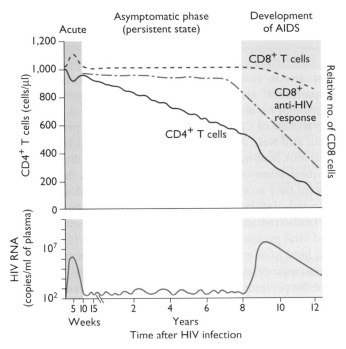

Figure 6.11 Schematic diagram of events occurring after HIV-1 infection. Adapted from Fig. 13.1 of J. A. Levy, *HIV and the Pathogenesis of AIDS*, 3rd ed. (ASM Press, Washington, DC, 2007), with permission.

new source of susceptible cells and their infection produces chronic immune stimulation.

During the period of acute infection, the virus population is relatively homogeneous. In most cases, it appears that the predominant virus was a minor variant in the population present in the source of the infection. These early-stage isolates are generally CCr5-macrophage tropic; the reason for such selective transmission is a topic of intense investigation, as it has great bearing on vaccine development.

The Asymptomatic Phase

By 3 to 4 months after infection, viremia is usually reduced to low levels, with small bursts of virus appearing from time to time. It is known that the degree of viremia at this stage of infection, the so-called **virologic set point**, is a direct predictor of how fast the disease will progress in a particular individual: the higher the set point, the faster the progression. During this time, CD4+ T cell numbers decrease at a steady rate, estimated to be approximately 60,000 cells/ml/year. Direct cytopathogenicity by the virus and apoptosis due to continued immune stimulation and inappropriate cytokine production seem likely explanations. In this protracted asymptomatic period, which can last for years, the CTL level remains slightly elevated, but virus replication continues at a low rate, mainly in the lymph nodes. In

lymphoid tissues a relatively large, stable pool of virions can be detected bound to the surface of follicular dendritic cells. Small numbers of infected T cells are also observed. During this asymptomatic phase of persistent infection, known also as **clinical latency**, only 1 in 300 to 400 infected cells in the lymph nodes may actually release virus. It is thought that, as in acute infection, virus propagation is suppressed at this stage by the action of antiviral CTLs. The number of these specific lymphocytes decreases toward the end of this stage. During the asymptomatic phase, the virus population becomes more heterogeneous, probably because of continual selection for specific mutations as a result of immunological pressures (Chapter 4).

The Symptomatic Phase and AIDS

The end stage of disease, when the infected individual develops symptoms of AIDS, is characterized by a CD4+ T-cell count below 200 per ml and increased quantities of virus. The total CTL count also decreases, probably owing to the precipitous drop in the number specific for HIV. In the lymph nodes, virus replication increases with concomitant destruction of lymphoid cells and of the normal architecture of lymphoid tissue. The cause of this lymph node degeneration is not clear; it may be due directly to virus replication or may be an indirect effect of chronic immune stimulation.

In this last stage, the virus population again becomes relatively homogeneous and, generally, CXCr4 T-cell tropic. Properties associated with increased virulence predominate, including an expanded cellular host range, ability to cause formation of syncytia, rapid replication kinetics, and CD4+ T-cell cytopathogenicity. Late-emerging virus also appears to be less sensitive to neutralizing antibodies and more readily recognized by antibodies that enhance infectivity. In some cases, strains that have enhanced neurotropism or increased pathogenicity for other organ systems emerge. Where analyzed, these changes can be traced to specific mutations, for example, in the viral envelope gene or in a regulatory gene (e.g., *tat*).

Variability of Response to Infection

Studies of large cohorts of HIV-1-infected adults show that approximately 10% progress to AIDS within the first 2 to 3 years of infection. Over a period of 10 years, approximately 80% of infected adults show evidence of disease progression and, of these, 50% have developed AIDS. Of the remainder, 10 to 17% are AIDS free for over 20 years; a very small percentage of these individuals are completely free of symptoms, with no evidence of progression to disease. What are the parameters that contribute to such variability?

One parameter is the degree to which an individual's immune system may be stimulated by infection with other antibodies against platelets, T cells, and peripheral nerves were detected in AIDS patients. Subsequently, autoantibodies to a large number of normal cellular proteins have been found in infected individuals (Table 6.6). The specific reason for the appearance of such antibodies is not clear, but their production might be stimulated in part by cellular proteins on the surface of viral particles or by viral proteins, regions of which may resemble cellular proteins (called molecular mimicry) (Chapter 4 and Table 6.7).

Immune Responses to HIV

Humoral Responses

Antibodies to HIV-1 can be detected shortly after acute infection, sometimes as early as a few days after exposure to the virus but generally within 1 to 3 months. These antibodies, which are secreted into the blood and are present on mucosal surfaces of the body, can be detected in genital and other body fluids. This phenomenon has been exploited in the design of home kits for detecting anti-HIV antibodies in the blood or urine. Among the various isotypes, IgG1 antibodies are known to play a dominant role at all stages of infection, giving rise to an antibody-dependent cellular cytotoxicity response, complement-dependent cytotoxicity, and neutralizing and blocking responses (Fig. 6.12; Chapter 4). Levels of other classes of antibody may vary at different clinical times, but there is no known correlation between the isotype and the clinical stage of disease.

Neutralizing antibodies are likely to play a role in limiting viral replication during the early, asymptomatic stage

Table 6.6 Some autoantibodies detected in HIV infection[a]

Antibodies to:	Associated clinical condition
CD4	CD4+ cell loss
Cellular components (Golgi complex, centriole, vimentin)	Immune disorder
Erythrocytes	Anemia
HLA	Lymphocyte depletion
Lymphocytes	Loss of CD4+, CD8+, and B lymphocytes
Myelin basic protein	Dementia, demyelination
Nerves (myelin)	Peripheral neuropathy
Neutrophils	Neutropenia
Nuclear protein (antinuclear antibody)	Autoimmune symptoms
Platelets	Thrombocytopenia
Sperm, seminal plasma	Aspermia

[a]Adapted from Table 10.4 (p. 253) of J. A. Levy, *HIV and the Pathogenesis of AIDS*, 3rd ed. (ASM Press, Washington, DC, 2007), with permission.

Table 6.7 Some regions of HIV that resemble normal cellular proteins and exhibit cross-reactivity[a]

Normal cellular protein	HIV protein(s)
Astrocyte protein	MA, TM
Brain cell protein	SU (V3 loop)
Epithelial cell protein	MA
Neuroleukin (phosphohexose)	SU
Platelet glycoprotein	SU
Platelet protein	CA
Thymosin	MA, TM
Vasoactive intestinal polypeptide	SU (peptide T)

[a]Adapted from Table 10.6 (p. 255) of J. A. Levy, *HIV and the Pathogenesis of AIDS*, 3rd ed. (ASM Press, Washington, DC, 2007), with permission.

of infection. However, the titers of these antibodies are generally very low. This property is consistent with recent structural analyses of the HIV-1 envelope protein, which indicate that many neutralizing epitopes are hidden from the immune system. The low titer may favor selection of resistant mutants. Indeed, many individuals produce antibodies that neutralize earlier virus isolates but not those present at the time of serum collection, suggesting effective immune "escape" by the virus. Some studies show loss of neutralizing antibodies with progression to AIDS, but the clinical relevance of such antibodies during the later stages of infection remains obscure.

Neutralizing antibodies generally bind to specific sites on the viral envelope complex, SU-TM. Variable region 3 (V3) in HIV-1 SU is one of the initial targets. Anti-V3 antibodies appear to block coreceptor interactions that occur after the virus attaches to the CD4 receptor. Because of the high sequence variation within the V3 loop (hence the name "variable"), neutralizing antibodies to this region are usually strain specific. Consequently, despite its relatively strong antigenicity, V3 is not a good target for the development of vaccines or broadly specific antiviral drugs. Members of another class of neutralizing antibodies block the binding of SU to the CD4 receptor. These antibodies bind to numerous conserved sites on SU and usually react with many strains of HIV-1. Other conserved or variable regions on both SU and TM can be targets for neutralizing antibodies. Broad neutralizing activity against carbohydrate-containing regions of the viral envelope protein has also been detected. Even less well understood is the relative importance of antibodies to other proteins on the surface of virions, such as the adhesion molecules Lfa-1 and Icam. Antibodies to these cell surface molecules inhibit the formation of syncytia following infection, perhaps because antibody-treated cells are less likely to form aggregates. In some instances, these antibodies seem to

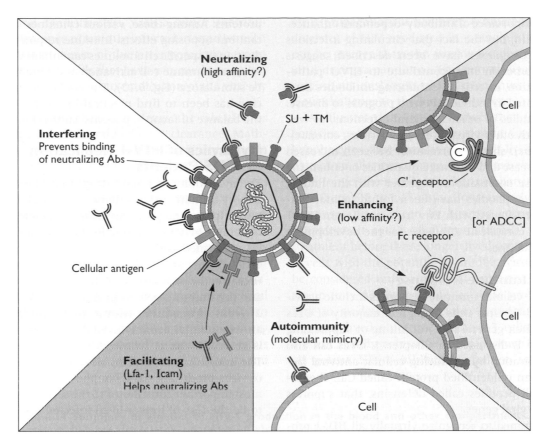

Figure 6.12 Antibody (Ab) responses to HIV infection. A summary of the various responses described in the text is presented. "Cellular antigen" refers to cellular membrane proteins that are incorporated in the virus envelope. One idea is that the relative affinities of the antibodies may be critical. According to this hypothesis, high-affinity antibodies neutralize the virus by binding tightly to SU, causing it to become detached from the virion; low-affinity antibodies bind to SU but not tightly enough to cause its detachment. Conformational changes in SU that might occur as a consequence of such low-affinity binding would then facilitate viral entry. C', complement; ADCC, antibody-dependent cellular cytotoxicity.

enhance neutralization of HIV-1 by antiviral antibodies (Fig. 6.12, bottom left).

Antibodies called **interfering antibodies** can bind to virions or infected cells and block interaction with neutralizing antibodies (Fig. 6.12, left). Others, called **enhancing antibodies**, can actually facilitate infection by allowing virions coated with them to enter susceptible cells (Fig. 6.12, right). In complement-mediated antibody enhancement, the complement receptors Cr1, Cr2, and Cr3 play a critical role in attaching virion-antibody complexes to susceptible cells. In Fc-mediated enhancement, attachment is via Fc receptors that are abundant not only on monocytes/macrophages and NK cells but also on other human cell types. As HIV-1 has been shown to replicate in cells that lack CD4 but express an Fc receptor, binding to the CD4 receptor is probably not required for Fc-mediated enhancement.

It is noteworthy that the same cellular receptors (for complement and Fc) are implicated in infection enhancement and the antibody-dependent cellular cytotoxicity response. In the case of enhancement, the receptors allow antibody-coated virions to enter susceptible cells bearing such receptors. In the case of antibody-dependent cellular cytotoxicity, such receptors on CTLs, NK cells, or monocytes/macrophages mediate the recognition and killing of antibody-coated infected cells.

Both neutralizing and enhancing antibodies recognize epitopes on SU and TM. Consequently it has been difficult to identify the features that specify either response. Indeed, polyclonal antibodies against SU possess both neutralizing and enhancing activities, and it has not been possible to decide how either effect might predominate. This idea and the distinctions between other classes of anti-HIV-1 antibodies are summarized in Fig. 6.12.

J. J. Goedert, T. R. O'Brien, M. W. Hilgartner, D. Vlahov, S. J. O'Brien, and M. Carrington. 1998. Genetic acceleration of AIDS progression by a promoter variant of *CCR5*. *Science* **282**:1907–1911.

Smith, M. W., M. Dean, M. Carrington, C. Winkler, G. A. Huttley, D. A. Lomb, J. J. Goedert, T. R. O'Brien, L. P. Jacobson, R. Kaslow, S. Buchbinder, E. Vittinghoff, D. Vlahov, K. Hoots, M. W. Hilgartner, and S. J. O'Brien. 1997. Contrasting genetic influence of *CCR2* and *CCR5* variants on HIV-1 infection and disease progression. *Science* **277**:959–965.

Winkler, C., W. Modi, M. W. Smith, G. W. Nelson, X. Wu, M. Carrington, M. Dean, T. Honjo, K. Tashiro, D. Yabe, S. Buchbinder, E. Vittinghoff, J. J. Goedert, T. R. O'Brien, L. P. Jacobson, R. Detels, S. Donfield, A. Willoughby, E. Gomperts, D. Vlahov, J. Phair, and S. J. O'Brien. 1998. Genetic restriction of AIDS pathogenesis by an SDF-1 chemokine gene variant. *Science* **279**:389–393.

Websites

http://www.unaids.org *UNAIDS, the Joint United Nations Programme on HIV/AIDS*

http://www.iavireport.org *Archives of the International AIDS Vaccine Initiative*

7

Introduction
   Properties of Transformed Cells
   Control of Cell Proliferation

Oncogenic Viruses
   Discovery of Oncogenic Viruses
   Viral Genetic Information
   in Transformed Cells
   The Origin and Nature of Viral
   Transforming Genes
   Functions of Viral Transforming Proteins

**Activation of Cellular Signal
Transduction Pathways by Viral
Oncogene Products**
   Viral Mimics of Cellular Signaling
   Molecules
   Alteration of the Production or Activity
   of Cellular Signal Transduction Proteins

**Disruption of Cell Cycle Control
Pathways by Viral Oncogene
Products**
   Abrogation of Restriction Point Control
   Exerted by the Rb Protein
   Production of Virus-Specific Cyclins
   Inactivation of Cyclin-Dependent Kinase
   Inhibitors

**Transformed Cells Must Also Grow
and Survive**
   Integration of Mitogenic and Growth-
   Promoting Signals
   Mechanisms That Permit Survival of
   Transformed Cells

**Tumorigenesis Requires
Additional Changes in the
Properties of Transformed Cells**
   Inhibition of Immune Defenses

**Other Mechanisms of
Transformation and Oncogenesis
by Human Tumor Viruses**
   Nontransducing, Complex Oncogenic
   Retroviruses: Tumorigenesis with Very
   Long Latency
   Oncogenesis by Hepatitis Viruses

Perspectives

References

Transformation and Oncogenesis

Introduction

Cancer is a leading cause of death in developed countries: about 500,000 individuals succumb each year in the United States alone. Consequently, efforts to understand and control this deadly disease have long been high priorities for public health institutions. Our general understanding of the mechanisms of **oncogenesis**, the development of cancer, as well as of normal cell growth, has improved enormously in the past half century. Such progress can be traced in large part to efforts to elucidate how members of several virus families cause cancer in animals. In fact, as we discuss in this chapter, study of such oncogenic viruses has led to a detailed understanding of the molecular basis of oncogenesis.

It is now clear that cancer (defined in Box 7.1) is a genetic disease: it results from the growth of successive populations of cells in which mutations have accumulated (Box 7.2). These mutations affect various steps in the regulatory pathways that control cell communication, growth, and proliferation, and lead to uncontrolled growth, increasing tissue disorganization, and ultimately cancer. One or more mutations may be inherited (Box 7.2), or they may arise as a consequence of endogenous DNA damage as well as exposure to environmental carcinogens or infectious agents, including viruses. It is estimated that viruses are a contributing factor in approximately 20% of all human cancers. For some, such as liver and cervical cancer, they are the major cause. However, it is important to understand that the induction of malignancy is **not** a requirement for the propagation of any virus. Rather, this unfortunate outcome for the host is a side effect of either infection or the host's response to the presence of the virus. From this perspective, viruses can be thought of as cofactors, or unwitting initiators of oncogenesis.

Understanding the development of cancer ultimately depends on knowledge of how individual cells behave within an animal. As described in Chapters 1, 3, and 4, analysis of viral pathogenesis must encompass a consideration of the organism as a whole, especially the body's immune defenses. However, understanding how members of several virus families cause cancer in animals

BOX 7.1 TERMINOLOGY
Some cancer terms

Benign An adjective used to describe a growth that does not infiltrate into surrounding tissues; opposite of malignant

Cancer A malignant tumor; a growth that is not encapsulated and that infiltrates into surrounding tissues, replacing normal with abnormal cells; it is spread by the lymphatic vessels to other parts of the body; death is caused by destruction of organs to a degree incompatible with their function, by extreme debility and anemia, or by hemorrhage

Carcinogenesis The complex, multistage process by which a cancer develops

Carcinoma A cancer of epithelial tissue

Endothelioma A cancer characterized by overproduction of erythrocytes

Fibroblast A cell derived from connective tissue

Fibropapilloma A solid tumor of cells of the connective tissue

Hepatocellular carcinoma A cancer of liver epithelial cells

Leukemia A cancer of white blood cells

Lymphoma A cancer of lymphoid tissue

Malignant An adjective applied to any disease of a progressive and fatal nature; opposite of benign

Neoplasm An abnormal new growth, i.e., a cancer

Oncogenic Causing a tumor

Retinoblastoma A cancer of cells of the retina

Sarcoma A cancer of fibroblasts

Tumor A swelling, caused by abnormal growth of tissue, not resulting from inflammation; may be benign or malignant

began with studies of cultured cells in the laboratory. In particular, early investigators noticed that the growth properties and morphologies of some cultured cells could be changed upon infection with certain viruses. We therefore consider such cells **transformed**. The advantages of these cell culture systems are many. The molecular virologist can focus attention on particular cell types, can manipulate their behavior in a controlled manner, and can easily distinguish effects specific to the virus. In many cases, cells transformed by viruses in culture can form tumors when implanted in animals. But tumors do not always form, and it is important to realize that transformed cultures are **not** tumors. The major benefit of cell culture systems is that they allow researchers to study the molecular events that establish an oncogenic potential in virus-infected cells.

Properties of Transformed Cells

Cellular Transformation

The proliferation of cells in the body is a strictly regulated process. In a young animal, total cell multiplication exceeds cell death as the animal grows to maturity. In an adult, the processes of cell multiplication and death are carefully balanced. For some cells, high rates of proliferation are required to maintain this balance. For example, human intestinal cells and white blood cells have half-lives of only a few days and need to be replaced rapidly. On the other hand, red blood cells live for over 100 days, and healthy neuronal cells rarely die. Occasionally, this carefully regulated process breaks down, and a particular cell begins to grow and divide even though the body has sufficient numbers of its type; such a cell behaves as if it were immortal. Acquisition of **immortality** is generally acknowledged to be an early step in oncogenesis. An immortalized cell may acquire one or more additional genetic changes to give rise to a clone of cells that is able to expand, ultimately forming a mass called a **tumor**. Some tumors are **benign**; they cease to grow in the body after reaching a certain size and generally do no great harm. Other tumor cells grow and divide indefinitely to form **malignant** tumors that damage and impair the normal function of organs and tissues. Occasionally, some cells in a tumor acquire additional mutations that confer the ability to escape the boundary of the mass, to invade surrounding tissue, and to be disseminated to other parts of the body, where the cells take up residence. There they continue to grow and divide, giving rise to secondary tumors called **metastases**. Such cells cause the most serious and life-threatening disease.

Many studies of the molecular biology of oncogenic animal viruses employed primary cultures of normal cells, for example, rat or mouse embryo fibroblasts. The advantage of these cells is that all the molecular changes necessary to convert them to the oncogenic state can be studied. Such primary cells, like their normal counterparts in the animal, have a finite capacity to grow and divide in culture. Cells from some animal species, such as rodents, undergo a spontaneous transformation when maintained in culture. Immortalized cells appear after a "crisis" period in which the great majority of the cells die (Fig. 7.1). As such cells are otherwise normal, and do not induce tumors when introduced into animals, they can be used to identify viral gene products needed for steps in transformation subsequent to immortalization. For reasons that are not fully understood, human and simian cells rarely undergo spontaneous transformation to immortality when passaged in culture. In fact, established lines of human cells generally can be derived only from tumors, or following exposure of primary cells to chemical carcinogens or to oncogenic

BOX 7.2

Genetic alterations associated with the development of colon carcinoma

Colorectal cancer is the fourth most common cancer worldwide, and the second most frequent in developed countries. The clinical stages in the development of this cancer are particularly well defined. Furthermore, as shown, several genes that are frequently mutated to allow progression from one stage to the next have been identified. The early adenomas or polyps that initially form are benign lesions. Their conversion to malignant metastatic carcinomas correlates with the acquisition of additional mutations in the *p53* and *dcc* ("deleted in colon carcinoma") genes. Inherited mutations in the genes listed can greatly increase the risk that an individual will develop colon carcinoma. For example, patients with familial adenomatous polyposis can inherit defects in the *apc* (adenomatous polyposis coli) gene that result in the development of hundreds of adenomatous polyps. The large increase in the **number** of these benign lesions increases the chance that some will progress to malignant carcinomas. In contrast, patients with hereditary nonpolyposis colorectal cancer develop polyps at the same rate as the general population. However, polyps develop to carcinomas more frequently in these patients, because defects in genes encoding proteins that correct mismatched bases in DNA lead to a higher mutation rate. Consequently, the **likelihood** that an individual polyp will develop into a malignant lesion increases from 5 to 70%.

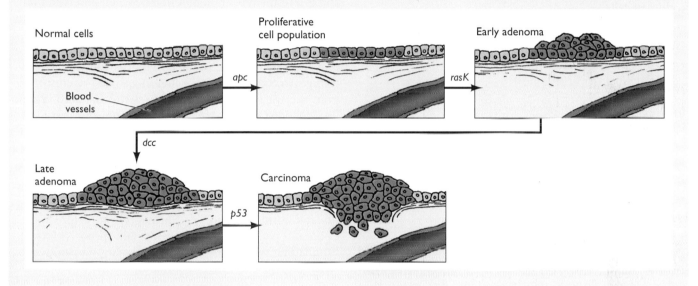

RNA or DNA viruses (or their transforming genes). The realization that such transformed cells share a number of common properties, regardless of how they were obtained, provided a major impetus for the investigation of viral transformation.

Properties That Distinguish Transformed from Normal Cells

The definitive characteristic of transformed cells is their lack of response to the signals or conditions that normally control DNA replication and cell division. This property is illustrated by the list of growth parameters and behaviors provided in Table 7.1. As described above, transformed cells are immortal: they can grow and divide indefinitely, provided that they are diluted regularly into fresh medium. Production of **telomerase**, an enzyme that maintains telomeric DNA at the ends of chromosomes,

has been implicated in immortalization. In addition, transformed cells typically exhibit a reduced requirement for the growth factors present in serum. Some transformed cells actually produce their own growth factors and the cognate receptors, providing themselves **autocrine growth stimulation**. Normal cells cease to grow and enter a quiescent state (called G_0, described in "Control of Cell Proliferation" below) when essential nutrient concentrations drop below a threshold value. Transformed cells are deficient in this capacity, and some may even kill themselves by trying to continue to grow in an inadequate environment.

Transformed cells grow to high densities. This characteristic is manifested by the cells piling up, over, or under each other. They also grow on top of untransformed cells, forming visually identifiable clumps called **foci** (Fig. 7.2). Transformed cells behave in this manner because they have lost **contact inhibition**, a response by which normal cells

A Mouse cells

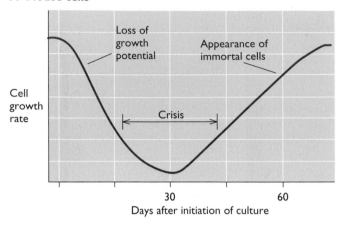

B Human cells

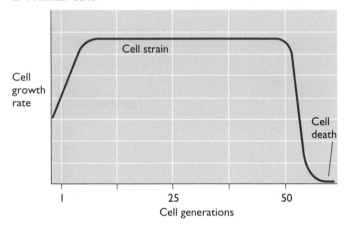

Figure 7.1 Stages in the establishment of a cell culture. **(A)** Mouse or other rodent cells. When mouse embryo cells are placed in culture, most cells die before healthy growing cells emerge. As these cells are maintained in culture, they begin to lose growth potential and most cells die (the culture goes into crisis). Very rarely cells do not die but continue growth and division until their progeny overgrow the culture. These cells constitute a cell line, which will grow indefinitely if it is appropriately diluted and fed with nutrients: the cells are immortal. **(B)** Human cells. When an initial explant is made (e.g., from foreskin), some cells die and others (mainly fibroblasts) start to grow; overall, the growth rate increases. If the surviving cells are diluted regularly, the cell strain grows at a constant rate for about 50 cell generations, after which growth begins to decrease. Eventually, all the cells die.

cease growth and movement when they sense the presence of their neighbors. Unlike normal cells, many transformed cells have also lost the need for a surface on which to adhere, and we describe them as being **anchorage independent**. Some anchorage-independent cells form isolated colonies in semisolid media (e.g., 0.6% agar). This property correlates well with the ability to form tumors in animals and often is used as an experimental surrogate for malignancy. Finally, transformed cells **look** different from

Table 7.1 Growth parameters and behavior of transformed cells

Immortal: can grow indefinitely
Reduced requirement for serum growth factors
Loss of capacity for growth arrest upon nutrient deprivation
Growth to high saturation densities
Loss of contact inhibition (can grow over one another or normal cells)
Anchorage independence (can grow in soft agar)
Altered morphology (appear rounded and refractile)
Tumorigenic

normal cells; they are more rounded, with fewer processes, and as a result appear more refractile when observed under a microscope (Fig. 7.2).

There are other ways in which transformed cells can be distinguished from their normal counterparts. These properties include metabolic differences and characteristic changes in cell surface and cytoskeletal components. However, the list in Table 7.1 comprises the standard criteria used to judge whether cells have been transformed.

Control of Cell Proliferation

Sensing the Environment

As noted previously, proliferation of cells in an organism is strictly regulated to maintain tissue or organ integrity and normal physiology. Normal cells therefore possess elaborate pathways that receive and process growth-stimulatory or growth-inhibitory signals transmitted by other cells in the tissue or organism. Much of what we know about these pathways comes from study of the cellular genes transduced or activated by oncogenic retroviruses. Signaling often begins with the secretion of a growth factor by a specific type of cell. The growth factor may enter the circulatory system, as in the case of many hormones, or may simply diffuse through the spaces around cells in a tissue. Growth factors bind to the external portion of specific receptor molecules on the surface of the same or other types of cells. Alternatively, signaling can be initiated by binding of a receptor on one cell to a specific protein (or proteins) present on the surface of another cell, or to components of the extracellular matrix (Volume I, Chapter 5). The binding of the ligand triggers a change, often via oligomerization of receptor molecules, that is transmitted to the cytoplasmic portion of the receptor. In the case illustrated in Figure 7.3, the cytoplasmic domain of the receptor possesses protein tyrosine kinase activity, and interaction with the growth factor ligand triggers autophosphorylation. This modification sets off a **signal transduction cascade**, a chain of sequential physical interactions among,

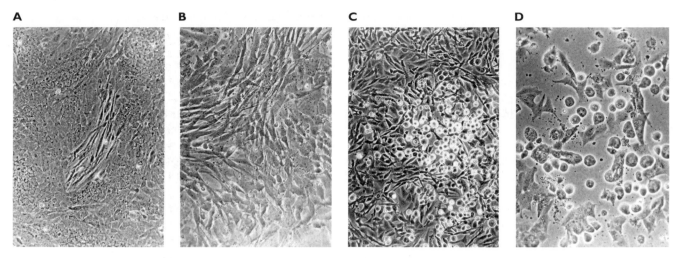

Figure 7.2 Foci formed by avian cells transformed with two strains of Rous sarcoma virus. Differences in morphology are due to genetic differences in the transduced *src* oncogene. **(A)** A focus of infected cells with fusiform morphology shown on a background of flattened, contact-inhibited, uninfected cells. **(B)** Higher magnification of a fusiform focus showing lack of contact inhibition of the transformed cells. **(C)** A focus of highly refractile infected cells with rounded morphology and reduced adherence. **(D)** Higher magnification of rounded infected cells, showing tightly adherent normal cells in the background. Courtesy of P. Vogt, The Scripps Research Institute.

and biochemical modifications of, membrane-bound and cytoplasmic proteins. Ultimately, the behavior of the cell is altered. Signaling proteins include small guanine nucleotide-binding proteins (G proteins), such as Ras, and protein kinases that phosphorylate serine or threonine residues. Signal transduction cascades can also include enzymes that produce small molecules (e.g., cyclic adenosine monophosphate AMP and certain lipids) that act as diffusible **second messengers** in the signal relay. Changes in ion flux across the plasma membrane, or in membranes of the endoplasmic reticulum, may also contribute to transmission of signals.

Relay of the signal can terminate at cytoplasmic sites to alter metabolism or cell morphology and adhesion. However, many signaling cascades culminate in the modification of transcriptional activators or repressors, and therefore alter the expression of specific cellular genes. The products of these genes either allow the cell to progress through another growth cycle or cause the cell to stop growing, to differentiate, or to die, whichever is appropriate to the situation. Errors in the signaling pathways that regulate these decisions can lead to cancer.

Regulation of the Cell Cycle

The capacity of cells to grow and divide is regulated by a molecular timer. The timer comprises an assembly of proteins that integrate stimulatory and inhibitory signals received by, or produced within, the cell. Eukaryotic cells do not attempt to divide until all their chromosomes have

been duplicated and are precisely organized for segregation into daughter cells. Nor do they initiate DNA synthesis and chromosome duplication until the previous cell division is complete, or unless the extra- and intracellular environments are appropriate. Consequently, the molecular timer controls a tightly ordered **cell cycle** comprising intervals, or **phases**, devoted to specific processes.

The duration of the phases in the cell cycle shown in Fig. 7.4 is typical of those of many mammalian cells growing actively in culture. However, there is considerable variation in the length of the cell cycle, largely because of differences in the **gap phases** (G_1 and G_2). For example, early embryonic cells of animals dispense with G_1 and G_2, do not increase in mass, and move immediately from the **DNA synthesis phase (S)** to **mitosis (M)** and again from M to S. Consequently, they possess extremely short cycles of 10 to 60 min. At the other extreme are cells that have ceased growth and division and have withdrawn from the cell cycle. The variability in duration of this specialized **resting state**, termed G_0, accounts for the large differences in the rates at which cells in multicellular organisms proliferate. As discussed in Volume I, Chapter 9, many viruses replicate successfully in cells that spend all or most of their lives in G_0, a state that has been likened to "cell cycle sleep." In many cases, synthesis of viral proteins in such resting or slowly cycling cells induces them to reenter the cell cycle and grow and divide rapidly. To describe the mechanisms by which these viral proteins induce such abnormal activity and transform cells, we first introduce

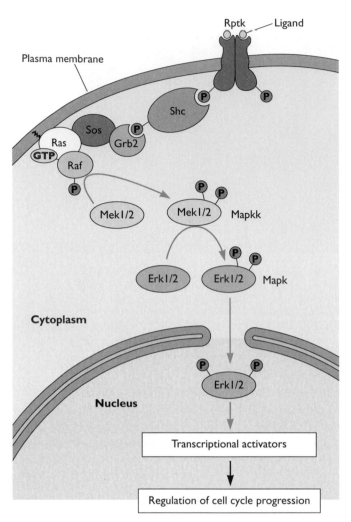

Figure 7.3 The mitogen-activated protein kinase signal transduction pathway. Signal transduction is initiated by binding of ligand to the extracellular domain of a receptor protein tyrosine kinase (Rptk), for example, the receptors for epidermal growth factor or platelet-derived growth factor. Binding of ligand (yellow circles) induces receptor dimerization and autophosphorylation of tyrosine residues in the cytoplasmic domain. Adapter proteins like Shc and the Grb2 component of the Grb2-Sos complex are recruited to the membrane by binding to these phosphotyrosine-containing sequences (or to a substrate phosphorylated by the activated receptor), along with Ras. Sos is the guanine nucleotide exchange protein for Ras and stimulates exchange of GDP for GTP bound to Ras by. The GTP-bound form of Ras binds to members of the Raf family of serine/threonine protein kinases. Raf then becomes autophosphorylated, and initiates the Map kinase (Mapk) cascade. The pathway shown contains Mek1 and Mek2 (Mek1/2) (Map kinase kinases [Mapkk]) and Erk1 and Erk2 (Erk1/2) (Map kinases). Phosphorylated Erk1/Erk2 can enter the nucleus, where they modify and activate transcriptional regulators. These kinases can also regulate transcription indirectly, by effects on other protein kinases.

the molecular mechanisms that control passage through the cell cycle.

The Cell Cycle Engine

The orderly progression of eukaryotic cells through periods of growth, chromosome duplication, and nuclear and cell division is driven by intricate regulatory circuits. The elucidation of these circuits must be considered a *tour de force* of contemporary biology. The first experimental hint that cells contain proteins that control transitions from one phase of the cell cycle to another came more than 35 years ago. Nuclei of slime mold (*Physarum polycephalum*) cells in early G_2 were found to enter mitosis immediately following fusion with cells in late G_2 or M. This crucial observation led to the conclusion that the latter cells must contain a **mitosis-promoting factor.** Subsequently, similar experiments with mammalian cells in culture identified an analogous S-phase-promoting factor. The convergence of many observations eventually led to the identification of the highly conserved components of the cell cycle engine (Fig. 7.5A).

Mitosis-promoting factor purified from amphibian oocytes proved to be an unusual protein kinase: its catalytic subunit is activated by the binding of an unstable regulatory subunit. Furthermore, the concentration of the regulatory subunit was found to oscillate reproducibly during each and every cell cycle. The regulatory subunit was therefore given the descriptive name **cyclin**, and the associated protein kinase was termed **cyclin-dependent kinase (Cdk)**. Similar proteins were implicated in cell cycle control in the yeast *Saccharomyces cerevisiae*, and it soon became clear that all eukaryotic cells contain multiple cyclins and Cdks, which operate in specific combinations to control progression through the cell cycle. The cyclins are related in sequence to one another, and they share such properties as activation of cyclin-dependent kinases and controlled destruction by the proteasome.

Figure 7.5A lists the various mammalian cyclin-Cdk complexes and shows the phases in the cell cycle in which they accumulate. The patterns of accumulation illustrate a critical feature of the cell cycle: individual cyclin-Cdks, the active protein kinases, accumulate in successive waves. The concentration of each increases gradually during a specific period in the cycle, but decreases abruptly as the cyclin subunit is degraded. In mammalian cells, proteolysis is important in resetting the concentrations of individual cyclins at specific points in the cycle, but production of cyclin mRNAs is also regulated. For example, the concentrations of mRNAs encoding cyclins E, A, and B, like those of the proteins, oscillate during the cell cycle. The orderly activation and inactivation of specific kinases govern passage through the cell cycle. For example, cyclin E synthesis is rate limiting for the transition from G_1 to S phase in

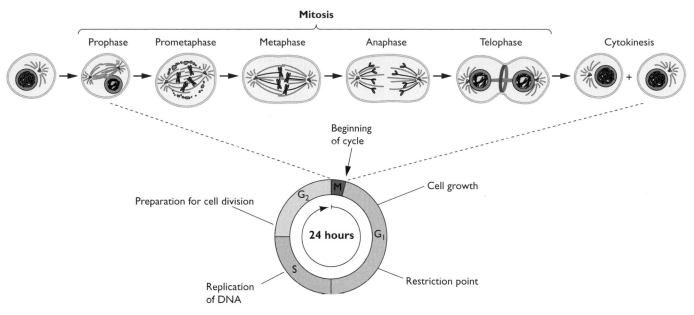

Figure 7.4 The phases of a eukaryotic cell cycle. The most obvious phase morphologically, and hence the first to be identified, is mitosis, or M phase, the process of nuclear division that precedes cell division. During this period, the nuclear envelope breaks down. Duplicated chromosomes become condensed and aligned on the mitotic spindle and are segregated to opposite poles of the cell, where nuclei reform upon chromosome decondensation (top). The end of M phase is marked by cytokinesis, the process by which the cell divides in two. Despite this remarkable reorganization and redistribution of cellular components, M phase occupies only a short period within the cell cycle. During the long interphase from one mitosis to the next, cells grow continuously. Interphase was divided into three parts with the recognition that DNA synthesis takes place only during a specific period, the synthetic or S phase, beginning about the middle of interphase. The other two periods, which appeared as "gaps" between defined processes, are designated the G_1 and G_2 (for gap) phases.

mammalian cells, and cyclin E-Cdk2 accumulates during late G_1. Soon after cells have entered S phase, cyclin E rapidly disappears from the cell; its task is completed until a new cycle begins.

While the oscillating waves of active Cdk accumulation and destruction are thought of as the ratchet that advances the cell cycle timer, it is important not to interpret this metaphor too literally. The orderly and reproducible sequence of DNA replication, chromosome segregation, and cell division is not determined solely by the oscillating concentrations of individual cyclin-Cdks. Rather, the cyclin-Cdk cycle serves as a device for integrating numerous signals from the exterior and interior of the cell into appropriate responses. The regulatory circuits that feed into and from the cycle are both many and complex (e.g., Fig. 7.5B). These regulatory signals ensure that the cell increases in mass and divides **only** when the environment is propitious or, in multicellular organisms, when the timing is correct. Many signal transduction pathways that convey information about the local environment or the global state of the organism therefore converge on the cyclin-Cdk integrators. In addition, various surveillance mechanisms monitor

such internal parameters as DNA damage, problems with DNA replication, and proper assembly and function of the mitotic spindle. Such mechanisms protect cells against potentially disastrous consequences of continuing a cell division cycle that could not be completed correctly. As we shall see, it is primarily these signaling and surveillance (**checkpoint**) mechanisms that are compromised during transformation by oncogenic viruses.

Oncogenic Viruses

Oncogenic viruses cause cancer by inducing changes that affect the control of cell proliferation. Indeed, the study of the mechanisms of viral transformation and oncogenesis laid the foundation for our current understanding of cancer, for example, with the identification of oncogenes that are activated or captured by retroviruses (originally known as RNA tumor viruses) and viral proteins that inactivate tumor suppressor gene products (Fig. 7.6). Specific members of a number of different virus families, as well as an unusual, unclassified virus (Box 7.3), have been implicated in naturally occurring or experimentally induced cancers in animals (Table 7.2). It has been estimated that

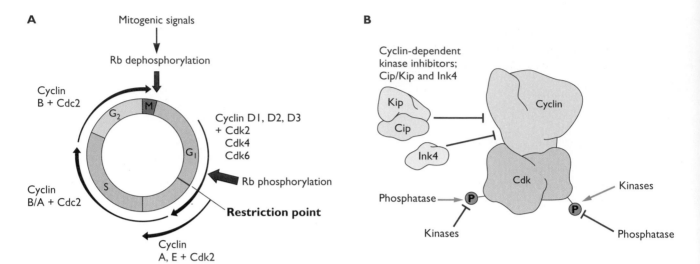

A

Mitogenic signals

Rb dephosphorylation

Cyclin
B + Cdc2

Cyclin D1, D2, D3
+ Cdk2
Cdk4
Cdk6

Cyclin
B/A + Cdc2

Rb phosphorylation

Restriction point

Cyclin
A, E + Cdk2

B

Cyclin-dependent
kinase inhibitors;
Cip/Kip and Ink4

Kip

Cip

Ink4

Cyclin

Cdk

Phosphatase

P

Kinases

Kinases

P

Phosphatase

Figure 7.5 The mammalian cyclin-Cdk cell cycle engine. (A) The phases of the cell cycle are denoted on the circle. The progressive accumulation of specific cyclin-Cdks is represented by the broadening arrows, with the arrowheads marking the time of abrupt disappearance. **(B)** The production, accumulation, and activities of both cyclins and cyclin-dependent kinases are regulated by numerous mechanisms. Activating and inhibitory reactions are indicated by green arrows and red bars, respectively. Activation of the kinases can require not only binding to the appropriate cyclin, but also phosphorylation at specific sites and removal of phosphate groups at others. The activities of the kinases are also controlled by association with members of two families of cyclin-dependent kinase-inhibitory proteins, which control the activities of only G_1 (Ink4 proteins) or all (Cip/Kip proteins) cyclin-Cdks. Both types of inhibitor play crucial roles in cell cycle control. For example, the high concentration of p27^{Kip1} characteristic of quiescent cells falls as they enter G_1, and inhibition of synthesis of this protein prevents cells from becoming quiescent.

Figure 7.6 A genetic paradigm for cancer. The pace of the cell cycle can be modulated both positively and negatively by different sets of gene products. Cancer arises from a combination of dominant, gain-of-function mutations in proto-oncogenes and recessive, loss-of-function mutations in tumor suppressor genes, which encode proteins that block cell cycle progression at various points. The function of either type of gene product can be affected by oncogenic viruses.

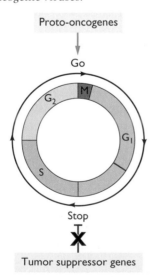

Proto-oncogenes

Go

Stop

Tumor suppressor genes

approximately 20% of all cases of human cancer are associated with infection by one of five viruses: Epstein-Barr virus, hepatitis B virus, hepatitis C virus, human T-lymphotropic virus type 1, and human papillomaviruses. In this section, we introduce oncogenic viruses and general features of their transforming interactions with host cells.

Discovery of Oncogenic Viruses

Retroviruses

Oncogenic viruses were discovered 100 years ago when Vilhelm Ellerman and Olaf Bang (1908) first showed that avian leukemia could be transmitted by filtered extracts (i.e., viruses) of leukemic cells or serum from infected birds. Because leukemia was not recognized as cancer in those days, the significance of this discovery was not generally appreciated. Shortly thereafter (in 1911), Peyton Rous demonstrated that solid tumors could be produced in chickens by using cell extracts from a transplantable sarcoma that had appeared spontaneously. Despite the viral etiology of this disease, the cancer viruses of chickens were thought to be oddities until similar murine malignancies, as well as mammary tumors, were found to be associated with infection by viruses. These oncogenic viruses all

**BOX
7.3**

EXPERIMENTS
A cancer virus with genomic features of both papillomaviruses and polyomaviruses

Efforts are under way to prevent the extinction of the western barred bandicoot, an endangered marsupial now found only on two islands in the UNESCO World Heritage Area of Shark Bay, Western Australia. Unfortunately, conservation has been hindered by a debilitating syndrome, in which wild and captive animals develop papillomas and carcinomas in several areas of the skin. The histological properties of the tumors suggested that a papillomavirus or a polyomavirus might contribute to development of the disease.

In fact, a previously unknown viral genome was discovered in tumor tissues from these animals by multiply primed amplification, cloning, and sequencing, and also by polymerase chain reaction (PCR) with degenerate primers specific for papillomavirus DNA. As summarized in the figure, this DNA genome exhibits features characteristic of both papillomaviruses and polyomaviruses and includes coding sequences related to those of both families. The papillomavirus-like and polyomavirus-like sequences were shown to be continuous with one another in the

viral DNA genome. This property excludes the possibility that the tumor tissues were coinfected with a member of each family, as well as artifacts such as laboratory contamination of samples.

The origin of this unique virus, which was named bandicoot papillomatosis carcinomatosis virus type 1, is not known. The virus might have arisen as a result of a recombination event between the genomes of a papillomavirus and a polyomavirus. Alternatively, it might represent the first known member of a new virus family that evolved from a common ancestor of

the *Papillomaviridae* and *Polyomaviridae*. Regardless, the viral genome was detected in 100% of bandicoots with papillomatosis and carcinomatosis syndrome, implicating the virus as a necessary factor in the development of this disease.

Woolford, L., A. Rector, M. Van Ranst, A. Ducki, M. D. Bennett, P. K. Nicholls, K. S. Warren, R. A. Swan, G. E. Wilcox, and A. J. O'Hara. 2007. A novel virus detected in papillomas and carcinomas of the endangered western barred bandicoot (*Perameles bougainville*) exhibits genomic features of both the *Papillomaviridae* and *Polyomaviridae*. *J. Virol.* **81:**13280–13290.

**Properties of
papillomaviruses**

Large genome (~7.3 kbp)

Coding sequences for
major capsid proteins
L1 and L2

**Properties of
polyomaviruses**

Putative early and late gene products
encoded in different strands of the genome

Coding sequences for nonstructural
proteins and large and small T antigens

---

proved to be members of the retrovirus family, which do not kill their host cells. We now know that retroviruses are endemic in many species, including mice and chickens. For example, most chickens in a flock will have been infected by avian leukosis virus within a few months of hatching. In the vast majority of cases, chickens experience a transient viremia but disease is rare. Embryos can also be infected before their immune systems have been developed, and eventually become tolerant to the virus.

Early researchers classified the oncogenic retroviruses into two groups depending on the rapidity with which they caused cancer (Table 7.3). The first group comprises rare, rapidly transforming **transducing oncogenic retroviruses**. These are all highly carcinogenic agents that cause malignancies in nearly 100% of infected animals in a matter of days. They were later discovered to have the ability to transform susceptible cells in culture. The second class, **nontransducing oncogenic retroviruses**, includes less carcinogenic agents. Not all animals infected with these viruses develop tumors, which appear only weeks or months after infection. In the late 1980s, a third type of oncogenic retrovirus, a **long-latency virus** was identified

in humans. Tumorigenesis is very rare, and occurs months or even years after infection.

Infection by each group of oncogenic retroviruses induces tumors by a distinct mechanism. As their name

Table 7.2   Oncogenic viruses and cancer

Family	Associated cancer(s)
RNA viruses	
Flaviviridae	
Hepatitis C virus	Hepatocellular carcinoma
Retroviridae	Hematopoietic cancers, sarcomas, and carcinomas
DNA viruses	
Adenoviridae	Various solid tumors
Hepadnaviridae	Hepatocellular carcinoma
Herpesviridae	Lymphomas, carcinomas, and sarcomas
Papillomaviridae	Papillomas and carcinomas
Polyomaviridae	Various solid tumors
Poxviridae	Myxomas and fibromas

BOX 7.5

EXPERIMENTS

A human herpesvirus 8 micro-RNA that may contribute to transformation of host cells

Micro-RNAs (miRNAs) are small (21- to 23-nucleotide) RNAs that base pair with complementary sequences in mRNAs to inhibit translation of the mRNA, or induce its degradation (see Volume I, Chapter 10). Viral miRNAs are made in cells infected by members of several families of DNA viruses, included herpesviruses. The genome of human herpesvirus 8, which is associated with Kaposi's sarcoma and certain B-cell lymphomas, encodes at least 10 miRNAs.

One of the human herpesvirus 8 miR-NAs, miR-K-11-12 has sequence homology to the human mi-RNA miR-155, particularly in the region critical for base pairing with target mRNAs, the so-called seed region (see the figure). The viral miR-K-11-12 miRNA was shown to target a set of cellular mRNAs that are also affected by miR-155.

- Cellular mRNAs with at least one match to the sequence complementary to the seed sequence of miR-155 were highly enriched in the 150 mRNAs that decreased in concentration to the greatest degree when the viral miRNA was made in

```
UUAAUGCU AAUCGUGAUAGGGU   miR-155
UUAAUGCU UAGCGUGUGUCCGA   K-miR-11-12
```

The sequences of human miR-155 and human herpesvirus type 8 K-miR-11-12 are shown aligned, with conserved nucleotides indicated in blue. The "seed" region important for base pairing of miRNAs with target mRNA is outlined in red.

human B cells in the absence of any other viral gene products
- The cellular miR-155 reduced the concentrations of a set of target mRNAs to the same degree as the viral miRNA, and both reduced the expression of reporter genes carrying 12 different mRNA 3' untranslated regions that were predicted to be complementary to the common miRNA seed sequence
- Reduction of the concentrations of 6 target mRNAs by miR-K-11-12 was prevented by mutation of the seed sequence
- The cellular and viral miRNAs both induced decreases in the concentrations of the proteins encoded by two predicted target mRNAs (encoding Fos and Bach1).

These observations strongly suggest that K-miR-11-12 is a functional analog of cellular miR-155. The latter is the product of the *bic* gene, which is overexpressed in several types of human B-cell lymphoma. When introduced into transgenic mouse, *bic* induced B-cell lymphomas, indicating that it is an oncogene. These properties suggest that the viral miRNA may contribute to the development of B-cell tumors associated with human herpesvirus 8.

Gottwein, E., N. Mukherjee, C. Sachse, C. Funsel, W. H. Majoros, J.-T. A. Chi, R. Braich, M. Manoharan, J. Soutschek, V. Ohler, and B. R. Cullen. 2007. A viral microRNA functions as an orthologue of cellular mi-R-155. *Nature* **450:**1096–1101.

Common Properties of Oncogenic Viruses

Although they are members of different families (Table 7.2), the majority of oncogenic viruses share several general features. In all cases that have been analyzed, transformation is observed to be a single-hit process (defined in Volume I, Chapter 2), in the sense that infection of a susceptible cell with a single virus particle is sufficient to cause transformation. In addition, all or part of the viral genome is usually retained in the transformed cell. With few exceptions, cellular transformation is accompanied by the continuous expression of specific viral genes. On the other hand, while specific viral genes are present and expressed, transformed cells need not and (except in the case of some retroviruses) **do not** produce infectious virus particles. Most importantly, viral transforming proteins alter cell proliferation by a limited repertoire of molecular mechanisms.

Viral Genetic Information in Transformed Cells

State of Viral DNA

Cells transformed by oncogenic viruses generally retain viral DNA in their nuclei. These DNA sequences correspond to all or part of the infecting DNA genome, or the proviral DNA made in retrovirus-infected cells. Viral DNA sequences are maintained by one of two mechanisms: they can be integrated into the cellular genome (Box 7.7) or persist as autonomously replicating episomes.

As discussed in Volume I, Chapter 7, integration of retroviral DNA by the viral enzyme integrase is an essential step in the viral life cycle. Although there are some virus-specific biases, integration can occur at essentially any site in cellular DNA, but the reaction preserves a fixed order of viral genes and control sequences in the provirus (see Volume I, Fig. 7.15). When the provirus carries a v-oncogene, the site at which it is integrated into the cellular genome is of no importance (provided that viral transcription is unimpeded). In contrast, integration of proviral DNA within specific regions of the cellular genome is a hallmark of the induction of tumors by nontransducing retroviruses.

The proviral sequences present in every cell of a tumor induced by nontransducing retroviruses are found in the same chromosomal location, an indication that all arose from a single transformed cell. Such tumors are, therefore, **monoclonal**. Although infection is initiated with a nondefective retrovirus, the proviruses in the resulting tumor cells

**BOX
7.6**

DISCUSSION
*Does a polyomavirus contribute to development
of Merkel cell carcinoma in humans?*

Mouse polyomavirus and simian virus 40 have been important models for studies of oncogenesis and transformation (see the text). Two human members of this family, BK and JC polyomaviruses, were discovered in 1971. These viruses commonly infect the urinary tract, and can be pathogenic in immunosuppressed patients (Appendix A, Fig. 20). A distantly related polyomavirus genome was detected recently in tumors from patients with Merkel cell carcinoma, a rare but rapidly metastasizing skin cancer.

Viral DNA sequences initially were identified in tumor tissue by a method based on high-throughput sequencing. Complementary DNA (cDNA) libraries were prepared from a single tumor and from a pool of three tumors. A total of 395,734 cDNAs were then sequenced. The vast majority (99.4%) were of human chromosomal or mitochondrial origin. Among the unassigned sequences, one from the single tumor exhibited significant homology to African green monkey lymphotropic polyomavirus and BK polyomavirus T antigen coding sequences. The 3′ end of this cDNA was shown to include sequences of the human receptor tyrosine phosphatase type G, suggesting that viral DNA sequences were integrated in the genome of tumor cells. Integration of the viral genome was subsequently confirmed by several methods, including Southern blotting (see Box 7.7). The organization of the viral genome is that typical of polyomaviruses, and includes sequences homologous to the early transforming gene products, large and small T antigens, of animal members of the family.

The genome of this virus, which was called Merkel cell polyomavirus, was detected in tumor tissue from 8 of 10 patients with Merkel cell carcinoma, but was present only at lower concentrations in 9 of 84 control samples. Furthermore, the pattern of viral DNA integration in the tissues examined indicated that the tumors were monoclonal in origin, implying that viral DNA integration proceeded proliferation of the cells. These observations suggest a causal association between virus infection and the development of Merkel cell carcinoma. More extensive epidemiological studies will be required to assess such a role. Furthermore, it is not yet known whether the predicted Merkel cell polyomavirus large and small T antigens are present in tumor cells, or possess transforming activity.

Feng, H., M. Shuda, Y. Chang, and P. S. Moore. 2008. Clonal integration of a polyomavirus in Merkel cell carcinoma. *Science* **319:**1096–1100.

The evolutionary relationship of Merkel cell carcinoma polyomavirus to some other mammalian polyomaviruses is shown schematically. Adapted from R. P. Viscidi and K. V. Slak, *Science* **319:**1049–1050, 2008, with permission.

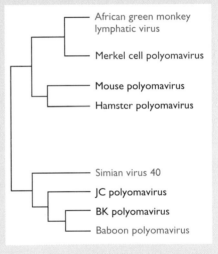

have usually lost some or most of the proviral sequences. Nevertheless, at least one long terminal repeat (LTR) containing the transcriptional control region is always present. Viral transcription signals, but not protein-coding sequences, are therefore required for transformation by nontransducing retroviruses. The significance of these properties became apparent when it was discovered that in several tumors proviruses were integrated in the vicinity of some of the same cellular oncogenes that are captured by transducing retroviruses. The study of nontransducing oncogenic retroviruses has also led to the identification of some additional proto-oncogenes, as illustrated for two murine retroviruses in Table 7.4. As integration of retroviral DNA into the host genome is essentially random, there is a limited probability that integration will occur in the vicinity of an oncogene. The long latency for tumor induction by these viruses can be explained by the need for multiple cycle of replication and integration for such an event to occur.

Integration of viral DNA sequences is not a prerequisite for successful propagation of **any** oncogenic DNA viruses. Nevertheless, integration is the rule in adenovirus- or polyomavirus-transformed cells. Such integration is the result of rare recombination reactions (catalyzed by cellular enzymes) between generally unrelated DNA sequences. Integration can therefore occur at many sites in the cellular genome. The great majority of cells transformed by these viruses retain only partial copies of the viral genome. The genomic sequences integrated can vary considerably among independent lines of cells transformed by the same virus, but a common, minimal set of genes is always present. The low probability that viral DNA will become integrated into the cellular genome, and the fact that only a fraction of these recombination reactions will maintain the integrity of viral transforming genes, are major factors contributing to the low efficiencies of transformation by these viruses. In the case of simian virus 40, infection of cultures with 10^2 to 10^4 plaque-forming units (PFU) is required to obtain one focus of transformed cells, while the ratio is even higher for human adenoviruses.

BOX 7.7 **METHODS**
Detection and characterization of integrated viral DNA sequences in transformed cells

Integration of viral DNA into the cellular genome was initially demonstrated by the cosedimentation of viral DNA with high-molecular-mass cellular DNA under strongly denaturing conditions. The development of restriction endonucleases as molecular tools and of the Southern hybridization assay allowed direct proof of such integration, as well as characterization of integrated viral DNA. These approaches are illustrated for a circular, double-stranded viral DNA genome.

Cleavage of genomic DNA isolated from transformed cells with restriction endonuclease A (no sites in the viral genome) yields one high-molecular-mass fragment of lower mobility than free viral DNA (F) for each site of integrated viral DNA (I). The number of integration sites can therefore be counted. When used with appropriate standards, the number of copies of viral sequences at each such site can also be estimated. Restriction endonuclease B (one site in the viral genome) generates two fragments differing in mobility from the free, linear viral DNA. The results obtained with enzymes A and B demonstrate that viral DNA must be covalently attached to cellular DNA. Restriction endonuclease C, with two cleavage sites

in the viral genome, produces three fragments from transformed cell DNA. The comigration of one fragment with the smaller product of digestion of free viral

DNA indicates that this segment of the viral genome is intact in the integrated viral DNA.

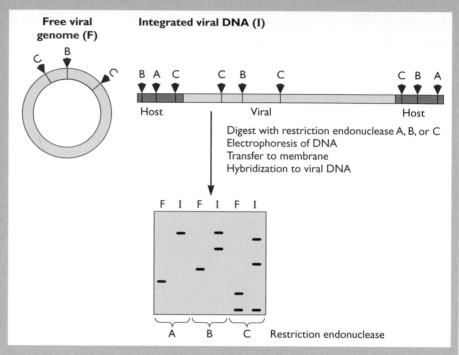

Table 7.4 New proto-oncogenes targeted by two nontransducing, simple retroviruses[a]

Virus	Neoplasia	Cellular gene	Normal function
Moloney murine leukemia virus	Pre-B lymphoma	*Ahi1*	Adapter protein
	T-cell lymphoma	*cyclin D2*	G_1 cyclin
		Lck	Nonreceptor protein tyrosine kinase
		Notch-1	Transmembrane receptor that functions in development
		Pim1	Serine/threonine kinase
		Tpl2	Serine/threonine kinase
	B-cell lymphoma	*Pim2*	Serine/threonine kinase
Mouse mammary tumor virus	Mammary tumors	*Int-1* (*wnt-1*)	Secreted glycoprotein important in pattern formation in early embryonic development
		Int-2 (*hst*)	Secreted proteins that may act as a growth factor
		Int-3	Protein presumed to function in development (*notch* family)
		IntH/Int5	Protein that converts androgens to estrogens

[a]Adapted from J. M. Coffin, S. H. Hughes, and H. E. Varmus (ed.), p. 482–484, *Retroviruses*, Cold Spring Harbor Laboratory Press, Cold Spring Harbor, NY, 1997, with permission.

A second mechanism by which viral DNA can persist in transformed cells is as a stable, extrachromosomal episome (Volume I, Box 1.4). Such episomal viral genomes are a characteristic feature of B cells immortalized by Epstein-Barr virus, and they can also be found in cells transformed by papillomaviruses. The viral episomes are maintained at concentrations of tens to hundreds of copies per cell, by both replication of the viral genome in concert with cellular DNA synthesis and orderly segregation of viral DNA to daughter cells (see Volume I, Chapter 9). Consequently, transformation depends on the viral proteins necessary for the survival of viral episomes, as well as those that modulate cell growth and proliferation directly.

Identification and Properties of Viral Transforming Genes

Transforming genes of oncogenic viruses have been identified by classical genetic methods, characterization of the viral genes present and expressed in transformed cell lines, and analysis of the transforming activity of viral DNA fragments directly introduced into cells (Box 7.8). For example, analysis of transformation by temperature-sensitive mutants of mouse polyomavirus established as early as 1965 that the viral early transcription unit is necessary and sufficient to initiate and maintain transformation. Of even greater value were mutants of retroviruses, in particular two mutants of Rous sarcoma virus isolated in the early 1970s. The genome of one mutant carried a spontaneous deletion of approximately 20% of the viral genome; this mutant could no longer transform the cells it infected, but it could still replicate. The second mutant was temperature sensitive for transformation, but the virus could replicate at both temperatures. These properties of the mutants therefore showed unequivocally that cellular transformation and viral replication are distinct processes. More importantly, the deletion mutant allowed preparation of the first nucleic acid probe specific for a v-oncogene, v-src (Box 7.9). This src-specific probe was found to hybridize to cellular DNA, providing the first conclusive evidence that v-oncogenes are of cellular and not viral origin. This finding, for which J. Michael Bishop and Harold Varmus received the 1989 Nobel Prize in physiology or medicine, had far-reaching significance, because it immediately suggested that such cellular genes might become oncogenes by means other than viral transduction.

The presence of cellular oncogenes in their genomes turned out to be the definitive characteristic of transducing retroviruses, as illustrated in Fig. 7.7. As noted earlier, the acquisition of these cellular sequences is a very rare event. In addition, with the exception of Rous sarcoma virus, the transducing retroviruses are replication defective, having lost all or most of the viral coding sequences during oncogene capture. Such defective transducing viruses can, however, be propagated in mixed infections with replication-competent "helper" viruses, which provide all the

---

BOX 7.8

Multiple lines of evidence identified the transforming proteins of the polyomavirus simian virus 40

Early-gene products are necessary and sufficient to initiate transformation.

1. Viruses carrying temperature-sensitive mutations in the early transcription unit (*ts*A mutants), but no other region of the genome, fail to transform at a nonpermissive temperature.
2. Simian virus 40 DNA fragments containing only the early transcription unit transform cells in culture; DNA fragments containing other regions of the genome exhibit no activity.

Early-gene products are necessary to maintain expression of the transformed phenotype.

1. Many lines of cells transformed by simian virus 40 *ts*A mutants at a permissive temperature revert to a normal phenotype when shifted to a nonpermissive temperature.
2. Integration of viral DNA sequences disrupts the late region of the viral genome but not the early transcription unit, and early-gene products are synthesized in all transformed cell lines.

Both LT and sT contribute to transformation.

1. Simian virus 40 mutants carrying deletions of sequences expressed only in sT fail to transform rat cells to anchorage-independent growth.
2. Introduction and expression of LT complementary DNA are sufficient for induction of transformation, but expression of sT can stimulate transformation (especially at low LT concentrations), is necessary for expression of specific phenotypes in specific cells, and is required for transformation of resting cells.

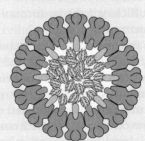

proteins, simian virus 40 LT, and the E7 proteins of onco-genic human papillomavirus (types 16 and 18) can induce DNA synthesis and cell proliferation. Such mitogenic activity requires regions of the viral proteins that are necessary for their binding to hypophosphorylated Rb. All three viral proteins make contacts with two noncontiguous regions by which Rb associates with E2f family members (regions A and B in Fig. 7.19A). In this way, they sequester the inhibitory form of Rb and disrupt Rb-E2f complexes. As a result, they induce transcription of E2f-dependent genes. The relief of Rb-mediated repression leads to inappropriate entry of cells into S phase (Fig. 7.18B).

Disruption of E2f-Rb complexes by the viral proteins appears to be an active process, rather than simply the result of passive competition for binding to Rb. Induction of cell cycle progression by simian virus 40 LT requires not only the Rb-binding site, but also the N-terminal J domain located nearby (Fig. 7.19B). Because the J domain func-tions as a molecular chaperone (see Volume I, Chapter 9), it has been suggested that LT actively dismantles the Rb-E2f complex. The CR1 sequence of the adenoviral E1A proteins fulfills a similar function. Exactly how these viral proteins disrupt the association of Rb with E2f remains to be determined, but conformational alteration of Rb seems likely to be important.

In some cases the viral proteins also alter Rb metabolism. For example, the human papillomavirus type 16 and 18 E7 proteins can bind directly to one subunit of the proteasome and induce polyubiquitinylation and proteasome-mediated degradation of Rb in transformed epithelial cells.

Inhibition of Negative Regulation by Rb-Related Proteins

The Rb protein is the founding member of a small family of related gene products, which also includes the proteins p107 and p130. The latter two proteins were discovered

Figure 7.19 Model for active dismantling of the Rb-E2f complex by simian virus 40 LT. (A) Functional domains of the human Rb protein are shown to scale. The A- and B-box regions form the so-called pocket domain, which is necessary for binding of Rb to both E2fs and the viral proteins described in the text. This segment is also sufficient to repress transcription when fused to a heterologous DNA-binding domain, and it is required for binding to histone deacetylases (Hdacs). Like the viral Rb-binding proteins, Hdacs contain the motif LXCXE within the region that binds to Rb. The N-terminal segment of the protein, which is also important for suppression of cell proliferation, binds to human Mcm-7, a component of a chromatin-bound complex required for DNA replication and control of initiation of DNA synthesis. **(B)** The LT protein binds to the Rb A- and B-box domains via the sequence that contains the LXCXE motif, designated R. The adjacent, N-terminal J domain of LT is not necessary for binding to Rb, but is required for induction of cell cycle progression. It has been proposed that the J domain recruits the cellular chaperone Hsc70. The chaperone then acts to release E2f–Dp-1 heterodimers from their association with Rb, by a mechanism that is thought to depend on ATP-dependent conformational change.

A

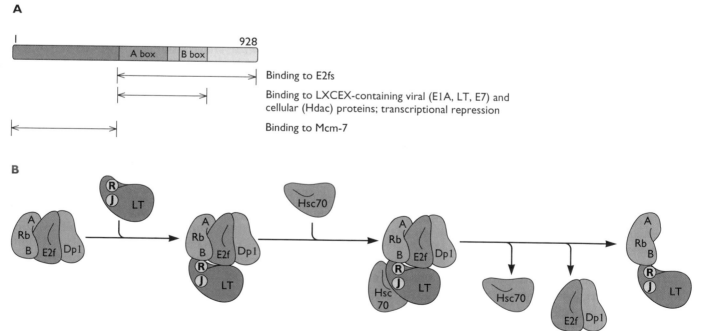

Binding to E2fs

Binding to LXCEX-containing viral (E1A, LT, E7) and cellular (Hdac) proteins; transcriptional repression

Binding to Mcm-7

B

by virtue of their interaction with adenoviral E1A proteins (Fig. 7.20), but they also bind to both simian virus 40 LT and human papillomavirus type 16 and 18 E7 proteins. The similarity of p107 and p130 to Rb is most pronounced in the A and B sequences needed for binding of Rb to both E2f and the viral transforming proteins (Fig. 7.19A). Indeed, the residues by which Rb contacts the common Rb-binding motif of the viral proteins (LXCXE) are invariant among the other family members. Binding of the viral proteins to p107 and p130 can make important contributions to transforming activities. For example, the LXCXE sequence of simian virus 40 LT is required for transformation of fibroblasts derived from *Rb*-null mice. The J domain of LT, which induces hypophosphorylation of p107 and p130, concomitant with their increased degradation, is also necessary.

The Rb, p107, and p130 proteins bind preferentially to different members of the E2f family during different phases of the cell cycle. Hypophosphorylated Rb binds primarily to E2f-1, E2f-2, or E2f-3 during the G_0 and G_1 phases. In contrast, p107 is largely associated with E2f-4 and E2f-5 during S phase, whereas p130 binds these same two E2fs in G_0. These properties, and the targeting of p107 and p130 by transforming proteins of the smaller DNA viruses, indicate that inhibition (or more subtle regulation) of the activity of E2f-4 and E2f-5 must be important for orderly progression through the cell cycle. Binding of p130 to these E2f family members appears to be critical for maintaining cells in the quiescent state, and such complexes predominate in mammalian cells in G_0. Their disruption by adenoviral, papillomaviral, or polyomaviral transforming

proteins is thought to allow such cells to reenter the cycle, in part via stimulation of the transcription of genes encoding both the E2f proteins and the cyclin-dependent kinase (Cdk2) needed for entry into S phase. For example, when quiescent monkey cells are productively infected by simian virus 40, LT disrupts the association of p130 with E2fs.

Production of Virus-Specific Cyclins

Human herpesvirus 8 and its close relative herpesvirus saimiri encode functional cyclins. The cyclin gene of human herpesvirus 8, designated v-*cyclin*, has 31% identity and 53% similarity to the human gene that encodes cyclin D2, and its product binds predominantly to Cdk6. Like its cellular counterpart, v-cyclin activates this protein kinase, which then phosphorylates the Rb protein. The viral cyclin also alters the substrate specificity of the kinase: the v-*cyclin*–Cdk6 complex phosphorylates proteins normally recognized by cyclin-bound Cdk2, but not by cyclin D-Cdk6. These targets include the cyclin-dependent kinase inhibitor p27 and the replication proteins Cdc6 and Orc1 (see Volume I, Fig 9.23). Normally, synthesis of the Cip/Kip and Ink4 proteins that inhibit cellular cyclin-Cdks (Fig. 7.5) blocks cell cycle progression. However, neither Cip/Kip nor Ink4 family members bind well to the viral cyclin. Synthesis of the viral cyclin can therefore overcome the G_1 arrest imposed when either type of inhibitory protein is made in human cells, and can induce cell cycle progression in quiescent cells and initiation of DNA replication. The specific advantages conferred by production of the viral cyclin during the infectious cycle have not been identified. However, it would be surprising if v-cyclin does not contribute to the oncogenicity of these herpesviruses in their natural hosts, as synthesis of v-cyclin in B cells of transgenic mice results in B-cell lymphoma (when the mice are p53-null [see below]).

Inactivation of Cyclin-Dependent Kinase Inhibitors

The production of viral cyclins in infected cells appears to be a unique property of certain herpesviruses, but other DNA viruses encode proteins that inactivate specific inhibitors of Cdks. For example, the E7 protein of human papillomavirus type 16 binds to the p21^{Cip1} protein and inactivates it. This member of the cellular Cip/Kip family inhibits G_1 cyclin-Cdk complexes (Fig. 7.5). The increase in intranuclear concentrations of p53 triggered by unscheduled inactivation of the Rb protein (see next section) results in accumulation of p21^{Cip1}. The ability of the papillomaviral E7 protein to inactivate both Rb and p21^{Cip1} would therefore appear to ensure entry of cells into S phase. Indeed, both these functions of the E7 protein are necessary to induce differentiated human epithelial cells to enter S phase.

Figure 7.20 Organization of the larger adenoviral E1A protein. Regions of the protein are shown to scale. Those designated CR1 to CR3 are conserved in the E1A proteins of human adenoviruses. The CR3 region, most of which is absent from the smaller E1A protein because of alternative splicing, is not necessary for transformation. The locations of the Rb-binding motif and of the regions required for binding to the other cellular proteins discussed in the text are indicated.

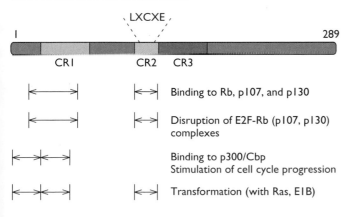

The adenoviral E1A proteins can also overcome G_1 arrest induced by p21^{Cip1} or p27^{Kip1}, probably by binding to the region of the inhibitors that mediates association with cyclin-dependent kinases. In addition, binding of E1A proteins to the transcriptional coactivators p300 and Cbp (Fig. 7.20) blocks activation of transcription of the p21^{Cip1} gene in response to DNA damage or differentiation.

Transformed Cells Must Also Grow and Survive

Integration of Mitogenic and Growth-Promoting Signals

Prior to division, cells must increase in size and mass, as they duplicate their components for division between the two daughter cells produced by cytokinesis (Fig. 7.4). Consequently, signals that induce cell proliferation also lead to the metabolic changes required to promote and sustain cell growth. Not surprisingly, the mechanisms that regulate growth of normal cells are integrated with those that lead to cell proliferation in response to mitogenic signals (Fig. 7.21A). The small G protein Ras and the protein kinase Akt are important components of the networks that achieve such integration: their activation leads to not only increased production of active D-type cyclins, but also stimulation of translation and regulation of many metabolic enzymes.

The rapid proliferation of cells transformed by viral proteins depends on high rates of metabolism and growth during each cell cycle. It seems likely that any viral oncogene product that results in activation of Ras (or Akt) promotes cell growth, as well as proliferation. How viral transforming proteins that impinge directly on the nuclear circuits that govern cell cycle progression induce altered cell growth is less clear. However, the actions of many of these proteins lead to changes in the transcription of numerous cellular genes, responses that might increase the concentrations of biosynthetic and other metabolic enzymes.

Mechanisms That Permit Survival of Transformed Cells

As discussed in Chapter 3, metazoan cells can undergo programmed cell death (apoptosis). This program is essential during development and serves as a powerful antiviral defense of last resort. Apoptosis can be activated not only by external cues, but also by intracellular events, notably damage to the genome or unscheduled DNA synthesis. Consequently, viral transforming proteins that induce cells to enter S phase and proliferate when they would not normally do so will also promote the apoptotic response. This potentially fatal side effect is foiled by a variety of mechanisms that allow survival of infected, and under appropriate circumstances, transformed cells.

Viral Inhibitors of the Apoptotic Cascade

Many viral genomes encode mimics of cellular proteins that hold apoptosis in check (see Table 3.9). Such viral inhibitors of apoptosis can contribute to transformation. For example, the human adenovirus E1B 19-kDa protein, one of the first viral homologs of cellular antiapoptotic proteins to be identified, allows survival, and hence transformation, of rodent cells that also contain the viral growth proliferation-promoting E1A protein (Table 7.5).

Integration of Inhibition of Apoptosis with Stimulation of Proliferation

Cells must continually interpret the numerous internal and external signals that impinge upon them to execute an appropriate response. Not all the mechanisms that integrate the many types of information that cells receive have been elucidated. However, it is well established that signal transduction cascades that induce cell proliferation in response to external signals can simultaneously promote cell survival by blocking the apoptotic response. For example, signaling via the small G protein Ras results in activation of the cyclin-dependent kinases that drive the G_1-to-S-phase transition (Fig. 7.18A). In addition, such signaling activates phosphoinositide-3-kinase and the protein kinase Akt. Akt induces transcriptional and posttranscriptional mechanisms that inhibit the production and activity of proapoptotic proteins, such as Bad and Bim, and stimulate synthesis of inhibitors of apoptosis, including Bcl-2 (Fig. 7.21B). The various retroviral transforming proteins that function in the receptor tyrosine kinase pathway that activates Ras (Fig. 7.3; Table 7.6) therefore also induce inhibition of apoptosis. Signaling via Src proteins also activates phosphoinositide-3-kinase (Fig. 7.11). It therefore seems likely that v-Src and viral proteins that operate via c-Src, such as polyomavirus mT (Table 7.7; Fig. 7.16), block apoptosis in the same way.

Inactivation of the Cellular Tumor Suppressor p53

Transformation by several DNA viruses requires inactivation of a second cellular tumor suppressor, the p53 protein. This transcriptional regulator, first identified by virtue of its binding to simian virus 40 LT, is a critical component of regulatory circuits that determine the response of cells to damage to their genomes, as well as to low concentrations of nucleic acid precursors or hypoxia. The importance of this protein in the appropriate response to such damage or stress is emphasized by the fact that *p53* is the most frequently mutated gene in human tumors.

The accumulation and activity of p53 are tightly regulated. The intracellular concentration of p53 is

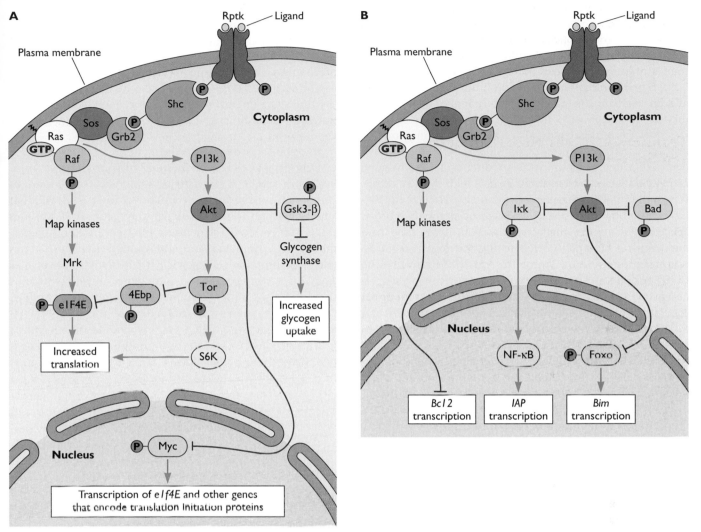

Figure 7.21 Some signaling pathways that promote cell growth and survival. (A) Cell growth. Upon activation, in this example by signaling initiated by binding of its ligand to a receptor protein tyrosine kinase (Rptk), signaling via Ras and the Map kinase cascade activates a Map kinase-dependent kinase (Mak), which phosphorylates and activates the translation initiation protein eIF4E. The activity of this initiation protein is also increased when signaling from Ras via phosphatidylinositol 3-kinase (PI3k) stimulates the protein kinase Akt. Phosphorylation of the inihibitory eIF4E-binding protein (4Ebp) by Akt suppresses its ability to inactivate eIF4E. The transcription of the genes encoding eIF4E and other translation initiation proteins is stimulated via effects on the transcription activator Myc. Akt-dependent phosphorylation of ribosomal protein S6 kinase increases the rate of translation elongation. The mechanisms increase the availability and activity of proteins crucial for protein synthesis, and allow cells to provide proteins at a rate that sustains cells growth. Signaling from Akt also regulates metabolism via phosphorylation and inactivation of glycogen synthase kinase (Gsk3-β). As a result, glycogen synthase cannot be phosphorylated and inactivated. **(B)** Cell survival. Activation of Ras also promotes cell survival by inhibition of synthesis or activity of proapoptotic proteins and by stimulation of production of inhibitors of programmed cell death. Substrates of activated Akt include the pro-apoptotic protein Bad and the transcriptional regulator Foxo, which are inactivated by phosphorylation. Akt also phosphorylates the inhibitor of Nf-κb (Iκk) to promote transcription of genes that encode other inhibitors of apoptosis. As shown in Fig. 7.18A, signaling via Ras and the Map kinase cascade induces cell proliferation. The PI3k-Akt pathway also promotes proliferation, for example by phosphorylation and inactivation of cyclin-dependent kinase inhibitors. Consequently, these (and other) signaling networks integrate cell proliferation and survival.

normally very low, because the protein is targeted for nuclear export and proteasomal degradation by binding of the Mdm-2 protein to an N-terminal sequence (Fig. 7.22 and 7.23). However, DNA damage, such as double-strand breaks produced by γ-irradiation or the accumulation of DNA repair intermediates following ultraviolet (UV) irradiation, leads to the stabilization of p53 and a substantial increase in its concentration (Fig. 7.23). The rate of translation of the protein may also increase. Various proteins that appear to be important for stabilization of p53 have been identified, including the product of a human gene called *Atm* (<u>a</u>taxia <u>t</u>elangiectasia <u>m</u>utated), which recognizes potentially genotoxic DNA damage. The genetic disease caused by mutation in *Atm* is associated with a broad spectrum of defects, including hypersensitivity to X rays and ionizing radiation. Cells lacking the Atm protein do not accumulate the p53 protein, and fail to arrest at the G_1/S boundary in response to DNA damage.

The p53 protein is a sequence-specific transcriptional regulator, containing an N-terminal activation domain and a central DNA-binding domain (Fig. 7.22). Its ability to stimulate transcription of p53-responsive genes is tightly regulated. For example, the DNA-binding activity of p53 is stimulated by association of the C-terminal domain with double-stranded DNA ends or sites of excision repair damage, and binding of the Mdm-2 protein to the activation domain inhibits p53-dependent transcription. The many mechanisms by which the accumulation or activity of p53 can be regulated (Fig. 7.23) provide the means to integrate the multiple signals that are monitored to ensure that this potent protein alters cell physiology only under extreme conditions.

In response to damage to the genome, or other inducing conditions, p53 can promote one of two major responses, leading to either G_1/S arrest or apoptosis. One important component of the former pathway is the p53-dependent stimulation of transcription of the gene that encodes p21^{Cip1}, the G_1 cyclin-dependent kinase inhibitor. The p53 protein also stimulates transcription of a number of genes encoding proteins that participate in apoptosis, such as Apaf-1, Bax, and Fas, as well as of genes encoding proteins that inhibit signaling pathways that promote cell survival. For example, increased production of the protein Pten leads to impaired signaling via phosphoinositide 3-kinase to Akt (Fig. 7.21B), as Pten is a phosphatase that dephosphorylates phosphoinositides. The ability of p53 to repress

Figure 7.22 The human p53 protein. (A) The functional domains of the protein are shown to scale. TafII70 and TafII31 are Tbp-associated proteins present in TfIId (Volume I, Chapter 8); other proteins are defined in the text. **(B)** Structures of the central, DNA-binding domain bound to DNA (left) and of the C-terminal domain that mediates tetramer formation (right) were determined by X-ray crystallography. From Y. Cho et al., *Science* **265**:346–355, 1994, and P. D. Jeffrey et al., *Science* **267**:1498–1502, 1996, with permission. Courtesy of P. Jeffrey and N. Pavletich, Memorial Sloan-Kettering Cancer Center.

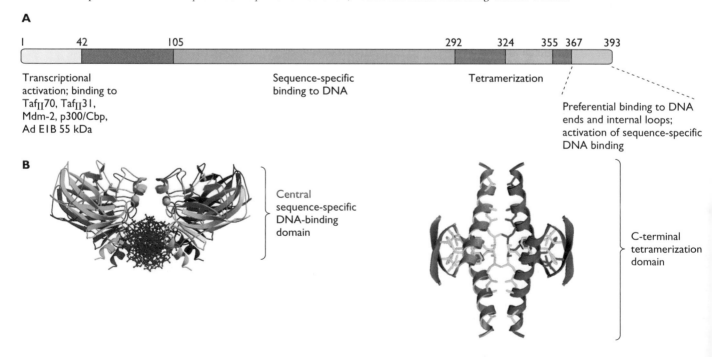

A

| 1 | 42 | 105 | | 292 | 324 | 355 | 367 | 393 |

Transcriptional activation; binding to Taf$_{II}$70, Taf$_{II}$31, Mdm-2, p300/Cbp, Ad E1B 55 kDa

Sequence-specific binding to DNA

Tetramerization

Preferential binding to DNA ends and internal loops; activation of sequence-specific DNA binding

B

Central sequence-specific DNA-binding domain

C-terminal tetramerization domain

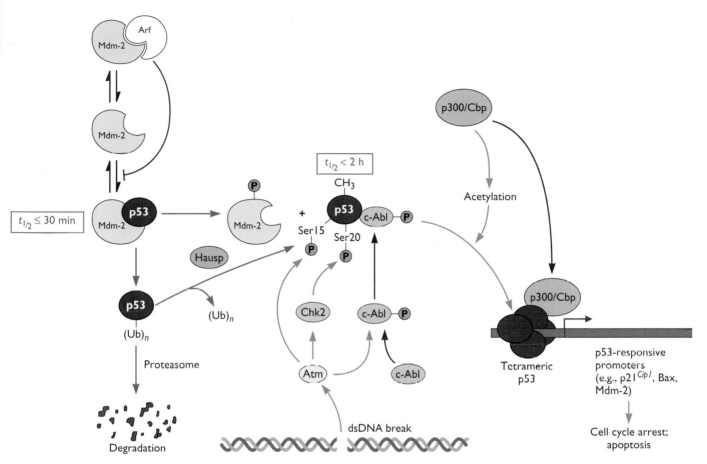

Figure 7.23 Regulation of the stability and activity of the p53 protein. Under normal conditions (left), cells contain only low concentrations of p53. This protein is unstable, turning over with a half-life of minutes, because it is targeted for proteasomal degradation by the Mdm-2 protein. Mdm-2 is a p53-specific E3 ubiquitin ligase that catalyzes polyubiquitinylation of p53, the signal that allows recognition by the proteasome. The Mdm-2 protein therefore maintains inactive p53 at low concentrations. The availability and activity of Mdm-2 are also regulated, for example, by Arf proteins encoded by the *ink4a/arf* tumor suppressor gene, and by stimulation of Mdm-2 transcription by the p53 protein itself. Signaling pathways initiated in response to damage to the genome or other forms of stress lead to stabilization of p53. Such posttranscriptional regulation is thought to allow a very rapid response to conditions that could be lethal to the cell. As illustrated with pathways operating in response to DNA damage (double-strand [ds] breaks) caused by ionizing radiation, p53 is stabilized in multiple ways. These mechanisms include phosphorylation of p53 at specific serines by Atm (see the text) and checkpoint kinase 2 (Chk2), binding to the c-Abl tyrosine kinase, sequestration of the Mdm-2 protein by Arf, and deubiquitinylation of p53 (in the presence of Mdm-2) by the herpesvirus-associated ubiquitin-specific protease (Hausp). Multiple mechanisms, including various modifications within the C-terminal domain (e.g., acetylation) also stimulate the sequence-specific DNA-binding activity of p53 or its association with the transcriptional coactivators p300/Cbp, and hence transcription from p53-responsive promoters.

transcription of genes for antiapoptotic proteins, such as *survivin*, may also be important. Whether p53 promotes cell cycle arrest or apoptosis is determined by numerous parameters, including the cell type, the nature of extracellular stimuli, and the concentration of the p53 protein itself. However, the apoptotic response prevails in many cell types under many circumstances: following DNA damage, in the absence of a hormone or growth factor necessary for

cell survival, and following expression of viral oncogenes that induce entry into S phase.

Inactivation of p53 by binding to viral proteins. The genomes of many viruses encode proteins that have been reported to interact with p53. However, the mechanisms by which the functions of this critical cellular regulator can be circumvented are best understood for the small DNA tumor viruses. As we have seen, the transforming proteins of these viruses induce release of E2f family members from association with Rb to promote cell cycle progression. Stabilization of p53 appears to be an inevitable consequence: E2f activates transcription from the promoter of the *Ink4/Arf* gene, which encodes a negative regulator of Mdm-2 (Fig. 7.24).

In contrast to the common mechanism of inactivation of Rb by the different viral proteins discussed above, adenoviral, papillomaviral, and polyomaviral proteins block p53 function in different ways (Fig. 7.25). The human papillomavirus type 16 or 18 E6 proteins bind to both p53 and a cellular ubiquitin protein ligase called the E6-associated protein, and thereby target p53 for proteasome-mediated destruction. Consequently, p53 is cleared from the infected or transformed cell. In conjunction with the viral E4 Orf6 protein, the adenoviral E1B 55-kDa protein also induces increased turnover of p53, but by directing it to a different ubiquitin ligase. In contrast, simian virus 40 LT actually stabilizes the p53 protein upon binding to it, but sequesters this cellular regulator

in inactive complexes. The adenoviral E1B 55-kDa protein also binds specifically to the N-terminal activation domain of p53 and can convert the cellular protein from an activator to a repressor of transcription. The results of mutational analyses have correlated the changes in concentration or activity in p53 induced by these viral proteins with their transforming activities.

Despite the mechanistic differences summarized in Fig. 7.25, cells infected by adenoviruses, papillomaviruses, or polyomaviruses all lack functional p53, and are therefore refractory to stimuli that would normally trigger p53-mediated cell cycle arrest or apoptosis.

Alteration of p53 activity via the p300/Cbp proteins. The nuclear phosphoprotein p300 was first identified by virtue of its binding to adenoviral E1A proteins. This protein, and the closely related transcriptional coactivator Creb-binding protein (Cbp), are required for stimulation of transcription by various sequence-specific DNA-binding proteins. It is thought that these coactivators

Figure 7.24 Stabilization of p53 by viral transforming proteins that bind to Rb. As described previously, binding of the adenoviral E1A (or polyomavirus LT or human papillomavirus E7) proteins to Rb allows transcription of E2f-responsive genes. This large set of genes includes the *ink4/arf* gene, and Arf therefore accumulates. Binding of Arf to Mdm2 sequesters this ubiquitin ligase, and hence leads to accumulation of the p53 protein. The E1A proteins also stabilize p53 via p300/Cbp-mediated acetylation of Rb. Acetylated Rb forms a ternary complex with p53 and Mdm2 and blocks p53 degradation. The N-terminal transcriptional activation domain of p53 remains blocked by Mdm-2, but in this form p53 can repress transcription and promote apoptosis.

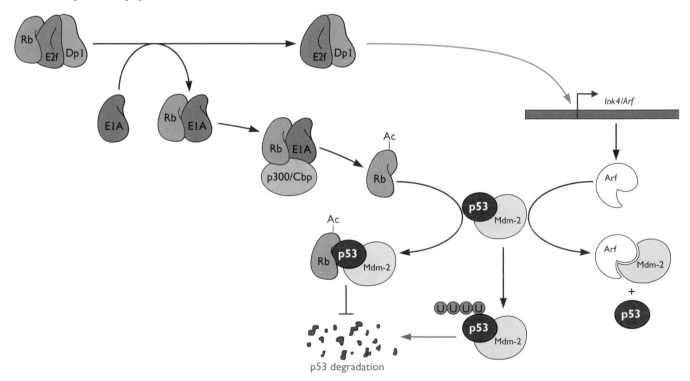

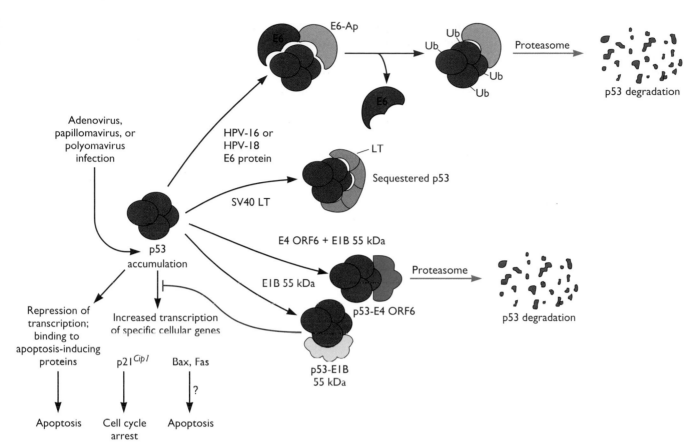

Figure 7.25 Inactivation of the p53 protein by adenoviral, papillomaviral, and polyomaviral proteins. The synthesis of transforming proteins in infected or transformed cells leads to accumulation of p53 (Fig. 7.24). Each of these viral genomes encodes proteins that interfere with the normal function of this critical cellular regulator. The E6 proteins of human papillomavirus types 16 and 18 bind to p53 via the cellular E6-associated protein (E6-Ap). The latter protein is a ubiquitin protein ligase that polyubiquitinylates p53 in the presence of the viral E6 protein, targeting p53 for degradation by the proteasome. Binding of simian virus 40 LT to p53, an interaction that is facilitated by sT, sequesters the cellular protein in inactive complexes. The adenoviral E1B 55-kDa and E4 Orf6 proteins bind to p53 at the N-terminal activation domain and a C-terminal region of p53, respectively. In experimental systems, the former interaction converts p53 from an activator to a repressor of transcription. In transformed rodent cells, it also induces relocalization of p53 from the nucleus to a perinuclear, cytoplasmic body. Binding of the E4 Orf6 and the E1B proteins increases the rate of degradation of p53 via recruitment of an ubiquitin ligase complex, that contains elongins B and C and cullin 5.

improve access of components of the transcriptional machinery to DNA in chromatin templates by acetylating nucleosomal histones. Because the E1A proteins bind to the regions of p300/Cbp that contact other transcriptional regulators, they disrupt complexes containing these cellular proteins. Binding of E1A proteins, and of simian virus 40 LT, to p300 correlates with induction of cell cycle progression and transformation, indicating that inhibition of the histone-modifying role of p300/Cbp proteins is important for the transforming activities of these viral proteins.

As mentioned above, the p53 protein blocks cell cycle progression by stimulating transcription of genes that encode cyclin-Cdk-inhibitory proteins, such as p21^{Cip1}.

Such stimulation of transcription is mediated by p300/Cbp, which binds to the activation domain of p53 via sequences that can also interact with the E1A proteins. Indeed, binding of E1A proteins to the coactivator inhibits p53-dependent transcription of *Cip1*. Consequently, this interaction blocks the induction of G_1 arrest by p53 following DNA damage.

Tumorigenesis Requires Additional Changes in the Properties of Transformed Cells

The mechanisms described in the preceding sections account for the sustained proliferation and survival of cells transformed by viral oncogenes. However, they are not necessarily sufficient for the induction of tumors or other types of cancer: tumorigenesis also generally depends on the ability of transformed cells to survive in the face of immune defenses. In some cases, induction of the growth of new blood vessels (angiogenesis) is also required (see "Other Viral Homologs of Cellular Genes" above).

Inhibition of Immune Defenses

The crucial contribution of mechanisms that protect transformed cells from immune defenses against tumorigenesis is illustrated by the properties of rodent cells transformed by oncogenic or nononcogenic human adenoviruses (Box 7.12). As discussed in Chapters 3 and 4, mechanisms that render infected cells refactory to immune defenses are important for the ability of many viruses to replicate in immunocompetent animals. How such mechanisms facilitate the survival of transformed cells and oncogenesis is best understood for herpesviruses associated with human cancers: Epstein-Barr virus and human herpesvirus 8.

Epstein-Barr virus is associated with Burkitt's lymphoma (a B-cell lymphoma) and nasopharyngeal carcinoma.

Although LMP-1 is the only viral gene product that can transform cells in culture, other viral proteins are made in such tumor cells. These products include EBNA-1, which is necessary for replication and maintenance of the episomal viral genome (Volume I, Chapter 9). This protein also contains a sequence that inhibits presentation of EBNA-1 epitopes by major histocompatibility complex (MHC) class I proteins on the cell surface. Consequently, tumor cells cannot be detected so readily by components of the adaptive immune system.

Similarly, several of the human herpesvirus 8 genes that have been implicated in transformation or tumorigenicity encode proteins that inhibit innate or adaptive immune responses. For example, the viral cytokine v-IL-6, which is a B-cell mitogen, also blocks the action of interferon by inhibiting phosphorylation of substrates of the interferon receptor, such as Stat2. In addition, the vFL1P protein, which can enhance the tumorigenicity of murine B cells, inhibits killing by natural killer (NK) cells. The viral genome also encodes proteins that reduce the cell surface expression of MHC class I proteins, block the transport of

---

BOX 7.12	**BACKGROUND**

Escape from immune surveillance and the oncogenicity of adenovirus-transformed cells

One of the earliest classifications of human adenovirus serotypes was on the basis of the ability of the viruses to induce tumors in laboratory animals (see the table). Rodent cells transformed with the viral E1A and E1B genes in culture exhibit the tumorigenicity characteristic of the virus. For example, cells transformed by the adenovirus type 12 genes form tumors efficiently when inoculated into syngeneic, immunocompetent animals, whereas cells transformed by adenovirus type 5 DNA induce tumors only in immunoincompetent animals, such as nude mice. This difference was exploited to map the ability of transformed cells to form tumors efficiently in normal animals to a small region of the E1A gene, unique to adenovirus type 12 (and other highly oncogenic adenoviruses). The tumorigenicity of transformed cells was also correlated with repression of transcription of MHC class I genes: the adenovirus type 12 proteins, but not those of adenovirus type 5, inhibit transcription of MHC class I genes by stimulating the binding of a translational repressor and histone deacetylases to the MHC class I enhancers.

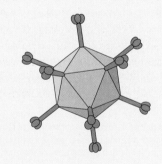

In addition, they inhibit the binding of the activator Nf-κb.

The inhibition of MHC class I transcription induced by adenovirus type 12 E1A proteins results in reduced expression of MHC class I proteins on the cell surface, and hence in impaired presentation of

antigens to cells of the immune system: recognition of transformed, tumor cells by cytotoxic T lymphocytes depends on MHC class I presentation of antigens. Adenovirus type 12-transformed cells therefore escape immune surveillance and destruction, whereas those transformed by adenovirus type 5 do not.

Yewdell, J. W., J. R. Bennink, K. B. Euger, and R. P. Ricciardi. 1998. CTL recognition of adenovirus-transformed cells infected with influenza virus: lysis by anti-influenza CTL parallels Ad12-induced suppression of class I MHC molecules. *Virology* **162:**236–238.

Zhao, B., S. Huo, and R. P. Ricciardi. 2003. Chromatin repression by COUP-TFII and HDAC dominates activation by NF-κB in regulating major histocompatibility complex class I transcription in adenovirus tumorigenic cells. *Virology* **306:**68–76.

Classification of human adenoviruses on the basis of oncogenicity

Subgroup	Representative serotypes	Oncogenicity in animals
A	12, 18, 31	High: induce tumors rapidly and efficiently
B	3, 7, 21	Low: induce tumors in only a fraction of infected animals, with a long latent period
C	1, 2, 5	None

B-cell receptors to the plasma membrane, or inhibit the complement cascade. Whether these proteins contribute to tumorigenesis remains to be established.

Other Mechanisms of Transformation and Oncogenesis by Human Tumor Viruses

The mechanisms by which some viruses associated with human cancers transform cells and contribute to tumor development cannot be subsumed within the general paradigms discussed in the preceding sections. Our current understanding or the development of these neoplastic diseases, which has taken many years, is described in this section.

Nontransducing, Complex Oncogenic Retroviruses: Tumorigenesis with Very Long Latency

The prototype for the nontransducing oncogenic retroviruses with complex genomes is human T-lymphotropic virus type 1, which is associated with adult T-cell lymphocytic leukemia (ATLL). This disease was first described in Japan in 1977, and has since been found in other parts of the world, including the Caribbean and areas of South America and Africa. The virus, which was isolated in 1980, is now classified as a deltaretrovirus, a group that includes other retroviruses with complex genomes, such as bovine leukemia virus and simian T-cell leukemia virus.

Human T-lymphotropic virus is transmitted via the same routes as human immunodeficiency virus; during sexual intercourse, by intravenous drug use and blood transfusions, and from mother to child. Infection is usually asymptomatic, but can progress to ATLL in about 5% of infected individuals over a period of 30 to 50 years (see Chapter 5). There is no effective treatment for the disease, which is usually fatal within a year of diagnosis. The mechanism(s) by which the virus induces malignancies is still uncertain, but some of the features of ATLL are consistent with a role for a viral regulatory protein. A provirus is found at the same site in all leukemic cells from a given case of ATLL, indicating clonal origin, but there are no preferred chromosomal locations for these integrations. Activation or inactivation of a specific cellular gene is not, therefore, a likely mechanism of transformation. As the genome of human T-lymphotropic virus type 1 does not contain **any** cell-derived nucleic acid, some viral sequences must be responsible for this activity. Surprisingly, however, proviral genes are not expressed in ATLL cells. But viral proteins are made if these cells are placed in culture. The last two features suggest an unusual mechanism of oncogenesis, in which expression of viral genes is necessary for the initiation of transformation but not for its maintenance.

In addition to the conserved genes of all retroviruses, the human T-lymphotropic virus type 1 genome contains a region, denoted X, which encodes a number of regulatory and accessory proteins (Fig. 7.26). One of the best-studied among these is the transcriptional activator Tax. A role for the *tax* gene in oncogenesis is suggested by the observation that transgenic mice carrying *tax* under the control of the viral LTR synthesize this viral protein in muscle tissue and develop multiple soft tissue sarcomas. A current model of how this virus might contribute to oncogenesis proposes that Tax alters the expression of cellular genes encoding proteins that regulate T-cell physiology. Infection by human T-lymphotropic virus type 1 does induce proliferation of T cells and increased synthesis of a number of cellular proteins, among them cytokines (e.g., IL-2 and its receptor) and the product of the *fos* proto-oncogene. Furthermore, the Tax protein inhibits apoptosis and activates Nf-κb, which is required for transcription of a number of genes that encode cellular regulators. Other proteins encoded in the X region may also play a role in oncogenesis, both by down-regulating the synthesis of viral proteins (Tax and Rex) late in infection (thereby avoiding immune recognition), and by affecting the expression of certain cellular proteins on the cell surface. For example, one of the accessory proteins has been found to decrease cell surface expression of MHC class 1 molecules.

A new gene, which is encoded in the **antisense** [(−)] strand of proviral DNA, was recently identified in the human T-lymphotropic virus type 1 genome. Transcription of this *hbz* gene is initiated in the 3′ LTR (Fig. 7.26), and the spliced mRNA encodes a basic leucine zipper protein. This protein forms a heterodimer with the cellular transcription factor Creb, rendering Creb incapable of binding to the Tax-responsive element in the 5′ LTR. Consequently, Tax-mediated transcription is inhibited, an event that may contribute to immune evasion. In addition to this protein function, the *hbz* RNA itself has been found to promote T-cell proliferation by a mechanism that is currently unknown but which could be mediated by RNA interference. The 3′ LTR is conserved in all ATLL cells, but the 5′ LTR is often deleted, or silenced by methylation. It has been proposed, therefore, that the *hbz* RNA may play a vital role in oncogenesis associated with this virus, even at late stages, when Tax is no longer synthesized. Other investigators have hypothesized that at late stages of the disease, no viral products might be required. Rather, genetic changes that accumulate in the DNA of the infected cells may drive oncogenesis. Because the virus-induced oncogenic events occur a long time before ATLL appears, it will be difficult to distinguish among these possibilities. Another challenge to the study of human T-lymphotropic virus has been the inefficiency of infection

The worldwide effort to eradicate poliomyelitis, launched by the World Health Assembly in 1988, has stalled. The goal for eradication was set to be the year 2000, but it was not to be. Many setbacks necessitated shifting the target date forward to the present target of 2010.

Enthusiasm was high during the first 12 years of the campaign, when the number of cases of the disease fell from an estimated 350,000 to 2,971. Now, the initial optimism is gone, replaced by doubt over whether eradication is realistic in light of the biological and political realities that have emerged in the course of the campaign. Furthermore, cessation of vaccination, which was to follow eradication, now seems ill advised. Why have the hopes engendered by this once popular program plummeted?

The strategy to eradicate polio makes use of large-scale immunization campaigns with live attenuated poliovirus vaccine. These vaccine strains were known to revert to neurovirulence and cause vaccine-associated poliomyelitis. However, it was thought initially that vaccine-derived poliovirus strains do not circulate efficiently in the population, and that once wild-type poliovirus was eliminated, cessation of vaccination would eradicate the cases of vaccine-associated disease. Unfortunately, the 2000 outbreak of poliomyelitis in Hispaniola shattered this incorrect assumption. In this outbreak, a total of 21 confirmed cases were reported, all but 1 of which occurred in unvaccinated or incompletely vaccinated children. However, subsequent analyses showed that the viruses responsible for the outbreak were derived from Sabin poliovirus type 1 administered in 1998 and 1999. The neurovirulence and transmissibility of these viruses were indistinguishable from those of wild-type poliovirus type 1.

Evidence of circulating vaccine-derived poliovirus was subsequently identified elsewhere. In Egypt, type 2 vaccine-derived poliovirus circulated from 1983 to 1993 and was associated with 32 reported cases. Vaccine-derived strains are also thought to have caused outbreaks of poliomyelitis in Africa and Asia in the 1980s.

Vaccine-derived polioviruses caused more than 100 cases of poliomyelitis in Nigeria in 2005 and 2006. Unchecked, and with continued circulation, vaccine-derived strains could eventually cause outbreaks as large as those caused by wild-type strains. The previously underestimated threat of vaccine-derived polioviruses now makes the plan to cease vaccination unacceptable.

During the eradication campaign, countries are certified as free from wild-type polioviruses when the virus cannot be isolated for a 3-year period. As of 2007, poliovirus is endemic in only four countries: Afghanistan, India, Nigeria, and Pakistan. Failure to eliminate transmission of wild polioviruses in these countries is probably a consequence of insufficient vaccine coverage due to politics and war. For example, it is difficult to deliver polio vaccine to the border of Pakistan and Afghanistan, where skirmishes occur regularly. In the north of India, children continue to contract polio despite multiple immunizations. In this case, poor sanitation, crowding, poverty, and infection with other microbes likely contribute to vaccine failure. In 2003, immunization was halted in

Nigeria due to fears that the vaccine was contaminated. This led to a resurgence of polio in Nigeria, which then spread across central Africa where immunization also had been reduced, and eventually spread to Yemen, Indonesia, and northern India. Although immunization in Nigeria resumed a year later, many children in that country still do not receive vaccine. The recent outbreak of vaccine-associated polio in Nigeria is a consequence of low immunization rates, which allow circulation and evolution of vaccine-derived polioviruses.

Certification of a region as polio-free means only that wild polioviruses cannot be detected, not that they are absent. Unfortunately, the surveillance network is not infallible. For example, orphan polioviruses, with genotypes unrelated to known strains have been identified, suggesting a failure of some point in the surveillance chain.

These considerations lend strength to the conclusion that polio eradication, followed by cessation of immunization, is not a realistic goal and that the program should be modified to ensure the protection of as many individuals as possible from

Globally reported incidence of poliomyelitis in 2008. The Americas, Western Pacific, and European regions have been declared poliomyelitis free by the WHO. The number of cases has declined from an estimated 350,000 in 1988 to ca. 1,300 in 2008. At the same time, the number of countries in which poliovirus is endemic has decreased from >125 to 4.

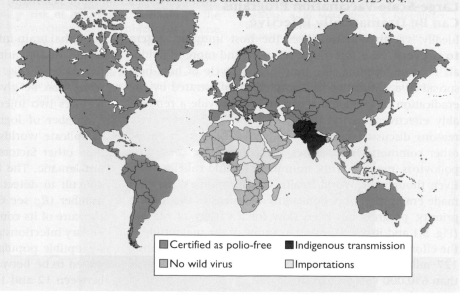

■ Certified as polio-free   ■ Indigenous transmission
□ No wild virus   □ Importations

poliomyelitis. The vaccine of choice for this purpose would be one that does not revert to neurovirulence. Furthermore, the use of a killed vaccine should bring the circulation of vaccine-derived polioviruses in check. This strategy, combined with surveillance to provide information on the extent of circulation of poliovirus strains, will provide protection of the population for the foreseeable future.

Chumakov, K., E. Ehrenfeld, E. Wimmer, and V. I. Agol. 2007. Vaccination against polio should not be stopped. *Nat. Rev. Microbiol.* **5:**952–958.

Dove, A. W., and V. R. Racaniello. 1997. The polio eradication effort: should vaccine eradication be next? *Science* **277:**779–780.

Minor, P. D. 2004. Polio eradication, cessation of vaccination and re-emergence of disease. *Nat. Rev. Microbiol.* **2:**473–482.

Eradicating a Viral Disease: Is It Possible?

The concept and possibility of eradicating a viral disease deserves careful contemplation. Viruses have survived countless bouts of selection during evolution, so the hubris of declaring a viral disease eradicated is obvious. Nevertheless, in 1978, the Director General of the WHO announced that smallpox was eradicated. Indeed, since that time no natural cases of smallpox have been reported (although accidental laboratory infections have occurred). What makes global elimination conceivable? A number of features are important, as summarized in Table 8.1, but two are absolutely essential: the viral infectious cycle must take place in a single host, and infection (or vaccination) must induce lifelong immunity. By definition, a vaccine that renders the host population immune to subsequent infection by a virus **that can grow only in that host** effectively eliminates the virus. In contrast, a virus with alternative host species in which to propagate cannot be eliminated by vaccination of a single host population: other means of blocking viral spread are required.

Table 8.1 Features of smallpox that enabled its eradication

Virology and disease aspects

No secondary hosts; this is a human-only virus

Long incubation period

Infectious only after incubation period

Low communicability

No persistent infection

Subclinical infections are not a source of infection

Easily diagnosed

Immunology

Infection confers long-term immunity

One stable serotype

Effective vaccine is available

Vaccine is stable and cheap

Social and political aspects

Severe disease with high morbidity and mortality

Considerable savings to developed countries where infection is not endemic

Eradication from developed countries demonstrated its feasibility

Few cultural or social barriers to case tracing and control

National Programs for Eradication of Agriculturally Important Viral Diseases Differ Substantially from Global Programs

National programs typically are established for economically important livestock diseases. The goals are to keep a country free of a particular viral disease even though the disease may still be present in other countries. For example, the United States and Canada have been declared to be free of foot-and-mouth disease, but infections still occur in parts of Europe and South America. Indeed, a devastating epidemic occurred in the United Kingdom in 2001 and a laboratory outbreak occurred in 2007. Such national programs can be successful only when supplemented with broad governmental enforcement and border security, as animals in the virus-free country are constantly exposed to sources from outside the country. Obviously, other principles of control must be implemented. For example, surveillance and containment strategies must be mobilized quickly and aggressively to identify and stop the spread from niche outbreaks. A common practice is to slaughter every host animal in farms at increasing distances surrounding an outbreak site (the so-called "ring-slaughter" program [Box 8.3]). Because acute infections spread rapidly from the outbreak site by many routes, and do so before identifiable symptoms are visible, the ring-slaughter containment often is breached unknowingly. To deal with this fact, preemptive slaughter of **all** animals on "at-risk" farms may be required. Obviously, the faster an outbreak is identified, the more likely the success of containment actions. Unfortunately, on-farm diagnostic tools that provide guaranteed identification of pathogens before symptoms are visible are simply not available. A false-positive identification of an outbreak will have serious consequences.

The strategies devised originally for national veterinary virus control take on new significance when the possibility that human and animal viruses may be used for nefarious purposes is considered. We cannot be sure that deadly, frightening, or economically devastating viruses are not hidden away for use as weapons, or to inflict terror on an unsuspecting and unprotected public (Box 8.4).

Vaccine Basics

Immunization Can Be Active or Passive

Ideally, **active immunization** with modified virions or purified viral proteins induces an immunologically mediated resistance to infection or disease. In contrast, **passive immunization** introduces the **products** of the immune response (e.g., antibodies or stimulated immune cells) obtained from an appropriate donor(s) directly into the patient. Passive immunization is a preemptive response, usually given when a virus epidemic is suspected. In 1997, consumption of contaminated fruit led to a widespread outbreak of hepatitis A infections in the United States. Pooled human antibodies (also called immune globulin) were administered in an attempt to block the spread of infection and reduce disease. Immune globulin contains the collective experience of many individual infections and provides

instant protection against some. The standard procedure for smallpox vaccination with live vaccinia virus requires that so-called "vaccine immune globulin" be available should disseminated vaccinia occur. When stimulated immune cells (e.g., T cells) are used, the process often is called **adoptive transfer**. Passive immunization is expected to produce short-term effects, depending on the biological half-lives of the antibodies or immune cells. Mothers passively immunize their babies through colostrum (antibody-rich first milk), or by transfer of maternal antibody to the fetus via the placenta, providing a protective umbrella against a number of pathogens (Fig. 8.2). This protective effect can be detrimental if active immunization of infants is attempted too early, as maternal antibody may block the vaccine from stimulating immunity in the infant.

Active Vaccines Stimulate Immune Memory

Vaccines work primarily because the immune system can recall the identity of a specific virus years after the initial encounter, a phenomenon called **immune memory** (Box 8.5). While the molecular aspects of the establishment and maintenance of memory continue to be active topics of research and debate, the resounding practical success of immunization

---

BOX 8.3

BACKGROUND
Stopping epidemics in agricultural animals by culling and slaughter

Vaccination of agriculturally important animals such as cattle and swine may not be cost-effective or may run afoul of government rules that block the shipping and sale of animals with antibodies to certain viruses. The 2001 foot-and-mouth disease epidemic in the United Kingdom provides a dramatic example of how viral disease is controlled when vaccination is not possible. The solution that stopped the epidemic was mass slaughter of **all animals** surrounding the affected areas and chemical decontamination of farms. It is estimated that 6,131,440 animals were destroyed in less than a year before the spread of foot-and-mouth disease virus was contained.

Animal slaughter is often the only alternative available to officials dealing with potential epidemic spread. For example, in recent years, millions of chickens in Hong Kong were killed to stop an influenza virus epidemic with potential to spread to humans. In 2002, millions of chickens in California were slaughtered to

stem the spread of Newcastle disease virus in major poultry factories.

Keeling, M. J., M. E. J. Woolhouse, R. M. May, G. Davies, and B. T. Grenfell. 2003. Modeling vaccination strategies against foot-and-mouth disease. *Nature* **421:**136–142.

Kitching, R., A. Hutber, and M. Thrusfield. 2005. Factors relevant to predictive modelling of the disease. *Vet. J.* **69:**197–209.

Woolhouse, M., and A. Donaldson. 2001. Managing foot-and-mouth. The science of controlling disease outbreaks. *Nature* **410:**515–516.

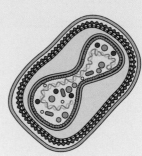
in stimulating long-lived immune memory stands as one of humankind's greatest medical achievements.

Immune memory is maintained by dedicated T and B lymphocytes that remain after an infection has waned. These cells provide a heightened ability to respond quickly to a subsequent infection (Fig. 8.3). Antiviral vaccines establish immunity and memory without the pathogenic events typical of the initial encounter with a virulent virus. Ideally, an effective vaccine is one that induces and maintains significant concentrations of specific antibodies (products of B cells) in serum and at points of viral entry. At the same time, T cells responsible for specific cellular immunity must be maintained in a precursor state, ready to make their lethal products (e.g., granzymes and perforins) when challenged by subsequent infection.

Two Central Concepts about Vaccines: Protection from Infection or Protection from Disease

It is critical to understand two fundamental ideas about vaccination and the potential outcomes. It is possible to immunize an individual so that a subsequent infection is stopped before it starts. In this case, the immune response from vaccination is strong enough such that sufficient humoral and cell-based immune effectors are produced and maintained for long periods. Second, it also is possible to immunize an individual so that infection is not blocked immediately, but the increased immune protection provided by vaccination delays or ameliorates disease. In this case,

the immune response in the vaccinated individual is said to be "primed." As a result, the subsequent cooperation of vaccine-induced immune effectors and infection-induced molecules may clear the invading virus, but **only** after infection begins. It also is possible that the pathogen may not be eliminated because the vaccine response or the natural response (or both) to the infection is not adequate. In this case, vaccination might only delay the appearance of disease. A further possibility is that immunized individuals may be infected, might mount an active immune response, and will have little or no disease, but the immune response does not eliminate the agent. These infections are said to be persistent or latent (see Chapter 5), and can occur naturally in the absence of vaccination.

Vaccines Must Be Safe, Efficacious, and Practical

The major prerequisites for an effective vaccine are presented in Table 8.2. The overriding requirement for any vaccine is safety: the vaccine cannot cause undue harm. For example, it is imperative that infectious particles and viral nucleic acids be undetectable in vaccines containing inactivated virions or viral proteins. If a live vaccine is used, virulent revertants must be exceedingly rare or undetectable. Contamination of vaccines with adventitious agents, such as other microbes introduced during production, must be avoided. These safety ideals are easy to state, but absolute safety is impossible to guarantee. Human error during vaccine production and testing, as well as limits of process biochemistry and cell biology, are simple facts of life. Rare

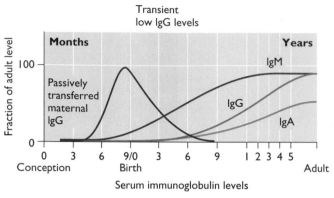

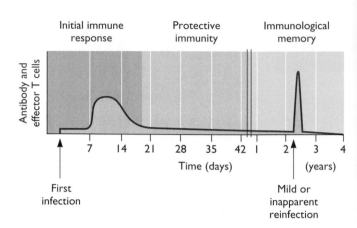

Figure 8.2 Passive transfer of antibody from mother to infant. The fraction of the adult concentration of various antibody classes is plotted as a function of time, from conception to adulthood. Newborn babies have high levels of circulating IgG antibodies derived from the mother during gestation (passively transferred maternal IgG), enabling the baby to benefit from the broad immune experience of the mother. This passive protection falls to low levels at about 6 months of age as the baby's own immune response takes over. Total antibody concentrations are low from about 6 months to 1 year after birth, which may lead to susceptibility to disease. Premature infants are particularly at risk for infections because the level of maternal IgG is lower and their immune system is less well developed. The time course of production of various isoforms of antibody (IgG, IgM, and IgA) synthesized by the baby is indicated. Adapted from C. A. Janeway, Jr., P. Travers, M. Walport, and M. Shlomchik, *Immunobiology: the Immune System in Health and Disease* (Current Biology Limited, Garland Publishing Inc., New York, NY, 2001), with permission.

Figure 8.3 Antibody and effector T cells are the basis of protective immunity. The relative concentration of antibody and T cells is shown as a function of time after first (primary) infection. Antibody levels and numbers of activated T cells decline after the primary viral infection is cleared (purple). Reinfections at later times (years later), even if mild or inapparent, are marked by rapid and robust immune response because of "memory." Adapted from C. A. Janeway, Jr., P. Travers, M. Walport, and M. Shlomchik, *Immunobiology: the Immune System in Health and Disease* (Current Biology Limited, Garland Publishing Inc., New York, NY, 2001), with permission.

side effects or immunopathology often can be identified only after millions of people have been vaccinated (Box 8.6). Furthermore, live vaccine agents have the potential to spread to nonvaccinated individuals in a population. For example, the smallpox vaccine is not given intentionally to

immunosuppressed individuals, because they cannot contain the infection and may die. These individuals obviously are at risk in a major vaccination program with the current smallpox vaccine (Box 8.1). Untoward side effects or other safety issues have been the death knell of many otherwise effective vaccines. Safety is paramount because vaccines are given to millions of people, some of whom may never be exposed to the virus targeted by the vaccine.

The next requirement is that the vaccine must induce protective immunity **in a significant fraction of the population.** Not every individual in the population need be

---

BOX 8.5

EXPERIMENTS

A natural "experiment" demonstrating immune memory

A striking example of immune memory is provided by a natural "experiment" in the 18th and 19th centuries on the Faroe Islands in the northern Atlantic Ocean. In 1781 a devastating measles outbreak drastically reduced the islands' population. For the next 65 years, the islands remained measles free and the surviving population flourished. In 1846 measles struck again, infecting over 75% of the population with similar devastating results. An astute Danish physician noted that none of the

aged people who survived the 1781 epidemic were infected. However, their age-matched peers who had not been infected earlier were ravaged by measles.

This natural experiment illustrates two important points: immune memory lasts a long time, and memory is maintained during this time without reexposure to the virus.

Ahmed, R., and D. Gray. 1996. Immunological memory and protective immunity: understanding their relation. *Science* **272:**54–60.

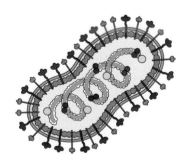

Table 8.2   Requirements of an effective vaccine

Safety	The vaccine must not cause disease
Side effects must be minimal	
Induction of protective immune response	Vaccinated individuals must be protected from illness due to pathogen
	Proper innate, cellular, and humoral responses must be evoked by vaccine
Practical issues	Cost per dose must not be prohibitive
	The vaccine should be biologically stable (no genetic reversion to virulence; able to survive use and storage in different surroundings)
	The vaccine should be easy to administer (oral delivery preferred to needles)
	The public must see more benefit than risk

immunized to stop viral spread, but a sufficient number must become immune to impede virus transmission. Virus spread stops when the probability of infection drops below a critical threshold. This effect has been called **herd immunity**. The actual threshold is agent and population specific, but it generally corresponds to 80 to 95% of the population acquiring vaccine-induced immunity. The herd immunity threshold is calculated as $1 - 1/R_0$. For smallpox this number is 80 to 85%, while for measles it is 93 to 95%. No vaccine is 100% effective, and as a result, the level of immunity is not equal to the number of people immunized. In fact, we know that when 80% of a population is immunized with measles vaccine, about 76% of the population actually is immune, clearly well below the 93 to 95% required for herd

immunity. Obviously, achieving such high levels of immunity by vaccination is a daunting task. Moreover, if the virus remains in other populations or in alternative hosts, reinfection is always possible. Closed populations (e.g., military training camps) can be immunized easily, but extended populations spread over large areas of a country present serious logistical and social problems. In addition, public complacency, or reluctance to be immunized, is dangerous to any vaccine program (Box 8.6). Unfortunately, epidemics can occur if the immunity of a population falls below a critical level because an available vaccine is not used.

The protection provided by a vaccine must also be long-term, lasting many years. Some vaccines cannot provide lifelong immunity after a single administration. In such cases, inoculations given after the initial vaccination (booster shots) to stimulate waning immunity are effective, but may be impractical to administer in large populations. Protective immunity also requires that the proper immune response be mounted. Primary infection by some viruses, such as poliovirus, can be blocked only when a robust **antibody response** is evoked by vaccination. On the other hand, a potent **cellular immune response** is required for protection against herpesviral disease. To maximize effectiveness, a vaccine must be tailored to fit its viral target.

Outbred populations always have varied responses to vaccination. Some individuals exhibit a robust response, while others may not respond as well (a "poor take"). While many factors are responsible for this variability, the age and health of the recipient are major contributors. For example, the influenza virus vaccine available each year is far more effective in young adults than in the elderly. Weak immune responses to vaccination pose several problems. Obviously, protection against subsequent infection may be inadequate, but another concern is that upon such subsequent infection, replication will occur in the presence of weak immune effectors. Mutants that can escape the host's immune response can then be selected, and may spread

BOX 8.6

DISCUSSION
The public's view of risk-taking is a changing landscape

Whooping cough used to be a major lethal disease of children until the introduction of the DPT vaccine (diphtheria, pertussis, and tetanus), which virtually eliminated the disease. Immunization resulted in some common, mild side effects: about 20% of children experienced local pain and a feeling of being tired. In addition, about 1 immunized child in 1,000 had more

severe side effects. Given that whooping cough was well known to be a child killer, these side effects were acceptable.

However, the vaccine is exceptionally effective and whooping cough was nearly eliminated. Now some parents assess the risk of immunization side effects to be unacceptable, and whooping cough is being seen again in clinics. The risk posed by

the vaccine has not changed, but in the face of reduced threat of natural disease, the perceived risk of vaccination was elevated.

Johnson, B. 2001. Understanding, assessing, and communicating topics related to risk in biomedical research facilities. *ABSA Anthology of Biosafety IV—Issues in Public Health*, Chapter 10. http://www.absa.org/0100johnson.html.

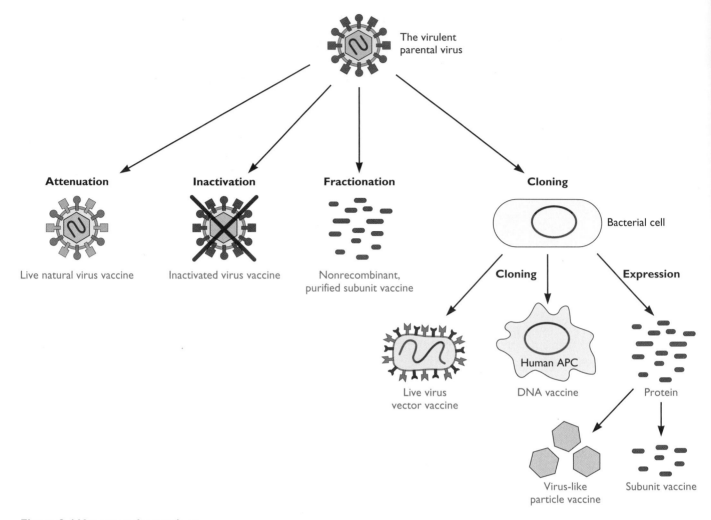

Figure 8.4 How to make vaccines.

in the immunized population. Indeed, vaccine "escape" mutants are well documented.

Important practical requirements for an effective vaccine include stability, ease of administration, and low cost. If a vaccine can be stored at room temperature, rather than refrigerated or frozen, it can be used where cold storage facilities are limited. When a vaccine can be administered orally rather than by injection, more people will accept the vaccination. The WHO estimates that a vaccine must cost less than $1 per dose if much of the world is to afford it. However, the research and development costs for a modern vaccine are in the range of hundreds of millions of dollars. Another, often prohibitive expense is covering the liability of the vaccine producer. Liability expenses can be astronomical in a litigious society and have forced many companies to forgo vaccine development completely. Unfortunately, there is an inherent conflict between providing a good return on investment to vaccine developers

and supplying vaccines to people and government agencies with a limited ability to bear the cost.

The Fundamental Challenge

Given the remarkable success of smallpox, measles, and polio vaccines, it might seem reasonable to prepare vaccines against all viral diseases. Unfortunately, designing and producing an effective vaccine are exceedingly difficult. Despite considerable progress in research, we cannot predict with confidence the efficacy or side effects of different vaccine preparations. We lack sufficiently detailed knowledge of the important mechanisms of immune protection against most viral infections, so the basic design of a vaccine is not always obvious. Even questions such as "Is a neutralizing antibody response important?" or "Is a cytotoxic T-lymphocyte (CTL) response essential?" cannot be answered with certainty even for our most common viral infections. Will a vaccine induce sufficient immunity to block infection completely, or will it delay disease onset? In fact, only when a vaccine is effective, or more often, when it fails, can we learn what actually does or does not constitute a protective response. The regulatory network of gene products required to reproduce an effective antiviral response is understood only in outline. To complicate the situation, even when an experienced vaccine manufacturer sets out to develop, test, and register a new vaccine, the process can take years. For example, it took 22 years to develop and license a relatively straightforward hepatitis A vaccine. The fundamental challenge is to find ways to capitalize on the discoveries in molecular virology and medicine to speed up the process of vaccine development. Despite many problems, those involved in vaccine research and development remain optimistic because of our remarkable history of success and our increasingly detailed knowledge of the molecular nature of viruses and viral disease.

The Science and Art of Making Vaccines

At present, we can outline four basic approaches to produce vaccines (Fig. 8.4). Each approach starts with a pathogenic virus of interest. A vaccine developer may produce large quantities of the virulent virus of interest and chemically inactivate it (**killed vaccine**), may attenuate the pathogenicity through laboratory manipulation (**live, attenuated vaccine**), may produce individual proteins free of the viral nucleic acid (**subunit vaccines**), or may clone all or portions of the viral genome to give rise to recombinant DNA vaccines of several types (**recombinant vaccines**). The most common, commercially successful vaccines simply comprise attenuated or inactivated virions. Their preparation is based on principles that would be understood by Pasteur (Table 8.2 and Boxes 8.1, 8.7, and 8.8).

Basic Approaches

Inactivated or "Killed" Virus Vaccines

The inactivated poliovirus, influenza virus, hepatitis A virus, and rabies virus vaccines are examples of effective inactivated vaccines administered to humans (Table 8.3). In addition, these vaccines are widely used in veterinary medicine. To prepare such a vaccine, virions of the virulent virus are isolated and inactivated by chemical or physical procedures. These treatments eliminate the infectivity of the virus, but do not compromise the antigenicity (i.e., the ability to induce the desired immune response). Common techniques include treatment with formalin or β-propriolactone, or extraction of enveloped virus particles with nonionic detergents.

Theoretically, inactivated vaccines are very safe, but accidents can and do happen. In the 1950s, a manufacturer of Salk polio vaccine did not inactivate it completely. As a result, more than 100 children developed disease after vaccination. Incomplete inactivation and contamination of vaccine stocks with potentially infectious viral nucleic acids have been singled out as major problems with this type of vaccine.

Vaccination is currently the most important measure for reducing influenza virus-induced morbidity and mortality. In the United States alone, influenza virus infections may cause as many as 50,000 deaths every year and cost at least $12 billion in health care. Every year, millions of citizens seeking to avoid infection receive their "flu shot," which contains several strains of influenza virus that have been predicted to reach the United States that year. The magnitude of this undertaking is noteworthy. For example, between 75 million and 100 million doses of inactivated vaccine must be manufactured annually. Typically, these vaccines are formalin-inactivated whole virions or

---

BOX 8.7

BACKGROUND
The response to infection and vaccination

After a natural infection, viral proteins that can be recognized by the immune system are made in the infected cell. The production of progeny virions and their subsequent spread to other cells amplifies the response. However, vaccination may not reproduce all aspects of this response. Several important variables include the following.

- Infection by live attenuated viruses may provoke an immune response that is qualitatively different from

that stimulated by infection with virulent virus.

- In the case of killed vaccines or subunit vaccines, no new viral proteins are made after injection. The immune system therefore **must** recognize only the input material.
- Inactivated virus particles and virion subunits often stimulate an antibody response, but they rarely stimulate an effective CTL response.

- Inactivated virus particles or virion subunits may not persist in the body long enough to establish an effective immune memory response.

Ada, G. L. 1994. Vaccines and the immune response, p. 1503–1507. *In* R. G. Webster and A. Granoff (ed.), *Encyclopedia of Virology*. Academic Press, San Diego, CA.

| BOX | BACKGROUND |
| 8.8 | *Our best vaccines are based on old technology* |

It seems ironic in this age of modern biology that the mutations in many of our common vaccine strains were not introduced by site-directed mutagenesis of genes known to be required for viral virulence, but rather were isolated by selection of mutants that could replicate in various cell types. The vaccines were produced with little bias as to how to reduce virulence.

Despite the old technology, the vaccines are relatively safe and remarkably effective. Consequently, their analysis has led to the identification of important attenuating mutations, as well as parameters affecting the protective immune response. The current vaccines not only provide protection but also are the foundation for future vaccines.

Painting of Louis Pasteur examining the dried spinal cord of an infected rabbit used to prepare an attenuated strain of rabies virus. Image courtesy of the Pasteur Institute (Photothèque/Relations Presse et Communication externe, Institut Pasteur, Paris, France).

detergent- or chemically disrupted virions (often called a "split" virus preparation). The viruses, which are mass-produced in embryonated chicken eggs, can be natural isolates or reassortant viruses constructed to express the appropriate hemagglutinin (HA) or neuraminidase (NA) gene from the virulent virus.

Currently, a typical influenza vaccine dose is standardized to comprise 15 µg of each viral HA protein, but it contains other viral structural proteins as well. The efficacy of these vaccines varies considerably. They are reportedly about 60 to 90% effective in healthy children and adults younger than 65 years exposed to virus strains in the vaccine. The influenza virus vaccines are all highly immunogenic in young adults, but less so in the elderly, immunosuppressed individuals, and people with chronic illnesses. Protection against illness correlates with the concentration of serum antibodies that react with viral HA and NA proteins produced after vaccination. Immunization

may also stimulate limited mucosal antibody synthesis and CTL activities, but these responses vary widely.

The envelope proteins of influenza viruses change by antigenic drift and shift as the virus replicates in various animal hosts around the world (see Chapter 5). Consequently, protection one year **does not guarantee protection the next.** To deal with this ever-changing agent, vaccine manufacturers must reformulate the vaccine **every year** so that it contains antigens from the predicted next generation of viruses. A committee of the Food and Drug Administration chooses the particular strains, in conjunction with WHO-designated laboratories that monitor influenza infections. Timing is critical, as the final decision for the virus composition in the vaccine must be made within the first few months of each year to allow sufficient time for production of the vaccine. Any delay or error in the process, from prediction to manufacture, has far-reaching consequences, given the millions of people who are vaccinated and expect safe protection. Even if the vaccine contains the appropriate viral antigens, and is made promptly and safely, inactivated influenza virus vaccines have the potential to cause side effects in some individuals who are allergic to the eggs in which the vaccine strains are grown. As an example of other problems, the H5N1 avian virus that first infected humans in Hong Kong in the 1990s was extraordinarily cytopathic to chicken embryos, making it difficult to propagate for vaccine purposes. Reassortants had to be constructed by placing the new H5N1 segments in less cytopathic viruses. The risk, of course, is that such reassortants will not provide the proper immune protection against the original strain.

Live Attenuated Virus Vaccines

Successful live attenuated vaccines are effective for at least three reasons: viral replication occurs and stimulates an immune response, progeny virions often are contained at the site of replication and do not spread to other sites, and the infection induces mild or inapparent disease (Fig. 8.5). Less virulent (attenuated) viruses can be selected by growth in cells other than those of the normal host, or by propagation at nonphysiological temperatures (Fig. 8.6). Mutants able to propagate better under these selective conditions arise during viral replication. When such mutants are isolated, purified, and subsequently tested for pathogenicity in appropriate models, some may be less pathogenic than their parent. Temperature-sensitive and cold-adapted mutants are often less pathogenic than the parental virus, because of reduced capacity for replication and spread in the warm-blooded host. Cold-adapted influenza viruses with mutations in almost every gene are licensed vaccines. In the case of viruses with segmented genomes (e.g., arenaviruses, orthomyxoviruses,

Table 8.3 Viral vaccines licensed in the United States

Disease or virus	Type of vaccine	Indications for use	Schedule
Adenovirus	Live attenuated, oral	Military recruits	One dose
Hepatitis A	Inactivated whole virus	Travellers, other high-risk groups	0, 1, and 6 mo
Hepatitis B	Yeast-produced recombinant surface protein	Universal in children, exposure to blood, sexual promiscuity	0, 1, 6, and 12 mo
Influenza	Inactivated viral subunits	Elderly and other high-risk groups	One dose seasonally
Influenza	Live attenuated	Children 2–8 yr old, not previously vaccinated with influenza vaccine	Two doses at least 1 mo apart
		Children 2–8 yr old, previously vaccinated with influenza vaccine	One dose
		Children, adolescents, and adults 9–49 yr old	One dose
Japanese encephalitis	Inactivated whole virus	Travelers to or inhabitants of high-risk areas in Asia	0, 7, and 30 days
Measles	Live attenuated	Universal vaccination of infants	12 mo of age; 2nd dose, 6 to 12 yr of age
Mumps	Live attenuated	Universal vaccination of infants	Same as measles, given as MMR
Papilloma (human)	Yeast- or SF9-produced virus-like particles	Females 9–26 yr old	Three doses
Rotavirus	Live reassortant	Healthy infants	2, 3, and 6 mo or 2 and 4 mo of age depending on vaccine
Rubella	Live attenuated	Universal vaccination of infants	Same as measles, given as MMR
Polio (inactivated)	Inactivated whole viruses of types 1, 2, and 3	Changing: commonly used for immunosuppressed where live vaccine cannot be used	2, 4, and 12–18 mo of age, then 4 to 6 yr of age
Polio (live)	Live, attenuated, oral mixture of types 1, 2, and 3	Universal vaccination; no longer used in United States	2, 4, and 6–18 mo of age
Rabies	Inactivated whole virus	Exposure to rabies, actual or prospective	0, 3, 7, 14, and 28 days postexposure
Smallpox	Live vaccinia virus	Certain laboratory workers	One dose
Varicella	Live attenuated	Universal vaccination of infants	12 to 18 mo of age
Varicella-zoster	Live attenuated	Adults 60 yr old and older	One dose
Yellow fever	Live attenuated	Travel to areas where infection is common	One dose every 10 yr

bunyaviruses, and reoviruses), attenuated, reassortant viruses may be obtained after mixed infections with pathogenic and nonpathogenic viruses.

Live oral poliovirus vaccine comprises three attenuated strains selected for their reduced neurovirulence. Type 1 and 3 vaccine strains were isolated by the passage of virulent viruses in different cells and tissues until mutants with reduced neurovirulence in laboratory animals were obtained (see Fig. 8.7 for a description of how type 3 was derived). The type 2 component was derived from a naturally occurring attenuated isolate. The mutations responsible for the attenuation phenotypes of all three serotypes are shown in Fig. 8.7. Curiously, each of the three serotypes contains a different mutation in the 5' noncoding region that may affect translation. Alterations in capsid proteins are thought to influence viral assembly.

The live measles virus vaccine currently in use was derived from a virulent virus isolated in 1954 by John Enders, called the Edmonston strain. Attenuated viruses were isolated following serial passages through various types of cells (Fig. 8.8). Even though this approach was empirical, the virions that were isolated replicated poorly at body temperature and caused markedly less disease in primates. The vaccine strain harbors a number of mutations, including several that affect the viral attachment protein.

The live varicella-zoster vaccine is currently the only licensed human herpesvirus vaccine. It has proved to be safe and effective in children and adults, providing significant protection against infection by this prevalent human herpesvirus. Recently, the vaccine has been licensed for use in previously infected adults (over 60 years of age) to protect against recurrent disease (herpes zoster, or shingles).

Live attenuated viruses are administered by injection (e.g., measles-mumps-rubella [MMR] and varicella-zoster vaccines) or by mouth (e.g., poliovirus, rotavirus, and adenovirus vaccines). The highly effective Sabin poliovirus

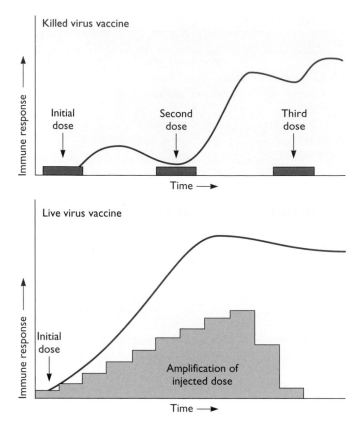

Killed virus vaccine

Immune response →

Initial dose

Second dose

Third dose

Time →

Live virus vaccine

Immune response →

Initial dose

Amplification of injected dose

Time →

Figure 8.5 Comparison of the predicted immune responses to live and inactivated viruses used in vaccine protocols. (Top) Immune responses plotted against time after injection of a killed virus vaccine (red curve). Three doses of inactivated virions were administered as indicated. **(Bottom)** Results after injection of a live attenuated virus vaccine. A single dose was administered at the start of the experiment. The filled histogram (lavender-colored area) under the curve displays the titer of infectious attenuated virus. Redrawn from C. A. Mims et al., *Mims' Pathogenesis of Infectious Disease*, 4th ed. (Academic Press, Inc., Orlando, FL, 1995), with permission.

vaccine is administered as drops to be swallowed, and enteric adenovirus vaccines are administered as virus-impregnated tablets. One virtue of the oral delivery method for enteric viruses is that it mimics the natural route of infection and, therefore, has the potential to induce the natural immune response. Another is that it bypasses the traditional need for hypodermic needles required for intramuscular or intradermal delivery. Live attenuated respiratory viruses (e.g., the live influenza virus vaccine) can also be delivered by nasal spray, simulating the natural route of infection and immune response.

Live attenuated virus vaccines have inherent problems. For example, despite reduced spread of attenuated viruses in the vaccinee, some viral shedding occurs and these virions are available to infect unvaccinated individuals (Box 8.2). Given the high rate of mutation associated with

RNA virus replication, a reversion to virulence should not surprise us. Such reversion is one of the main obstacles to developing effective live attenuated vaccines and is formally equivalent to the emergence of drug-resistant mutants (see "Resistance to Antiviral Drugs" in Chapter 9). While virulent revertants are a serious problem, considerable insight into viral pathogenesis and the immune response can be obtained by determining the changes responsible for increased virulence.

Because our knowledge about virulence is so limited, it is difficult to predict how a live attenuated virus will behave in individuals and in the population. The attenuating mutations may lead to unexpected diversions from the natural infection and expected host response. For example, the attenuated virus may be eliminated from the vaccinated individual before it can induce a protective response, it may infect new niches in the host with unpredictable effects, or it may initiate atypical infections (e.g., slow or chronic infections) that can trigger immunopathological responses of unknown etiology, such as **Guillain-Barré syndrome.** This syndrome may follow viral illness or vaccination and is characterized by rapidly progressing, symmetric weakness of the extremities. It is the most frequent cause of acute generalized paralysis. Vaccine side effects, whether real or not, often have a detrimental effect on public acceptance of national vaccine programs (Box 8.9).

Reversion is not the only problem for live attenuated vaccines, as illustrated by the following informative example. The varicella-zoster live vaccine generally is quite stable and not prone to genetic drift or reversion. Indeed, reversion of the attenuating mutations has been exceedingly difficult to document. However, in studies where viral genomes were isolated and sequenced from individuals who developed vaccine-associated rash, scientists found an unexpected, nonrandom selection of mutations in the isolated viruses. These mutations were not reversions of attenuating mutations, but, rather, had preexisted at a very low level in stocks of the vaccine virus. They were selected in the vaccinated individuals who experienced rashes. The implication was that viral, rather than host, factors were responsible for the rare vaccine side effect of severe rash. This unexpected discovery points to the importance of defining the genetic purity of live-vaccine stocks.

Ensuring purity and sterility of the product is a problem inherent in the production of biological reagents on a large scale. If the cultured cells used to propagate attenuated viruses are infected with unknown viruses, the vaccine may well contain these adventitious agents. For example, in the 1950s, early lots of poliovirus vaccine were grown in monkey cells that were unknowingly infected with the polyomavirus simian virus 40. It is estimated that 10 million to 30 million persons received one or more doses of live simian virus 40

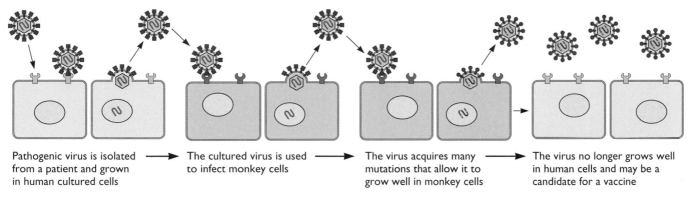

Pathogenic virus is isolated from a patient and grown in human cultured cells ➔ The cultured virus is used to infect monkey cells ➔ The virus acquires many mutations that allow it to grow well in monkey cells ➔ The virus no longer grows well in human cells and may be a candidate for a vaccine

Figure 8.6 Viruses specific for humans may become attenuated by passage in nonhuman cell lines. The four panels show the process of producing an attenuated human virus by repeated transfers in cultured cells. The first panel depicts isolation of the virus from human cells (yellow). The second panel shows passage of the new virus in monkey cells (lavender). During the first few passages in nonhuman cells, virus yields may be low. Viruses that grow better can be selected by repeated passage, as shown in the third panel. These viruses usually have several mutations, facilitating growth in nonhuman cells. The last panel shows one outcome in which the monkey cell-adapted virus now no longer grows well in human cells. This virus may also be attenuated (have reduced ability to cause disease) after human infection. Such a virus may be a candidate for a live vaccine if it will induce immunity but not disease. Adapted from C. A. Janeway, Jr., P. Travers, M. Walport, and M. Shlomchik, *Immunobiology: the Immune System in Health and Disease* (Current Biology Limited, Garland Publishing Inc., New York, NY, 2001), with permission.

in contaminated vaccines. Many even developed antibody to simian virus 40 virion proteins. Some concern exists that rare tumors may be linked to this inadvertent infection, but proving cause and effect has been difficult (see Chapter 7).

Alternatives to the classical empirical approach to attenuation based on modern virology and recombinant DNA technology can now be applied. Viral genomes can be cloned, sequenced, and synthesized. Their genetic information can be analyzed using animals and cultured cells. Viral genes required for pathogenesis in model systems can be identified in systematic fashion. In many cases, attenuation can be deliberately achieved by genetic manipulation. For example, deletion mutations with exceedingly low probabilities of reversion can be created (Fig. 8.9). In another approach that relies on genome segment reassortment of influenza viruses and reoviruses, genes contributing to virulence are replaced with those from related but nonpathogenic viruses. No matter which technology is applied to achieve attenuation, the genetic engineer and the classical virologist must satisfy the same fundamental requirements: isolation or construction of an infectious agent with low pathogenic potential that is, nevertheless, capable of inducing a long-lived, protective immune response.

Subunit Vaccines

A vaccine may consist only of a subset of viral proteins, as demonstrated by successful hepatitis B and the "split" influenza vaccines. Vaccines formulated with purified components of viruses, rather than the intact virion, are called subunit vaccines. We can deduce what viral proteins to include in a vaccine by determining which are recognized by the antibodies and CTLs found in individuals recovering from disease. Although the most obvious proteins would be those present on the virion surfaces, in fact, any viral protein could be a target.

Synthetic peptides of about 20 amino acids or more in length can induce specific antibody responses when chemically coupled to certain protein carriers that can be taken up, degraded, and presented by major histocompatibility complex (MHC) class II proteins. In principle, synthetic peptides should be the basis for an extremely safe, well-defined vaccine. To date, however, peptide vaccines have had little success, mainly because synthetic peptides are expensive to make in sufficient quantity, and the antibody response they elicit is often weak and short-lived. A weak immune response can be far worse than no response at all, because viral escape mutants may be selected. Given the simplicity of the antipeptide response (usually a single epitope is represented by the peptide), selection of escape mutants is highly probable.

Recombinant DNA Approaches to Subunit Vaccines

Recombinant DNA methods enable cloning of appropriate viral genes into nonpathogenic viruses, bacteria, yeasts, insect cells, or plant cells to produce the immunogenic protein(s). As only a portion of the viral genome is required for such production, there can be no contamination of the resulting vaccine with the original virus, solving a major safety problem inherent in inactivated whole-virus vaccines. Viral proteins can be made inexpensively in large quantities by engineered organisms

A **Derivation of Sabin type 3 attenuated poliovirus**

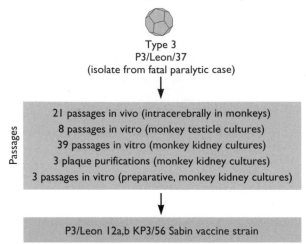

Type 3
P3/Leon/37
(isolate from fatal paralytic case)

Passages

21 passages in vivo (intracerebrally in monkeys)
8 passages in vitro (monkey testicle cultures)
39 passages in vitro (monkey kidney cultures)
3 plaque purifications (monkey kidney cultures)
3 passages in vitro (preparative, monkey kidney cultures)

P3/Leon 12a,b KP3/56 Sabin vaccine strain

B **Determinants of attenuation in the Sabin vaccine strains**

Virus	Mutation (location/nucleotide position)
P1/Sabin	5'-UTR nt 480 VP1 aa 1106 VP1 aa 1134 VP3 aa 3225 VP4 aa 4065
P2/Sabin	5'-UTR nt 481 VP1 aa 1143
P3/Sabin	5'-UTR nt 472 VP3 aa 3091

C **Reversion of P3/Sabin**

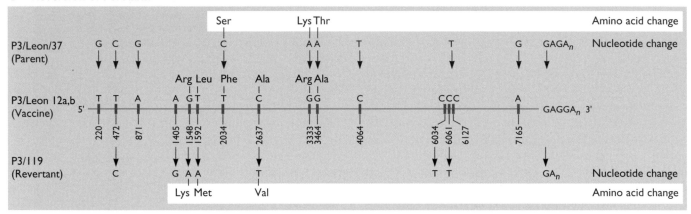

Figure 8.7 Live attenuated Sabin oral poliovirus vaccine. (A) All three viral serotypes may cause poliomyelitis. Therefore, the Sabin vaccine is administered as a mixture of three different strains that are representatives of poliovirus serotypes 1, 2, and 3. Shown is the derivation of the type 3 vaccine strain, called P3/Sabin (the letter P means poliovirus). The parent of P3/Sabin is P3/Leon, a virus isolated from the spinal cord of an 11-year-old boy named Leon, who died of paralytic poliomyelitis in 1937 in Los Angeles. P3/Leon virus was passaged serially as indicated. At various intervals, viruses were cloned by limiting dilution, and the virulence of the virus was determined in monkeys. An attenuated strain was selected to be the final P3/Sabin strain included in the vaccine. **(B)** Determinants of attenuation in all three strains of the Sabin vaccine. The mutations responsible for the reduced neurovirulence of each serotype of the live poliovirus vaccine are indicated (5'-UTR is the 5' untranslated region, VP1 to VP4 are the viral structural proteins; nt, nucleotide). **(C)** Reversion of the P3/Sabin vaccine strain. Differences in the nucleotide sequences of the virulent P3/Leon strain and the attenuated P3/Sabin vaccine strain are shown above the cartoon of the parental viral RNA (green) and of a virulent virus (P3/119) isolated from a case of vaccine-associated poliomyelitis (below).

under conditions that simplify purification and quality control. For example, problems with egg allergies after vaccination can be eliminated completely when influenza virus proteins are synthesized in *Escherichia coli* or yeasts.

Unfortunately, many candidate subunit vaccines fail because they do not induce an immune response sufficient to protect against infection. Protection against infection

(often called "challenge") is the "gold standard" of any vaccine. To achieve this standard, many variables must be assessed: the nature, dose, and virulence of the challenging virus, as well as the route of immunization and the age and health of the host, come into play. The immune repertoire evoked by a live-virus infection may be only partially represented in a response to a subunit vaccine. In particular,

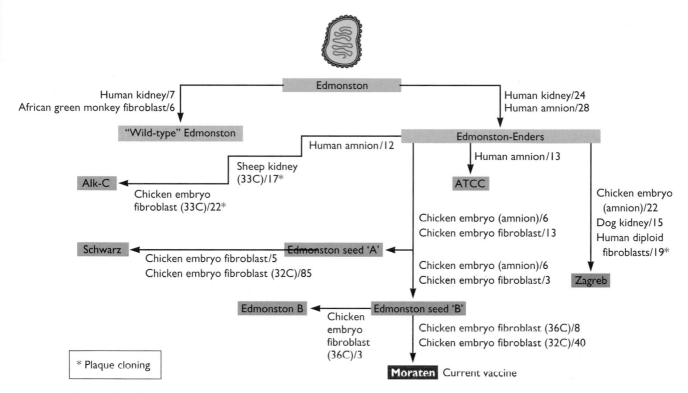

Figure 8.8 Live attenuated measles virus vaccine. Passage histories of live attenuated measles virus vaccines derived from John Enders' original isolate of Edmonston virus (top, blue). The current vaccine is called the Moraten strain (bottom, red). The cells used in passaging the virus to select attenuated mutants are indicated. The temperature during growth is given in parentheses, and the number of passages of virus in the particular cell or cell line follows the slash. An asterisk indicates that a single plaque was picked for further propagation. Adapted from D. D. Richman, R. J. Whitley, and F. G. Hayden (ed.), *Clinical Virology,* 2nd ed. (ASM Press, Washington, DC, 2002), with permission.

purified protein antigens rarely simulate the appearance of mucosal antibodies, particularly immunoglobulin A (IgA).

Virus-Like Particles

Capsid proteins of nonenveloped and some enveloped virions may self-assemble into virus-like particles. Virus-like particles have virtually identical capsid structure to virions, but unlike authentic virions, the capsids are empty; they contain no nucleic acid. These empty capsids retain most of the conformational epitopes not found on purified or unstructured proteins. Consequently, unlike simple subunit vaccines of pure proteins, virus-like particle vaccines often induce authentic neutralizing antibodies and other protective responses after injection. Importantly, because the particles are completely noninfectious, inactivation with formalin or other agents typically used to inactivate live-virus vaccines are not required. This fact presents at least two more advantages: immunogenicity is not affected (formalin and other alkylating chemicals often alter the immunogenicity of the inactivated virions), and concerns about efficiency of inactivation are avoided. Virus-like particle vaccines have proven to be particularly attractive for viruses that replicate poorly in cell culture.

The highly successful hepatitis B subunit vaccine comprises virus-like particles produced in yeast. This vaccine contains a single viral structural protein (the surface antigen) that assembles spontaneously into virus-like particles, whether made in yeast, *E. coli,* or mammalian cell lines. Formation of particles is critical, as purified monomeric capsid protein does not induce a protective immune response. In mice, as little as 0.025 µg of virus-like particles can elicit antibody production without the presence of an adjuvant. Typically, in the human vaccine, 10 to 20 µg of virus-like particles per dose is administered in three doses over a 6-month period, and more than 95% of recipients develop antibody against the surface antigen.

The virus-like particle vaccine effective against human papillomavirus infections is attracting considerable attention. More than 80% of sexually active women will be infected with several serotypes of human papillomavirus during their lifetime. As a result, many of them will develop genital warts and also cervical carcinoma. There are many serotypes of

BOX 8.9 DISCUSSION
National vaccine programs depend on public acceptance of their value

November 2001: Doctors warn of possible measles, mumps, or rubella epidemic

Doctors in Devon [United Kingdom] are warning of a possible outbreak of measles, mumps or rubella as the number of children immunised with MMR vaccine drops.

Update: July 2006

Thirty British doctors call for responsible media coverage amid doubts about the MMR vaccine in the United Kingdom. The number of vaccinated children in the United Kingdom declined from 93% in 1995 to 83% in 2005. Doctors cite a dramatic rise of measles in 2005.

From the BBC News Service, November 2001 and July 2006

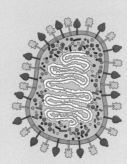

The measles-mumps-rubella (MMR) vaccine has proved to be effective in reducing the incidence of these highly contagious and serious diseases. The economic benefit in the United States from use of the MMR vaccine has been estimated to exceed $5 billion per year. However, in the past few years, the press has reported anecdotal studies that link the MMR vaccine to autism. Infants can receive many immunizations early in life, and the symptoms of autism often first appear at the time of these immunizations. The MMR vaccine was singled out as a potential link to autism in some studies. Of prime concern was that thimerosal, a mercury compound with potential neurological effects, was used as a preservative in vaccines. As a result of the publicity and parental concern, public confidence has been shaken in most developed countries. In some, people are refusing to have their children vaccinated. For example, the immunization rate has fallen significantly in the United Kingdom and, consequently, the incidence of measles infection is on the rise.

In the United States, the Centers for Disease Control and Prevention and the National Institutes of Health recognized the need for an independent group to examine the hypothesized MMR-autism link and address other vaccine safety issues. The committee concluded that "The evidence favors rejection of a causal relationship at the population level between MMR vaccine and autistic spectrum disorders (ASD)," and that "a consistent body of epidemiological evidence shows no associ-

ation at a population level between MMR and ASD." Other leading medical groups, the American Academy of Pediatrics, the WHO, and British health authorities have come to similar conclusions for largely the same reasons. In 1999, the Food and Drug Administration ordered the elimination of thimerosal from children's vaccines, despite evidence that the mercury exposure was too low to be responsible for neurological defects.

Unfortunately, public confidence in government proclamations is not always high. Moreover, members of the lay public are not well trained to analyze data collected from complex studies that seek to find correlations, or to assess cause and effect. It makes no difference if the public perceives that cancer is caused by high-tension lines or that autism is caused by vaccination: proving (or disproving) cause and effect to a disbelieving public is an exceedingly difficult task.

Information on the Immunization Safety Review Committee can be found at http://www.iom.edu/ImSafety. The full report can be purchased or read online at http://www.nap.edu/catalog/ 10101.html. Copies of *Immunization Safety Review: Measles-Mumps-Rubella Vaccine and Autism* are available for sale from the National Academy Press; call (800) 624-6242 or (202) 334-3313 (in the Washington, DC, metropolitan area), or visit the National Academy Press home page at http://www.nap.edu.

---

the virus, but serotypes 6, 11, 16, and 18 cause 70% of cervical cancers and 90% of genital warts. It had been known for some time that the human papillomavirus L1 capsid protein forms virus-like particles when synthesized in a variety of heterologous systems. Upon testing, these empty capsids proved to be exceptional inducers of a protective immune response. As a result, a quadrivalent, virus-like particle vaccine effective against the four major serotypes of the virus was formulated. In 2006, the Food and Drug Administration approved this formulation as the first vaccine to be developed to prevent cervical cancer induced by a virus.

DNA Vaccines

In 1992, scientists developed a variation of the subunit vaccine approach that is showing exceptional promise. The

DNA vaccine consisted simply of a DNA plasmid encoding a viral gene that can be expressed inside cells of the animal to be immunized. In the simplest case, the plasmid encodes only the immunogenic viral protein under the control of a strong eukaryotic promoter. The plasmid DNA, produced in bacteria, can be prepared free of contaminating protein and has no capacity to replicate in the vaccinated host (Fig. 8.10). Remarkably, unlike the requirements for standard protein subunit vaccines, no adjuvants or special formulations are necessary to stimulate an immune response. The vaccine can be delivered by injection of an aqueous solution containing a few micrograms of the plasmid DNA into muscle or skin tissue. Another effective delivery method uses a "gene gun" that literally shoots DNA-coated microspheres through the skin to introduce the plasmid into dermal tissue. Because

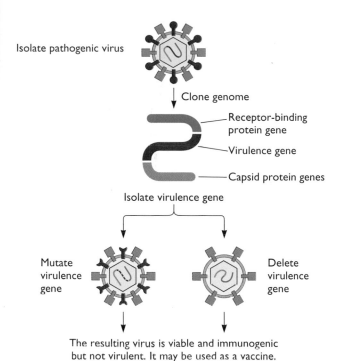

Isolate pathogenic virus

Clone genome

Receptor-binding protein gene

Virulence gene

Capsid protein genes

Isolate virulence gene

Mutate virulence gene

Delete virulence gene

The resulting virus is viable and immunogenic but not virulent. It may be used as a vaccine.

Figure 8.9 Construction of attenuated viruses by using recombinant DNA technology. Once the genome of a pathogenic virus is cloned in a suitable system, deletions, insertions, and point mutations can be introduced by standard recombinant DNA techniques. If the cloned genome is infectious or if mutations in plasmids can be transferred to infectious virus, it is possible to mutate viral genes systematically to find those required for producing disease. The virulence gene can then be isolated and mutated, and attenuated viruses can be constructed. Such viruses can be tested for their properties as effective vaccines. The mutations in such attenuated viruses may be point mutations (e.g., temperature-sensitive mutations) or deletions. Multiple point mutations or deletions are preferred to reduce or eliminate the probability of reversion to virulence. Adapted from C. A. Janeway, Jr., P. Travers, M. Walport, and M. Shlomchik, *Immunobiology: the Immune System in Health and Disease* (Current Biology Limited, Garland Publishing Inc., New York, NY, 2001), with permission.

the viral protein is made *de novo* inside a cell, a fraction of it is presented on the surfaces of producing cells by MHC class I molecules, and the cells are recognized by T cells. Both the humoral response and CTLs can be stimulated by DNA vaccination, and, most important, the vaccine protects against challenge in some animal models. The striking property of DNA vaccination is that a relatively low dose of immunizing protein seems sufficient to induce long-lasting immune responses: the quantity of encoded protein produced is perhaps in the nanogram range or less, as determined by sensitive reporter proteins such as luciferase.

The type of immune response is dictated by the method of inoculation. A Th1 response predominates after injection of an aqueous DNA solution into muscle. In this method,

DNA rapidly disappears from the site of injection and is found in the spleen, where the immune response occurs. Animals vaccinated in this way usually synthesize high concentrations of gamma interferon (IFN-γ), interleukin-2 (IL-2), and IL-12, but only low levels of IL-4 or IL-10. In contrast, after DNA immunization by gene gun, the DNA either enters cells directly or is taken up by skin cells and keratinocytes to produce the viral protein. Langerhans dendritic cells in the skin then acquire the protein and move to the draining lymph node. The response induced often is more typical of a Th2 response, with synthesis of IL-4 and antibodies, not IFN-γ and CTLs.

DNA vaccines are effective without addition of adjuvants. The reason is that plasmid DNA itself has intrinsic adjuvant activity in mammalian cells, presumably due to its recognition by Toll-like receptors such as Tlr9. When DNA with unmethylated CpG sequences binds Tlr9 on dendritic cells and monocytes, a burst of IL-12 and IFN-γ ensues and activates Th1 cells. These findings explain why techniques that bypass the dendritic cell Tlr9 protein, such as gene guns that place DNA directly into target cells, often activate the Th2 antibody response rather than the Th1 CTL response.

From a basic research point of view, much remains to be investigated. A technique called **gene shuffling**, which can be applied to produce diverse coding sequences, may have utility in DNA vaccine technology (Box 8.10). Another variation on the single-gene DNA vaccine is a **genomic vaccine**: a library of all the genes of a particular pathogen is prepared in multiple DNA vaccine vectors. The entire plasmid mixture is injected into an animal. Such a vaccine has the potential to present every gene product of the pathogen to the immune system. Because each plasmid encodes a unique mRNA, it should be straightforward to determine the gene(s) necessary and sufficient to induce a protective immune response.

We cannot be sure if DNA vaccine technology has a commercial future. Safety of DNA vaccines is a prime concern. Some possible dangers include integration of plasmid DNA leading to insertional mutagenesis, induction of autoimmune responses such as anti-DNA antibodies, and induction of immune tolerance.

Live Attenuated Viral Vectors and Foreign Gene Expression

Genomes of nonpathogenic viruses can be constructed to produce selected viral proteins that can immunize a host against the pathogenic virus (see Volume I, Chapter 3). This approach merges subunit and live attenuated virus vaccine technologies. Genes from a pathogenic virus are cloned in a nonpathogenic viral vector and used to infect the animals to be immunized. In theory, the vector provides the benefits of a viral infection with respect to stimulating an immune response to the expressed proteins, but with none

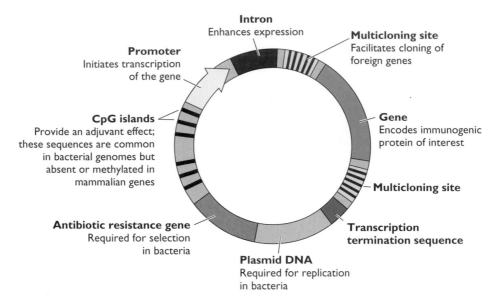

Figure 8.10 Functional components of a DNA vaccine expression vector. The system is based on a plasmid, typically from *E. coli*, carrying its own origin of replication and an antibiotic resistance gene as a selectable marker. A critical component is a strong eukaryotic promoter to initiate transcription of the cloned gene. Expression may be enhanced by including an intron sequence (to facilitate transport of messenger RNA [mRNA] from the nucleus). A multicloning site offers restriction enzyme sites to facilitate cloning of the antigen-encoding gene downstream of the strong promoter and intron. A short untranslated termination sequence provides a polyadenylation signal as well as stability to the antigen-encoding mRNA. CpG-rich sequences can be added and are often found naturally in the bacterial plasmid DNA. Such sequences function as an adjuvant to stimulate the immune response. Adapted from M. Oyaski and H. Ertl, *Sci. Med.* **7:**30–39, 2000, with permission.

of the pathogenesis associated with a virulent virus. However, any replicating viral vector has the potential to produce pathogenic side effects, particularly if injected directly into organs or the bloodstream (see Box 4.7). The immune response to such hybrid viruses is not always predictable, particularly if children, the elderly, and immunocompromised individuals are to be treated.

Poxviruses, such as vaccinia virus, often are used as vaccine vectors. Highly attenuated vaccinia virus mutants have been isolated and may be the basis for the next generation of vaccinia virus-based vaccines. A wide variety of systems are available for the construction of vaccinia virus recombinants that cannot replicate in mammalian cells, but allow the efficient synthesis of cloned gene products that retain their immunogenicity. Attenuated vaccinia virus vectors can accommodate more than 25 kb of new genetic information. Vaccinia virus recombinants can also be used to dissect the immune response to a given protein from a pathogenic virus. This application is illustrated in Fig. 8.11. Other poxviruses, including raccoonpox, canarypox, and fowlpox viruses, offer additional possibilities because they are able to infect, but not replicate in, humans.

The successful use of an oral rabies vaccine for wild animals in Europe and the United States demonstrates that recombinant vaccinia virus vaccines have considerable potential. Recombinant vaccinia virus genomes encoding the major envelope protein of rabies virus yield virions that are formulated in edible pellets to be spread in the wild. The pellets are designed to attract the particular animal to be immunized (e.g., foxes or raccoons). Ideally, the animal eats the pellet, is infected by the recombinant virus, and responds with a protective immune response to subsequent rabies virus infection. While effective, this vaccine approach must be applied with care. Given the rare but serious side effects possible when vaccinia virus infects humans, inadvertent human infection by these wildlife vaccines must be avoided.

The rhabdovirus vesicular stomatitis virus is a particularly promising vaccine vector (see Volume 1, Chapter 3). This virus causes an acute infection of short duration producing vesicular lesions in tongue, teats, and hooves of cattle, pigs, and horses. It also can infect humans, causing mild symptoms. Molecular virologists have studied this virus for many years and have developed techniques to construct attenuated mutants and recombinants that direct the synthesis of almost any protein. Infection of mice and rhesus monkeys induces strong cellular and humoral responses to any foreign protein that has been made after viral infection. The virus is attractive for vaccine technology, because

E X P E R I M E N T S
DNA shuffling: directed molecular evolution

DNA shuffling enables scientists to assemble and analyze new combinations of DNA fragments not found in nature. The utility of this approach for vaccine development is illustrated by the following example. Many different clades of HIV exist, and neutralizing antibodies often recognize the Env protein from one clade, but not others. Some scientists suggest that an important component of a vaccine would be a protein that induces an antibody response capable of recognizing the Env proteins from all known viral clades.

DNA shuffling may provide technology to achieve this goal. DNA encoding Env proteins from various clades would be mixed, digested with restriction enzymes, denatured, reannealed, and amplified by the polymerase chain reaction (PCR). Some of the single-stranded DNA fragments from the *env* gene of one clade will anneal to similar (but not identical) DNA sequences derived from another *env* gene. The 3′ ends of these hybrids are extended in the PCR, and single-stranded regions are copied by DNA polymerase. A mixture of chimeric double-stranded DNA fragments is produced, representing not only all combinations of *env* genes, but also many not found naturally. Expression libraries created by insertion of the new

shuffled DNA segments into suitable plasmids would then be screened for production of new proteins that bind antibodies specific for Env proteins, or that induce antibodies capable of neutralizing a broad spectrum of virus isolates.

Optimists assert that DNA shuffling can be used to make effective, safe, multivalent DNA vaccines, and also can be used to boost the potency of known vaccines.

Stemmer, W. P. C. 1994. Rapid evolution of a protein *in vitro* by DNA shuffling. *Nature* **370**:389–391.
Whalen, R. G., R. Kaiwar, N. W. Soong, and J. Punnonen. 2001. DNA shuffling and vaccines. *Curr. Opin. Mol. Ther.* **3**:31–36.

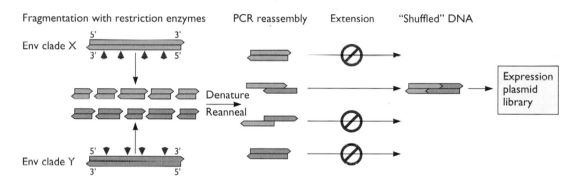

high titers can be obtained easily and it can be delivered by a mucosal route without the need for injection. Initial experiments are promising: when rhesus monkeys were vaccinated with recombinant vesicular stomatitis viruses expressing the human immunodeficiency virus type 1 (HIV) *env* and *gag* genes, and then challenged with a strain of human virus that causes AIDS in monkeys, all animals remained healthy for over a year, with low or undetectable viral loads. In contrast, all nonvaccinated, challenged monkeys progressed to AIDS in less than 5 months.

Two general problems arise with live-virus vector vaccines. First, the host usually is immunized against the viral vector, as well as the vaccine antigen. As a result, subsequent uses of the particular vector may result in a weak response, no response, or an immunopathological response. Second, introducing live viral vectors, even though they are attenuated, into the population at large may have long-term effects. For example, immunocompromised individuals may be infected, with adverse consequences.

Vaccine Technology

Most Killed and Subunit Vaccines Rely on Adjuvants To Stimulate an Immune Response

Charles Janeway revealed what he called "immunologists' dirty little secret": inactivated virions or purified proteins often do not induce the same immune response as live attenuated preparations, unless mixed with a substance that stimulates early processes in immune recognition, particularly the inflammatory response. Such immunostimulatory substances are called **adjuvants**. Development of these substances has been largely empirical, although as our understanding of the various regulators of immune responses increases, more specific and powerful molecules are being discovered. The fundamental understanding of adjuvant action is that adjuvants stimulate early intrinsic and innate defense signals, and therefore can shape subsequent adaptive responses. Adjuvants work in at least three different ways: by presentation of antigen as particles,

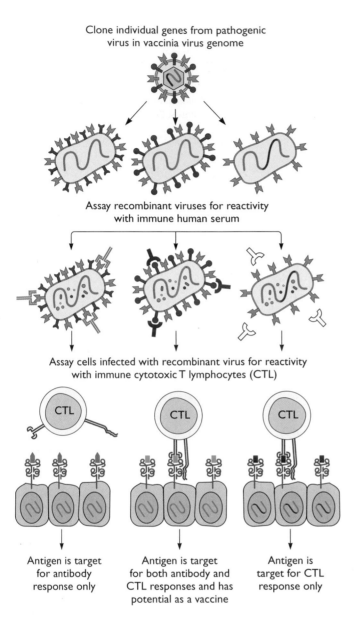

Clone individual genes from pathogenic
virus in vaccinia virus genome

Assay recombinant viruses for reactivity
with immune human serum

Assay cells infected with recombinant virus for reactivity
with immune cytotoxic T lymphocytes (CTL)

| Antigen is target for antibody response only | Antigen is target for both antibody and CTL responses and has potential as a vaccine | Antigen is target for CTL response only |

Figure 8.11 Use of recombinant vaccinia viruses to identify and analyze T- and B-cell epitopes from other viral pathogens. As illustrated, it is possible to determine if a particular viral protein contains a B-cell epitope (binds antibodies), a T-cell epitope (recognized by CTLs), or both. Subsequent site-directed mutational analysis of the viral genes enables precise localization of these epitopes on the viral protein. Adapted from C. A. Janeway, Jr., P. Travers, M. Walport, and M. Shlomchik, *Immunobiology: the Immune System in Health and Disease* (Current Biology Limited, Garland Publishing Inc., New York, NY, 2001), with permission.

by localization of antigen to the site of inoculation, and by direct stimulation of the intrinsic and innate immune responses. The immune system can be stimulated directly when adjuvants mimic or induce cellular damage or alter homeostasis (sometimes called "danger" signals), or when they engage intrinsic cellular defense receptors.

Adjuvants vary in composition, from complex mixtures of killed mycobacteria and mineral oil (the classic complete Freund's adjuvant) to lipid vesicles or mixtures of aluminum salts (Table 8.4). Some adjuvants, like alum (microparticulate aluminum hydroxide gel), are widely used for human vaccines such as hepatitis A and B vaccines. Others, such as complete Freund's adjuvant, are used only in research. Freund's adjuvant is extremely potent, but is not licensed for use in humans because it causes tissue damage and other undesirable effects. Two of the active components in Freund's adjuvant have been identified as muramyl dipeptide and lipid A, both potent activators of the inflammatory response. We now understand that the strong adjuvant effects of this complex adjuvant are due, at least in part, to the mycobacterial DNA present in the emulsion that activates the Tlr9 pathogen recognition protein. Less toxic derivatives, along with saponins and so-called block copolymers (linear polymers of clustered hydrophobic and hydrophilic monomers), are promising adjuvants.

As noted above, virus-like particles do not require adjuvants to stimulate strong immune responses. It may be possible to produce other viral protein complexes with adjuvant activity. For example, liposomes and lipid micelles can be prepared with viral proteins exposed on the surface (immune-stimulating complexes [ISCOMs]). These preparations are being evaluated in human trials (Table 8.4).

Some new adjuvants under study enable exogenous proteins to enter the major MHC class I pathway normally reserved for proteins synthesized *de novo*. Preparations with saponins or muramyl tripeptide-phosphatidylethanolamine induce macrophages and dendritic cells to synthesize IL-12, which stimulates Th1 immune responses. Others include molecules that stimulate the pattern recognition receptors that provide the initial responses to infection (Chapter 3). Administration of a natural or synthetic ligand for a particular Tlr, triggers a rapid and targeted immune response to a coinjected foreign protein in laboratory animals.

Vaccine researchers speak of "tailoring" a vaccine by using different combinations of adjuvant and vaccine to induce a protective immune response. The principle is promising, but much work remains to be done to demonstrate the safety and efficacy of this approach.

Delivery

Improvement of administration or delivery is an important goal of vaccine developers. At present, vaccines can be administered by a variety of methods, including the traditional hypodermic needle injection and the "air gun" injection of liquid vaccines under high pressure through the skin. Other formulation methods under consideration include new emulsions, artificial particles, and direct injection of fine powders through the skin. Oral delivery of vaccines can be effective in stimulating appearance of

Table 8.4 Vaccine delivery systems and adjuvants[a]

Adjuvant	Formulation and properties
Aluminum salts	Aluminum hydroxide or phosphate (alum). Form precipitates with soluble antigen, making the complexes more immunogenic; form antigen "depot" at site of injection; complement activation.
Emulsions	Freund's complete adjuvant: antigen suspended in water/mineral oil emulsion with killed *M. tuberculosis* bacteria or muramyl di- or tripeptide to stimulate strong T-cell responses. Freund's incomplete adjuvant: antigen suspended in water-in-mineral oil emulsion.
Microspheres	Antigen encapsulated in polymers of lactic and glycolic acids. They are biodegradable and cause slow release of antigen.
ISCOMs	Immune stimulating complexes composed of glycosides, a purified saponin from the plant *Quillaja saponaria*, cholesterol, phospholipids, and antigens. Form spheres of 30–40 nm in diameter that incorporate antigen.
Nucleic acid vaccines	Genes encoding antigens expressed from strong promoters are introduced directly to muscle or skin by using physical methods or liposomes, leading to intracellular protein production and presentation of antigen to the immune system.
Engineered viruses	Genes encoding foreign antigens are introduced into a viral genome (the vector) such that the new protein made following infection. Common viral vectors are vaccinia virus, adenovirus, and baculovirus. Many other viruses can also be modified to express foreign genes.

[a]See B. Guy, *Nat. Rev. Microbiol.* **5**:505–517, 2007, for more information.

IgA antibodies at mucosal surfaces of the intestine, and in inducing a more systemic response. However, oral delivery is not always possible, because the protective surfaces and enzymes of the oral cavity and alimentary tract block or destroy many vaccines. Specially constructed edible plants that make viral proteins represent an attractive, new approach. Transgenic plants can be engineered, or plant viruses with genomes encoding immunogenic proteins can be used to infect food plants. Early experiments are promising: when such a plant is eaten, antibodies to the viral structural protein can be demonstrated in the animal's serum. Another oral vaccine formulation technology is based on principles discovered by research on enteric viruses that survive passage through the stomach. Biopolymers that mimic the protective action of capsids can be formulated with vaccine components to resist the low pH and digestive enzymes of the stomach, but dissolve in the more hospitable regions of the upper intestinal tract. Such coating technology has been used for oral delivery of adenovirus vaccines and the vaccinia virus-based wildlife rabies vaccine discussed previously.

Immunotherapy

Patients already infected with viruses that cause persistent infections, or that reactivate from latency in the face of an immune response, present special problems. **Immunotherapy** provides the already-infected host with antiviral cytokines, other immunoregulatory agents, antibodies, or lymphocytes **over and above those provided by the normal immune response** in order to reduce viral pathogenesis. Immunotherapy can be administered by introduction of the purified compounds, or of a gene encoding the immunotherapeutic molecule (a gene therapy approach).

In at least the case of the varicella-zoster vaccine, a large boost of the same live, attenuated vaccine given to children will stimulate the immune response of latently infected adults such that recurrent disease (e.g., zoster or shingles) is markedly reduced. A live attenuated virus or a DNA vaccine can be modified to synthesize cytokines that stimulate the appropriate immune response. If a live attenuated vaccine is used, care must be taken as it is possible that the altered immune response will have unexpected effects, such as increased virulence, persistence, and pathogenesis.

It is possible to isolate lymphocytes from patients, infect these cells with a defective virus vector (e.g., a retrovirus) encoding an immunoregulatory molecule, and then introduce the transduced cells back into the patient (*ex vivo* approach). If these cells survive and synthesize the protein, the patient's immune response may be bolstered. If stem cells are transduced, a long-term effect may be achieved, as these cells replicate and propagate the transgene.

We know that immunotherapy using cytokines can be effective. Historically, it was pioneered by using interferons to treat a variety of diseases, including those caused by persistent viruses. For example, IFN-α is approved in the United States for treatment of chronic hepatitis caused by hepatitis B and C viruses. Its effect on chronic hepatitis B virus infection is remarkable: as many as 50% of treated patients have no detectable infection. This result illustrates the potential of immunotherapy. However, similar treatment of hepatitis C virus infection has been much less successful for reasons that are not clear. Limitations of IFN therapy (and probably cytokine therapy in general) are that the biological activity of IFN is not sustained for a prolonged time, side effects are significant, patients older than 50 tend to be less responsive, and treatment is expensive.

Immunomodulating agents, including IFN, cytokines that stimulate the Th1 response (e.g., IL-2), and certain chemokines, are being studied individually and in combinations for their ability to reduce virus load and complications of persistent infections caused by papillomavirus and human immunodeficiency virus. Cytokines that stimulate natural killer cells (e.g., IL-12 and IFN-γ) may have promise as well. An alternative to using cytokines or immunoregulatory molecules is the introduction of more antibodies or activated lymphocytes into an infected individual. Care must be taken when introducing immune-modulating genes into viral genomes. For example, cloning of the IL-4 gene into mousepox virus, with the hope that this recombinant would produce a strong antibody response, led to a hypervirulent virus. A fundamental problem with these approaches is that immune stimulation can lead to pathology or provide new avenues for virus spread.

The Quest for an AIDS Vaccine

In 1984, several years after HIV was identified, officials in the U.S. government predicted that an AIDS vaccine would be available within 3 years. Although research has continued for nearly 25 years with unprecedented intensity, we still do not know how to make a vaccine that will protect against AIDS. Such ill-founded optimism has a precedent in the history of vaccination. Poliovirus was isolated in 1908, and 3 years later Simon Flexner of the Rockefeller Institute announced that a vaccine would be prepared in 6 months. Almost 50 years of research on basic poliovirus biology was necessary to provide the knowledge of pathogenesis and immunity that allowed the development of effective poliomyelitis vaccines (Table 8.5).

Formidable Challenges

Despite an investment of more than 1.5 billion dollars toward the development of an HIV vaccine since 1996, it is safe to say that the prospect of a licensed vaccine appearing even by the next decade is remote. The reasons for this lack of progress lie deep in HIV biology and the interaction of the virus with the host immune system. Although a vigorous immune response is induced after infection, this response is not effective in clearing it. Not only that, but viral proteins actually derail subsequent immune responses to the extent that the host often dies due to other microbial infections. As a consequence, HIV pathogenesis is almost completely one sided, in that patients never spontaneously recover. At this time, we do not know what constitutes protective immunity and have few, if any, clues to guide development of a prophylactic or therapeutic vaccine. Vaccine developers are resigned to testing their hypotheses in surrogate systems or even in the clinic before any progress can be made. A number of approaches, including inactivated virions, subunit vaccines based on single viral proteins, and passive immunization, have already been tested with no obvious success. In particular, subunit vaccines, although capable of inducing strong antibody responses, are markedly inefficient in eliciting a CTL response. Therapeutic vaccines administered to infected individuals may have value, if administered in combination with a drug regimen that protects the immune system. Live attenuated HIV vaccines, modeled after the successful live poliovirus vaccine, present difficult scientific and ethical problems: the risks associated with injecting thousands of healthy uninfected volunteers with a living (albeit attenuated) virus are currently considered unacceptable. More promising, however, are vaccines based on recombinant adenovirus, poxvirus, vesicular stomatitis virus, or alphavirus genomes that encode various HIV proteins. Such recombinant vaccines have proved to be potent activators of CTLs as well as of humoral responses. A promising area of research is combination of a DNA vaccine to provide strong cellular and humoral responses (DNA priming) and a live attenuated virus vaccine (vector boosting). Nevertheless, truth be told, in the last 15 years, more than 30 vaccine candidates have been tested and all have failed. As a sobering example of the paucity of basic knowledge, a recent book on HIV vaccines stated that "… there is a general, but unproven notion, that both humoral and cellular immune responses will ultimately be required for a successful AIDS vaccine."

Table 8.5 When can we expect an HIV vaccine?

Viral vaccine	Yr when etiologic agent was discovered	Yr when vaccine was developed in the United States	No. of yr elapsed
Polio	1908	1955	47
Measles	1953	1983	30
Hepatitis B	1965	1981	16
Rotavirus	1970	1998	28
Hepatitis A	1973	1995	22
HIV	1983	None yet	>25

The Central Issues

Lentivirus replication requires integration of a DNA copy of the viral genome into the host genome. Such a close relationship with the host often results in a true latent infection where cells with integrated proviral DNA produce no proteins. These reservoirs of infection remain invisible to immune recognition. Consequently, the infection is maintained in the face of a vigorous immune response. Some scientists consider this the crux of the challenge: how do we induce an immune response **superior** to the normal response that apparently is not good enough to clear the infection? How do we eliminate immune escape mutants?

At a minimum, we can make a short list of our expectations for an HIV vaccine (Table 8.6). It should be safe, of course, but it also must be effective in preventing infection in most of the vaccinated people. The protection should last for many years and be effective against as many of the diverse HIV strains as possible. The vaccine should not be so complicated that it cannot be produced on a large scale for a reasonable price. It should be stable with a significantly long shelf life so that it can be distributed, stored, and delivered when needed. Finally, it should be easy to administer. If anything has been learned over the past 25 years of managing HIV infections in the clinics around the world, it is that a predictive relationship exists between the viral load (as measured by the number of viral RNA copies in the circulation) and disease progression: patients with low viral loads progress more slowly to disease. The hope is that a vaccine that will mimic the response resulting in low viral load can be developed. What this immune response is remains a matter of conjecture. The window of opportunity to block a primary infection, integration, and dissemination within a host is very small.

There are complicated social, ethical, and political issues that must be addressed, when vaccines are to be tested in humans (Table 8.7). For example, can vaccine trials be conducted so that appropriate placebo controls can be evaluated? Vaccine developers face costly litigation if unanticipated reactions occur. On a larger scale, significant political issues arise when decisions are made by Western leaders about vaccines to be used in developing countries. Resolution of these problems will not be easy.

Despite the formidable challenges, scientists in academia and industry must work with public health scientists and government policy makers to muster the political and scientific might to solve the problems. The HIV pandemic is already extracting an inestimable cost and, sadly, shows no signs of diminishing.

Perspectives

The goal of modern vaccine research is to formulate vaccines that can be tailored to particular infections, and that will produce a safe and protective defense. We have had noteworthy success with new vaccine formulations such as the virus-like particle preparations effective in protecting against hepatitis B and papillomavirus infections. However, our lack of progress with others reminds us that viral infection and pathogenesis are complex and poorly understood processes. Because of the complexity of host-virus interactions, the problems of controlling pathogenesis are daunting. What we don't know about both processes is humbling.

Even when available, vaccines can have unexpected side effects, people may refuse to accept them, and societies may not use or be able to pay for them. In addition, we have learned that it is difficult to intervene in any complex

Table 8.6 Challenges of developing an HIV vaccine[a]

1. Provirus is integrated in host genome
2. Infected cells transmit the infection (cell-cell spread)
3. Must deal with high mutation rate/immune selection
4. Virus infects areas of reduced immune surveillance (brain, testes)
5. Infection compromises immune function
6. Must induce the proper balance of Th1 and Th2 responses

[a]For more details, see Y. Bhattacharjee, *Science* **318:**28–29, 2007.

Table 8.7 Three phases of vaccine trials[a]

Phase	Time	Action
I	12–18 mo	Safety. A small group of seronegative individuals is given the vaccine and observed for adverse affects and tested for their immune responses.
II	1–2 yr	Safety, dose optimization, immunogenicity. Several hundred seronegative individuals are given the vaccine. Initial and long-term immune responses at various doses are measured.
III	Several years	Safety and efficacy. Several thousand individuals are enrolled. One group receives the vaccine, and the other receives a placebo control. The extent of virus transmission and the effect of the vaccine on virus replication are determined.

[a]Adapted from J. A. Levy, *HIV and the Pathogenesis of AIDS*, 3rd ed. (ASM Press, Washington, DC, 2007).

host-parasite interaction without unanticipated effects. Indeed, we often find out how little we know when we test vaccines in the real world: formulations that worked in the lab fail in the field. Furthermore, it is inevitable that viral mutants able to escape immune defenses will arise.

An important principle is that the built-in survival mechanisms of virus and host provide both opportunities and problems in our quest to control viral disease. However, it is telling that essentially all successful vaccines on the market today have been developed empirically. We anticipate that the situation will change as we learn more about the molecular mechanisms of antiviral immune defense and the epidemiology of infections.

References

Books

Koff, W., P. Kahn, and I. Gust. 2007. *AIDS Vaccine Development: Challenges and Opportunities.* Caister Academic Press, Wymondham, United Kingdom.

Levy, J. A. 2007. *HIV and the Pathogenesis of AIDS,* 3rd ed. ASM Press, Washington, DC.

Mims, C., A. Nash, and J. Stephan. 2001. *Mims' Pathogenesis of Infectious Disease,* 5th ed. Academic Press, Inc., Orlando, FL.

Murphy, K., P. Travers, and M. Alpert. 2007. *Janeway's Immunobiology,* 7th edition. Garland Science, Garland Publishing Inc. New York, NY.

Plotkin, S. A., W. A. Orenstein, and P. A. Offit (ed). 2004. *Vaccines,* 4th ed. The W. B. Saunders Co., Philadelphia, PA.

Richman, D. D., R. J. Whitley, and F. G. Hayden (ed.). 2009. *Clinical Virology,* 3rd ed. ASM Press, Washington, DC.

Historical Papers and Books

Bodian, D. 1955. Emerging concept of poliomyelitis infection. *Science* **122:**105–108.

Enders J. F., T. Weller, and F. Robbins. 1949. Cultivation of the Lansing strain of poliomyelitis virus in cultures of various human embryonic tissues. *Science* **109:**85–87.

Enders, J. F., and T. C. Peebles. 1954. Propagation in tissue cultures of cytopathic agents from patients with measles. *Proc. Soc. Exp. Biol. Med.* **86:**277–286.

Jenner, E. 1788. Observations on the natural history of the cuckoo. *Philos. Trans. R. Soc. Lond. Ser. B* **78:**219–237.

Jenner, E. 1798. *An Inquiry into the Causes and Effects of the Variolae Vaccinae, a Disease Discovered in Some of the Western Counties of England, especially Gloucestershire, and Known by the Name of Cow Pox.* Sampson Low, London, England.

Metchnikoff, E. 1905. *Immunity in the Infectious Diseases.* Macmillan Press, New York, NY.

Salk, J., and D. Salk. 1977. Control of influenza and poliomyelitis with killed virus vaccines. *Science* **195:**834–847.

Woodville, W. 1796. *The History of the Inoculation of the Smallpox in Great Britain; Comprehending a Review of all the Publications on the Subject: with an Experimental Inquiry into the Relative Advantages of Every Measure Which Has Been Deemed Necessary in the Process of Inoculation.* James Phillips, London, England.

Reviews

Ahmed, R., and D. Gray. 1996. Immunological memory and protective immunity: understanding their relation. *Science* **272:**54–60.

Angel, J., M. Franceo, and H. Greenberg. 2007. Rotavirus vaccines: recent developments and future considerations. *Nat. Rev. Microbiol.* **5:**529–539.

Arvin, A., and H. Greenberg. 2006. New viral vaccines. *Virology* **344:**240–249.

Baltimore, D. 1988. Intracellular immunization. *Nature* **335:**395–396.

Bukreyev, A., M. Skiadopoulos, B. Murphy, and P. Collins. 2006. Nonsegmented negative-strand viruses as vaccine vectors. *J. Virol.* **80:**10293–10306.

Buonaguro, L., M. Tornesello, and F. Buonaguro. 2007. Human immunodeficiency virus type 1 subtype distribution in the worldwide epidemic: pathogenetic and therapeutic implications. *J. Virol.* **81:** 10209–10219.

DesJardin, J., and D. Snydman. 1998. Antiviral immunotherapy. A review of current status. *Biodrugs* **9:**487–507.

Doms, R., and J. Moore. 2000. Human immunodeficiency virus-1 membrane fusion: targets of opportunity. *J. Cell Biol.* **151:**9–14.

Donnelly, J., B. Wahren, and M. Liu. 2005. DNA vaccines: progress and challenges. *J. Immunol.* **175:**633–639.

Gay, N. J. 2004. The theory of measles elimination: implications for the design of elimination strategies. *J. Infect. Dis.* **189:**S27–S35.

Guy, B. 2007. The perfect mix: recent progress in adjuvant research. *Nat. Rev. Microbiol.* **5:**505–517.

Ishii, K., S. Koyama, A. Nakagawa, C. Coban, S. Akira. 2008. Host innate immune receptors and beyond: making sense of microbial infections. *Cell Host Microbe* **3:**352–365.

Letvin, N. 2006. Progress and obstacles in the development of an AIDS vaccine. *Nat. Rev. Immunol.* **6:**930–939.

Liu, M., B. Wahren, and G. Hedestam. 2006. DNA vaccines: recent developments and future possibilities. *Hum. Gene Ther.* **17:**1051–1061.

Moss, B. 1996. Genetically engineered poxviruses for recombinant gene expression, vaccination and safety. *Proc. Natl. Acad. Sci. USA* **93:**11341–11348.

Moss, W., and D. Griffin. 2006. Global measles elimination. *Nat. Rev. Microbiol.* **4:**900–908.

Nathanson, N., and B. Mathieson. 2004. AIDS vaccine: can the scientific community outwit 10,000 nucleotides? *ASM News* **70:** 406–411.

Neutra, M., and P. Kozlowski. 2006. Mucosal vaccines: the promise and the challenge. *Nat. Rev. Immunol.* **6:**148–158.

Restifo, N. 2000. Building better vaccines: how apoptotic cell death can induce inflammation and activate innate and adaptive immunity. *Curr. Opin. Immunol.* **12:**597–603.

Reyes-Sandoval, A., and H. C. Ertl. 2001. DNA vaccines. *Curr. Mol. Med.* **1:**217–243.

Robinson, H., and K. Weinhold. 2006. Phase I clinical trials of the National Institutes of Health vaccine research center HIV/AIDS candidate vaccines. *J. Infect. Dis.* **194:**1625–1627.

Schijns, V. 2000. Immunological concepts of vaccine adjuvant activity. *Curr. Opin. Immunol.* **12:**456–463.

Schoenly, K. and D. Weiner. 2008. Human immunodeficiency virus type 1 vaccine development: recent advances in the cytotoxic T-lymphocyte platform "spotty business." *J. Virol.* **82:**3166–3180.

Stewart, P., and G. Nemerow. 1997. Recent structural solutions for antibody neutralization of viruses. *Trends Microbiol.* **5:**229–233.

Ulmer, J., B. Wahren, and M. Liu. 2006. Gene-based vaccines: recent technical and clinical advances. *Trends. Mol. Med.* **12:**216–222.

Walker, B., and B. T. Korber. 2001. Immune control of human immunodeficiency virus: the obstacles of HLA and viral diversity. *Nat. Immunol.* **2:**473–475.

Wherry, E., and R. Ahmed. 2004. Memory CD8 T-cell differentiation during viral infection. *J. Virol.* **78:**5535–5545.

Whitehead, S., J. Blaney, A. Durbin, and B. Murphy. 2007. Prospects for a dengue virus vaccine. *Nat. Rev. Microbiol.* **5:**518–528.

Selected Papers

Baden, L., G. Curfman, S. Morrissey, and J. Drazen. 2007. Human papillomavirus vaccine—opportunity and challenge. *N. Engl. J. Med.* **356:**1990–1991.

Bolker, B. M., and B. T. Grenfell. 1996. Impact of vaccination on the spatial correlation and persistence of measles dynamics. *Proc. Natl. Acad. Sci. USA* **93:**12648–12653.

Brochier, B., M. P. Kieny, F. Costy, P. Coppens, B. Baudilin, J. P. Lecocq, B. Languet, G. Chappuis, P. Desmettres, K. Afiademanyo, R. Lilbois, and P. P. Pastoret. 1991. Large scale eradication of rabies using recombinant vaccinia-rabies vaccine. *Nature* **354:**520–522.

Chen, D., R. Endres, C. Erickson, K. Weis, M. McGregor, Y. Kawaoka, and L. Payne. 2000. Epidermal immunization by a needle-free powder delivery technology: immunogenicity of influenza vaccine and protection in mice. *Nat. Med.* **6:**1187–1190.

Chu, R. S., O. S. Targoni, A. M. Krieg, P. V. Lehmann, and C. V. Harding. 1997. CpG oligodeoxynucleotides act as adjuvants that switch on T helper 1 (Th1) immunity. *J. Exp. Med.* **186:**1623–1631.

Clark, H., P. Offit, R. Ellis, J. Eiden, D. Krah, A. Shaw, M. Pichichero, J. Treanor, F. Borian, L. Bell, and S. Plotkin. 1996. The development of multivalent bovine rotavirus (strain WC3) reassortant vaccine for infants. *J. Infect. Dis.* **174**(suppl):S73–S80.

Eisenbarth, S., O. Colegio, W. O'Connor, F. Sutterwaa, and R. Flavell. 2008. Crucial role for the Nalp3 inflammasome in the immunostimulatory properties of aluminium adjuvants. *Nature.* **453:**1122–1126.

Eo, S., S. Lee, S. Chun, and B. Rouse. 2001. Modulation of immunity against herpes simplex virus infection via mucosal genetic transfer of plasmid DNA encoding chemokines. *J. Virol.* **75:**569–578.

Gans, H. A., A. M. Arvin, J. Galinus, I. Logan, R. DeHovitz, and Y. Maldonado. 1998. Deficiency of the humoral immune response to measles vaccine in infants immunized at age 6 months. *JAMA* **280:**527–532.

Hassett, D. E., J. Zhang, M. Slifka, and J. Whitton. 2000. Immune responses following neonatal DNA vaccination are long-lived, abundant, and qualitatively similar to those induced by conventional immunization. *J. Virol.* **74:**2620–2627.

Kimbauer, R., F. Booy, N. Cheng, D. R. Lowy, and J. T. Schiller. 1992. Papillomavirus L1 major capsid protein self-assembles into virus-like particles that are highly immunogenic. *Proc. Natl. Acad. Sci. USA* **89:**12180–12184.

Modelska, A., B. Dietzschold, and V. Yusibov. 1998. Immunization against rabies with plant-derived antigen. *Proc. Natl. Acad. Sci. USA* **95:**2481–2485.

Mortara, L., F. Letourneur, H. Gras-Masse, A. Venet, J.-G. Guillet, and I. Bourgault-Villada. 1998. Selection of virus variants and emergence of virus escape mutants after immunization with an epitope vaccine. *J. Virol.* **72:**1403–1410.

Murphy, B., D. Morens, L. Simonsen. R. Chanock, J. LaMontagne, and A. Kapikian. 2003. Reappraisal of the association of intussusception with the licensed live rotavirus vaccine challenges initial conclusions. *J. Infect. Dis.* **187:**1301–1308.

Quinlivan, L., A. Gershon, R. Nichols, and J. Breuer. 2007. Natural selection for rash-forming genotypes of the varicellal-zoster vacine virus detected within immunized hosts. *Proc. Natl. Acad. Sci. USA* **104:**208–212.

Rodriguez, F., J. Zhang, and J. L. Whitton. 1997. DNA immunization: ubiquitination of a viral protein enhances cytotoxic T-lymphocyte induction and antiviral protection but abrogates antibody induction. *J. Virol.* **71:**8497–8503.

Rose, N., P. Marx, A. Luckay, D. Nixon, W. Moretto, S. Donahoe, D. Montefiori, A. Roberts, L. Buonocore, and J. Rose. 2001. An effective AIDS vaccine based on live attenuated vesicular stomatitis virus recombinants. *Cell* **106:**539–549.

Taboga, O., C. Tami, E. Carrillo, J. I. Nunez, A. Rodriguez, J. C. Saiz, E. Blanco, M. L. Valero, X. Roig, J. A. Camarero, D. Andreu, M. G. Mateu, E. Giralt, E. Domingo, F. Sobrino, and E. Palma. 1997. A large-scale evaluation of peptide vaccines against foot-and-mouth disease: lack of solid protection in cattle and isolation of escape mutants. *J. Virol.* **71:**2606–2614.

Xu, L., A. Sanchez, Z. Yang, S. R. Zaki, E. G. Nabel, S. T. Nichol, and G. Nabel. 1998. Immunization for Ebola virus infection. *Nat. Med.* **4:**37–42.

Zhou, J., X. Sun, D. Stenzel, and I. Frazer. 1991. Expression of vaccinia recombinant HPV 16 L1 and L2 ORF protein in epithelial cells is sufficient for assembly of HPV virus-like particles. *Virology* **185:**251–257.

9

Introduction
    Paradox? So Much Knowledge, So Few
    Antivirals
    Historical Perspective

Discovering Antiviral Compounds
    The New Lexicon of Antiviral Discovery
    Screening for Antiviral Compounds
    Designer Antivirals and Computer-Based
    Searching
    The Difference between "R" and "D"
    Examples of Some Approved Antiviral
    Drugs
    The Search for New Antiviral Targets
    Antiviral Gene Therapy and
    Transdominant Inhibitors
    Resistance to Antiviral Drugs

**Human Immunodeficiency Virus
and AIDS**
    Examples of Anti-HIV Drugs
    The Combined Problems of Treating
    a Persistent Infection and Emergence
    of Drug Resistance
    Combination Therapy
    Strategic Treatment Interruption
    Challenges and Lessons Learned

Perspectives

References

Antiviral Drugs

Introduction

Some viral infections can be controlled effectively by public health measures and vaccines. However, for many others, these measures have no effect, are not available, or cannot be applied. Antiviral drugs are intended to fill a portion of this void. However, despite almost 50 years of research, our armamentarium of such drugs remains surprisingly small. The current arsenal comprises fewer than 50 drugs, and most of these are directed against human immunodeficiency virus (HIV) and herpesviruses (Table 9.1). It is important to understand why we are in this precarious situation.

Paradox? So Much Knowledge, So Few Antivirals

Because we know so much about some viruses, this dearth of antiviral drugs is unexpected. There are many reasons for the lack of therapies, but a primary reason is that antiviral drugs **must** be safe. This simple goal often is unattainable, as viruses are parasites of cellular mechanisms, and compounds interfering with viral growth often have adverse effects on the host. Another reason is that antiviral compounds must be extremely potent, virtually 100% efficient in blocking viral growth. Even modest replication in the presence of an inhibitor provides the opportunity for resistant mutants to prosper. Achieving the potency to block viral replication **completely** is remarkably difficult. Additionally, many medically important viruses cannot be propagated conveniently in the laboratory (e.g., hepatitis B virus and papillomaviruses), and for some human viruses, there are no available small-animal models (e.g., measles virus and hepatitis C virus). The testing and development of safe, potent, and efficacious compounds that block infection by these viruses pose a significant challenge. Finally, as noted in Chapter 5, many acute infections are of short duration, and by the time the individual feels ill, viral replication is completed and infected cells are being cleared. Antiviral drugs must be given early in the infection, or prophylactically to populations at risk. It takes time, not only to identify the specific viral pathogen but also to obtain and dispense the drug (Box 9.1). The lack of rapid

Table 9.1   The antiviral repertoire[a]

Targets	Viruses[b]	Examples of compounds approved
Virion uncoating	Influenza A	Amantadine, rimantadine
DNA polymerase	Herpesviruses (HSV-1, HSV-2, VZV, CMV, EBV, HHV-6, HHV-7, HHV-8)	Nucleosides: acyclovir, valacyclovir, ganciclovir, valganciclovir, penciclovir famciclovir, brivudin,[c] foscarnet
	Herpesvirus (CMV)	Acyclic nucleoside phosphonates: cidofovir, tenofovir
	HIV	
Reverse transcriptase	HIV	Nucleosides: zidovudine, didanosine, zalcitabine, stavudine, lamivudine,[d] abacavir
		Nonnucleosides: nevirapine, delavirdine, efavirenz
Viral protease	HIV	Saquinavir, ritonavir, indinavir, nelfinavir, amprenavir, lopinavir
Viral neuraminidase	Influenza A and B virus	Zanamivir, oseltamivir
Inosine monophosphate dehydrogenase	HCV, RSV	Ribavirin[e]

[a]Data from E. De Clercq, *Nat. Rev. Drug Discov.* **1:**13–25, 2002.

[b]Abbreviations: CMV, cytomegalovirus; EBV, Epstein-Barr virus; BCAR, 5-ethynyl-1-β-D-ribofuranosylimidazole-4-carboxamide; HBV, hepatitis B virus; HCV, hepatitis C virus; HHV, human herpesvirus; HIV, human immunodeficiency virus; HSV, herpes simplex virus; IMP, inosine 5′-monophosphate; NNRTI, nonnucleoside reverse transcriptase inhibitor; NRTI, nucleoside reverse transcriptase inhibitor; RSV, respiratory syncytial virus; VZV, varicella-zoster virus.

[c]Birivudin is approved in some countries, for example, Germany.

[d]Lamivudine is also approved for the treatment of HBV.

[e]Ribavirin is used in combination with interferon-α for HCV.

BOX 9.1

DISCUSSION

We can put a person on the moon, but we cannot cure the common cold

While the title phrase of this box is time-worn and trite, it contains elements of truth that underlie some basic problems of finding treatments to common, acute infections. The common cold is a syndrome caused by many different viruses, including rhinoviruses (about 50% of all common colds), adenoviruses, coronaviruses, and others. Several fundamental problems confound the quest for a cure for the common cold.

- The common cold is not a life-threatening disease, but a mild, inconvenient illness. Antiviral drugs typically are not sold over the counter. Will people see their physicians to obtain a prescription every time they experience a drippy nose, congestion, or cough?
- By the time the symptoms of a common cold are evident, viral replication has reached its peak. Therefore, by the time an antiviral is given, it may be too late for significant effect.
- Drugs and vaccines may have to be given to millions of healthy people

to be effective. Short-term and long-term safety of any treatment must be ensured.

In 2001, a new antiviral compound called pleconaril was tested in humans. This compound is effective against rhinoviruses, but not coronaviruses or adenoviruses. It binds to the hydrophobic pocket under the canyon in the receptor-binding sites of many rhinoviruses (Volume I,

Chapter 5). Pleconaril is thought to act by stabilizing the particle so that uncoating and genome release cannot occur. In human trials, symptoms were reduced and the time course of disease was shortened by a day or two. Consequently, the compound has yet to be approved for widespread use.

An effective treatment of the common cold simply may be one that targets mechanisms of symptom production, not the viral proteins that promote infection and replication. As the symptoms are the result of immunopathology, targeting immune cells or soluble mediators may be a solution. We have incomplete knowledge about the mediators of immune defense against cold viruses, but it may be that there will be overlap among the host responses to different viruses giving rise to the common symptoms.

Hayden, F. G., T. Coats, K. Kim, H. Hassman, M. M. Blatter, B. Zhang, and S. Liu. 2002. Oral pleconaril treatment of picornavirus-associated viral respiratory illness in adults: efficacy and tolerability in phase II clinical trials. *Antiviral Ther.* **7:**53–65.

diagnostic reagents alone has hampered the development and marketing of antiviral drugs to treat many acute viral diseases, despite the existence of effective therapies.

Historical Perspective

The first large-scale effort to find antiviral compounds began in the early 1950s and focused on inhibitors of smallpox virus replication. At that time, virology was in its infancy, and smallpox was a worldwide scourge. In the 1960s and 1970s, drug companies expanded efforts because of increased knowledge and understanding of the viral etiology of common diseases, as well as their remarkable progress in the discovery of antibiotics to treat bacterial infections. They launched massive screening programs to find chemicals with antiviral activities. Despite much effort, there was relatively little success. One notable exception was amantadine (Symmetrel), approved in the late 1960s for treatment of influenza A virus infections. These antiviral discovery programs were called **blind screening**, because random chemicals and natural-product mixtures were tested for their ability to block replication of a variety of viruses in cell culture systems. "Hits" were then purified, and fractions were tested in various cell and animal models for safety and efficacy. Promising molecules, called "leads," were modified systematically by medicinal chemists to reduce toxicity, increase solubility and bioavailability, or improve biological half-life. As a consequence, hundreds if not thousands of molecules were made and screened before a specific antiviral compound was tested in humans. Moreover, the mechanism by which these compounds inhibited the virus was often unknown. For example, the mechanism of action of amantadine was not deduced until the early 1990s, almost 30 years after its discovery!

Discovering Antiviral Compounds

With the advent of modern molecular virology and recombinant DNA technology, the random, blind-screening procedures described above have been all but discarded. Instead, viral genes essential for growth can be cloned and expressed in genetically tractable organisms, and their products can be purified and analyzed in molecular and atomic detail. The life cycles of many viruses are known, revealing numerous targets for intervention (Table 9.2). Inhibitors of these processes can be found, even for viruses that cannot be propagated in cultured cells.

The New Lexicon of Antiviral Discovery

Mechanism- and cell-based assays, high-throughput screens, small interfering RNA screens, *in silico* screens, rational drug design, and combinatorial chemistry are the approaches of modern antiviral discovery (Fig. 9.1). Despite all the modern technology, this process does not find drugs (compounds approved and licensed for use in humans), but, rather, detects hits and highlights strategies for developing them as leads. The hard work begins when a lead compound is found. Will the compound get to the right place in the body at the appropriate concentration (**bioavailability**)? Will it

Table 9.2 Some viral targets for antiviral drug discovery[a]

Function	Lead compound or example	Virus
Attachment	Peptide analogs of attachment protein	HIV
Penetration and uncoating	Dextran sulfate, heparin	HIV, herpes simplex virus
mRNA synthesis	Interferon	Hepatitis A, B, and C viruses; papillomavirus
	Antisense oligonucleotides	Papillomavirus, human cytomegalovirus
Protein synthesis		
Initiation	Interferon	Hepatitis A, B, and C viruses; papillomavirus
IRES elements	Ribozymes; antisense oligonucleotides	Flaviviruses and picornaviruses
DNA replication		
Polymerase	Nucleoside, nonnucleoside analogs	Herpesviruses, HIV, hepatitis B virus
Helicase/primase	Thiazole ureas	Herpes simplex virus
Processing/packaging	Benzimadolazoles	Herpesviruses
Nucleoside scavenging		
Thymidine kinase	Nucleoside analog	Herpes simplex virus, varicella-zoster virus
Ribonucleotide reductase	Inhibitors of protein-protein interaction of large and small subunits	Herpes simplex virus
Assembly		
Protease	Peptidomimetics	HIV
Virion integrity	Nonoxynol-9	
Lipid raft disruption	β-Cyclodextrins	HIV, herpes simplex virus

[a] Data from E. De Clercq, *Nat. Rev. Drug Discov.* **1:**13–25, 2002; S.-L. Tan, A. Pause, Y. Shi, and N. Sonenberg, *Nat. Rev. Drug Disc.* **1:**867–881, 2002.

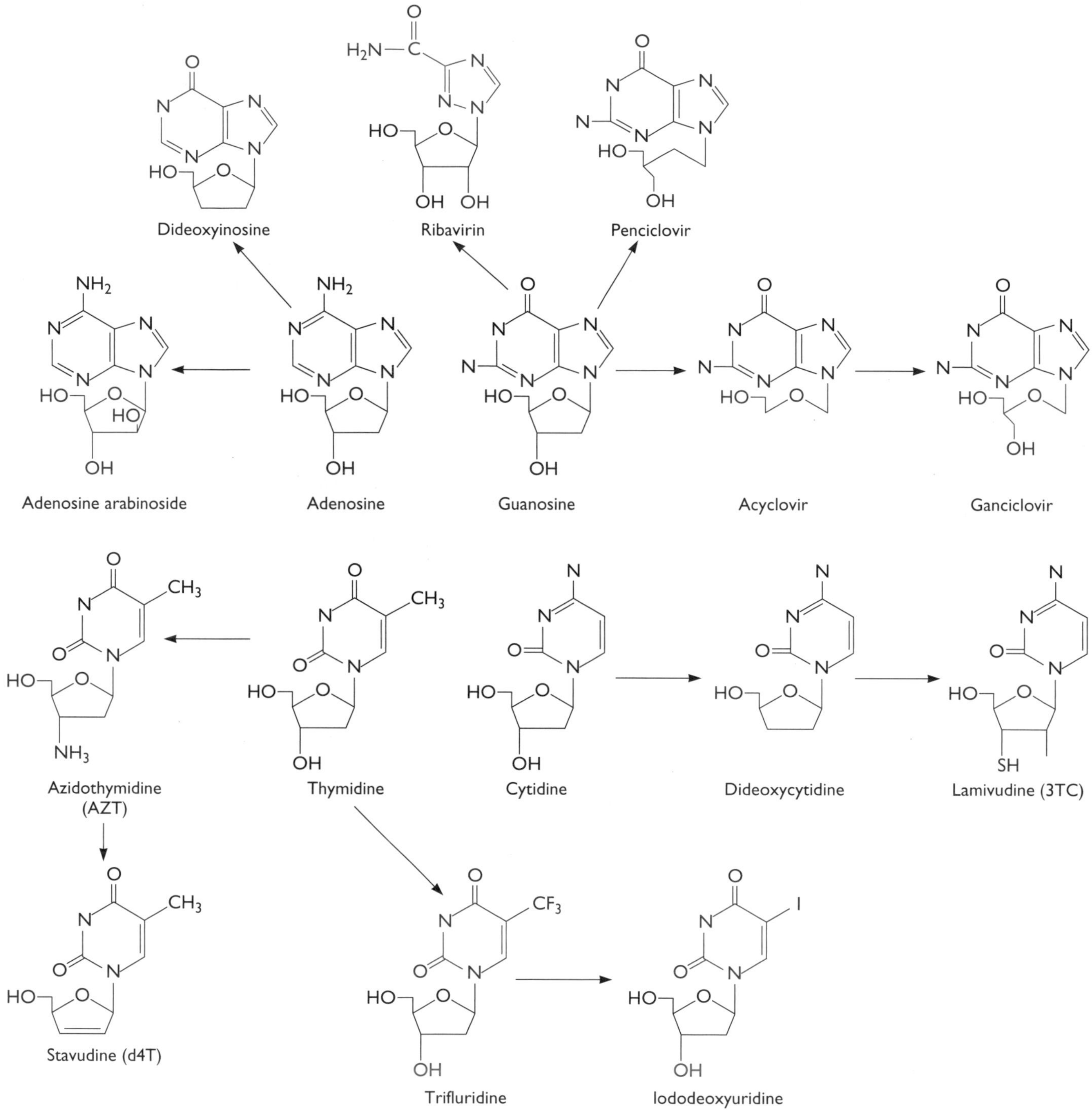

Figure 9.9 Many well-known antiviral compounds are nucleoside and nucleotide analogs. The four natural deoxynucleosides are highlighted in the yellow box. The chemical distinctions between the natural deoxynucleosides and antiviral drug analogs are highlighted in red. Arrows connect related drugs. Adapted from E. De Clercq, *Nat. Rev. Drug Discov.* **1:**13–25, 2002, with permission.

methylated derivative, cannot cross the blood-brain barrier and therefore has fewer central nervous system side effects. For this reasons, the drug often replaces amantadine in the treatment of influenza A virus infections.

An unusual property of amantadine is the concentration dependence of its antiviral activity. The drug has broad antiviral effects at high concentrations, but at low concentrations it is specific for influenza virus A, with no effect on B or C strains or other viruses. Analysis of resistant mutants provided insight into the apparent complex mechanism of action of amantadine. At concentrations of 100 mM or higher, the compound acts as a weak base and raises the pH of endosomes so that pH-dependent membrane fusion is blocked. Any virus with a pH-dependent fusion mechanism could be affected by high concentrations of amantadine. Resistant mutants of influenza A virus selected under these conditions in cultured cells harbor amino acid substitutions in HA that destabilize the protein and enable fusion at higher pH. Influenza virus mutants selected at concentrations of 5 mM or lower carried mutations in the M2 gene. Specifically, these mutations affected amino acids in the membrane-spanning region of the M2 ion channel protein.

Zanamivir and Oseltamivir

Zanamivir and oseltamivir are inhibitors of the neuraminidase enzyme synthesized by influenza A and B viruses (Box 9.3). Zanamivir is delivered via inhalation, while oseltamivir can be given orally. When used within 48 h of onset of symptoms, the drugs reduce the median time to alleviation of symptoms by about 1 day compared to placebo. When used within 30 h of disease onset, the drugs reduce the duration of symptoms by about 3 days.

The Search for New Antiviral Targets

Entry and Uncoating Inhibitors

The first step of viral replication has long been an attractive target, as virus-receptor interactions offer the promise of high specificity. Early enthusiasm for entry inhibitors came from experiments with monoclonal antibodies that inhibited virion attachment or entry in cultured cells. Passive immunization with these antibodies often can protect animals from challenge, suggesting that small-molecule inhibitors of entry may be useful leads. At present, it is impractical, or not cost-effective, to develop antibodies as antiviral agents; these molecules have been most valuable in identifying proteins required for entry and their mechanisms.

The binding site of antibodies that block entry provides one starting point for screening chemical libraries, or for design of small molecules that block viral entry. It is important that such compounds interfere with virus-host interactions, but not with the normal function of the cellular receptor. Obviously, an important assumption in this strategy is that no other entry routes exist. However, alternative receptors are available for many viruses (e.g., herpesviruses and HIV), and blocking a single virion-receptor interaction may not be effective.

Membrane fusion, the ubiquitous process by which enveloped virions enter cells, is an attractive target for chemotherapeutic intervention because fusion mechanisms are conserved among enveloped viruses. To identify inhibitors of influenza virus fusion, a computer program was first used to predict which small molecules might bind into a pocket of the HA molecule and possibly prevent low-pH-induced conformational change. From the molecules identified in this way, several benzoquinone- and hydroquinone-containing compounds were tested and found to inhibit HA-mediated membrane fusion at low pH. One of these inhibits influenza virus replication. Although it is not known how binding of the compound prevents fusion, it has been suggested that it acts as a "molecular glue" to prevent movement of the fusion peptide. Inhibitors of HIV entry are discussed in more detail below.

Considerable effort is being expended on **microbicides**, creams or ointments that either inactivate or block virions before they can attach and penetrate tissues. The concept is that these treatments either would inactivate virions directly or would bind to many potential receptors on the cell surface, and therefore may act as broad spectrum antivirals. Particular attention is focused on vaginal microbicides to prevent sexually transmitted infections. For example, certain polyanions act as competitive inhibitors of both human immunodeficiency virus and herpes simplex virus binding to cell surface receptors and are easily formulated as inexpensive topical creams. The primary problem is attaining full protection. Many of these compounds are competitive inhibitors of binding, and thus are dependent on the concentration of virions and microbicide.

Proteases

Viral proteases often cleave protein precursors to form functional units or to release structural components during ordered particle assembly (maturational proteases). As discussed in Volume I, Chapter 13, all herpesviruses encode a serine protease that is absolutely required for formation of nucleocapsids. Many features of these enzymes and their substrates are conserved among the members of the family *Herpesviridae*. The structure of the human cytomegalovirus protease is shown in Fig. 9.12. Interest in this enzyme as a target is based on the unusual serine protease fold and the mechanism of catalysis, which bodes well for the safety and efficacy of high-specificity inhibitors.

Hepatitis C virus infects more people than HIV in the United States, and infects millions more individuals around the world. As a result, an intensive antiviral drug discovery effort is in progress. The essential viral protease called NS3 is a major target (Fig. 9.13). Until recently, hepatitis C virus could not be propagated in cultured cells, and so the effect of inhibiting NS3 protease on virus replication was tested initially by using minireplicons in cell-based assays. The crystal structure of NS3 protease has been solved by several companies, as have the structures of NS3 helicase, the NS5B polymerase, and other proteins. Major efforts in structure-based design, as well as mechanism-based

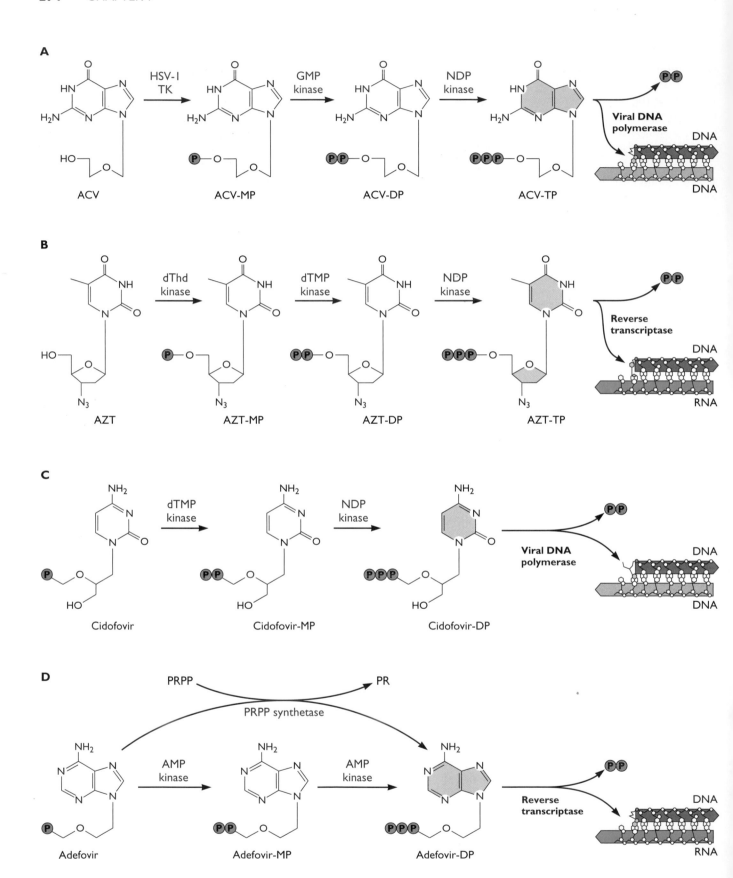

screening, have identified several NS3 and NS5B inhibitors that currently are in clinical trials.

Virus-Specific Nucleic Acid Synthesis and Processing

Viral replication enzymes are primary targets for antiviral drugs. From genetic analyses, we know that DNA polymerase accessory proteins, such as those that promote processivity or bind to viral origins, are essential for viral replication and also represent attractive targets. The RNA-dependent RNA polymerases of any RNA virus appear to be unique to the virus world. Their varied activities, which include synthesis of primers, cap snatching, and recognition of RNA secondary structure in viral genomes, can be exploited for drug discovery. The unique helicases encoded by many RNA viruses are already in many companies' high-throughput screens, but no lead compounds have yet progressed to the market.

Newly replicated, concatemeric herpesviral DNA is cleaved by viral enzymes into monomeric units during the packaging/assembly process. These processes, which are essential for viral replication, are carried out by specific enzymes. These enzymes represent promising targets for antiviral drugs. For example, 5-bromo-5,6-dichloro-1-β-D-ribofuranosyl benzimidazole binds to the human cytomegalovirus UL89 gene product that is a component of the "terminase" complex responsible for cleaving and packaging replicated concatemeric DNA. Members of the UL89 gene family are highly conserved among all herpesviruses, and have homology to a known ATP-dependent endonuclease encoded in the genome of bacteriophage T4. This class of compounds may therefore be the basis for discovery of broad-based inhibitors of herpesvirus replication. Another class of compounds, the 4-oxo-dihydroquinolines, also have potential as broad-spectrum herpesvirus inhibitors because they interact with a sequence conserved in many herpesvirus polymerases.

Regulatory Proteins

Viral proteins that control transcription often are essential for viral growth and are prime targets for antiviral screens. Fomivirsen is the first licensed compound designed to inhibit the function of a viral regulatory protein. The drug is a phosphorothioate, antisense oligonucleotide and is approved to treat retinitis, a disease caused by human cytomegalovirus. Inhibition depends on binding of the 21-nucleotide antisense molecule to the cytomegalovirus immediate-early 2 mRNA. Host proteins that play central roles in stimulating damaging intrinsic or immune responses are also attractive targets (Box 9.4).

Regulatory RNA Molecules

Micro-RNAs (miRNAs) control the activity of a substantial number of genes and are encoded in both host and viral genomes. Often one class of miRNA regulates the expression of an entire network of genes. One such host miRNA, miR-122, is expressed only in the liver and regulates expression of more than 400 genes including those involved in cholesterol metabolism. Remarkably, when expression of miR-122 is inhibited, not only do levels of cholesterol in circulation drop, but also the liver is protected from hepatitis C virus infection. The mechanism of viral inhibition is not understood. Nevertheless, this finding demonstrates that small RNA molecules may be targets for antiviral compounds.

Antiviral Gene Therapy and Transdominant Inhibitors

In 1988, scientists constructed a cell line that produced an altered VP26 protein from herpes simplex virus type 1. The cell line was subsequently shown to be highly resistant to infection by herpes simplex virus type 1. The mutation in the VP16 gene conferred a dominant-negative phenotype: the altered protein interfered with the function of the

---

Figure 9.10 Chain termination by antiviral nucleoside analogs. (A) Acyclovir (ACV) is a prodrug that must be phosphorylated in the infected cell. The viral thymidine kinase (herpes simplex virus type 1 [HSV-1] TK), but not the cellular kinase, adds one phosphate (orange circle labeled P) to the 5'-hydroxyl group of acyclovir. The monophosphate is a substrate for cellular enzymes that synthesize acyclovir triphosphate. The triphosphate compound is recognized by the viral DNA polymerase and incorporated into viral DNA. As acyclovir has no 3'-hydroxyl group, the growing DNA chain is terminated, and viral replication ceases. **(B)** AZT targets the HIV reverse transcriptase. It must be phosphorylated by cellular kinases in three steps to the triphosphate compound, which is incorporated into the viral DNA to block reverse transcription. **(C)** Cidofovir [S-1-(3-hydroxy-2-phosphonylmethoxypropyl) cytosine] is an acyclic nucleoside analog. In contrast to acyclovir and AZT, cidofovir requires only two phosphorylations by cellular kinases to be converted to the active triphosphate chain terminator. **(D)** Adefovir [9-(2-phosphonylmethoxyethyl)adenine] is an acyclic nucleoside analog and also requires only two phosphorylations by cellular AMP kinases. Through the action of phosphoribosyl pyrophosphate synthetase, which forms the triphosphate from the monophosphate in one step, both cidofovir and adefovir bypass the nucleoside-kinase reaction that limits the activity of dideoxynucleoside analogs such as AZT. DP, diphosphate; dThd, (2'-deoxy)-thymidine; MP, monophosphate; NDP, nucleoside 5'-diphosphate; PR, 5-phosphoribose; TP, triphosphate. Adapted from E. De Clercq, Nat. Rev. Drug Discov. 1:13–25, 2002, with permission.

BOX 9.3

EXPERIMENTS
Designer drugs: inhibitors of influenza virus neuraminidase

Influenza virus neuraminidase (NA) protein cleaves terminal sialic acid residues from glycoproteins, glycolipids, and oligosaccharides. It plays an important role in the spread of infection from cell to cell, because in cleaving sialic acid residues, the enzyme releases virions bound to the surfaces of infected cells and facilitates viral diffusion through respiratory tract mucus. Moreover, the enzyme can activate transforming growth factor β by removing sialic acid from the inactive protein. Because the activated growth factor can induce apoptosis, NA may influence the host response to viral infection. NA is a particularly intriguing target for drug hunters.

NA is a tetramer of identical subunits, each of which consists of six four-stranded antiparallel sheets arranged like the blades of a propeller. The enzyme active site is a deep cavity lined by identical amino acids in all strains of influenza A and B viruses that have been characterized. Because of such invariance, compounds designed to fit in this cavity would be expected to inhibit the NA activity of all A and B strains of influenza virus, a highly desirable feature in an influenza antiviral drug. Moreover, as NA inhibitors are predicted to block spread, they may be effective in reducing the transmission of infection to other individuals.

Sialic acid fits into the active site cleft in such a fashion that there is an empty pocket near the hydroxyl at the 4 position on its sugar ring. On the basis of computer-assisted analysis, investigators predicted that replacement of this hydroxyl group with either an amino or a guanidinyl group would fill the empty pocket and therefore would increase the binding affinity by contacting one or more neighboring glutamic acid residues.

Currently, two antiviral drugs, zanamivir and oseltamivir, that inhibit influenza A and B virus NA are licensed for use. Significantly, these drugs are inhibitors of both influenza A and B viruses in cultured cells. They do not inhibit other nonviral NAs, an important requirement for safety and lack of potential side effects.

Varghese, J. N., V. C. Epa, and P. M. Colman. 1995. Three dimensional structure of the complex of 4-guanidino-Neu5Ac2en and influenza virus neuraminidase. *Protein Sci.* **4:**1081–1087.

von Itzstein, M., W. Y. Wu, G. B. Kok, M. S. Pegg, J. C. Dyason, B. Jin, T. Van Phan, M. L. Smythe, H. F. White, S. W. Oliver, P. M. Coleman, J. N. Varghese, D. M. Ryan, J. M. Woods, R. C. Bethell, V. J. Hotham, J. M. Cameron, and C. R. Penn. 1993. Rational design of potent sialidase-based inhibitors of influenza virus replication. *Nature* **363:**418–423.

Structure of influenza A virus NA. (A) A ribbon diagram of influenza A virus NA with α-sialic acid bound in the active site of the enzyme. The molecule is viewed down the fourfold axis of the tetramer. The molecule is an N2 subtype from A/Tokyo/3/67. The C terminus is on the outside surface near a subunit interface. The six β-sheets of the propeller fold are indicated in colors. **(B)** The packing of the N2 NA tetramer viewed from above and down the symmetry axis. Adapted from J. N. Varghese, p. 459–486, *in* P. Verrapandian (ed.), *Structure-Based Drug Design* (Marcel Dekker, New York, NY, 1997), with permission. Courtesy of J. Varghese.

A

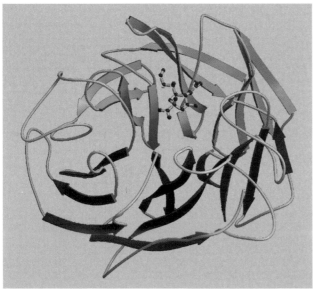

B

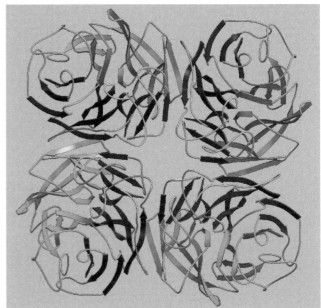

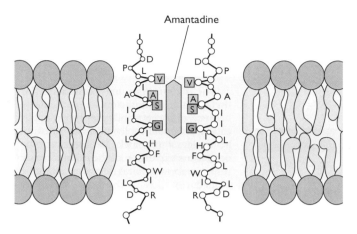

Amantadine

Figure 9.11 Interaction of amantadine with the transmembrane domain of the influenza A virus M2 ion channel. It is thought that at low concentrations (5 μM), amantadine exerts its antiviral effect by blocking the virus-encoded M2 ion channel activity in infected cells. M2 protein is a tetramer with an aqueous pore in the middle of the four subunits. The four locations of single-amino-acid changes in different amantadine-resistant mutants are boxed. A diagram of the putative interaction of amantadine with two diagonally located α-helices of the M2 tetramer in a lipid bilayer is shown. Adapted from A. J. Hay, *Semin. Virol.* **3**:21–30, 1992, with permission. For new data, see D. A. Steinhauer and J. J. Skehel, *Annu. Rev. Genet.* **36**:305–332, 2002.

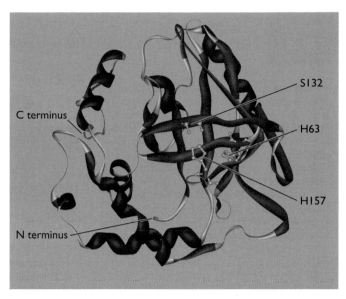

Figure 9.12 Human cytomegalovirus protease. The β-strands are shown as blue ribbons, α-helices are red, and the connecting loops are indicated in white. The active site residues (serine 132, histidine 63, and histidine 157) are indicated in yellow. The amino and carboxy termini are indicated. Adapted from L. Tong et al., *Nature* **383**:272–275, 1996, with permission.

normal VP16 protein. When transgenic mice were produced that synthesized the altered VP16 protein, they too were resistant to infection by herpes simplex virus type 1. Dominant-negative mutations are not new in genetics, but the idea that such an altered protein could be used to protect an animal against infection was novel. A particularly ingenious implementation of this strategy for antiviral drug discovery is described in Box 9.5.

Many natural and laboratory produced gene products that should block intracellular viral growth when present in an infected cell can be imagined, and some of them may yield effective antiviral therapies. For example, RNA-binding proteins, DNA-binding proteins, ribozymes, and antisense oligonucleotides offer interesting possibilities. Some clever ideas are based on hybrid proteins composed of a target domain and a killer domain. The former brings the latter

Figure 9.13 The hepatitis C virus polyprotein is cleaved by several proteases. Hepatitis C virus is a human flavivirus with a (+) strand RNA genome. The viral proteins are encoded in one large open reading frame that is translated into a polyprotein. The polyprotein is processed by cellular and viral proteases to release the viral proteins. The numbers above the polyprotein indicate the amino acid marking the start of each viral protein. The last number, 3010, indicates the amino acid number of the C terminus of the polyprotein. The proteases and their cleavage sites are indicated by arrows. Signal peptidase is the host enzyme in the endoplasmic reticulum that cleaves signal peptides from membrane proteins. The red open arrows and dotted lines represent cleavage sites that are inefficiently processed or not well studied. The viral metalloprotease is an autoprotease comprising NS2 and the amino-terminal domain of NS3. The viral serine-type protease is the NS3 protein bound to NS4A and is essential for the assembly of virus particles.

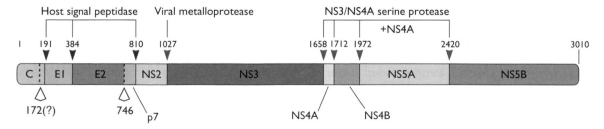

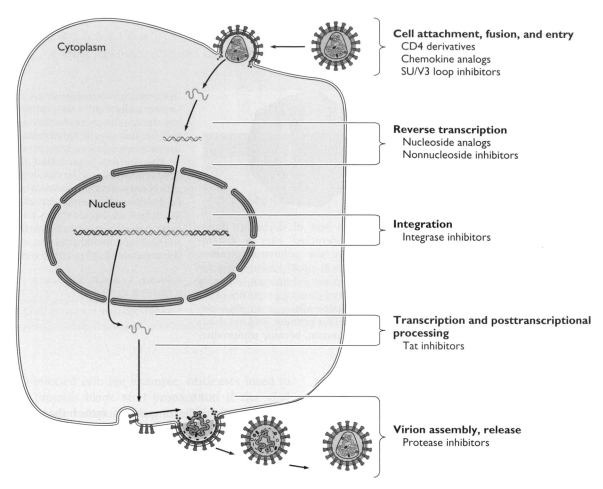

Figure 9.14 Important steps in the replication of human immunodeficiency virus. Five steps in the life cycle of the virus are highlighted as potential targets for antiviral compounds. Known or hypothetical inhibitory compounds are listed under each step. Adapted from D. D. Richman et al., *Clinical Virology* (Churchill Livingstone, New York, NY, 1997), with permission.

treatment for accidental needle sticks, and for treatment of infected pregnant women, when it can reduce considerably the probability of delivering an HIV-infected baby.

AZT toxicity is of great concern when the drug must be given for long periods. AZT treatment can damage bone marrow, resulting in a reduction in the number of neutrophils. Treatment of this problem requires multiple transfusions of red blood cells. Muscle wasting, nausea, and severe headaches are but a few of AZT's other side effects that must be endured. Despite these problems, the drug was used extensively because, until recently, there simply was no alternative.

Considerable effort has been devoted to discovering alternatives to AZT. Particular emphasis has been placed on finding reverse transcriptase inhibitors that are effective against AZT-resistant mutants. Several nucleoside analogs that have therapeutic value are now available (Fig. 9.9).

Nonnucleoside Inhibitors of Reverse Transcriptase

Nonnucleoside inhibitors of viral reverse transcriptase do not bind at the nucleotide-binding site of the enzyme (Fig. 9.15). Examples of these compounds are nevirapine and the tetrahydroimidazobenzodiazepinone (TIBO) class of compounds. Initially, they offered the hope of complementing nucleoside analog inhibitors, but the rapid emergence of resistant mutants dampened enthusiasm. Although they are effective inhibitors of reverse transcriptase, a substitution in any of seven residues that line their binding site on the enzyme confers resistance. Because resistant mutants are selected rapidly, nonnucleoside inhibitors cannot be used by themselves for the treatment of AIDS (monotherapy). However, as with AZT, nevirapine has value for treatment of pregnant women before delivery to prevent infection of the newborn. This use is now the preferred

BOX 9.6

DISCUSSION

A heroic effort: 19 new drugs, 3 targets, 9 companies, and 15 years

We must never forget the daunting task that faced the scientific and medical community in the 1980s when HIV was first identified and every infection was a death sentence. There was no experi- ence with such infections in the clinics, and the drug hunters had nothing in the pipeline that was proven to be effective against retroviruses. In fact, there were few scientists with any experience at all with lentiviruses. Yet as the data in this table demonstrate, a truly heroic effort was mounted over the first 15 years of the pandemic, but it took time, money, and unprecedented cooperation.

Target or mechanism	Generic name	Brand name	Manufacturer	Yr approved
Nucleoside reverse transcriptase inhibitors	Zidovudine (AZT, ZDV)	Retrovir	GlaxoSmithKline	1987
	Didanosine (ddI)	Videx	Bristol-Myers Squibb	1991
	Zalcitabine (ddC)	Hivid	Roche	1992
	Stavudine (d4T)	Zerit	Bristol-Myers Squibb	1994
	Lamivudine (3TC)	Epivir	GlaxoSmithKline	1995
	AZT/3TC	Combivir	GlaxoSmithKline	1997
	Abacavir (ABC)	Ziagen	GlaxoSmithKline	1998
	AZT/3TC/ABC	Trizivir	GlaxoSmithKline	2000
	Tenofovir (TDF)	Viread	Gilead	2001
Nonnucleoside reverse transcriptase inhibitors	Nevirapine	Viramune	Roxane	1996
	Delavirdine	Rescriptor	Agouron	1997
	Efavirenz	Sustiva	Dupont	1998
	Saquinavir (hard gel)	Invirase	Roche	1995
Protease inhibitors	Saquinavir (soft gel)	Fortovase	Roche	1997
	Ritonavir	Norvir	Abbott	1996
	Indinavir	Crixivan	Merck	1996
	Nelfinavir	Viracept	Agouron	1997
	Amprenavir	Agenerase	GlaxoSmithKline	1999
	Lopinavir/ritonavir	Kaletra	Abbott	2000
Summary				
Three enzyme targets	**16 unique compounds**	**19 approved drugs**	**9 companies**	**15 years**

treatment in underdeveloped countries. These drugs have also proved valuable in combination therapy (see below).

Protease Inhibitors

HIV protease, which is encoded in the *pol* gene, is essential for production of mature infectious viral particles. The enzyme cleaves itself from the Gag-Pol precursor polyprotein and then cleaves at seven additional sites in Gag-Pol to yield nine proteins and three enzymes. Active protease has been produced in high yields in many recombinant organisms and has even been synthesized chemically. It was the first HIV enzyme to be crystallized and studied at the atomic level (Fig. 9.5). The active enzyme is a small (only 99-amino-acid) dimeric aspartyl protease, similar to renin and pepsin.

The seven cleavage sites in Gag-Pol are similar but not identical. In the early stages of research to find protease inhibitors, it was essential to understand how the enzyme recognized and cleaved these sites. In pursuing this goal, an important discovery was made: the enzyme can recognize and cleave small peptide substrates in solution. It was subsequently determined that the active site is large enough to accommodate seven amino acids. The parameters of peptide binding and protease activity were determined by screening synthetic peptides containing variations of the seven cleavage sites for their ability to be recognized and cleaved by the enzyme. The first inhibitor leads were peptide mimics (peptidomimetics) modeled after inhibitors of other aspartyl proteases, such as human renin, an enzyme implicated in hypertension (Fig. 9.16). Subsequent screens for mechanism-based and structure-based inhibitors designed *de novo* have yielded several powerful peptidomimetic inhibitors of the protease (Box 9.7).

Integrase

Integrase protein is particularly attractive, because biochemical and structural data are available. In the absence of

A

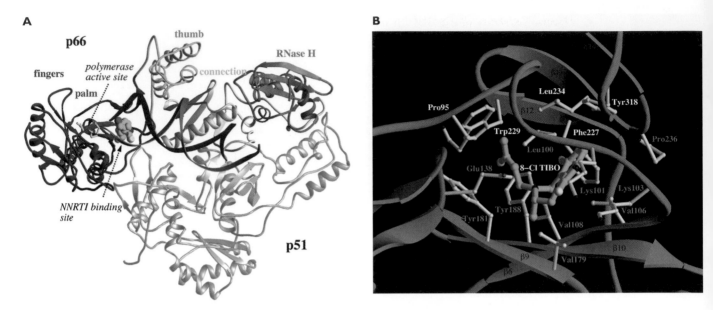

Figure 9.15 Structure of HIV type 1 reverse transcriptase highlighting the polymerase active site and the nonnucleoside reverse transcriptase inhibitor binding site. (A) Structure of the reverse transcriptase p66-p51 heterodimer bound to a double-stranded DNA template-primer, showing the relative locations of the polymerase active site and the site for binding nonnucleoside reverse transcriptase inhibitors (NNRTIs). Data from A. Jacobo-Molina et al., *Proc. Natl. Acad. Sci. USA* **90**:6320–6324, 1993. **(B)** Structure of 8-CL TIBO (a prototype NNRTI; green) bound to reverse transcriptase (green molecule). Amino acid side chains corresponding to sites of drug resistance mutations are shown in orange; residues with cyan side chains are sites at which NNRTI resistance mutations have not been detected. Data from J. Ding et al., *Nat. Struct. Biol.* **2**:407–415, 1995. Courtesy of K. Das and E. Arnold, Center for Advanced Biotechnology and Medicine, Rutgers University.

integrase, viral DNA cannot be inserted into the host genome. As a result, the viral genome is unable to be expressed efficiently and to propagate in dividing cells. Several companies have lead compounds under analysis, and rational design programs based on structures of integrase complexed with inhibitors are in progress. One drug, called raltegravir (Isentress), was approved in late 2007. All current integrase inhibitors target the joining step in the reaction (Volume I, Fig. 7.16). The inhibitors stabilize the DNA-protein intermediate so that the reaction cannot be completed.

Inhibitors of Fusion and Entry

Some neutralizing antibodies that block viral attachment, bind the third variable domain (the so-called V3 loop) of the SU protein (Volume I, Fig. 5.5). A variety of natural and synthetic molecules interfere with V3 loop activity. These compounds include specific antibodies and polysulfated or polyanionic compounds such as dextran sulfate and suramin. These compounds were identified early in the search for antiviral agents, but were discarded because of intolerable side effects such as anticoagulant activity. Although considerable effort was expended in the development of inhibitors of the SU-CD4 receptor interaction, including a "soluble CD4" that theoretically would act as a competitive inhibitor of infection, no effective antiviral agents have been found using this

strategy. This lack of success can be attributed, in part, to the high concentration of SU on the virion, as well as to alternative mechanisms for spread in an infected individual.

As often happens, the early research with failed Env inhibitors provided much insight into how virions enter cells, and has focused attention on other targets in the process. For example, it was curious that mutants resistant to neutralizing antibodies have clustered mutations in the V3 loop, yet virus-cell fusion is not affected. The implication was that CD4-V3 interactions were not involved in entry. As described in Volume I, Chapter 5, entry is a multistep process requiring that the target cells synthesize not only CD4, but also any one of several chemokine receptors, such as CCr5 present on macrophages or CXCr4 found on T cells. Chemokine receptors are attractive targets because individuals homozygous for mutations in one such receptor (CCr5) are partially resistant to infection and suffer no apparent ill effects. Recent studies have shown that infected individuals carrying a mutation in the CCr5 chemokine receptor experience a delay of 2 to 4 years in the progression to AIDS. Development of a safe chemokine receptor antagonist that delays the onset of AIDS for a long period, if not indefinitely, may be feasible.

We now know that the V3 loop of SU interacts with the chemokine receptor, exposing previously buried SU sequences required for membrane fusion. These transiently

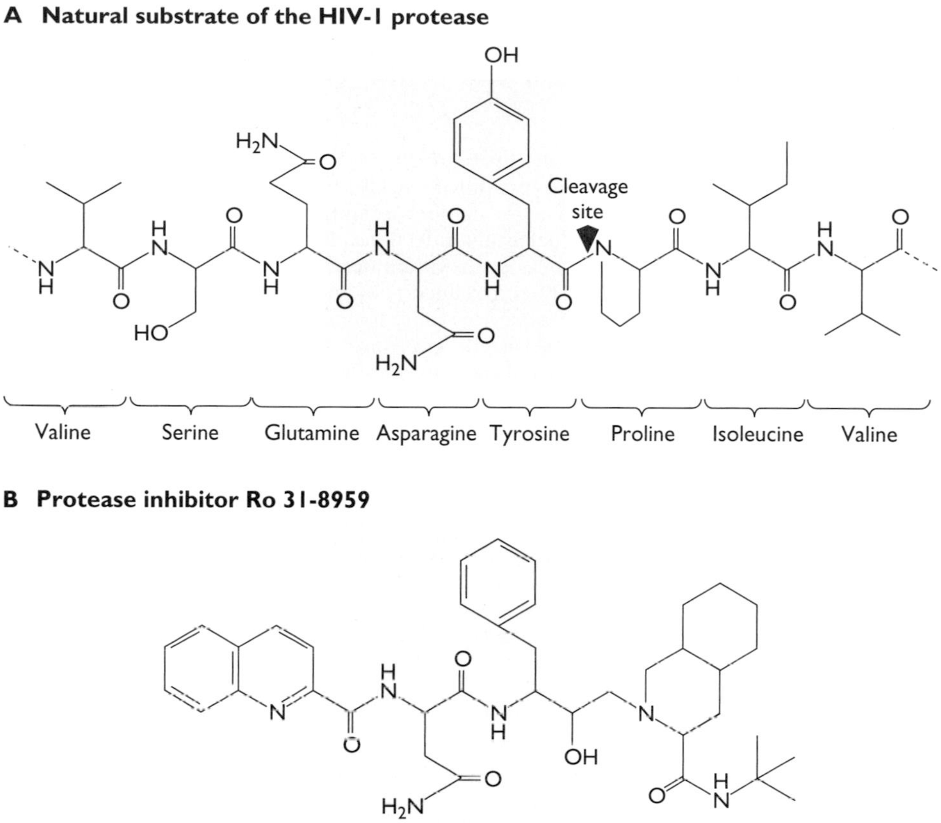

A Natural substrate of the HIV-1 protease

B Protease inhibitor Ro 31-8959

Figure 9.16 Comparison of one natural cleavage site for HIV protease with a peptidomimetic protease inhibitor. (A) The chemical structure of eight amino acids comprising one of the cleavage sites in the Gag-Pol polyprotein. The cleavage site between the tyrosine and proline is indicated by a red arrow. **(B)** The chemical structure of an inhibitory Roche peptide mimic (Ro 31-8959). The dotted box indicates the region of similarity. Adapted from D. R. Harper, *Molecular Virology* (Bios Scientific Publishers Ltd., Oxford, United Kingdom, 1994), with permission.

exposed surfaces are excellent targets for antiviral agents. A 36-amino-acid synthetic peptide, termed T20, derived from the second heptad repeat of SU, binds to the exposed grooves on the surface of a transient triple-stranded coiled-coil and perturbs the transition of SU into the conformation active for fusion. T20 (enfuvirtide) was the first drug approved with this mode of action. enfuvirtide is a peptide and was difficult to develop as a drug. Large-scale synthesis is expensive, and patients must actually prepare the peptide formulation so that they can inject it. Nevertheless, this drug is remarkably effective in reducing HIV titers in the blood.

The Combined Problems of Treating a Persistent Infection and Emergence of Drug Resistance

The Heart of the Problem

The slow pattern of infection exemplified by HIV is characterized by an asymptomatic period that lasts for years to decades after primary infection (see Volume I,

Chapters 5 and 6). This asymptomatic phase was initially thought to reflect a quiescent period of no viral replication and unfortunately was called "latency." We now understand that there is extensive viral replication throughout this period and that such relentless replication is the heart of the problem facing effective antiviral therapy.

The Critical Parameters

About 0.1% of an infected individual's CD4$^+$ T cells are replicating HIV at any time during the asymptomatic period, and at least 10% of the total CD4$^+$ population contains viral nucleic acid. The steady-state level of HIV RNA detected in blood depends primarily on the rate of virion production in CD4$^+$ T cells. Ultimately, in the untreated individual, viral RNA concentrations increase, CD4$^+$ T-cell numbers decrease, and death is inevitable. The late stage of infection is characterized by high levels of replication (often more than 10^9 particles produced per day), a high turnover of viral particles (50% turnover in less than a day), and

EXPERIMENTS
Highly specific, designed inhibitors may have unpredicted activities

The discovery and development of structure-based inhibitors of HIV protease have been pronounced a triumph of rational drug design. Structural biology and molecular virology came together to provide the protease inhibitors that anchor today's highly active antiretroviral therapy. However, patients receiving protease inhibitors often respond in unexpected ways, and the extent of immunological recovery with treatment is being debated. One study showed that the protease inhibitor ritonavir unexpectedly inhibits the chymotrypsin-like activity of the proteasome. As a result, the protease inhibitor blocks the formation and subsequent presentation of peptides to cytotoxic T lymphocytes by major histocompatiblity complex class I proteins. In another study, the saquinavir protease inhibitors were found to inhibit Zmpste24, a protease involved in conversion of farnesylprenlamin A to lamin A, a structural component of the nuclear lamina.

The challenge is to determine if these secondary activities help or hinder AIDS therapy. As discussed in Chapter 4, CTLs not only kill virus-infected cells, but also are responsible for significant immunopathology in persistent infections. Perhaps ritonavir blocks such immunopathology. On the other hand, reduction in immunosurveillance by cytotoxic T lymphocytes potentiates persistent infections. In this case, the secondary activity of ritonavir may presage long-term problems. We now know that ritonavir and saquinavir inhibit proteasome activity, while indinavir and nelfinavir do not. The inhibition of Zmpste24 by the saquinavir class of compounds may be involved in the debilitating partial lipodystrophy side effect (redistribution of adipose tissue from the face, arms, and legs to the trunk). Genetic data indicate that individuals with missense mutations in *LmnA*, the gene encoding prelaminA and lamin C, have a significant loss of adipose tissues.

Therefore, it will be of some interest to monitor the antiviral effects, lymphocyte functions, and accumulation of prelamin A in patients under treatment with these different protease inhibitors. Tailoring HIV protease inhibitors to limit their action to the intended target is an important goal. As noted by the investigators who found these surprising activities, the human genome carries approximately 400 genes encoding proteases. About 70 of these proteases are targets for new drugs, and the unexpected side effects of antiviral protease inhibitors may be useful in finding new therapies.

Andre, P., M. Groettrup, P. Klenerman, R. DeGiuli, B. L. Booth, Jr., V. Cerundolo, M. Bonneville, F. Jotereau, R. M. Zinkernagel, and V. Lotteau. 1998. An inhibitor of human immunodeficiency virus-1 protease modulates proteasome activity, antigen presentation, and T-cell responses. *Proc. Natl. Acad. Sci. USA* **95:**13120–13124.

Coffinier, C., S. Hudon, E. Farber, S. Chang, C. Hrycyna, S. Young, and L. Fong. 2007. HIV protease inhibitors block the zinc metalloproteinase ZMPSTE24 and lead to an accumulation of prelamin A in cells. *Proc. Natl. Acad. Sci. USA* **104:**13432–13437.

massive loss of CD4$^+$ T cells. The quantity of HIV RNA and the number of CD4$^+$ T cells are therefore the most useful monitors of clinical status and effectiveness of antiviral therapy. More than 500 CD4$^+$ lymphocytes per ml is considered a normal value; fewer than 200 is the definition of AIDS. Viral load is estimated from the number of copies of viral RNA per milliliter of serum or plasma. The lowest level detectable with current methodologies is about 50 copies, and values greater than 10^6 have been recorded.

The First Drug-Resistant Mutants

Viral mutants that grew with impunity in the presence of AZT appeared almost immediately after the drug was approved for general use. The genomes of the mutants harbor single-base-pair changes at one of at least four sites in the reverse transcriptase gene. Reverse transcriptase enzymes bearing these substitutions no longer bind phosphorylated AZT, but they retain enzymatic activity. Mutants resistant to other nucleoside analogs, as well as to protease inhibitors, also arose with disheartening frequency. Resistance to protease inhibitors is found only after acquisition of several distinct amino acid substitutions, but mutants resistant to these drugs have been isolated from patients. Such drug-resistant mutants were transmitted to new hosts and threatened to undermine the entire antiviral effort.

Because HIV replicates extensively during the asymptomatic period, mutations accumulate. On average, every new viral genome can be expected to carry at least one new mutation, and a single patient will harbor many viral genomes in various tissues (Fig. 9.17). If an antiviral drug is given long after the primary infection, selection of drug-resistant mutants is inevitable (Fig. 9.18). Even genomes with detrimental mutations are maintained in the population, if their defects can be complemented. Drug resistance does not have to be absolute, because any increase in fitness, no matter how subtle, can result in large changes in the virus population. The reality is that we do not know the principal factors controlling viral load in an infected patient. Future drug and vaccine trials providing quantitative data on immune responses will go far to solve this problem.

Antiviral Therapy Has the Potential To Promote or Prevent the Emergence of Resistant Viruses

Mutations appear only when the viral genome is replicated. Accordingly, when replication is blocked, no

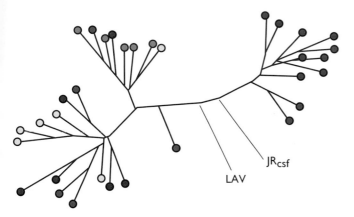

Figure 9.17 HIV-1 evolution within a single patient. An unrooted phylogenetic tree shows relationships among 33 viral variants coexisting within a single infected patient. Two prototype viruses, LAV and JR$_{csf}$, are shown in the center of the tree. Colors denote tissue sources for virus isolation: blue, lymph nodes; red, peripheral blood; yellow, spleen; green, lung; purple, spinal cord; brown, dorsal root ganglion. Adapted from M. Ait-Khaled et al., *AIDS* **9**:675–683, 1995, with permission.

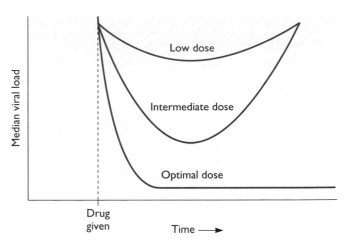

Figure 9.18 Viral load depends on the dose of antiviral drug. If virus replication is allowed in the presence of an antiviral drug, mutants resistant to that drug will be selected. This phenomenon is illustrated by plotting median virus load in relative units on the *y* axis as a function of time after exposure to a drug on the *x* axis as indicated (Drug given). In the top curve (Low dose), the concentration of antiviral drug is insufficient to block virus replication, and the viral load is reduced transiently, if at all. Viruses that replicate may be enriched for resistant mutants. In the middle curve (Intermediate dose), the concentration of antiviral drug appears to be successful in lowering the viral load initially, indicating that some replication was blocked. In this example, replication was not blocked completely, and resistant viruses overwhelmed the patient. In the bottom curve (Optimal dose), the concentration of the antiviral drug is such that all virus replication is blocked. As a consequence, no drug-resistant mutants can arise, and the viral load drops dramatically. Redrawn from J. H. Condra and E. A. Emini, *Sci. Med.* **4**:14–23, 1997, with permission.

drug-resistant mutants can arise. If an individual harboring a small number of viral genomes with no relevant preexisting mutations is given sufficient drug to block all viral replication, the infection will be held in check (Fig. 9.18). In contrast, if the same antiviral drug is given after the viral population has expanded, or if the drug concentration is insufficient to block replication entirely, genomes that harbor mutations will survive and will continue to replicate and evolve. When the viral genome numbers are small, the infection may still be cleared by the host's immune system before resistant mutants take over. If resistance to an antiviral drug requires multiple mutations (e.g., resistance to protease inhibitors), the chance that all mutations preexist in a single genome is much lower than if only a single mutation is required. But if replication is allowed in the presence of the inhibitor, resistant mutants will accumulate.

Cross-Resistance to Similar Inhibitors

HIV mutants resistant to different nucleoside analogs often carry different amino acid substitutions in the reverse transcriptase. Furthermore, a mutation conferring resistance to one inhibitor may suppress resistance to another (Table 9.4). Consequently, combinations of nucleoside analogs were tested with the expectation that double-resistance mutants would be rare, perhaps nonviable, or at least severely crippled. While initially promising, many combinations failed miserably, with mutants resistant to both drugs appearing after less than a year of therapy. Furthermore, the barrier to resistance to many pairwise combinations of nucleoside and nonnucleoside

inhibitors was higher than that for any single drug, but not high enough.

Experience with protease substrate analog inhibitors has been similar; resistance to two inhibitors emerges almost as fast as resistance to either one alone. As current protease inhibitors are all peptide mimics that bind to the substrate pocket of the enzyme, a change in residues lining this pocket can affect the binding of more than one inhibitor.

It is clear that treatment of a patient with one antiviral drug at a time is of limited clinical value. Furthermore, the use of a single drug must be contemplated carefully because this same drug may be used for combination therapy in the future.

Combination Therapy

The Use of Two or More Antiviral Drugs To Combat the Resistance Problem

The use of two or more treatments simultaneously is well known in tuberculosis therapy and cancer treatment. Combining two mechanistically different treatments often leads to more effective killing of the bacterium or tumor cells, thereby circumventing the appearance of cells

Table 9.4 Unpredicted drug resistance and susceptibility patterns

Compound	Substitution conferring resistance	Drug sensitivity phenotypes (amino acid substitution)
Zidovudine	T215F in reverse transcriptase	Didanosine resistance (L74V) restores zidovudine susceptibility
		Lamivudine resistance (M184V) restores zidovudine susceptibility
		Nevirapine and loviride resistance (Y181C) restores zidovudine susceptibility
		Foscarnet resistance (W88G) restores zidovudine susceptibility
VX-478	M46I + I47V + I50V in protease	Saquinavir resistance (G48V + I50V + I84L); restores VX-478 susceptibility
		Indinavir resistance (V32I, A71V); restores VX-479 resistance
Delavirdine	P236L in reverse transcriptase	Increased susceptibility to nevirapine; R82913 (TIBO); and L-697,661 (pyridinone)
Foscarnet	E89K + L92I + S56A + Q161L, H208Y in reverse transcriptase	Increased susceptibility to zidovudine, nevirapine, and R82150 (TIBO)

resistant to one treatment or the other. This principle also applies to antiviral therapy. In theory, if resistance to one drug occurs once in every 10^3 genomes, and resistance to a second occurs once in every 10^4, then the likelihood that a genome carrying both mutations will arise is the product of the two probabilities, or one in every 10^7.

Problems and Promise

Combination therapy does not always clear an infection. One obvious reason for resurgence of virion production is the appearance of drug-resistant mutants if therapy is given late in the progression of disease, or if the combination is not sufficiently potent. Combination therapy can be demanding for physician and patient. For example, if other infections are being treated, as they almost always are in AIDS patients, then many pills a day may be required. Other problems arise because storing and keeping track of different medications are daunting tasks for an ill patient. To compound the problems, every drug has side effects, and some are severe. For example, the gastrointestinal problems that accompany many protease inhibitors are particularly stressful for patients. Some side effects, such as the wasting of limbs and face with fat accumulation in the gut (lipodystrophy), may appear only after months of continuous use of current antiprotease drugs (Box 9.7). Because of these problems, some patients simply do not take their medication. The most insidious failure lies in wait when the patient begins to feel better and stops taking the pills. Viral replication resumes when the inhibitors are removed. Replication means mutation, and in such cases combination therapy may be ineffective if ever reinstated. Finally, combination therapy is very expensive, currently costing thousands of dollars per year for some regimens. In addition to the costs of medication, tests for viral load and CD4 counts must be performed regularly to monitor therapy. Such tests also are costly. Clearly, combination therapy is not accessible to everyone.

Despite these seemingly formidable issues, combination therapy has been effective, particularly for long-term control of infection. Many think that the clinical success of combination therapy is truly remarkable and represents one of the high points in the battle against AIDS (Box 9.8).

BOX 9.8

DISCUSSION
The first triple-drug combination, once-a-day pill

Twelve years ago, it was common for an HIV infected individual to take more than 20 pills a day to treat not only his or her HIV infection, but also the variety of other infections and complications inherent in AIDS. As more potent inhibitors were discovered, patients could take fewer pills to control their infection. More importantly, combination therapy developed as a therapeutic concept to deal with drug-resistant viruses. In 2006, the first triple drug, once-a-day pill for HIV therapy was approved. Atripla consists of three active antiviral drugs: a nucleotide reverse transcriptase inhibitor (tenofovir), a nucleoside reverse transcriptase inhibitor (emtricitabine), and a nonnucleoside reverse transcriptase inhibitor (efavirenz). It took over 20 years to develop this combination pill and required the cooperation of two pharmaceutical companies. This achievement is a landmark in the battle against HIV.

De Clercq, E. 2006. From adefovir to Atripla™ via tenofovir, Viread™ and Truvada™. *Fut. Virol.* **1:**709–715.

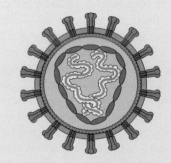

Strategic Treatment Interruption

The underlying premise of strategic treatment interruption is that combination drug treatment will stop replication and, as a consequence, the immune system will recover. The hope is that cycles of drug therapy, bolstered by drug-free periods, will enable the patient's own immune defenses to clear the infection. There are many reasons why such a hypothesis is attractive. A primary one is that "drug holidays" can be very attractive to patients. Initial studies with monkeys were encouraging when drug therapy was given very early in the acute infection. At these early times, the population of viral genomes in infected individuals is probably not as diverse, because it has not been subjected to many rounds of selection by the immune system. Consequently, replication could be held in check by the recovering immune system. However, studies with infected humans have not been as promising. Most AIDS patients have been infected for years, and the genetic diversity of viral genomes in their bodies is enormous. Indeed, the emergence of drug-resistant mutants, the small number of antiviral drugs available to treat resistant infections, and the appearance of cytotoxic T-lymphocyte and antibody escape mutants present serious problems for widespread use of strategic treatment interruption.

Challenges and Lessons Learned

Of utmost importance is that drugs must be potent. Such potency, which must be achieved to avoid selection of resistant mutants, can be accomplished only with combinations of drugs. Potency must be attained despite unpredictable patient adherence, individual differences in drug metabolism, and inevitable toxicity. A potent drug is of no use if the patient will not take it because the pills are too large, or too numerous or cause side effects.

At the moment, drug therapy is our only proven weapon. Therefore, it is prudent to avoid therapies in which resistance can be conferred by a single mutation in the viral genome. As a corollary, therapies that require multiple mutations for drug resistance to emerge should be used. Combinations of drugs will be most effective if the patient has not been treated previously with any drug.

With aggressive use of potent antiviral drugs, HIV replication can be suppressed, but the infection **cannot** be cured. Even when viral RNA has been undetectable for years during drug therapy, as soon as the drug is removed, replication begins again. We simply have no way to eliminate every last viral genome from the body of an infected individual. As years go by during this insidious, slow infection, the diversity of viral genomes increases as a result of mutation, recombination, and selection by drugs and an active immune system. While we understand in principle how this diversity arises, we have little insight about how to eliminate it.

The hope of any therapy is that as the cells of the immune system die and are replaced, proviral copies will be reduced to a point at which the infection cannot be sustained. Even if a cure is wishful thinking, combination therapies, where available, have converted HIV infection to a chronic treatable disease rather than a death sentence.

Perspectives

The world's surprisingly small arsenal of antiviral drugs is directed against a minor subset of viral diseases, notably those caused by HIV and the herpesviruses. Few drugs in the arsenal are effective against some of the most deadly viral diseases, which are caused by RNA viruses and poxviruses. There are many reasons for this state of affairs. One formidable problem is that we are unable to diagnose acute viral infections accurately and within sufficient time for effective intervention with antiviral drugs. Another arises from economic considerations. Many debilitating viral infections affect people in the developing world, a population that lacks the means to pay as well as the infrastructure to deliver therapy.

Persistent infections such as those caused by HIV, herpes simplex virus, the hepatitis viruses, and the papillomaviruses present a different set of challenges. At present, these infections are controlled by drugs, but not cured. Patients often must take the drug, or more likely a combination of drugs, for the rest of their lives, a difficult and expensive proposition. New approaches have been undertaken, and many promising lead compounds and therapies for treatment, and even cure, of persistent infections are being investigated. For example, it may be possible to reduce viral load by antiviral drugs and then promote clearance of the remaining infection by treating with drugs that bolster immune responses.

Regardless of what comes out of the antiviral drug pipeline, selection of resistant mutants is inevitable. This certainty is a specter that continues to haunt antiviral research and public health. As the search for effective therapies to treat HIV infection demonstrates, we still have much to learn.

Fortunately for our species, antiviral drugs and vaccines are not the only means to prevent viral infections. Effective public health measures, proper nutrition, and simple personal hygiene remain fundamental contributors to the prevention of viral infections. Indeed, in some underdeveloped countries, these nontechnical solutions are the most effective and often the only defense against viral diseases. The principles underlying the high- and low-technology approaches are remarkably similar: the sources and avenues of viral spread must be eliminated. Availability of clean drinking water, adequate sewage disposal, insect control, good medical practice, and an uncontaminated blood supply can be of paramount importance in reducing the incidence of viral infection. Simple personal actions such as washing hands, seeking protection from insects and rodents, and using condoms can be remarkably effective, low-cost preventative measures. Because malnourished individuals have reduced innate and acquired immune defenses, proper nutrition can be a major though unappreciated defense.

Unfortunately, even these fundamental practices and policies cannot be put in place without basic infrastructure and social policies that promote public awareness of infectious disease, understanding of potential hazards and risks, appreciation of the importance of early detection, and reporting of the incidence of infections.

References

Books

Blair, E., G. Darby, G. Gough, E. Littler, D. Rowlands, and M. Tisdale. 1998. *Antiviral Therapy*. Bios Scientific Publishers, Oxford, United Kingdom.

Levy, J. A. 2007. *HIV and the Pathogenesis of AIDS*, 3rd ed. ASM Press, Washington, DC.

Veerapandian, P. (ed.). 1997. *Structure-Based Drug Design: Diseases, Targets, Techniques and Developments*, vol. 1. Marcel Dekker, Inc., New York, NY.

Reviews

Air, G. M., A. Ghate, and S. Stray. 1999. Influenza neuraminidase as target for antivirals. *Adv. Virus Res.* **54:**375–402.

Balzarini, J. 2007. Targeting the glycans of glycoproteins:a novel paradigm for antiviral therapy. *Nat. Rev. Microbiol.* **5:**583–597.

Coffin, J. 1995. Human immunodeficiency virus population dynamics in vivo: implications for genetic variation, pathogenesis and therapy. *Science* **267:**483–489.

Condra, J., and E. Emini. 1997. Preventing human immunodeficiency virus-1 drug resistance. *Sci. Med.* **4:**14–23.

Craigie, R. 2001. Human immunodeficiency virus integrase, a brief overview from chemistry to therapeutics. *J. Biol. Chem.* **276:**23213–23216.

De Clercq, E., and A. Holy. 2005. Acyclic nucleoside phosphonates: a key class of antiviral drugs. *Nat. Rev. Drug Discov.* **4:**928–940.

De Clercq, E. 2004. Antivirals and antiviral strategies. *Nat. Rev. Microbiol.* **2:**704–720.

De Clercq, E. 2007. Three decades of antiviral drugs. *Nat Rev. Drug Discov.* **6:**941.

Ding, S.-W. 2007. RNAi restricts virus infections. *Microbe* **2:**296–301.

Drews, J. 2000. Drug discovery: a historical perspective. *Science* **287:**1960–1964.

Elion, G. B. 1986. History, mechanisms of action, spectrum and selectivity of nucleoside analogs, p. 118–137. *In* J. Mills and L. Corey (ed.), *Antiviral Chemotherapy: New Directions for Clinical Application and Research*. Elsevier Science Publishing Co., New York, NY.

Evans, J., K. Lock, B. Levine, J. Champness, M. Sanderson, W. Summers, P. McLeish, and A. Buchan. 1998. Herpesviral thymidine kinases: laxity and resistance by design. *J. Gen. Virol.* **79:**2083–2092.

Fields, B. 1994. AIDS: time to turn to basic science. *Nature* **369:**95–96.

Hajduk, P., R. Meadows, and S. Fesik. 1997. Discovering high-affinity ligands for proteins. *Science* **278:**497–499.

Harrison, S., B. Alberts, E. Ehrenfeld, L. Enquist, H. Fineberg, S. McKnight, B. Moss, M. O'Donnell, H. Ploegh, S. Schmid, K. Walter, and J. Theriot. 2004. Discovery of antivirals against smallpox. *Proc. Natl. Acad. Sci. USA* **101:**11178–11192.

Ikuta, K., S. Suzuki, H. Horikoshi, T. Mukai, and R. Luftig. 2000. Positive and negative aspects of the human immunodeficiency virus protease: development of inhibitors versus its role in AIDS pathogenesis. *Microbiol. Mol. Biol. Rev.* **64:**725–745.

Jackson, H., N. Roberts, Z. Wang, and R. Belshe. 2000. Management of influenza: use of new antivirals and resistance in perspective. *Clin. Drug Investig.* **20:**447–454.

Neutra, M., and P. Kozlowski. 2006. Mucosal vaccines: the promise and the challenge. *Nat. Rev. Immunol.* **6:**148–158.

Nolan, G. 1997. Harnessing viral devices as pharmaceuticals: fighting human immunodeficiency virus-1's fire with fire. *Cell* **90:**821–824.

Pang, Y. 2007. In silico drug discovery: solving the "target rich and lead poor" imbalance using the genome-to-drug-lead paradigm. *Clin. Pharm, Ther.* **81:**30–34.

Persaud, D., Y. Zhou, J. M. Siliciano, and R. F. Siliciano. 2003. Latency in human immunodeficiency virus type 1 infection: no easy answers. *J. Virol.* **77:**1659–1665.

Pulendran, B., and R. Ahmed. 2006. Translating innate immunity into immunological memory: implications for vaccine development. *Cell* **124:**849–863.

Richman, D. 2001. Human immunodeficiency virus chemotherapy. *Nature* **410:**995–1001.

Smith, K. 2001. To cure chronic human immunodeficiency virus infection, a new therapeutic strategy is needed. *Curr. Opin. Immunol.* **13:**617–624.

Wade, R. C. 1997. "Flu" and structure-based drug design. *Structure* **5:**1139–1145.

Zivin, J. 2000. Understanding clinical trials. *Sci. Am.* **282**(4):69–75.

Selected Papers

Bonhoeffer, S., R. May, G. Shaw, and M. Nowak. 1997. Virus dynamics and drug therapy. *Proc. Natl. Acad. Sci. USA* **94:**6971–6976.

Coen, D., and P. Schaffer. 1980. Two distinct loci confer resistance to acycloguanosine in herpes simplex virus type 1. *Proc. Natl. Acad. Sci. USA* **77:**2265–2269.

Crotty, S., D. Maag, J. Arnold, W. Zhong, J. Lau, Z. Hong, R. Andino, and C. Cameron. 2000. The broad-spectrum antiviral ribonucleoside ribavirin is an RNA virus mutagen. *Nat. Med.* **6:**1375–1379.

Crute, J., C. Grygon, K. Hargrave, B. Simoneau, A. Faucher, G. Bolger, P. Kibler, M. Liuzzi, and M. Cordingley. 2002. Herpes simplex virus helicase-primase inhibitors are active in animal models of human disease. *Nat. Med.* **8:**386–391.

DeLucca, G., S. Erickson-Viitanen, and P. Lam. 1997. Cyclic human immunodeficiency virus protease inhibitors capable of displacing the active site structural water molecule. *Drug Discov. Today* **2:**6–18.

Herold, B. C., N. Bourne, D. Marcellino, R. Kirkpatrick, D. M. Strauss, L. J. D. Zaneveld, D. P. Waller, R. A. Anderson, C. J. Chany, B. J. Barham, L. R. Stanberry, and M. D. Cooper. 2000. Poly(sodium 4-styrene sulfonate): an effective candidate topical antimicrobial for the prevention of sexually transmitted diseases. *J. Infect. Dis.* **181:**770–773.

Kim, J., K. Morgenstern, C. Lin, T. Fox, M. Dwyer, J. Landro, S. Chambers, W. Markland, C. Lepre, E. O'Malley, S. Harbeson, C. Rice, M. Murcko, P. Caron, and J. Thompson. 1996. Crystal structure of the hepatitis C virus NS3 protease domain complexed with a synthetic NS4A cofactor peptide. *Cell* **87:**343–355.

Kumar, P., H. Wu, J. McBride, K. Jung, M. Kim, B. Davidson, S. Lee, P. Shankar, and N. Manjunath. 2007. Transvascular delivery of small interfering RNA to the central nervous system. *Nature* **448:** 39–43.

Pearce-Pratt, R., and D. Phillips. 1996. Sulfated polysaccharides inhibit lymphocyte-to-epithelial transmission of human immunodeficiency virus-1. *Biol. Reprod.* **54:**173–182.

Perelson, A., P. Essunger, Y. Cao, M. Vesanen, A. Hurley, K. Saksela, M. Markowitz, and D. Ho. 1997. Decay characteristics of human immunodeficiency virus-1 infected compartments during combination therapy. *Nature* **387:**188–191.

Thomsen, D. R., N. L. Oien, T. A. Hopkins, M. L. Knechtel, R. J. Brideau, M. W. Wathen, and F. L. Homa. 2003. Amino acid changes within conserved region III of the herpes simplex virus and human cytomegalovirus DNA polymerases confer resistance to 4-oxo-dihydroquinolines, a novel class of herpesvirus antiviral agents. *J. Virol.* **77:**1868–1876.

Underwood, M., R. Harvey, S. Stanat, M. Hemphill, T. Miller, J. Drach, L. Townsend, and K. Biron. 1998. Inhibition of human cytomegalovirus DNA maturation by a benzimidazole ribonucleoside is mediated through the UL89 gene product. *J. Virol.* **72:**717–725.

van Zeijl, M., J. Fairhurst, T. Jones, S. Vernon, J. Morin, J. LaRocque, B. Feld, B. O'Hara, J. Bloom, and S. Johann. 2000. Novel class of thiourea compounds that inhibit herpes simplex virus type 1 DNA cleavage and encapsidation: resistance maps to the UL6 gene. *J. Virol.* **74:**9054–9061.

Varghese, J. N., V. C. Epa, and P. M. Colman. 1995. Three dimensional structure of the complex of 4-guanidino-Neu5Ac2en and influenza virus neuraminidase. *Protein Sci.* **4:**1081–1087.

von Itzstein, M., W. Wu, G. Kok, M. S. Pegg, J. Dyason, B. Jin, T. Van Phan, M. Smythe, H. White, S. Oliver, P. Colman, J. Varghese, D. Ryan, J. Woods, R. Bethell, V. Hotham, J. Cameron, and C. Penn. 1993. Rational design of potent sialidase-based inhibitors of influenza virus replication. *Nature* **363:**418–423.

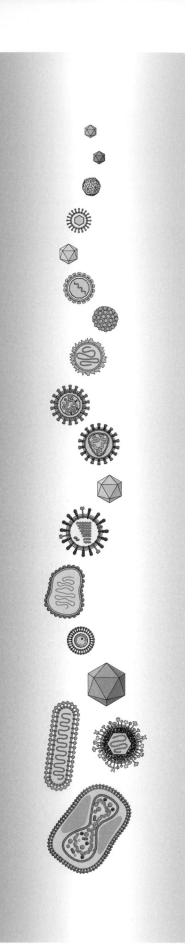

10

Virus Evolution

The Classic Theory of Host-Parasite
Interactions

How Do Viral Populations Evolve?

The Origin of Viruses

The Fundamental Properties of Viruses
Constrain and Drive Evolution

Emerging Viruses

The Spectrum of Host-Virus Interactions

Encountering New Hosts: Fundamental
Problems in Ecology

Expanding Viral Niches: Snapshots of
Selected Emerging Viruses

Host Range Can Be Expanded
by Mutation, Recombination, or
Reassortment

Some Emergent Viruses Are Truly Novel

A Revolution in Diagnostic Virology

Perceptions and Possibilities

Infectious Agents and Public Perceptions

What Next?

Perspectives

References

Evolution and Emergence

Virus Evolution

The word "evolution" conjures up images of fossils, dusty rocks, and ancestral phylogenetic trees covering eons. Modern virology shatters this staid image and reminds us that evolution not only is contemporary (we are made aware of it every day), but also carries profound implications for the future. Indeed, as host populations grow and adapt, virus populations that can infect them are selected. It also works the other way: viral infections can be significant selective forces in the evolution of host populations.

The public is made aware of virus evolution by frequent, often sensationalized reports in the press of new viral diseases such as severe acute respiratory syndrome (SARS), killer viruses like Ebola and Marburg, and real pandemics like the acquired immunodeficiency syndrome (AIDS) caused by human immunodeficiency virus. In addition to these more spectacular events, everyone experiences regular bouts with influenza and common cold viruses. Scientists are challenged regularly to devise new antiviral drugs, vaccines, diagnostics, and treatments for these emerging diseases. In truth, we are only beginning to develop an understanding of the selective forces that drive virus evolution, the emergence of new and old infections, and the corresponding implications for our own survival.

The Classic Theory of Host-Parasite Interactions

To put the process of evolution of viruses in perspective, it is important to have a basic understanding of host-parasite interactions. A fundamental principle of virus biology and evolution is that viral genomes must spread from host to host to maintain the viral population. An infection will spread if each infected individual infects, on average, more than one new individual before dying or clearing the infection. The infection rate is related to the population size: infections can spread only if population density exceeds a minimal value.

In 1983, these concepts were incorporated into a comprehensive theory of host-parasite interactions that is well known in ecological circles, but not always appreciated among molecular virologists. Briefly, this work suggests that spread of viral infection can be described in quantitative terms such as transmission rate, host survival/immunity, and virulence. The original work assumed well-mixed, homogeneous host populations in which each individual has the same probability of becoming infected. This theory states that selection maximizes R_0, the average number of secondary infections resulting from one infected host in an otherwise uninfected population. If R_0 is less than 1, it is impossible to sustain an epidemic. In fact, if R_0 is less than 1, it may be possible to eradicate the disease. If R_0 is slightly greater than 1, an epidemic is possible, but random fluctuations in early stages of infection in a susceptible population can lead to extinction or explosion of the infection. If R_0 is high, an epidemic is almost certain. A large R_0 value is typical of a disease that features "super-spreaders" (e.g., the individual who transmitted SARS coronavirus so successfully to others in the Hong Kong Hotel Metropole). A low R_0 value is typical of an infection in which the exposure time is reduced, the yield of infectious virus is low, or the duration of replication is short. In the standard model, the infection rate is predicted to be proportional to population size and virulence is related to transmissibility.

The original host-parasite theory remains robust, but additional terms and constraints must be added to the classic equations as we learn more about viral infections in the wild (Box 10.1, see also Fig. 5.2). For example, disease dynamics are oscillatory, but are not simple predator-prey cycles (e.g., foxes and rabbits). Actions of the immune system affect the appearance of immune-resistant viral mutants with differences in virulence and transmissibility. These events all are manifestations of complex variables that are difficult to evaluate. Not only are viral populations more complicated than first thought, but also the structure of host populations affects the evolution of viral agents. For example, as humans alter ecosystems and expand in numbers, viruses and hosts are mixing to an extent never before experienced. As a result, more viral diseases are appearing, precisely as predicted. It is possible to understand the basics of how these changes occur from first principles, as we discuss below.

How Do Viral Populations Evolve?

The harsh realities presented by a large, genetically variable host population dispersed in ever-changing environments appear to present insurmountable barriers to the survival of inanimate, submicroscopic, obligate intracellular pathogens. Obviously, such thinking must be flawed, as viruses abound everywhere. The primary reason for the remarkable success of viruses is that despite a minimal set of genes, viral populations display spectacular diversity. It is such diversity, manifested in the large collection of genomic permutations that are present in viral populations at any given time, that provides constant opportunities for survival. The sources of this diversity are **mutation, recombination and reassortment,** and **selection.** The constant change of a viral population in the face of selective pressures is the definition of **virus evolution**.

Positive and negative selection for particular preexisting mutants can occur at any step in a viral life cycle. The selective forces acting on viral populations are imposed not only by the environment, but also by the limitations of the information encoded in the viral genome. The requirement to spread within an infected host, as well as among individuals in the population, exposes virions to a variety of host antiviral defenses. The population density, the social behavior, and the health of potential hosts represent but a few of the powerful selective forces determining the survival of viral populations.

Is virulence a positive or negative trait for selection? One idea is that increased virulence reduces transmissibility (hosts die faster, reducing exposure to uninfected hosts). Debilitating disease may actually reduce transmission because the infected individual may not interact with other susceptible hosts.

If everything were simple, one might expect that all viruses would evolve to be maximally infectious and completely avirulent. A different view appears when real-life infections are studied. Persistent infections often lie dormant for years with respect to symptoms and then kill the host at the end stage of disease. Virulent viruses for one species may be maintained as asymptomatic infections in another. The interplay of contextual terms such as severity of disease and transmissibility is complicated. In fact, for some diseases, a strong case may be made that increased virulence actually increases R_0 and is strongly selected for in natural viral infections.

Boots, M., and M. Mealor. 2007. Local interactions select for lower pathogen infectivity. *Science* **315:**1284–1286.

Weiss, R. 2002. Virulence and pathogenesis. *Trends Microbiol.* **10:**314–317.

We define virus evolution in terms of a population, **not** an individual viral particle. Viral populations comprise diverse arrays of mutants that are produced in prodigious quantities. In most infections, thousands of progeny are produced after a single cycle of replication in one cell, and when copying is error prone, almost every new genome can differ from every other. Accordingly, it is confusing, and even misleading, to think of an individual particle as representing an average virion for that population. Virologists are population biologists whether they know it or not. As Stephen Jay Gould put it, the median is **not** the message when it comes to evolution. The diversity of the population provides surprising avenues for survival, yet every individual is a potential winner. Occasionally, even the most rare genotype in a current population may be the most common after a single selective event.

It is important to make a distinction between viral strategies that produce large numbers of progeny (high reproductive output or **r-replication strategies**) and those that result in a lower reproductive output but better competition for resources (**K-replication strategies**) (see Chapter 3). The former are characterized by short reproductive cycles with production of many progeny and are effective when resources are in short supply. The K-replication strategy produces persistent or latent infections with little pathogenesis. Fewer progeny are produced, but they have a high probability of surviving. Viruses that replicated via K-replication strategies survive as long their hosts survive.

The notations r and K come from the following equation:

$$dN/dt = rN(1 - N/K)$$

where r is the growth rate, N is the population size, and K is the carrying capacity in the current environment (see also Fig. 5.2).

The trajectory of evolution has long been a subject of debate. Scientists and philosophers have considered many questions such as the following. Is there a predictable direction for evolution? If so, what is the path? Are there really evolutionary dead ends? Virology provides a productive area for research into these questions. A primary lesson is that we must avoid judging outcomes as "good or bad." While making such judgments is a common human activity, anthropomorphic assessments and the comfort of analogies must be avoided, particularly when virus evolution is considered. Furthermore, from the first principle that there is no goal but survival, we can deduce that evolution does not move a viral genome from "simple" to "complex," or along some trajectory aimed at "perfection." Rather, change is effected by elimination of the ill adapted of the moment, not by the prospect of building something better for the future.

Virus-Infected Cells Produce Large Numbers of Progeny

The r-replication strategy (high reproductive output) is common among viruses. A single cell infected with poliovirus yields about 10,000 virus particles, and, in theory, three or four cycles of replication at this rate could produce enough virions to infect every cell in a human. Such overreplication does not happen for a variety of reasons, including a vigorous host defense and tropism for certain tissues. Nevertheless, high rates of replication over short periods and the resulting accumulation of large quantities of infectious particles are hallmarks of many common acute and persistent viral infections. This feature is illustrated in Table 10.1 for human immunodeficiency virus and hepatitis B virus. These high rates of particle production can continue for years. In the case of human immunodeficiency virus, the time from release of an infectious virion to infection and lysis of another target cell is estimated to be 2.6 days during the later stages of infection. What is more striking is the 50 to 90% turnover rate (replication minus elimination by host defenses) of this virus in plasma. Such measurements reveal the prodigious, relentless production of new virus particles in the face of a vigorous host

Table 10.1 *In vivo* dynamics of human immunodeficiency virus and hepatitis B virus

Characteristic	Value in:	
	Hepatitis B virus	**Human immunodeficiency virus**
Virus in plasma		
Half-life	24 h	6 h
Daily turnover	50%	90%
Total production in blood	>10^11	>10^9
Virus in infected cell		
Half-life	10–100 days	2 days
Daily turnover	1–7%	30%

defense. The interface of host defense and virus replication is fertile ground for selection and evolution.

Large Numbers of Mutants Arise When Viral Genomes Replicate

Two simple principles are that evolution is possible only when mutants arise in a population, and that mutations are introduced during copying of any nucleic acid molecule. When viral genomes replicate, mutations invariably accumulate in their progeny (Box 10.2).

RNA virus evolution. Most viral RNA genomes are replicated with considerably less fidelity than those comprising DNA (see Volume 1, Chapters 6 and 7). The average error frequencies reported for copying of RNA genomes are about one misincorporation in 10^4 or 10^5 nucleotides polymerized, more than 10^6 times greater than the rate for a host genome. Given a typical RNA viral genome of 10 kb, a mutation frequency of 1 in 10^4 corresponds to an average of 1 mutation in every replicated genome. Not all viral RNA genomes have the same mutation rate: there is some evidence that the replication machinery of viruses with larger RNA genomes (e.g., the 30,000-base human SARS coronavirus genome) may operate with higher replication fidelity than the polymerase complexes of smaller RNA viral genomes.

DNA Virus Evolution

The error rate of viral DNA replication is estimated to be about 300-fold lower than that for most RNA genomes described above. Many RNA polymerases lack error-correcting mechanisms, while most DNA polymerases can excise and replace misincorporated nucleotides (Volume 1, Chapter 9). Experimental data indicate that replication of small single-stranded DNA virus genomes (e.g., *Parvoviridae* and *Circoviridae*) is more error prone than is replication of the double-stranded DNA genomes of larger viruses. The best evidence for this conclusion comes from studies of canine parvovirus, first observed to cause disease in dogs in 1978, that has become a ubiquitous pathogen worldwide. When mutation rates were estimated by sequencing multiple isolates of canine parvovirus over the time of the pandemic, the rate of nucleotide substitution was closer to that of RNA viruses than to that of double-stranded DNA viruses. The mechanism for this increased misincorporation of nucleotides remains to be elucidated.

The Quasispecies Concept

A 1978 paper described a detailed analysis of an RNA bacteriophage population (phage Qβ). The authors made a startling conclusion:

> A Qβ phage population is in a dynamic equilibrium with viral mutants arising at a high rate on the one hand, and being strongly selected against on the other. The genome of Qβ cannot be described as a defined unique structure, but rather as a weighted average of a large number of different individual sequences.
>
> E. Domingo, D. Sabo, T. Taniguchi, and C. Weissmann,
> *Cell* **13**:735–744, 1978

This conclusion has been validated for many virus populations. Indeed, we now understand that virus populations

---

BOX 10.2

DISCUSSION
Error rates are difficult to quantify

Estimates of mutation rates must be viewed with caution. Absolute error rates (measured as the number of misincorporations per nucleotide polymerized) for any nucleic acid polymerase are difficult, if not impossible, to determine. A variety of technical issues must be considered, including sampling problems and potential artifacts of experimental design. Estimates of error rate can vary substantially, depending on the experimental method by which they are assessed. For example, PCR technology is commonly used to sample viral genomes, but the polymerase used may itself introduce copying errors that must be factored into the analysis.

Another popular method makes use of reporter genes (e.g., the *lacZ* gene, which encodes β-galactosidase). These genes can be inserted into a viral genome so that errors in the reporter gene can be scored by inspection or analysis of virus plaques. The error rate for the viral genome is then extrapolated from that determined for the reporter gene. While this method is relatively simple, it can yield misleading data, because errors of incorporation are not uniformly distributed as each genome is copied and are often dependent upon the particular sequence analyzed. For example, the reverse transcriptase from human immunodeficiency virus is inaccurate when measured *in vitro*, with an average error rate per nucleotide incorporated of 1 in 1,700. However, certain positions in this genome can be hot spots for mutation at which the error rate can be as high as 1 per 70 polymerized nucleotides.

Drake, J. 1992. Mutation rates. *Bioessays* **14**:137–140.

Drake, J. 1991. A constant rate of spontaneous mutation in DNA-based microbes. *Proc. Natl. Acad. Sci. USA* **88**:7160–7164.

Drake, J. 1993. Rates of spontaneous mutation among RNA viruses. *Proc. Natl. Acad. Sci. USA* **90**:4171–4175.

Hwang, Y. T., and C. B. C. Hwang. 2003. Exonuclease-deficient polymerase mutant of herpes simplex virus type 1 induces altered spectra of mutations. *J. Virol.* **77**:2946–2955.

exist as dynamic distributions of nonidentical but related replicons, often called **quasispecies**, a concept developed by Manfred Eigen. The classical definition of a species (an interbreeding population of individuals) has little meaning when considering viruses. For example, most viral infections are initiated not by a single virion, but rather by a population of particles. The large number of progeny produced after such infections are complex products of intense selective forces that operate **inside an infected host.** The relatively few virions that successfully infect another host have been subjected to an entirely new set of **external** selective forces. Therefore, a steady-state, equilibrium population of a given viral quasispecies must

comprise vast numbers of particles. Such an equilibrium cannot be attained in the small populations typically found after isolated infections in nature or in the laboratory. Consequently, when small populations of virus particles replicate, extreme fluctuations in genotype and phenotype are possible (Fig. 10.1).

For a given RNA virus population, the genome sequences cluster around a consensus or average sequence, but virtually every genome can be different from every other. A rare genome with a particular mutation may survive a selection event, and this mutation will be found in all progeny genomes. However, the linked but unselected mutations in that genome get a free ride. Consequently, the product of

Figure 10.1 Viral quasispecies, population size, bottlenecks, and fitness. Genomes are indicated by the horizontal lines. Mutations are indicated by different symbols. When an RNA virus replicates, mutations accumulate in the progeny genomes. A population of genomes, each member containing a characteristic set of mutations, is shown in the center. The consensus sequence for this population is shown as a single line at the bottom. Note that there are **no** mutations in the consensus sequence, despite their presence in every genome in the population (every genome is different). A population of genomes that emerges after passage of one genome through one bottleneck is depicted on the right. The consensus sequence for this population is shown as a single line at the bottom. Note that in this example, three mutations selected to survive the bottleneck are found in every member of the population, and these appear in the consensus sequence. If the large population is propagated without passage through bottlenecks (situation on the left), repeated passage enriches for mutant genomes that **improve replication and increase the fitness** of the population. If the population continues to propagate through serial bottlenecks, mutations accumulate that result in **reduced fitness**. Adapted from E. Domingo et al., p. 144, *in* E. Domingo, R. Webster, and J. Holland (ed.), *Origin and Evolution of Viruses* (Academic Press, Inc., San Diego, CA, 1999), with permission.

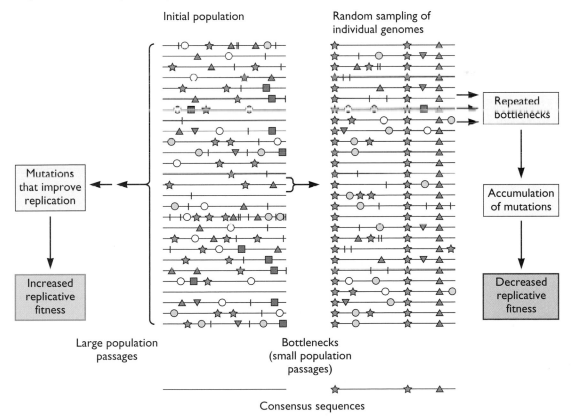

selection after replication is a new, diverse population that shares only the selected mutation. The ramifications of this phenomenon in virus evolution and pathogenesis are only now being appreciated.

Quasispecies theory predicts that viral quasispecies are not simply a collection of diverse mutants, but rather a group of interactive variants that characterize the particular population. Diversity of the population, therefore, is critical for survival. It has been possible to test the idea that viral populations, **not** individual mutants, are the target of selection by limiting diversity. Previously, we made the assertion that creation of diversity is the primary reason for virus survival. This conclusion follows in part from the identification of certain polymerase substitutions that reduce the frequency of incorporation errors during growth in cultured cells. Such mutants are called **antimutators**. Certain spontaneous mutants of human immunodeficiency virus that are resistant to the reverse transcriptase inhibitor lamivudine exhibit a 3.2-fold reduction in error frequency. The seemingly modest increase in fidelity was associated with a significant growth disadvantage in infected individuals. Poliovirus replication is notoriously error prone, producing a remarkably diverse population. Certain ribavirin-resistant poliovirus mutants have increased fidelity of about sixfold (~0.3 mutation per genome compared to ~2 mutations per genome for the parental virus). Importantly, the mutant was much less pathogenic in animals than was the wild-type virus: reduced diversity led to attenuation and loss of neurotropism. Further studies showed that in a diverse quasispecies, isolated viral mutants complemented each other, providing proof that it is the population, not the individual, that is evolving. As the wild-type viruses have maintained high mutation rates, we can infer that lower rates are neither advantageous nor selected in nature.

Sequence Conservation in Changing Genomes

Not all is in flux during viral genome replication. Despite high mutation rates, the so-called *cis*-acting sequences of RNA viruses change very little during propagation. These sequences are required for genome replication, messenger RNA (mRNA) synthesis, and genome packaging. They are often the binding sites for one or more viral or cellular proteins. Any genome with mutations in such sequences, or in the gene that encodes the corresponding binding protein, is likely to be less fit, or may not replicate at all. Changes must occur in both partners for restoration of function. The tight, functional coupling of binding protein and target sequence may be a marked constraint for evolution. In some instances, these sequences are stable enough to represent lineage markers for molecular phylogeny.

The Error Threshold, Lethal Mutagenesis, and Extinction

The capacity to sustain prodigious numbers of mutations is a powerful advantage. Yet, at some point, selection and survival must balance genetic fidelity and mutation rate. Many mutations are detrimental, and if the mutation rate is high, accumulating mutations can lead to a phenomenon called **lethal mutagenesis** when the population is driven to extinction. Intuitively, mutation rates higher than one error in 1,000 nucleotides incorporated must begin to challenge the very existence of the viral genome, as precious genetic information can be irreversibly lost. The **error threshold** is a mathematical parameter that measures the complexity of the information that must be maintained to ensure survival of the population. RNA viruses tend to evolve close to their error threshold (Box 10.3), while DNA viruses have evolved to exist far below it. We can infer these remarkable properties from experiments with mutagens. After treatment of cells infected by an RNA virus (such as vesicular stomatitis virus or poliovirus) with a base analog such as 5-azacytidine, virus titers drop dramatically. The error frequency per surviving genome increases only two- to threefold at best. In contrast, a similar experiment performed with a DNA virus, such as herpes simplex virus or simian virus 40, reveals an increase in single-site mutations of several orders of magnitude among survivors.

The error threshold concept is more complicated than it might first appear. In fact, one line of argument indicates that theory predicting error catastrophe, while mathematically rigorous, cannot represent a viral infection realistically. If this is so, other explanations for the antiviral behavior of mutagens must be considered. In addition, many important biological parameters contribute to virus survival, including a complex property called **fitness**, the replicative adaptability of an organism to its environment. Fitness depends on the context of the experiment and what outcome is measured. In the laboratory setting, fitness may be measured by comparison of growth rates or virus yields. Fitness is far more difficult to measure under more natural conditions, such as infection of complex organisms that live in large, interacting populations. Another essential component, equally difficult to measure, is the stability or predictability of the environment as it affects propagation of a virus. Host population dynamics and seasonal variation are but two examples of the many complicated environmental variables that exist. Finally, given the diversity in any viral population, determining the fitness of one population versus that of another depends on the mathematics of population genetics, a subject beyond the scope of this text.

EXPERIMENTS
Lethal mutagenesis: pushing human immunodeficiency virus over the "error threshold"?

At the end stage of AIDS, individuals infected with human immunodeficiency virus produce more than 1010 particles per day, and, on average, each genome contains one mutation. Diverse viral populations accumulate in the various tissues of every infected individual. Most of the virions appear to be nonviable, suggesting that the population exists at its error threshold and cannot tolerate many additional mutations. What would happen if one could intentionally push replicating genomes over the error threshold with a mutagenic nucleoside analog?

In 1990, scientists first reported that it is difficult to increase the mutation rate for two RNA viruses by using mutagens. Nine years later, another group selected antiviral nucleoside analogs that lacked toxicity for human cells, but were incorporated into human immunodeficiency virus DNA by the viral reverse transcriptase. These are not chain-terminating compounds, but rather they promote base misincorporations when copied. One compound, 5-hydroxydeoxycytidine, was promising

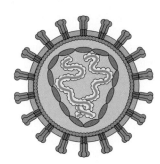

in that it was not toxic to human cells, but was efficiently incorporated into replicating viral genomes.

The experiments comprised serial passage of virions in human cells in the presence of the nucleoside analog. In seven of nine experiments, there was a precipitous loss of viral replication after 9 to 24 serial passages in the presence of the compound. Sequence analysis showed that the progeny genomes accumulated G-to-A substitutions. Loss of viral replication was not observed in 28 control infections in which virions were serially passaged without the analog.

Similar results have been obtained after infection of cultured cells infected with a variety of RNA viruses.

Bull, J., R. Sanjuan, and C. Wilke. 2007. Theory of lethal mutagenesis for viruses. *J. Virol.* **81**:2930–2939.

Crotty, S., C. E. Cameron, and R. Andino. 2001. RNA virus error catastrophe: direct molecular test by using ribavirin. *Proc. Natl. Acad. Sci. USA* **98**:6895–6900.

Eigen, M. 2002. Error catastrophe and antiviral strategy. *Proc. Natl. Acad. Sci. USA* **99**:13374–13376.

Holland, J. J., E. Domingo, J. C. de la Torre, and D. A. Steinhauer. 1990. Mutation frequencies at defined single codon sites in vesicular stomatitis virus and poliovirus can be increased only slightly by chemical mutagenesis. *J. Virol.* **64**:3960–3962.

Loeb, L. A., J. M. Essigmann, F. Kazazi, J. Zhang, K. D. Rose, and J. I. Mullins. 1999. Lethal mutagenesis of HIV with mutagenic nucleoside analogs. *Proc. Natl. Acad. Sci. USA* **96**:1492–1497.

Pariente, N., S. Sierra, P. R. Lowenstein, and E. Domingo. 2001. Efficient virus extinction by combinations of a mutagen and antiviral inhibitors. *J. Virol.* **75**:9723–9730.

Summers, J., and S. Litwin. 2006. Examining the theory of error catastrophe. *J. Virol.* **81**:20–26.

Genetic Shift and Drift

In Chapter 5, we discussed the process of antigenic variation and its contribution to modulating the immune response. Selection of mutants resistant to elimination by antibodies or cytotoxic T lymphocytes is inevitable when sufficient virus replication occurs in an immunocompetent individual. The terms **genetic drift** and **genetic shift** describe distinct mechanisms of diversity production. Diversity arising from copying errors and immune selection (drift) is contrasted with diversity arising after recombination or reassortment of genomes and genome segments (shift). Drift is possible every time a genome replicates. Shift can occur only under certain circumstances and is relatively rare. The episodic pandemics of influenza (Appendix B) provided strong evidence for this conclusion. For example, there are only six established instances of genetic shift for the influenza virus hemagglutinin gene since 1889. The combination of rapid drift and slow shift contributes to yet more diversity in populations (Boxes 10.4 and 10.5). When retrovirus infection results in integration of multiple proviral genomes in a single cell, genetic shift

may be a frequent event, with ramifications that are only now being appreciated.

Genetic Bottlenecks

Unlike lethal mutagenesis, which is deterministic and can cause extinction of large populations, the **genetic bottleneck** represents extreme selective pressure on small populations that results in loss of diversity, accumulation of nonselected mutations, or both (Fig. 10.1). A simple experiment illustrating this principle is easily done in the laboratory. A single RNA virus plaque formed on a monolayer of cultured cells is picked and expanded. Next, a single plaque is picked from the expanded stock, and the process is repeated many times. The bottleneck is the consequence of restricting further viral replication to the progeny found in a single plaque that contains a few thousand virions derived from a single infected cell. The perhaps surprising result is that after about 20 or 30 cycles of single-plaque amplification, many virus populations are barely able to propagate. They are markedly less fit than the original population (Table 10.2). The

BOX
10.4

DISCUSSION
Postulated evolution of human influenza A viruses from 1889 to 1977

The major influenza pandemics are characterized by viral reassortants. The emerging reassortant carries H and N genes that had not been in circulation in humans for some time, and consequently immunity is low or nonexistent. Six times since 1889, an influenza virus H subtype that had not been seen for years entered the human population. Three influenza virus H subtypes display a cyclic appearance, with the sequential introduction of H2 in 1889, H3 in 1900, H1 in 1918, H2 again in 1957, H3 again in 1968, and H1 again in 1977. With each H subtype introduction, the world experienced an influenza pandemic characterized by a new combination of H and N.

The figure depicts the appearance and transmission of distinct serotypes of influenza A virus in humans. In 1889 and 1900, only the H and N serology can be deduced from historical data. The bottom part shows the nature of the avian influenza viruses that reassort with human viruses. The color of the genome segments represents a particular viral genotype. Segments of the predominant influenza virus genome and its gene products are indicated in each human silhouette for each year. The number next to the arrow indicates how many segments of the viral genome are known to have been transmitted.

Adapted from R. G. Webster and Y. Kawaoka, *Semin. Virol.* **5:**103–111, 1994, with permission.

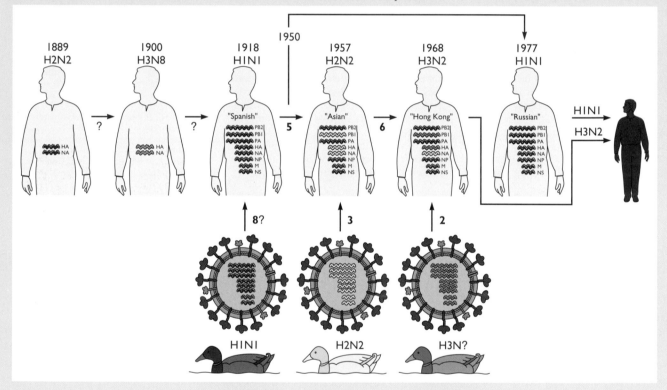

environment is constant, and the only apparent selection is that imposed by the ability of the population of viruses from a single plaque to replicate. Why does fitness plummet?

The answer lies in a phenomenon dubbed **Muller's ratchet**: small, asexual populations decline in fitness over time if the mutation rate is high. The genomes of replicating RNA viruses accumulate many mutations. As noted previously, they survive close to their error threshold. By restricting population growth to serial single founders (the bottleneck) under otherwise nonselective conditions, so many mutations accumulate that fitness decreases.

The ratchet metaphor should be clear: a ratchet on a gear allows the gear to move forward, but not backward. After each round of replication, mutations accumulate but are not removed. Each round of error-prone replication works like a ratchet, "clicking" relentlessly as mutations accumulate at every replication cycle. Each mutation has the potential to erode the fitness of subsequent populations. Simple studies such as the serial plaque transfer experiment indicate

BOX
10.5

DISCUSSION
Reassortment of influenza virus genome segments

Pandemic influenza results from shifts in H and N serotypes due to the exchange of genome segments by mammalian and avian influenza viruses. Virologists have demonstrated that certain combinations of H and N are better selected in avian hosts than in humans. An important observation was that both avian and human viruses replicate well in certain species such as pigs, no matter what the H-N composition. Indeed, the lining of the throats of pigs contains receptors for both human and avian influenza viruses, providing an environment in which both can flourish. As a result, the pig is a good

nonselective host for mixed infection of avian and human viruses, in which reassortment of H and N segments can occur, creating new viruses that can reinfect the human population.

One might think that this combination of human, bird, and pig infections must be extremely rare. However, the dense human populations in Southeast Asia that come in daily contact with domesticated pigs, ducks, and fowl remind us that these interactions are likely to be frequent. Indeed, epidemiologists can show that the 1957 and 1968 pandemic influenza A virus strains originated in China and that

the human H and N serotypes are circulating in wildfowl populations.

Studies of Italian pigs provide evidence for reassortment between avian and human influenza viruses. The figure shows how the avian H1N1 viruses in European pigs reassorted with H3N2 human viruses. The color of the segments of the influenza genome indicates the origin of the avian and human viruses. The host of origin of the influenza virus genes was determined by sequencing and phylogenetic analysis. These studies support the hypothesis that pigs can serve as an intermediate host in the emergence of new pandemic influenza viruses.

Adapted from R. G. Webster and Y. Kawaoka, *Semin. Virol.* **5:**103–111, 1994, with permission. For more information, see J. S. Peiris et al., *J. Virol.* **75:**9679–9686, 2001.

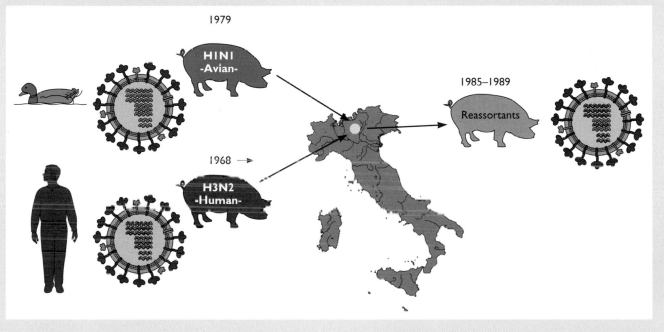

that Muller's ratchet can be avoided if a more diverse viral population is replicated by serial passage. One such study showed that pools of virus from 30 individual plaques were required in serial transfer to maintain the culture's original fitness. This observation can be explained as follows: more diversity in the replicating population facilitates the construction of a mutation-free genome by recombination or reassortment, removing or compensating for mutations that affect propagation adversely. Even if such a recombinant is rare, it has a powerful selective advantage in this

experimental paradigm. Indeed, its progeny ultimately will predominate in the population. The message is simple but powerful: diversity of a viral population is important for the survival of individual members; remove diversity, and the population suffers.

While the particular bottleneck of single-plaque passage is obviously artificial, infection by a small virus population and subsequent amplification are often found in nature. Examples include the small droplets of suspended virus particles during transmission as an aerosol, the activation

Table 10.2 Fitness decline compared to initial virus clone after passage through a bottleneck[a]

Virus	No. of bottleneck passages	% Decrease in fitness (avg)
φ6 (bacteriophage)	40	22
Vesicular stomatitis virus	20	18
Foot-and-mouth disease virus	30	60
Human immunodeficiency virus	15	94
MS2 (bacteriophage)	20	17

[a]A. Moya, S. Elena, A. Bracho, R. Miralles, and E. Barrio. *Proc. Natl. Acad. Sci. USA* **97**:6967–6973, 2000.

of a latent virus from a limited population of cells, and the small volume of inoculum introduced in infection by insect bites. An important question is, how do viruses that spread in nature by these routes escape Muller's ratchet? They do so by exchanging genetic information.

Exchange of Genetic Information

Genetic information is exchanged by recombination or by reassortment of genome segments (Volume I, Chapters 6, 7, and 9). In one step, recombination creates new combinations of many mutations that may be essential for survival under selective pressures. As discussed above, this process allows the construction of viable genomes from debilitated ones and can avoid Muller's ratchet. Recombination occurs when the polymerase changes templates (copy choice) during replication, or when nucleic acid segments are broken and rejoined. The former mechanism is common among RNA viruses, whereas the latter is more

typical of double-stranded DNA virus recombination. Another mechanism for exchange of genetic information is reassortment among genomic segments when cells are coinfected with segmented RNA viruses. It is an important source of variation, as exemplified by orthomyxoviruses and reoviruses (Boxes 10.4 and 10.5).

Insertion of nonviral nucleic acid into a viral genome (sequence-independent recombination) is well documented and makes a central contribution to virus evolution. Incorporation of cellular sequences can lead to defective genomes, or to more pathogenic viruses. Examples of such recombination include the appearance of a cytopathic virus in an otherwise nonpathogenic infection by the pestivirus bovine viral diarrhea virus (see Volume I, Chapter 6) or the sudden appearance of oncogenic retroviruses in nononcogenic retroviruses. The acquisition of activated oncogenes from the cellular genome is the hallmark of acutely transforming retroviruses such as Rous sarcoma virus (Chapter 7). Poxvirus and gammaherpesvirus genomes carry virulence genes with sequence homology to host immune defense genes. These genes are usually found near the ends of the genome. One explanation for their location is that the process of DNA packaging (gammaherpesviruses) or initiation of DNA replication (poxviruses) stimulates virus-host recombination when viral DNA is cleaved.

Information can be exchanged in a variety of unexpected ways during viral infections. For example, a host can be infected or coinfected by many different viruses during its lifetime (Box 10.6). In fact, serial and concurrent infections are commonplace. Both have a profound effect on virus evolution. In the simplest case, propagation of a virus quasispecies in an infected individual allows coinfection of single cells, phenotypic mixing, and genetic

---

BOX 10.6

EXPERIMENTS
Virus evolution by host switching and recombination

Viruses can be transmitted to completely new host species that have not experienced any prior infection. Usually host defenses stop the infection before any replication and adaptation can take place. On rare occasions, a novel population of viruses arises in the new host. Past interspecies infections can occasionally be detected by sequence analyses. The data provide a glimpse of the rather amazing and unpredictable paths of virus evolution.

Circoviruses infect vertebrates and have small, circular, single-stranded genomes (Volume I, Chapter 3). Nanoviruses have the same genome structure, but infect plants. The genes encoding the Rep protein of these viruses appear to be hybrids: they share significant sequence similarity in the 5′ coding sequences. They also exhibit homology in the 3′ sequences with an RNA-binding protein encoded by caliciviruses (RNA viruses that infect vertebrates).

The scientists who analyzed the DNA sequences suggested that two remarkable events occurred during the evolution of circoviruses and nanoviruses. At some time, a nanovirus was transferred from a plant to a vertebrate, perhaps when a vertebrate was exposed to sap from an infected plant. The virus adapted to vertebrates, and the circovirus family was established. As all known caliciviruses infect vertebrates, the recombination between circovirus and calicivirus sequences would have occurred after the host switch of nanoviruses to vertebrates.

Gibbs, M. J., and G. F. Weiller. 1999. Evidence that a plant virus switched hosts to infect a vertebrate and then recombined with a vertebrate-infecting virus. *Proc. Natl. Acad. Sci. USA* **96**:8022–8027.

complementation. As a result, recessive mutations are not immediately eliminated, despite the haploid nature of most viral genomes. Of course, such coinfection also provides an opportunity for physical exchange of genetic information.

Viral infections occur as host defenses are modulated or bypassed. Infections that suppress the immune response have a marked effect on concurrent or subsequent infections of the same host by very different viruses. Indeed, human immunodeficiency virus and measles virus infections lead to serious disease and death of their hosts by facilitating subsequent infections by diverse pathogens. Animals infected with poxviruses are more susceptible to infection by a variety of pathogens, because infected cells secrete soluble inhibitors of interferons and other cytokines required for innate defense. Conversely, activation of host defenses by one viral infection can impair subsequent infection by a different virus. The existence of such multifactorial interchange among diverse viruses reinforces the idea that selection acts not only on one population of viruses, but also on interacting populations of different viruses (Box 10.7).

Two General Pathways for Virus Evolution

An overarching principle is that viruses and their hosts exist along a continuum of *r*- and *K*-selection. *r*-selected viruses have high yields and short generation times and often kill their hosts, while *K*-selected viruses coexist with their hosts for long periods. Since viruses are absolutely dependent on their hosts for replication, viral evolution tends to take one of two general pathways. In one, viral populations coevolve with their hosts so that they share a common fate; as the host prospers, so does the viral population. However, given no other host, a bottleneck now exists: the entire viral population can be eliminated with antiviral measures or by loss of the host. In the other pathway, viral populations occupy broader niches and infect multiple host species. When one host species is compromised, the viral population can replicate in another. In general, as discussed below, the first pathway is typical of DNA viruses, whereas the second is common for RNA viruses.

The Origin of Viruses

Where did viruses come from? Certainly, the mists of time and lack of a fossil record cloud their origins. We can be sure of one basic fact: viruses cannot exist without living cells. Soon after the discovery of viruses, many articulated the idea that viral genomes may be very ancient and even predecessors to modern cells. Despite early interest, the origin of the first viruses remains a matter of conjecture and debate. Technological advances in nucleic acid chemistry, sequencing, and genomics/computer analyses have provided a wealth of data engendering some provocative speculation.

BOX 10.7

DISCUSSION

Evolution by nonhomologous recombination and horizontal gene transfer

In the early 1970s, scientists working with bacteriophage lambda and related viruses formed heteroduplexes between various pairs of viral DNA and visualized them in the electron microscope. The images were striking and revealed that the genomes of this group of lambdoid phages were mosaics; that is, the genomes contained blocks of genes (modules) that were shuffled during evolution by recombination. Further analyses of bacteriophages that had picked up host genes by nonhomologous recombination revealed that horizontal gene transfer among bacteria by bacteriophages was a central feature evolution of both. With large-scale genome sequencing, scientists know that bacteriophage genomes have ancestral connections to viruses of eukaryotes and archaea.

Murray, N., and A. Gann. 2007. What has phage lambda ever done for us? *Curr. Biol.* **17**:R305–R312.

Electron microscope image of phage lambda (negative stain). Courtesy of Robert Duda, University of Pittsburgh, Pittsburgh, PA (http://www.pitt.edu/~duda).

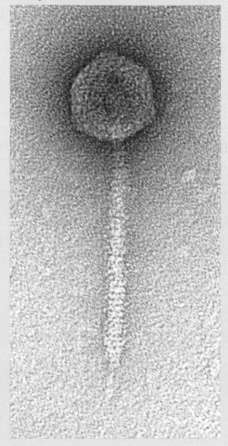

One challenge to viral genomics is that the "tree of life" as conceived by biologists does not include viral genomes. Their consensus is that the viruses are not alive and thus cannot be considered organisms. Obviously, such thinking is short sighted. Virions indeed are inanimate; they are complicated biomolecules incapable of replicating on their own. However, the infected cell has novel properties, clearly distinct from those of uninfected cells.

Theories about the origins of viruses center around three nonexclusive ideas. The **regressive theory** suggests that viruses are derived from intracellular parasites that have lost all but the most essential genes, those encoding products required for replication and maintenance. The **cellular origin theory** proposes that viruses arose from cellular components that gained the ability to replicate autonomously within the host cell. The **independent-entity theory** postulates that viruses coevolved with cells from the origin of life itself. As there is no fossil record and there are few viral stocks more than 80 years old, we cannot test the three hypotheses for the primordial origins of viruses. Nevertheless, we can gain some insight into viral evolution by examining contemporary viruses. One cannot help but conclude that nothing looks like the world of viral genomes (the virosphere). Viruses abound in the three domains of life (bacteria, eukaryotes, and archaea) and define three distinct arms of the virosphere. However, viral genomes in each arm share homologous features, providing a tantalizing hint that the viral genomes indeed are ancient and that they pre-date the last universal cellular ancestor. For example, the DNA viruses of green algae may be the oldest eukaryotic viruses, because sequence analysis places their genomes near the root of most eukaryotic sequences (Box 10.8).

A common but puzzling fact is that as more viral genomes are sequenced, more and more genes with no obvious homology to genes of known hosts are found. In addition, the very large viral DNA genomes often have **sequence coherence;** that is, these genomes do not appear to be mosaics (e.g., mimivirus [Box 10.9; see also Box 10.7]). The homogeneity of genomes within a family and the lack of any obvious homology among families are difficult to explain using the model that they arose by the sequential acquisition of exogenous genes by a primordial, precursor viral genome.

The existence of RNA and DNA viral genomes presents an interesting conundrum for any theory of virus origin. Which genome type came first? Some speculate that DNA genomes were a viral invention that was shared later with cells harboring RNA genomes. Given that viruses are obligate intracellular parasites, they are in a position to drive significant evolutionary change. The bringing together of two distinct genomes in a common cell is thought to have driven major evolutionary leaps such as the acquisition

of mitochondria by eukaryotic cells. It has been proposed that the eukaryotic nucleus arose from an infection of a primordial cell (perhaps with an RNA genome) with DNA viruses that replicate in the cytoplasm (Fig. 10.2). One critical issue inherent in such a model is the source of the deoxyribonucleoside triphosphates required for DNA replication. Can an organism with RNA biology produce these essential nucleotides?

To account for the shared gene pool for DNA viruses in the three domains of life, at least three independent transfers of DNA to the last universal cellular ancestor would be required to give rise to the modern viruses. The origin of RNA viruses remains even more speculative. They could be escapees from an ancient RNA world, but their distribution among the eukaryotes and bacteria is decidedly nonuniform. For example, at this time, no RNA viral genomes have been found in archaea. Given the extreme environments populated by these organisms, perhaps RNA genomes just cannot survive (or perhaps investigators have methodological problems). On the other hand, maybe the ancestral RNA protoviruses never infected the common archaeal ancestor. A better guess might be that we just have not looked hard enough.

DNA Virus Relationships Deduced by Nucleic Acid Sequence Analysis

We know the sequences of more than 1,600 viral genomes (roughly equal numbers of RNA and DNA genomes are in the databases). A "virochip" has been constructed with oligonucleotides representing the distinctive features of all 1,600 genomes. It is now possible to sample the ecosystem to determine the nature and diversity of viral genomes without having to propagate the viruses in the laboratory. Viral ecology is now described by the results of large-scale sequencing efforts. One type of study purifies all capsids from water samples and sequences all the DNA released from these capsids. This type of **metagenomic analysis** has revealed remarkable diversity. In fact, the vast majority of viral sequences determined so far by these technologies represent unknown viral genomes.

The origins of herpesviruses. Nucleic acid sequence analyses have identified many relationships among different viral genomes, providing considerable insight into the origin of viruses. The herpesvirus family exhibits a unique complex virion structure found in no other virus family (see Volume I, Appendix). The three main subfamilies of the family *Herpesviridae* (*Alphaherpesvirinae*, *Betaherpesvirinae*, and *Gammaherpesvirinae*) are easily distinguished by genome sequence analysis even though the original taxonomic separation of these families was based on general, often arbitrary, biological properties. Researchers have related the timescale of herpesviral genome evolution

BOX
10.8

DISCUSSION
Chlorella viruses: clues to viral and eukaryotic origins?

Paramecium bursaria chlorella virus (PBCV-1) is one of the oldest eukaryotic viruses. This remarkable virus is a large, icosahedral, plaque-forming, double-stranded DNA virus that replicates in certain unicellular, eukaryotic chlorella-like green algae (family *Phycodnaviridae*, genus *Chlorovirus*). In nature, the chlorella host is a hereditary endosymbiont of the ciliated protozoan *Paramecium bursaria*. The alga host can be grown in the laboratory in liquid and on solid media. Chloroviruses have been found in freshwater sources throughout the world, and many genetically distinct isolates usually can be found within the same sample. The titer of the viruses within a single water source can reach as high as 40,000 plaque-forming units (PFU) per ml.

The 330,744-bp PBCV-1 genome is predicted to harbor ~375 protein-encoding genes and 10 transfer RNA genes. The predicted products of ~50% of these genes resemble proteins of known function. Besides their large genome, the chloroviruses have other unusual features, including multiple DNA methyltransferases and DNA site-specific endonucleases, the entire machinery to glycosylate viral glycoproteins, at least two types of introns (a self-splicing intron in a transcription regulatory gene and a splicesome-processed intron in the viral DNA polymerase gene), a potassium ion channel, and the smallest known topoisomerase. Phylogenetic analyses based on DNA sequences and protein motifs indicate that these viral sequences lie near the

root of most eukaryotic sequences. The implication is that the earliest eukaryotes were exchanging information with ancient members of the *Phycodnaviridae*.

Plugge, B., S. Gazzarrini, M. Nelson, R. Cerana, J. L. Van Etten, C. Derst, D. DiFrancesco, A. Moroni, and G. Thiel. 2000. A potassium channel protein encoded by chlorella virus PBCV-1. *Science* **287**:1641–1644.

Van Etten, J. L., and R. H. Meints. 1999. Giant viruses infecting algae. *Annu. Rev. Microbiol.* **53**:447–494.

Van Etten, J. L., M. V. Graves, D. G. Muller, W. Boland, and N. Delaroque. 2002. Phycodnaviridae—large DNA algal viruses. *Arch. Virol.* **147**:1479–1516.

Villarreal, L. P., and V. R. DeFilippis. 2000. A hypothesis for DNA viruses as the origin of eukaryotic replication proteins. *J. Virol.* **74**:7079–7084.

Cryo-electron microscopic images courtesy of Tim Baker (University of California, San Diego [http://cryoem.ucsd.edu/]) and J. Van Etten (University of Nebraska).

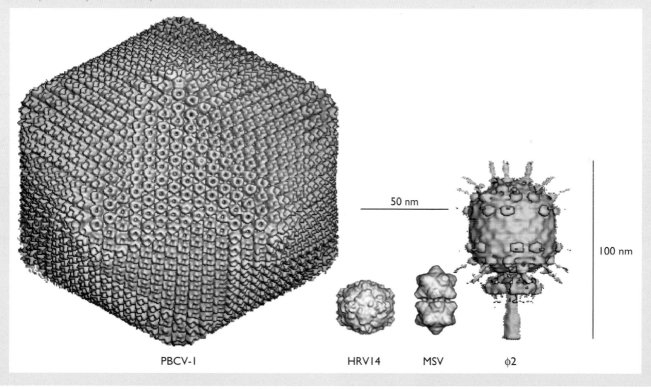

50 nm

100 nm

PBCV-1   HRV14   MSV   φ2

to that of the hosts (based on fossil records). For most herpesviruses, points of sequence divergence coincide with well-established points of host divergence. The conclusion is that an early herpesvirus infected an ancient host progenitor, and subsequent viruses developed by cospeciation with their hosts. Consistent with this conclusion, the genomes of **all** *Alphaherpesvirinae, Betaherpesvirinae,* and *Gammaherpesvirinae* that have been sequenced contain a core block of genes, often organized in similar clusters in the genome.

BOX 10.9

Mimivirus: the largest known virus particle and genome

A 1992 pneumonia outbreak in Bradford, England, led to the isolation of the world's largest virus. Investigators attempted to isolate *Legionella*-like pathogens of amoebae from hospital cooling towers and isolated what appeared to be a small gram-positive bacterium. All attempts to identify it using universal bacterial 16S ribosomal RNA (rRNA) PCR amplification failed. Transmission electron microscopy of infected *Acanthamoeba polyphaga* revealed 400-nm icosahedral virus particles in the cytoplasm. The virus was named "mimivirus" because it mimicked a microbe. This giant virus of amoebae challenges many preconceived notions about the nature of viruses and their origin.

Virion: 750 nm in diameter

Genome: 1.2×10^6 base pairs (bp)

Genes: 911 protein-coding genes

Still a virus: Despite having a number of genes predicted to be involved in protein synthesis, the genome does not encode a complete translation system.

Amazing sequence conservation: While it is difficult if not impossible to do genetics with mimivirus, bioinformatics has been used to probe the mimivirus DNA sequence. Mimivirologists have pointed out some rather provocative features that have

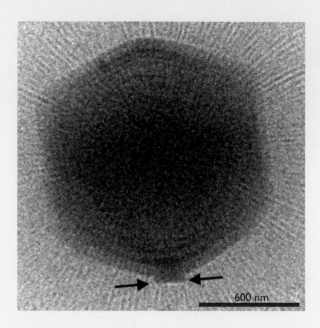

600 nm

implications for mimivirus evolution. For example, an AAAATTGA motif is found in more than half the mimivirus genes. This motif is proposed to be the structural motif of the TATA box core promoter element of unicellular eukaryotes, particularly the amoeba hosts of mimivirus. The motif is specific to the mimivirus lineage and may correspond to an ancestral promoter structure predating the radiation of the eukaryotes.

Raoult, D., S. Audic, C. Robert, C. Abergel, P. Renesto, H. Ogata, B. LaScola, M. K., Suzan, and J. Claverie. The 1.2Mb genome sequence of Mimivirus. *Science* **306:**1344–1350.

Suhre, K. I., S. Audic, and J.-M. Claverie. 2005. Mimivirus gene promoters exhibit an unprecedented conservation among all eukaryotes. *Proc. Natl. Acad. Sci. USA* **102:**14689–14693.

Our current best estimate is that the three major groups of herpesviruses arose approximately 180 million to 220 million years ago. These three subfamilies of viruses must therefore have been in existence before mammals spread over the Earth 60 million to 80 million years ago. Perhaps surprisingly, fish, oyster, and amphibian herpesviruses have virtually identical virion architecture, but little or no sequence homology to the *Alphaherpesvirinae*, *Betaherpesvirinae*, and *Gammaherpesvirinae*. They are related only tenuously to the mammalian and avian herpesviruses by common virion architecture and must represent a very early branch of this ancient family.

Origins of papillomaviruses and polyomaviruses. Coevolution with a host also is a characteristic of small DNA viruses, the parvoviruses, polyomaviruses, and

papillomaviruses. Here, the evidence for coevolution comes not from comparison of host and viral genes, but rather from finding close association of a given viral DNA sequence with a particular host group. The linkage of host to virus was particularly striking when human papillomavirus types 16 and 18 were compared: the distribution of distinct viral genomes is congruent with the racial and geographic distribution of the human population. Another example of the same phenomenon is provided by JC virus, a ubiquitous human polyomavirus associated with a rare, fatal brain infection of oligodendrocytes. This virus exists as five or more genotypes identified in the United States, Africa, and parts of Europe and Asia. Recent polymerase chain reaction (PCR) analyses of these subtypes indicate that JC virus not only coevolved with humans, but also did so within specific human subgroups. Probably the most striking finding was

Viral eukaryogenesis

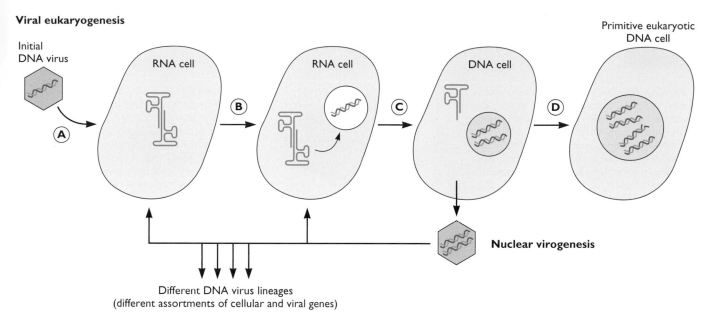

Figure 10.2 Primordial encounter of a DNA virus with an RNA cell: hypothetical origin of the nucleus. (A) An early DNA virus (perhaps a bacteriophage ancestor) engages a cell with an RNA genome. **(B)** The DNA virus is sequestered within a vesicle in the "cytoplasm" and replicates in this compartment. **(C)** Cellular genes are recruited to the enlarging nucleus; new DNA chemistry provides selective advantages. **(D)** This unstable situation may produce novel virions better adapted to infection of and cytoplasmic replication in cells with RNA genomes as well as the evolution of a stable eukaryotic cell with a nucleus and DNA replication machinery. See J.-M. Claverie, *Genome Biol.* **7:**110–114, 2006, and C. Zimmer, *Science* **312:**870–872, 2006.

that the JC virus of a particular group provides a convenient marker for human migrations from Asia to the Americas in both prehistoric and modern times.

How can virus evolution be linked to specific human populations in a manner akin to vertical transmission of a host gene? We can begin to appreciate this perhaps counterintuitive phenomenon from the unusual biology of human papillomaviruses (see also Box 10.10). Infection of the basal keratinocytes of adult skin leads to viral replication, as the cells differentiate. Virus particles are assembled only as cells undergo terminal differentiation near the skin surface. Mothers infect newborns with high efficiency, because of close contact or reactivation of persistent virus during pregnancy or birth. The infection therefore appears to spread vertically, in preference to the more standard horizontal spread between hosts. This mode of transmission was the predominant mechanism for papillomavirus and polyomavirus. It stands in contrast to that observed for most acutely infecting viruses, which are spread by aerosols, contaminated water, or food.

Origins of smallpox virus.   An analysis of smallpox virus genomes provides unexpected insights into the evolution of this scourge of humanity. The genome sequences of 45 epidemiologically different smallpox virus isolates are

remarkably similar. The genome sequences can be organized into three clades, which cluster according to the origin of the isolates (West Africa, South America, and Asia, respectively), but gene content is remarkably constant. Lack of diversity among diverse isolates indicates a recent introduction into humans. Interestingly, the only member of the large *Poxviridae* family with credible sequence homology to smallpox virus was a gerbil poxvirus. Perhaps human smallpox virus arose after a zoonotic infection from infected gerbils.

RNA Virus Relationships Deduced by Genome Sequence Analysis

The relationships among RNA viruses can also be deduced from sequence analyses, but the high rates at which mutations accumulate impose some difficulties. Moreover, genomes of RNA viruses are often small and carry few if any nonessential genes that might be useful for comparative studies. And, in contrast to the large DNA viruses, viral RNA genomes contain few, if any, genes in common with a host that might be used to correlate virus and host evolution. Nevertheless, when nucleotide sequences of many (+) and (−) strand RNA viral genomes are compared, blocks of genes that encode proteins with

BOX 10.10 EXPERIMENTS
The modular nature of papillomavirus genomes

Papillomaviruses infect vertebrate strati-fied squamous epithelia. Some are associ-ated with certain benign lesions and some are associated with cancers. Sequence analysis of host and viral genomes has revealed a close association of host and viral genomes. In general, papillomavirus genomes have been classified according to the sequence of the L1 capsid protein gene and a rather elaborate phylogenetic scheme as been proposed.

Garcia-Vallve et al. measured the sequence divergence of all the viral open reading frames and found five well-defined regions in the viral genome with appar-ently different evolutionary histories. They suggest that the primordial papillo-mavirus genome (protovirus) comprised

the E1, E2, L1, and L2 open reading frames, while the E5, E6, and E7 genes were acquired later. Given that all the early (E) viral gene products interact with host proteins that participate in cell cycle control and DNA replication, it is pos-sible to speculate about the host for the protovirus whose progeny now infect

essentially all warm-blooded vertebrates. The authors suggest that the low-diver-gence genes (E1, E2, L1, and L2) define the protopapillomavirus genome that infected an early land-dwelling vertebrate. The E6 and E7 genes have more diversity, consistent with their later addition to the protovirus genome. When the mamma-lian lineage appeared 150 million years ago, the protoviral genomes were exposed to various selective pressures, resulting in a rapid diversification of the sequences. The E5 open reading frame was the most recent addition to the protoviral genome about 65 million years ago.

Garcia-Vallve, S., A. Alonso, and I. Bravo. 2005. Papillomaviruses: different genes have different histories. *Trends Microbiol.* **13**:514–521.

similar functions can be defined. Common coding strategies can also be deduced. These groups are often called "super-groups" because the similarities suggest a common ances-try (Fig. 10.3 and 10.4). Alternatively, similarities may result from convergent evolution with no implications of shared lineages.

If one examines the sequences of many (–) strand RNA genomes, the first obvious common feature is the limited number of encoded proteins (as few as 4 and not more than 13). These proteins can be placed in one of three functional classes: core proteins that interact with the RNA genome, envelope glycoproteins that are required for attachment and entry of virus particles, and a polymerase required for replication and mRNA synthesis (Fig. 10.3).

The (+) strand RNA viruses (excluding the retroviruses) are the largest and most diverse subdivision of viruses: the genome sequences of more than 100 viruses representing at least 30 distinct groups are available for analysis. The number of proteins encoded by (+) strand RNA viruses ranges from 3 to more than 12, and, like the (–) strand RNA virus proteins, they can be divided into three groups by function: those required for RNA replication, encapsida-tion, and accessory functions. This comparison has resulted in the identification of three virus supergroups (Fig. 10.4). A unifying feature is that the RNA polymerase gene appears to be the most highly conserved, implying that it arose once in the evolution of these viruses. As each of the super-groups contains members that infect a broad variety of ani-mals and plants, an ancestor present before their separation

might have provided the primordial RNA polymerase gene. Alternatively, the ancestral (+) strand virus could have radiated horizontally among plants, animals, and bacteria (RNA viral genomes are not found in archaea).

It is instructive to consider the postulated evolution of supergroup 3, which contains the alphaviruses. Earlier work had indicated that they arose less than 5,000 years ago, but certain assumptions were made about the rate of RNA genome divergence and the constancy of virus evolution in different hosts and environments. More recent analyses comparing protein and RNA sequences indicate that the assumption of uniform nucleotide substitution rates was invalid; nucleotide changes are far from uniform. With new data from E1 protein and gene sequences, a phylogenetic tree can be constructed, but no time estimates for the appear-ance of the alphavirus progenitor are possible (Fig. 10.5). The data do suggest some interesting hypotheses about radi-ation of the species. The Old World viruses (e.g., the Semliki Forest virus complex) and the New World viruses (e.g., Ven-ezuelan, eastern, and western equine encephalitis viruses) are clearly distinct. The mosquito-borne alphaviruses could have arisen in either the Old World or the New World, but at least two transoceanic introductions are required to account for their current distribution.

Predictive Powers of Sequence Analyses
Phylogenetic dendrograms relating nucleic acid sequences depict the relationships as if founder and inter-mediary sequences were on a trajectory to the present

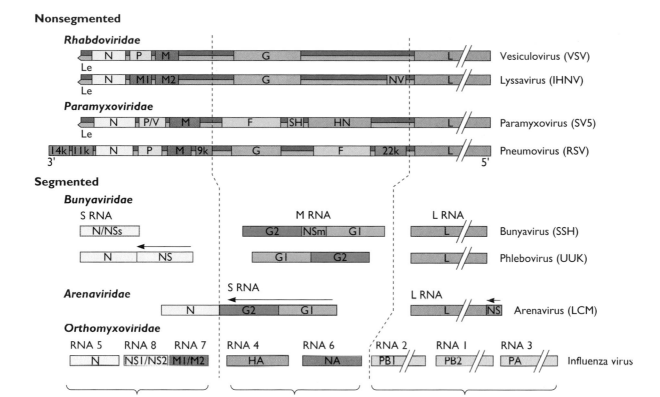

Figure 10.3 The genetic maps of selected (–) strand RNA viral genomes. Maps of the genes of *Rhabdoviridae*, *Paramyxoviridae*, *Bunyaviridae*, *Arenaviridae*, and *Orthomyxoviridae* are aligned to illustrate the similarity of gene products. The individual gene segments of the Orthomyxoviridae are arranged according to functional similarity to the two other groups of segmented viruses. Within a given genome, the genes are approximately to scale. For segmented genomes, blue-outlined genes are those in which multiple proteins are synthesized from different open reading frames. Red-outlined genes are expressed by the ambisense strategy. Virus abbreviations: VSV, vesicular stomatitis virus; IHNV, infectious hematopoietic necrosis virus; RSV, respiratory syncytial virus; SV5, simian virus 5; SSH, snowshoe hare virus; UUK, Uukuniemi virus; LCM, lymphocytic choriomeningitis virus. Le is a nontranslated leader sequence. Gene product abbreviations: N, nucleoprotein; P, phosphoprotein; M (M1 and M2), matrix proteins; G (G1 and G2), membrane glycoproteins; F, fusion glycoprotein; HN, hemagglutinin/neuraminidase glycoprotein; L, replicase; NA, neuraminidase glycoprotein; HA, hemagglutinin glycoprotein; NS (NV, SH, NSs, and NSm), nonstructural proteins; PB1, PB2, and PA, components of the influenza virus replicase. Figure derived from J. H. Strauss and E. G. Strauss, *Microbiol. Rev.* **58**:491–562, 1994, with permission.

sequences. This deduction is a gross oversimplification. Certainly, only extant genomes can be sequenced. Any intermediate that was lost during evolution will not contribute to the dendrogram. In addition, any recombination or gene exchange by coinfection with similar viral genomes will scramble ordered lineages (see also Box 10.7). A fair question is, can we predict the future trajectory of the dendrogram? What will comprise future branches of a given lineage? We can never answer these straightforward questions for two reasons: we cannot describe the diversity of any given virus population in an ecosystem, and we cannot we predict the selective pressures that will be imposed.

Origins of Viral Groups Suggested by Sequence Analysis

Japanese encephalitis virus. This flavivirus is the most important cause of epidemic encephalitis in the world. The virus was first isolated in 1935 and has subsequently been found across most of Asia. The origins of the virus are uncertain, but one idea has been entertained. Sequence analysis of all known Japanese encephalitis virus isolates indicates an origin a few centuries ago in the Indonesia-Malaysia region. Evolutionary biologists have long recognized that this area of the world is a unique environment with remarkable examples of divergent evolution of plants and animals. The number of insect species and their

Supergroup 1

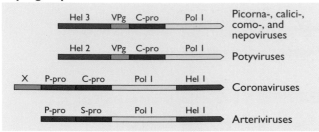

Supergroup 2

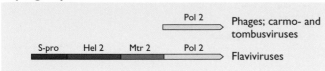

Supergroup 3

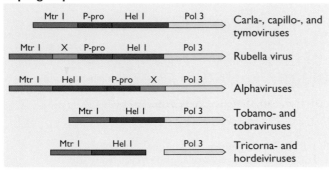

Figure 10.4 RNA virus genomes and evolution. Organization of (+) strand RNA genomes. The genomes of (+) strand RNA viruses comprise several genes for replicative functions that have been mixed and matched in selected combinations over time. These functions include a helicase (purple), a genome-linked protein (orange), a chymotrypsinlike protease (red), a polymerase (yellow), a papainlike protease (brown), a methyltransferase (dark blue), and a region of unknown function, X (green). Differences in the polymerase gene define the three supergroups. In this figure, the genes are not shown to scale and the structural proteins have been omitted for clarity. Derived from J. H. Strauss and E. G. Strauss, *Microbiol. Rev.* **58**:491–562, 1994, with permission.

viruses is enormous. The remarkable but unexplained variation among Murray Valley encephalitis virus isolates in New Guinea compared with those from Australia indicates that tropical Southeast Asia has a virus ecology worthy of detailed investigation.

Influenza virus. The ecology and biology of influenza have provided much food for thought about virus evolution. The same viral population can infect many different species. Each host species imposes new selections for replication and spread of the infection. As a result, the influenza virus gene pool is immense, with a dynamic ebb and flow

of genetic information as infection spreads among many different animals. Large-scale sequencing has provided a view of the state of the viral gene pool at various points in time and space, as infections are transmitted from human to animal, animal to human, and human to human. In one analysis alone, a consortium of scientists sequenced more than 200 human influenza virus genomes and collected almost 3 million bases of sequence. One salient finding was that a given influenza virus population in circulation contains multiple lineages at any time. In addition, alternative minor lineages exchange information with the dominant lineage. As selection pressures change, the numbers of distinct immune escape mutants rise and fall, as do the numbers of mutants with alterations in receptor-binding affinity. These studies even provide answers about gene function. A newly discovered open reading frame called PB1-F2 is preserved in almost all of the sequenced genomes, putting to rest the idea that it was not a functional gene.

Important clues to the epidemiology of influenza virus came from the sequencing and analysis of more than 1,300 influenza A virus isolates from various geographic locations. It was clear that the viral genome changes by frequent gene reassortment and occasional bottlenecks of strong selective "sweeps." More importantly, the study suggests that new antigenic subtypes have different dynamics but that all follow a classical "sink-source" model of viral ecology. In this model, antigenic variants emerge at intervals from a persisting reservoir in the tropics (the source) and spread to temperate regions, where they have only a transient existence before disappearing (the sink). Similar large-scale sequencing of avian influenza viruses is under way to expand our knowledge of the dynamic and rapid evolution of influenza virus genomes.

Measles virus. Sequence analyses indicates that measles virus, a human virus, is closely related to rinderpest virus, a bovine pathogen. It is thought to have evolved from a zoonotic ancestral rinderpest virus when humans first began to domesticate cattle. The best estimates indicate that measles virus is a relatively new human pathogen and probably became established in the Middle East about 5,000 years ago, when human populations began to congregate in cities. Measles virus spread around the world by colonization and migration, reaching the Americas in the 16th century (with disastrous effects on the native Americans).

The Protovirus Theory for Retroviruses

The origins of retroviruses may be more accessible than any other virus group. Their unique life cycle centers on the enzymes reverse transcriptase and integrase. These enzymes ensure that the RNA genome in virions is converted to a

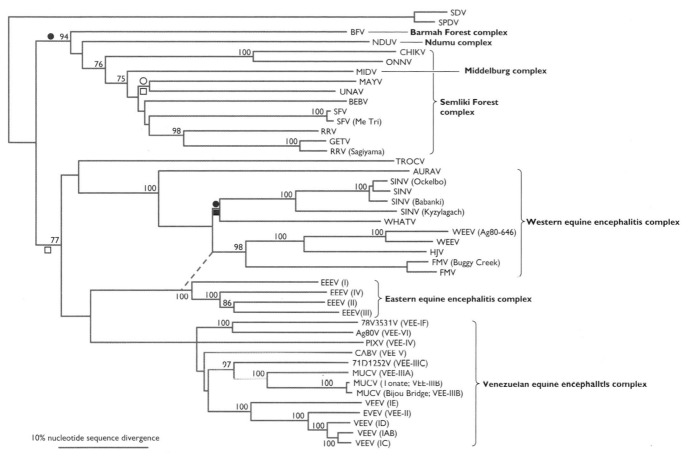

Figure 10.5 Dendrogram depicting a proposed phylogenetic tree of modern alphaviruses based on nucleic acid sequence comparisons of the E1 gene. (Top) Semliki Forest complex, **(middle)** western equine encephalitis complex; **(bottom)** Venezuelan equine encephalitis complex. The western equine encephalitis virus (WEEV) subgroup arose by a recombination event between a member of the eastern equine encephalitis virus (EEEV) lineage and the Sindbis virus lineage (dashed line). The bar indicates 10% nucleotide divergence. The open red circle adjacent to a branch indicates a hypothetical Old World-to-New World introduction, and the closed red circle indicates New World-to-Old World introduction, assuming a New World origin. The open red square indicates Old World-to-New World introduction, and the closed red square indicates New World-to-Old World introduction, assuming an Old World origin of the non-fish alphavirus clade. Redrawn from A. M. Powers et al., *J. Virol.* **75:**10118–10131, 2001, with permission.

DNA copy permanently integrated in the host DNA genome (the provirus). Howard Temin, who shared the Nobel Prize for the discovery of reverse transcriptase, first proposed the "protovirus theory" for the origin of this virus family. This theory posits that a cellular reverse transcriptase-like enzyme copied segments of cellular RNA into DNA molecules that were then inserted into the genome to form retroelements. These DNA segments in turn acquired more sequences, including those encoding integrase, ribonuclease H (RNase H) domains, regulatory sequences, and structural genes (Fig. 10.6). This theory predicts that evidence for this process might exist in the genomes of mammals and other species. Indeed, many of the predicted intermediates are found in abundance, including pseudogenes,

retrotransposons, and a variety of endogenous retroviruses (see Volume I, Chapter 7). In humans, such endogenous retroviruses are surprisingly abundant, comprising several percent of the human genome, and may represent footprints of ancient infection of germ cells.

Contemporary Virus Evolution

Although we cannot describe the origins of viruses, we should be able to define modern precursors of new ones by studying current virus ecology. Even with powerful technology, the task is daunting. It is probably safe to assume that every living thing is infected with viruses, and that we have only scratched the surface to identify them all. Nevertheless, studying the processes that result in the

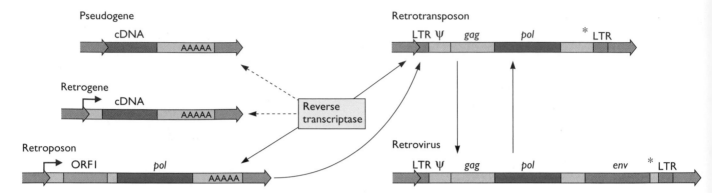

Figure 10.6 The action of a primordial reverse transcriptase may drive evolution of retroelements. A speculative scheme shows the evolution of various retroelements through the action of a primordial reverse transcriptase. Dashed arrows depict the emergence of genetic elements by reverse transcription; the complementary DNA copies are shown integrated into the host genome (gray). Solid arrows show the acquisition of new elements by recombination. cDNA, double-stranded DNA copied by reverse transcriptase; ORF, open reading frame; *gag,* capsid protein gene; *pol,* polymerase gene coding for reverse transcriptase, RNase H, integrase, and other enzymes; *env,* envelope protein gene; LTR, long terminal repeat; red arrow, promoter, initiation of transcription; AAAAA, poly(A) tail; ψ, packaging signal; *, polypurine tract. Direct repeats of cellular DNA are indicated in gray. Adapted from R. J. Loewer et al., *Proc. Natl. Acad. Sci. USA* **93:**5177–5184, 1996, with permission.

emergence of new viruses seems likely to prove more helpful in divining the present and future evolution of viruses than in explaining their origins.

As viruses are not likely to arise *de novo,* modern and future viruses must arise from progenitors that already exist. Hence, even human immunodeficiency virus, which appeared within the past few decades, has many relatives in nature and descended from one, or a combination, of them. Therefore, the sources of new viruses are strictly limited to two possibilities: a mutant virus already existing in an infected host can be selected, or a virus can enter a naive population from an entirely different infected species. These two simple possibilities notwithstanding, the interactions of host and virus required to establish a stable relationship are remarkably complex.

The Fundamental Properties of Viruses Constrain and Drive Evolution

We can recognize a herpesvirus, an adenovirus, a retrovirus, or an influenza virus genome by sequence analysis, despite many rounds of replication, mutation, and selection. This fact is emphasized by sequence analyses of human immunodeficiency virus strains isolated from patients around the world. As much as 10% of the total viral genome can vary from isolate to isolate, yet viral isolates fall into consistent subgroups called **clades.** Each clade differs from the others in amino acid sequence by at least 20% for Env proteins and 15% for Gag proteins. Differences within a clade can be as much as 8 to 10%, emphasizing the rather arbitrary delineation. As discussed in Chapter 6, these viruses have

replicated in widely dispersed geographic locations and have very different histories, yet each sequence is clearly recognizable as the human immunodeficiency virus genome.

The important message is that viral populations frequently maintain quite stable master or consensus sequences, despite opportunities for extreme variation (Fig. 10.1). Diversity exists—indeed is necessary for virus survival—but the consensus sequence remains, despite many years of selection and growth. How is stability maintained in the face of mutation, recombination, and selection? One answer is that all viruses share fundamental characteristics that define and constrain them. Those that can function within the constraints survive. Comprehending the evolution of viruses requires an understanding of these shared properties.

Constraining Viral Evolution

The very characteristics that enable us to define and classify viruses are the primary barriers to major genetic change; that is, extreme alterations in the viral consensus genome obviously do not survive selection. Certainly, some changes are simply impossible. One obvious constraint is the viral genome itself: DNA genomes cannot mutate to become RNA genomes and vice versa. Once a replication and expression strategy has evolved, there can be no turning back, because solutions to replication or the decoding of viral information are limited. Every step in viral replication requires interactions with host cell machinery. Consequently, any change in a viral component without a compensating change in the interacting host component may compromise replication. Similarly, inappropriate

DISCUSSION
A constraint on evolution? Selection for survival inside a host

A thought-provoking finding from studies of human immunodeficiency virus infections indicates that additional constraints on virus evolution must be considered. Virions that initiate infections typically are macrophage tropic and engage the CCr5 chemokine receptor. At the end stage of disease, the infected individual is producing billions of virions that survive in the face of host defenses and antiviral therapy. Invariably, these virions are T-cell tropic and engage the CXCr4 chemokine receptor. The diversity in this final population is a result of evolution **inside** a single individual.

Amazingly, when virions from end-stage disease infect a new host, the first replicating viral genomes that can be detected are macrophage tropic, and engage the CCr5 receptor. The progeny genomes have passed through a bottleneck, and only a few of the diverse variants are passed on. The processes that select these variants from the previous T-cell-tropic population are not well understood. However, one conclusion is clear: the virions that ultimately devastate the immune system after years of replication and selection within a host are not the most fit for infection of new hosts.

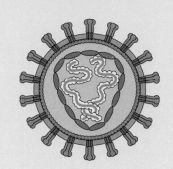

synthesis, concentration, or location of a viral component is likely to be detrimental.

A second constraint is the physical nature of the capsid required for transmission of the genome. For example, icosahedral capsids have a defined internal space that fixes the size of packaged nucleic acids. Once the genes encoding assembly of an icosahedral capsid are selected, genome size is essentially fixed; only very limited duplication or acquisition of sequences is allowed without compensating deletion of other sequences. A final constraint is that selection occurs during host-to-host spread of infection, as well as during spread of infection within a single individual (Box 10.11). All viral genomes encode products capable of modulating a broad spectrum of host defenses, including physical barriers to viral access and the vertebrate immune system. A mutant that is too efficient in bypassing host defenses will kill its host and suffer the same fate as one that does not replicate efficiently enough: it will be eliminated. These general constraints define the viruses we see today, as well as the further evolution of new viruses.

Finite Strategies To Replicate Viral Genomes

In Volume I, Chapter 4, we describe seven genome replication strategies that are likely to represent all possible solutions. We also outline a small number of expression strategies for protein production from these genomes. That the provenance of all viruses can be described by a short list of replication and expression strategies is extraordinary. Understanding how protein function and gene expression strategies evolved represents a new research frontier for which we have few data to guide us. Some initial studies have been provocative. For example, RNA

virus replication complexes described for different families have fundamental similarities (Box 10.12). Localization of genomes to membrane sites or to assembling capsids leads to precise temporal and spatial organization of viral compartments important for gene expression, replication, and particle assembly. Are these overtly similar mechanisms products of convergent evolution and coincidence, or do they imply a common evolutionary origin for this abundant group of viral genomes? One thought is that similar mechanisms were selected because they sequestered viral nucleic acid from the cytoplasmic intrinsic defense proteins such as RigI/Mda5, Pkr, and Tlr proteins.

Evolution of New Viruses

Even in the seemingly constrained context of a given virus, the number of all possible viable mutants is astronomical, if not inconceivable (Box 10.13). In fact, the number of possible mutants is so large that all the possibilities can never be tested in nature. Sequence comparisons of several viral RNA genomes have demonstrated that well over half of all nucleotides can accommodate mutations. This property means that, for a 10-kb viral RNA genome, more than $4^{5,000}$ sequence permutations define all possible mutants. If one considers deletions, recombination, and reassortment, the numbers become even larger. Considering that there are roughly 4^{135} atoms in the visible universe, this is a large number indeed. Even with the high rates of replication and mutation characteristic of viruses, we can be sure that only a minuscule fraction of all possible viral genomes have arisen since life began: even the most efficiently replicating virus will fail to spawn all the possible permutations for a run through the gauntlet

**BOX
10.12**

D I S C U S S I O N
*Parallels in replication of (+) strand and double-stranded
RNA genomes*

The mRNA templates of viruses with double-stranded RNA (dsRNA) and (+) strand RNA genomes (including retroviruses) are sequestered in a multisubunit protein core that directs synthesis of the RNA or DNA intermediate from which more viral mRNA is made. Similarities in how the mRNA template and core proteins are assembled suggest that all three

virus groups may share evolutionary history, despite a complete lack of genome sequence homology. It is possible that this ancient replicative strategy provides RNA genomes with increased template specificity and retention of negative-strand products in the core or vesicle for template use. In addition, by sequestering RNA in vesicles or capsids, host defenses

such as RNA interference, dsRNA-activated protein kinase, and RNase L are avoided.

Schwartz, M., J. Chen, J. Janda, M. Sullivan, J. den Boon, and P. Ahlquist. 2002. A positive-strand RNA virus replication complex parallels form and function of retrovirus capsids. *Mol. Cell* **9:**505–514.

Similarities in replication and budding reactions. In the case of retroviruses, specific sequences on the RNA genome bind to Gag proteins that define the budding site. Gag proteins encapsidate viral RNA and reverse transcriptase with plasma membrane. Similarly, (+) strand RNA genomes are replicated on intracellular membrane vesicles that form in response to a viral protein that binds to membranes. Polymerase complexes and viral RNA templates are recruited to these vesicles. Replication of dsRNA genomes occurs in compartments formed by assembling capsid proteins that sequester single-stranded genome templates via specific protein-RNA interactions. Blue circles are Gag proteins (retrovirus), 1A protein for (+) strand RNA virus, and inner capsid protein for dsRNA virus. The polymerase protein (Pol or 2APol) interacts with Gag or 1A, respectively. The polymerase is part of the assembling capsid of dsRNA viruses. The RNA genome is indicated in green, and the binding sites for interaction with Gag, 1A, or capsid protein are ψ, RE, or PS, as indicated for each virus. The final reaction for retroviruses is the release of an enveloped particle with the genome and polymerase; in the case of (+) strand RNA viruses, the product is not an enveloped virion but, rather, an involuted vesicle or the surface of a membrane vesicle where mRNA synthesis, (–) strand genome template synthesis, and (+) strand genome synthesis occur. In the case of dsRNA viruses, the product is a capsid compartment within which mRNA synthesis and complementary strand genome replication occur.

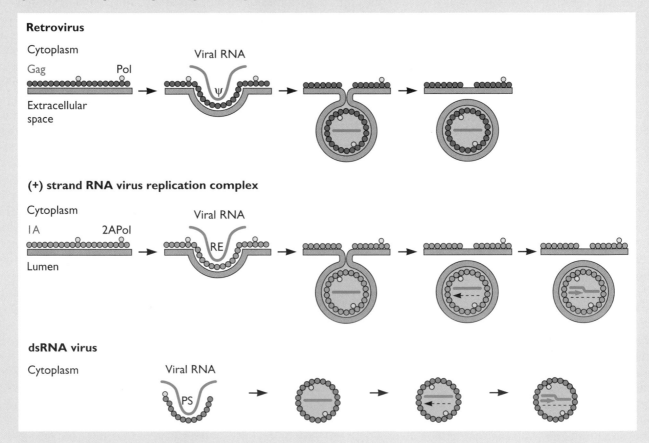

**BOX
10.13**

BACKGROUND
*The world's supply of human immunodeficiency virus genomes
provides remarkable opportunity for selection*

Tens of millions of humans are infected by human immunodeficiency virus. Before the end stage of disease, each infected individual produces billions of viral genomes per day. As a result, more than 10^{16} genomes are produced each day on the planet. Almost every genome has a mutation, and every infected human harbors viral genomes with multiple changes resulting from recombination and selection. Practically speaking, these large numbers provide an amazing pool of diversity. For example, mutants resistant to **every combination** of anti-reverse transcriptase and protease drugs in use, or in the pipeline, arise thousands of times each day, simply by chance.

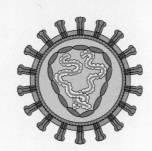

of selection in nature. The conclusions are inescapable: virus evolution is relentless, new mutants will always arise, and the possibilities are literally unimaginable. Those who seek to predict the trajectory of virus evolution face enormous challenges, as mutation rates are probabilistic and viral quasispecies are indeterminate. Nevertheless, we can be confident that all future viruses will arise from those now extant: they will be mutants, recombinants, and reassortants.

Emerging Viruses

As far as we know, humans have suffered for millions of years from infectious diseases. However, since the rise of agriculture (the past 11,000 years), new infectious agents have invaded human populations primarily because these infections (e.g., measles and smallpox) can be sustained only in large, dense populations that were unknown before agriculture and commerce. The source of these emerging infectious agents is a popular topic of research, debate, and concern.

We define an **emerging virus** as the causative agent of a new or hitherto unrecognized infection in a population. Occasionally, emerging infections are manifestations of expanded host range with an increase in disease that was not previously obvious. More generally, emerging infections of humans reflect transmission of a virus from a wild or domesticated animal with attendant human disease (**zoonotic infections**). Occasionally, a cross-species infection will establish a new virus in a population (e.g., human immunodeficiency virus moving from chimpanzees to humans). On the other hand, a similar cross-species infection will emerge in certain human populations, but the infection cannot be sustained (e.g., Ebola and Marburg viruses moving from bats to humans).

While the term "emerging virus" became part of the popular press in the 1990s (usually with dire implications ["killer viruses on the loose"]), emerging viruses are not

new to virologists, public health officials, and epidemiologists. These infections have long been recognized as an important manifestation of virus evolution. The most important factors driving the emergence of infectious diseases include unprecedented human population growth and large-scale change occurring in all ecosystems brought about by human occupation of almost every corner of the planet. The convergence of these factors drives viral emergence (Fig. 10.7). In recent years, emerging infections not only have been increasing in absolute frequency, but also are more easily detected because of advances in technology and better communication of disease outbreaks. Indeed, global communication has brought some emerging viral infections to center stage on the local news. Unfortunately, the lay public often know the name of the virus, but can not pinpoint the location of the outbreak on a map or articulate its significance. Anyone with access to television, radio, the Internet, or newspapers knows something about SARS, West Nile virus, Ebola virus, and certainly H5N1 avian influenza virus. Some examples of less well known emerging virus outbreaks are given in Table 10.3. Despite the many different viruses and geographical locations of these outbreaks, some common parameters do exist. These parameters define the rules of engagement for viruses and their potential hosts.

The Spectrum of Host-Virus Interactions

The spectrum of possible interactions among hosts and viruses may appear too complex for analysis. However, it is constructive to simplify the variables and consider only the general domains of these interactions. To illustrate this principle, we define four hypothetical interactions: **stable, evolving, dead-end,** and **resistant** (Fig. 10.8). These definitions are arbitrary snapshots of the extremes of dynamic host-virus interactions designed to emphasize essential concepts. The arrows suggest the hypothetical

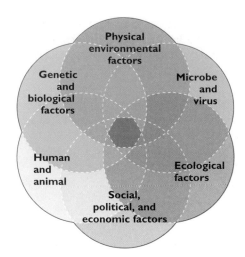

Figure 10.7 The convergence model for emerging viral infections. The provenance of six factors leads to convergence and the potential emergence of an infectious disease. The overlapping territories are obvious, but the dark center represents maximum convergence of these factors plus other unpredictable interactions. The point is that in the ecology of virus-host interactions, many factors are interlocking and interconnected. From the Institute of Medicine study *Microbial Threats to Health, Emergence, Detection and Response*, 2003.

transmission of infection; they stress the continuity of viral interactions in nature. In addition, it is important to understand that these definitions apply to large populations and **not** to a single virus-host interaction. In this hypothetical set of interactions, emerging viral infections are defined as human infections that derive from stable host-virus interactions preexisting in nonhuman hosts.

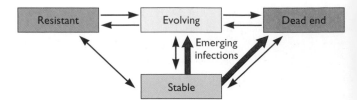

Figure 10.8 The general interactions of hosts and viruses. Four hypothetical host-virus interactions are indicated in the boxes. The **stable interaction** maintains the virus in the ecosystem. The **evolving interaction** describes the passage of a virus from "experienced" populations to naive populations in the same or other host species. The **dead-end interaction** represents one-way passage of a virus to different species. The host usually dies, or if it survives, the virus is not transmitted efficiently to the new host species. The **resistant host interaction** represents situations in which the host completely blocks infection. The arrows indicate possible transmission of infection from one situation to another or the possible transformation of one interaction into another. The red filled arrows indicate the sources of zoonotic (emerging) infections.

Stable Interactions

Stable host-virus interactions are those in which both participants survive and multiply. Such interactions are essential for the continued existence of the virus, and may influence host survival as well. This state is optimal for a host-parasite relationship, but the interaction need be neither benign nor permanent in an outbred population. Infected individuals can become ill, recover, develop immunity, or die, yet in the long run, both populations survive. While this situation is often described as an equilibrium, the term is misleading. The interactions

Table 10.3 Some examples of viruses that cause zoonotic infections

Virus	Family	Emergence factors
Dengue virus	*Flaviviridae*	Urban population density; open water storage favors mosquito breeding (e.g., millions of used tires)
Ebola virus	*Filoviridae*	Human contact with unknown natural host in Africa; importation of monkeys in Europe and the United States
Hantaan virus	*Bunyaviridae*	Human contact with rodents as a result of agricultural techniques
Human immunodeficiency virus	*Retroviridae*	Transfusions and blood products; sexual transmission; needle transfer during drug abuse
Human T-lymphotropic virus	*Retroviridae*	Transfusions and blood products; contaminated needles; social factors
Influenza virus	*Orthomyxoviridae*	Integrated pig-duck agriculture; mobile population
Junin virus	*Arenaviridae*	Agriculture techniques favor human contact with rodents
Noroviruses	*Caliciviridae*	New methods for detection; infectious diarrhea
Machupo virus	*Arenaviridae*	Agriculture techniques favor human contact with rodents
Marburg virus	*Filoviridae*	Unknown; importation of monkeys in Europe
Rift Valley virus	*Bunyaviridae*	Dams, irrigation
Sin Nombre virus	*Bunyaviridae*	Natural increase of deer mice and subsequent human/rodent contact
West Nile virus	*Flaviviridae*	Unknown introduction into United States

are dynamic and fragile, and certainly are rarely reversible. Viral populations may come to be more or less virulent if such a change enables them to be maintained in the population, while hosts may evolve mechanisms that attenuate the more debilitating effects of the viruses that infect them.

Some stable interactions are effectively permanent. For example, humans are the sole natural host for a several viruses, including measles virus, herpes simplex virus, human cytomegalovirus, and smallpox virus. Similarly, simian cytomegalovirus, monkeypox virus, and simian immunodeficiency virus infect only certain species of monkeys. Stable interactions can also include infection of more than one host species with the same virus. For example, influenza A virus, flaviviruses, and togaviruses, are capable of propagating in a variety of species. Indeed, many of the flaviviruses and togaviruses replicate efficiently in some insects as well as in mammals and birds. In these instances, the host that maintains the virus may not be apparent or obvious. Influenza A virus infects humans, wild birds, and pigs, but birds may be its natural host. This conclusion is discussed below.

Establishment of a stable host-virus interaction is not necessarily the optimal solution for survival. The trajectory of evolution is unpredictable: what is successful today may be suicidal at another time. Once a virus population becomes completely dependent on one, and only one, host, it has entered a potential bottleneck that may constrain its further evolution. If the host becomes extinct for whatever reason, the virus is also likely to be exterminated. If humans disappeared, many virus populations, including poliovirus, measles virus, and several herpesviruses, would cease to exist. Eradication of natural smallpox virus was possible because humans are the only hosts and worldwide immunization was achieved.

The Evolving Host-Virus Relationship

This relationship emphasizes that stability is unlikely to be established instantly (Fig. 10.9). We highlight this interaction to illuminate the dynamic consequences of a virus spreading among populations of the same or perhaps closely related species. The hallmarks of this interaction are instability and unpredictability. These properties are to be expected, as selective forces are applied to both host and virus, and are magnified when host populations are small. For example, some host subpopulations may experience high infection rates, while others are unaffected. The outcome of infection may range from relatively benign symptoms to death. Such an interaction is typified by the introduction of smallpox and measles to natives of the Americas by Old World colonists and slave traders. Europeans previously had experienced the same horror when these diseases spread to Europe from Asia. Other opportunities to enter the evolving host-virus interaction may arise if the virus in a stable interaction acquires a new property that increases its virulence or spread, or if the host population suffers a far-reaching catastrophe that reduces resistance (e.g., famine or mass population changes during wars). The introduction of West Nile virus into the Western Hemisphere in 1999 provides a contemporary example of an evolving host-virus interaction (Box 10.14).

Figure 10.9 Sources of three well-known emerging infections. (Left) SARS human coronavirus (SARS-CoV) is endemic in bats. Bats can spread infection to certain wild mammals that can, in turn, spread infection to humans. **(Right)** The progenitors of human immunodeficiency virus types 1 and 2 cause natural infections of chimpanzees and sooty mangabeys, respectively.

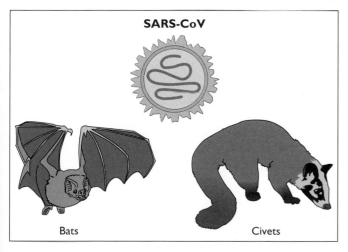

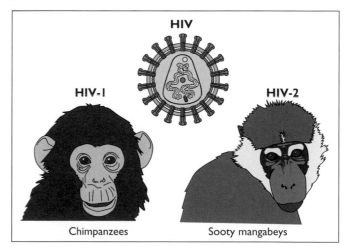

BOX
10.14

DISCUSSION
An evolving virus infection: the West Nile virus outbreak

In 1999, virologists had the opportunity to observe a virus population establish a new geographic niche. The West Nile virus, an Old World flavivirus discovered in 1937 in the West Nile district of Uganda, had never been isolated in the Western Hemisphere. In August 1999, six people were admitted to Flushing Hospital in Queens, NY, with similar symptoms of high fever, altered mental status, and headache. They were subsequently discovered to be infected with West Nile virus. The virus has now spread from the Atlantic to the Pacific, as well as to Canadian provinces and territories. In the summer of 2002, it reached epidemic status, causing encephalitis in hundreds of cases.

The New York isolate of West Nile virus is nearly identical to a virus isolated in 1998 from a domestic goose in Israel during an outbreak of the disease. The close relationship between these two isolates suggests that the virus was brought to New York City from Israel in the summer of 1999. How it crossed the Atlantic will probably never be known for sure, but it might have been via an infected bird, mosquito, human, horse, or other vertebrate host. These events mark the first introduction in recent history of an Old World flavivirus into the New World. A fascinating, and yet unanswered, question is why the infection was established in New York City. The summer of 1999 was particularly

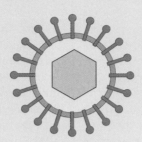

hot and dry. Similar conditions spawn outbreaks of West Nile virus encephalitis in Africa, the Middle East, and the Mediterranean basin of Europe. Such conditions may promote mosquito breeding in polluted, standing water.

The infection spreads via many species of mosquitoes and is now circulating in wild birds. The virus or virus-specific antibodies have been found in more than 157 species of birds, but crows and jays appear to be particularly sensitive. Many zoos are reporting deaths of their exotic birds from West Nile virus infections. The virus has been found to infect at least 37 kinds of mosquitoes and 18 other vertebrates.

Humans and other animals acquire infections by mosquito bites after the insect has fed on infected birds. Human infections may be spread by transfusion from an infected donor, a possibility with far-reaching implications for our blood supply. Horses develop a lethal encephalitis, hundreds of cases of which have been

reported. A vaccine for horses is available.

About 20% of infected humans experience flu-like symptoms, but only 1 in 150 of these individuals develop meningitis, encephalitis, or poliomyelitis-like symptoms. West Nile virus infection claimed a total of 564 lives in the United States in the 5 years from 1999 to 2003. In 2007, the Centers for Disease Control and Prevention reported 906 cases of West Nile virus infection with 26 fatalities.

By 2007, the North American epidemic was resolving as the virus became established. Sequence analysis indicates that viral populations were more diverse at the start of the epidemic than they are now. Until we understand the complex ecology of this viral infection, the consequences for public health are difficult to predict.

For an update on West Nile Virus, see http://www.cdc.gov/ncidod/dvbid/westnile/index.htm.

Brinton, M. A. 2002. The molecular biology of West Nile virus: a new invader of the Western Hemisphere. *Annu. Rev. Microbiol.* **56:**371–402.

Briese, T., and K. Bernard. 2005. West Nile virus—an old virus learning new tricks? *J. Neurovirol.* **11:**469–475.

Despommier, D. 2001. *West Nile Story.* Apple Trees Productions LLC, New York, NY.

Snapinn, K. W., E. C. Holmes, D. S. Young, K. A. Bernard, L. D. Kramer, and G. D. Ebel. 2007. Declining growth rate of West Nile virus in North America. *J. Virol.* **81:**2531–2534.

The Dead-End Interaction

A commonly encountered host-virus relationship is described rather vividly as a **dead-end interaction**. It has much in common with the evolving host-virus interaction, as both represent departures from a stable relationship, often with lethal consequences. A dead-end interaction is a frequent outcome of cross-species infection (but not of intraspecies infections). A **zoonosis** is a disease or infection that is naturally transmitted between vertebrate animals and humans. Unanticipated viral zoonoses often are classic cases of emerging viral infections. In many cases, the host is killed so quickly that there is little or no subsequent transmission of the virus to others. In other cases, the newly infected host is incapable of transmitting the infection to other individuals of the same species

The dead-end interaction is often observed with the many viruses carried by arthropods like ticks and mosquitoes that cycle in the wild between insects and a vertebrate host in a stable relationship. Occasionally the infected insect bites a new species (e.g., humans) and transmits the virus (Fig. 10.10). As a consequence, the infection results in severe pathogenic effects in the infected human, but is not transmitted further; i.e., it reaches a dead end. As the human is not part of the natural, stable host-virus relationship, these infections have little if any effect on the evolution of the virus and its natural host. To a first approximation, dead-end hosts are transparent participants. However, such logic may be too simplistic; infection by a less virulent mutant or infection of a more resistant individual may be the first step in establishing a new host-virus interaction.

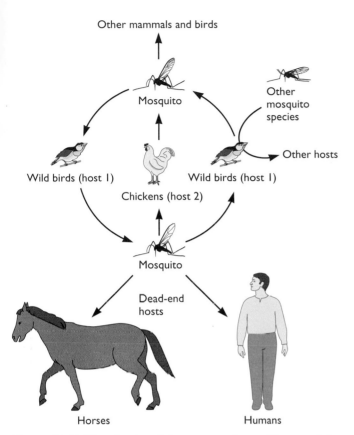

Figure 10.10 The dead-end host scenario as illustrated by a complex host-virus relationship. This arbovirus infection illustrates how multiple host species can maintain and transmit a virus in the ecosystem. In this example, the virus population is maintained in two different hosts (wild birds and domestic chickens) and is spread among individuals by a mosquito vector. The virus replicates in both species of bird and in the mosquito. Disease is likely to be nonexistent or mild in these species, as the hosts have adapted to the infection. A third host (in this example, horses or humans) occasionally is infected when bitten by a mosquito that previously fed on an infected bird. Horses and humans are dead-end hosts and contribute little to the spread of the natural infection, but they may suffer from serious, life-threatening disease. Occasionally another species of biting insect (e.g., other mosquito species) can feed on an infected individual (bird, horse, or human) and then transmit the infection to another species not targeted by the original mosquito vector.

Human diseases like yellow fever and dengue fever provide excellent examples of the dead-end interaction (Box 10.15). The viruses causing these diseases are endemic in the tropics and maintain a stable relationship with their natural host and insect vectors. When humans begin to develop a tropical area by clearing forests and building roads, dams, canals, and towns, they are at risk for being bitten by mosquitoes and other insects. The natural host-virus interaction that existed before the intrusion changes, and humans may experience new viral infections spread by insect vectors. A similar situation exists for some flaviviruses transmitted by ticks, which in turn are carried by rodents. Humans often experience European tick-borne encephalitis as dead-end hosts (Fig. 10.11).

The lethality of the Marburg and Ebola filoviruses in humans is typical of dead-end host infection. Filoviruses are single-stranded, (−) RNA viruses causing severe hemorrhagic manifestations in infected humans (Appendix A). Disease onset is sudden, with 25 to 90% fatality rates reported. Virus spreads through the blood and replicates in many organs, causing focal necrosis of the liver, kidneys, lymphatic organs, ovaries, and testes. Capillary leakage, shock, and acute respiratory disorders are observed in fatal cases. Patients usually die rapidly of intractable shock without evidence of an effective immune response. Even when recovery is under way, survivors do not have detectable neutralizing antibodies. The infection clearly overwhelms a particular individual, but apparently does not spread widely; human infections tend to cluster in local areas. Such viruses can be transmitted to other humans only by close personal contact with infected blood and tissue. Although the natural hosts for many filoviruses remain elusive, these viruses must have established a stable interaction with an animal host, and infect humans only inadvertently. Some investigators have argued that the natural host is responsible for the apparent "geographic containment" of Ebola virus in central Africa. Humans are not the only dead-end hosts for Ebola virus: gorillas are susceptible, and large numbers have died from Ebola virus infection.

Many animal models of disease might be considered examples of dead-end interactions. For example, herpes simplex virus is a human virus, but when it is introduced into mice, rabbits, or guinea pigs in the laboratory, these animals become infected and show pathogenic effects that mimic some aspects of the human disease. However, in their natural environment, these animals contribute nothing to the transmission or survival of the virus, and therefore provide experimental models of the dead-end interaction.

Rodent vectors play critical roles in the introduction of new viruses into populations in areas where these animals abound. Most hemorrhagic disease viruses, including Lassa, Junin, and Sin Nombre viruses, are endemic in rodents, their natural hosts. The viruses establish a persistent infection, and the rodents show few if any ill effects. Substantial numbers of virus particles are excreted in urine, saliva, and feces to maintain the virus in the rodent population. However, infection by such rodent viruses can cause local, often lethal, outbreaks in humans as dead-end hosts. Humans become infected only because they happen to come in contact with rodent excretions containing infectious virus particles.

DISCUSSION
Yellow fever virus: humans change the pattern and pay the price

In tropical forests, yellow fever virus is maintained by a monkey-mosquito cycle. Neither the monkey nor the mosquito is the worse for wear. Various mosquito species serve as vectors, including *Aedes* species in Africa and *Haemagogus* species in the Americas. If humans blunder into mosquito-infected areas, they stand a chance of being bitten by the infected insects and contracting the disease. Yellow fever came to the New World with the colonists and the slave trade, where it wreaked havoc among the indigenous human populations.

Humans were not the only species to suffer from the invasion. New World monkeys also died when infected with yellow fever virus, indicating that they were as unequipped to handle this new infection as were the indigenous humans.

Yellow fever virus and its mosquito vector spread by ship to the burgeoning populations of the U.S. South and East Coast with ease. Cities hit hard were New Orleans; Charleston, SC; Philadelphia; New York; and Boston. In 1800 Thomas Jefferson lamented, "Yellow Fever will discourage the growth of great cities in our nation." In a striking example of an emerging disease, 15% of the population of Philadelphia died of yellow fever in 1793. Neighboring New Jersey and Maryland attempted to bar panicked Philadelphians from entering their states.

The contribution of mosquitoes to the transmission of yellow fever was first glimpsed by Carlos Finlay in Cuba in 1880 and was established firmly by Walter Reed in 1900 when the disease was a problem during construction of the Panama Canal in the Central American jungle. A vaccine was developed in 1937 to contain the disease. Even with the vaccine, more than 10,000 people die of yellow fever every year in South America.

Yellow fever deaths in Philadelphia in the summer and early fall of 1793. From N. Nathanson, *ASM News* **63**:83–88, 1997, with permission.

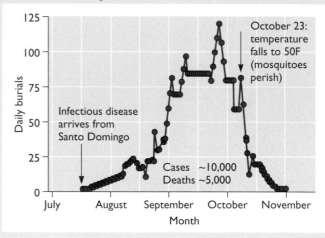

Bats are vectors of several dead-end, zoonotic infections (Fig. 10.9 and 10.12). Hendra virus, Nipah virus, rabies virus, and the progenitor of the SARS coronavirus are known to enter the human population via bats. The Old World fruit bats (genus *Pteropus*), commonly called flying foxes, are widely distributed in southeast Asia, Australia, and the Indian subcontinent. These fruit bats naturally harbor Hendra and Nipah viruses. Despite having high antibody titers, the animals exhibit no obvious disease. High virulence in humans and our complete lack of therapeutic interventions require that these viruses be studied only under the highest biological and physical containment (biosafety level 4 [BSL4]). Accordingly, we know very little about their biology and pathogenesis. In addition, we know next to nothing about fruit bat ecology and how humans and livestock become infected. Considerably more effort is required in the field before we can begin to understand the epidemiology of many zoonotic infections that produce the dead-end host scenario.

The Resistant Host

All living things are exposed continuously to viruses of all types, yet the vast majority of these interactions are uneventful. This may be because the host cells are not susceptible, not permissive, or both. A more likely possibility is that the primary physical, intrinsic, and innate defenses are so strong that potential invaders are diverted or destroyed upon contact. In other cases, organisms may become infected and produce some virions, but the infection is cleared rapidly without activation of the host's acquired immune system (Chapter 5). This outcome is in contrast to an inapparent infection, in which an immune response is mounted but the individual exhibits no signs of disease.

We include the resistant host in our analysis because the transition between the inability and ability of a host to support viral replication need not be insurmountable. Indeed, **xenotransplantation** (the use of animal organs in humans) is thought provoking. Not only does transplantation bypass physical and innate defenses by surgery,

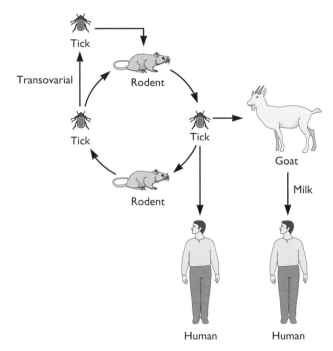

Fruit bats (flying foxes): a source of zoonotic infections

- Four previously unknown paramyxoviruses isolated from flying foxes since 1995, including Nipah and Hendra viruses

- Three cause severe disease in domestic animals (horses and pigs) and are known to infect humans

Figure 10.11 Replicative cycle of the central European tick-borne flavivirus may involve zoonotic infections. European tick-borne flavivirus infection is maintained and spread by multiple host infections. Congenital transmission in the tick maintains the virus in the tick population as they feed upon rodents. The infected newborn ticks have adapted to the infection and thrive. The virus is transmitted from the tick to a variety of animals, including cows, goats, and humans. Humans can also be infected by drinking milk from an infected goat, sheep, or cow. This zoonotic infection is another example of the dead end host interaction.

Figure 10.12 Emergence of Nipah and Hendra viruses from natural infections of fruit bats. All of the *Pteropus* species are considered flying foxes. *Pteropus vampyrus* is the Malayan flying fox (found in peninsular Malaysia), and *Pteropus conspicillatus* is the spectacled flying fox (found in far northern Queensland, Australia, and Papua New Guinea). *P. vampyrus* is one of the species that carries Nipah virus, and *P. conspicillatus* carries Hendra virus. Bat photograph courtesy of Juliet Pulliam, Princeton University. See B. Eaton et al., *Nat. Rev. Microbiol.* **4:**23–35, 2006.

but also drugs suppress the immune response. As a result, virus particles or genomes in xenografts have direct access to the once-resistant host in the absence of crucial antiviral defenses. As many of these viruses can infect human cells or have close, human-adapted relatives, the xenotransplantation patient represents a source of new viral diversity. The outcome of such an experiment in viral evolution is not predictable, and the experiment should not be attempted.

Encountering New Hosts: Fundamental Problems in Ecology

All virus-host interactions are governed by the concentrations of the participants and the probability of productive encounters. Rare chance encounters of viruses with new hosts may give rise to infections that are never seen, or at least never appreciated. These rare single-host infections may not be transmitted among humans for any number of reasons including insufficient quantity of progeny virus shed, limited duration of shedding, and small numbers of new human hosts exposed to the infected individual. In addition, the progeny virus produced in the new host may not have the genetic repertoire to facilitate high levels of replication and transmission.

Several ecological and social parameters facilitate the transmission of infection to new hosts in natural populations (Tables 10.4 and 10.5). Living together and sharing resources facilitates inter- and intraspecies transmission. Droughts concentrate many species at water holes; destruction of habitat forces new interactions. Predators eat their prey and become unwitting "test tubes" for cross-species infection by viruses found in tissues of the prey, no matter

Table 10.4 Human actions can promote large-scale changes in virus ecology

Air travel

Dams and water impoundments

Irrigation

Rerouting of wildlife migration patterns

Wildlife parks

Hot tubs

Air conditioning

Blood transfusion

Xenotransplants

Long-distance transport of livestock and birds

Moral and societal changes with regard to drug abuse and sex

Massive deforestation

Millions of used tires

Uncontrolled urbanization

Day care centers

Table 10.5 Ecological and social parameters facilitate transmission of infection to new hosts

Transmission parameter	Action or example
Contact with bodily fluids of infected hosts	Predation, hunting, and consumption of wild game; intimate contact with infected animals in the wild, at zoos, or in the home
Sharing of a resource with different species	Infected fruit bats, pigs, and humans share the same space
Sharing of insect or rodent vectors	Japanese encephalitis virus infection is spread by mosquitoes that feed on herons, people, and pigs
Encroachment by one species into the habitat of another	Humans enter the jungle and are bitten by mosquitoes that are part of a virus-bird infectious cycle

if aerosols, excretion, or close mucosal contact normally transmits the infection.

Successful Encounters Requires Access to Susceptible and Permissive Cells

Potential new hosts must have cells with accessible receptors that can engage virion ligands. The influenza virus hemagglutinin protein has a high affinity for sialosaccharides found on the cell surfaces of many different host species. The linkage of the terminal sialic acid/galactose residues plays a crucial role in tropism. Avian influenza virions bind sialic acid $\alpha(2,3)$-galactose-terminated oligosaccharides, whereas the human influenza virus hemagglutinin proteins bind tightly to oligosaccharides carrying a terminal $\alpha(2,6)$-linked galactose. Cells of the human respiratory tract do display the $\alpha(2,3)$-galactose-terminated oligosaccharides, but they lie deep in respiratory tissues. Conversely, sialic acid with terminal $\alpha(2,6)$ linkages abounds in the more accessible regions of the upper respiratory tract. This anatomical fact appears to be a prime reason why humans cannot be infected easily with avian influenza viruses.

Nipah virus was first identified during an outbreak in swine and humans in Malaysia in 1998 and 1999 (Fig. 10.12). Nipah virus infection of bats apparently is nonpathogenic, but copious quantities of virions are secreted in urine and feces. Two salient facts are that pig farmers often plant mangoes and durian trees next to pig pens, and fruit bats are messy eaters. When pigs come in contact with partially eaten contaminated fruit, they suffer a respiratory disease, and efficiently spread virions in the environment by sneezing and via mucous secretions.

For reasons that only now are becoming clear, bats and pigs established a one-way conduit for a zoonotic infection of humans. In rural Indonesian communities, humans often share accommodation with domestic swine, facilitating zoonotic infection. In addition, slaughterhouse workers are exposed to infected animals. Remarkably, when Nipah virus infects humans, it causes encephalitis, not respiratory infection. While often lethal for the infected human (Nipah virus killed 105 of 265 infected people in the Malaysian outbreak mentioned above), the infection is contained in infected brain tissue and does not spread. This state of affairs is apparently changing: Nipah virus isolated in India can infect the human upper respiratory tract, and these strains spread efficiently among humans in close contact.

Population Density and Health Are Important Factors

Two predominant parameters influencing the spread of infection are the population density and the health of individuals in that population. Close personal contact, either by direct methods (e.g., aerosols or sexual contact) or by indirect exposure (e.g., water or sewage), is also required. Variables such as duration of immunity and the quantity of virions produced and shed from each individual have marked effects on spread of infection. As discussed in Chapter 5, at least half a million people in a more or less confined urban setting are required to ensure a large enough annual supply of susceptible hosts to maintain measles virus in a human population. When this large population of interacting hosts is not available, measles virus dies out. One can reduce the effective population by splitting the half million hosts into small groups physically separated from social discourse, as exemplified by quarantine or the use of sanitariums to isolate infected patients. Another way is by immunizing the group such that the large majority are immune and cannot propagate the virus. If these or similar actions are not taken, our ever-expanding and increasingly interactive human population will ensure the maintenance and continued evolution of measles virus.

The age distribution of any potential host population is also important. For example, babies and the elderly are commonly more susceptible to a given virus than is the general population and, consequently, serve as sources of transmission. Predictably, prevention of infection in these groups tends to reduce the overall infection rate in the population at large. The distribution of poor and wealthy individuals in a population can influence infection rates. Malnourished individuals are more susceptible to disease than those who are well fed.

The age, health, and genetic variation in any given outbred host population contribute to the unpredictable nature of viral infection. Perhaps not so obvious are the other variables that modulate infection, notably, seasonal variations. Respiratory infections caused by adenoviruses and rhinoviruses often occur in the spring, but respiratory infections caused by coronaviruses and respiratory syncytial viruses tend to occur in the winter. Most arbovirus infections are experienced in the summer. While it is clear that insect-mediated infections are not likely to occur when temperatures drop below freezing for months at a time, the seasonal variations for other viral diseases are not easy to explain.

The Need for Experimental Analysis of Host-Virus Interactions

The definitions of different host-virus interactions listed above highlight some of the problems of coming to grips with virus evolution in natural populations. How do we do experiments and test hypotheses? Unfortunately, such experiments are complicated by a variety of issues, including safety, cost, and social and political issues, as well as by the uncertainty of the relevance of laboratory models to natural disease and virus spread. Natural infections occur in outbred populations in complicated settings with unknown ecological parameters, features that are difficult to model in the laboratory. To understand the mechanics of viral evolution, we must be able to quantify the relationship between viral virulence (ability to cause disease) and the rate of transmission. We need facts and figures to determine the importance of interspecies transmission in the establishment of new viruses. More quantitative and qualitative data are required to understand the variables that maintain quasispecies. Certainly, if we are to understand viral evolution and emerging viral infections, we must be knowledgeable about basic viral ecology.

We understand in principle that most virus populations survive in nature only because of **serial infections** among individuals (a chain of transmission). Quantitative measures of these interactions, determination of the molecular mechanisms responsible for them, and development of model systems that can predict them are sorely needed. Mathematical models are being developed to describe patterns of disease transmission in complex groups. Such models should help to determine the critical population size necessary to support the continual transmission of viruses with differing incubation periods, and the dynamics of persistent viral infections. The value of such models remains to be seen, simply because it is difficult to perform a controlled experiment in nature.

Learning from Accidental Natural Infections

Our understanding of the dynamics of a viral infection in a large outbred human population is rudimentary, at best. What we do know is based predominantly on a limited number of accidental "experiments." Two classic examples are provided by hepatitis B virus and poliovirus infections. During World War II, large doses of infectious hepatitis B virus were accidentally introduced into approximately 45,000 soldiers when they were injected with a contaminated yellow fever vaccine. Surprisingly, only 900 (2%) came down with clinical hepatitis, and fewer than 36 developed severe disease. Similarly, in 1955, 120,000 school-aged children were vaccinated with an improperly inactivated poliovirus vaccine. About half had preexisting antibodies to poliovirus thanks to inapparent infections by wild virus. Of the remainder, about 10 to 25% were infected by the vaccine, as determined by the appearance of antibodies. More than 60 cases of paralytic poliomyelitis were documented among these infected children, and the remainder escaped disease. These two experiments tell us that even when a large number of individuals are infected with a virulent virus, the outcome cannot be predicted; we have only a rudimentary understanding of why this is so. One of the more classic cases of deliberate release of a virus and the resulting effects on a wild host population is the attempt to use viral infection to rid Australia of rabbits (Box 10.16).

Expanding Viral Niches: Snapshots of Selected Emerging Viruses

Poliomyelitis: a Disease of Modern Sanitation

Host populations change with time, and each change can have unpredictable effects on virus evolution. Analysis of poliomyelitis, a disease caused by poliovirus infection, provides an instructive example. The disease is ancient, postulated by some to be present over 4,000 years ago (see Volume I, Chapter 1). For centuries, the host-virus relationship was stable, and infection was endemic in the human population. Poliomyelitis epidemics were unheard of (or, at least, not written about), but we imagine that occasional bouts of disease were obvious in scattered areas. This state of affairs changed radically in the first half of the 20th century, when large annual outbreaks of poliomyelitis appeared in Europe, North America, and Australia. Retrospective analysis established that the viral genome did not change or evolve substantially (Fig. 10.13A). How can the emergence of epidemic poliomyelitis be explained?

The answer is that humans changed their lifestyle on an unprecedented scale. Poliomyelitis is caused by an enteric virus spread by oral-fecal contact. As a consequence, endemic disease was characteristic of life in rural communities, which generally had poor sanitation and small populations. Because virions circulated freely, most children were infected at an early age and developed antibodies to at least one of the serotypes. Maternal

EXPERIMENTS

*A classic experiment in virus evolution: deliberate release
of rabbitpox virus in Australia*

In 1859, 24 European rabbits were introduced into Australia for hunting, and, lacking natural predators, the friendly rabbits reproduced to plague proportions. In 1907, the longest unbroken fence in the world (1,139 miles long) was built to protect portions of the country from invading rabbits. Such heroic actions were to no avail. As a last resort, the rabbitpox virus, myxoma virus, was released in Australia in the 1950s in an attempt to rid the continent of rabbits. The natural hosts of myxoma virus are the cottontail rabbit, the brush rabbit of California, and the tropical forest rabbit of Central and South America. The infection is spread by mosquitoes, and infected rabbits develop superficial warts on their ears. However, European rabbits are a different species and are killed rapidly by myxoma virus. In fact, the infection is 90 to 99% fatal!

In the first year, the infection was amazingly efficient in killing rabbits, with a 99.8% mortality rate. However, by the second year the mortality dropped dramatically to 25%. In subsequent years, the rate of killing was lower than the reproductive rate of the rabbits, and hopes for 100% eradication were dashed. Careful epidemiological analysis of this artificial epidemic provided important information about the evolution of viruses and hosts.

The infection spread rapidly during spring and summer, when mosquitoes are abundant, but slowly in winter, as expected. Given the large numbers of rabbits and virus particles, and the almost 100% lethal nature of the infection, advantageous mutations were quickly selected. Within 3 years, less-virulent viruses appeared, as did rabbits that survived the infection. The host-virus interaction observed was that predicted for an evolving host coming to an equilibrium with the pathogen. A balance is struck: some infected rabbits die, but they die more slowly, and many rabbits survive.

What was learned? To paraphrase the Rolling Stones: you always get what you select, but you don't often get what you want. Probably the most obvious lesson was that the original idea to eliminate rabbits with a lethal viral infection was flawed. Powerful selective forces that could not be controlled or anticipated were at work.

In this experiment, the viral genomes acquired mutations resulting in an attenuated infection: fewer rabbits were killed, infected rabbits were able to survive over the winter, and in the spring mosquitoes spread the infection. Moreover, rabbits that were more resistant to, or tolerant of, the infection were selected.

Surprisingly, more experiments in the biological control of rabbits are under way in Australia. One line of experimentation uses a lethal rabbit calicivirus, while another employs a genetically engineered myxoma virus designed to sterilize, but not kill, rabbits. The latter viruses encode a rabbit zona pellucida protein, and infected rabbits synthesize antibodies against their own eggs (so-called immunocontraception).

antibodies, which protect newborns, were also prevalent, as most mothers had experienced a poliovirus infection at least once. An important consideration is that most infected children do not develop paralysis, the most visible symptom of poliomyelitis. Paralysis is a more frequent result when older individuals are infected. Even the most virulent strains of poliovirus cause 100 to 200 subclinical infections (inapparent infections) for every case of poliomyelitis. These inapparent infections in children provided a form of natural vaccination. As childhood disease and congenital malformations were not uncommon in rural populations, the few individuals who developed poliomyelitis were not seen as out of the ordinary. No one noticed endemic poliovirus.

However, during the 19th and 20th centuries, industrialization and urbanization changed the pattern of poliovirus transmission. Improved sanitation broke the normal pattern and effectively stopped natural vaccination. In addition, dense populations and increased travel provided new opportunities for rapid spread of infection. As a result, children tended to encounter the virus for the first time at a later age, without the protection of maternal antibodies, and were at far greater risk for developing paralytic disease. Consequently, epidemic poliovirus

A

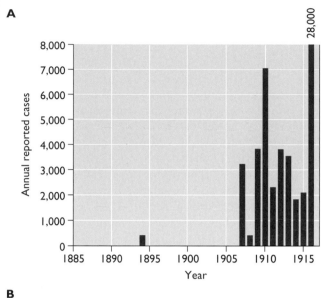

B

Figure 10.13 Poliovirus in the early 20th century. (A) The emergence of paralytic poliomyelitis in the United States, 1885 to 1915. From N. Nathanson, *ASM News* **63**:83–88, 1997, with permission. **(B)** Board of Health quarantine notice, San Francisco, CA, circa 1910.

infections emerged time and time again in communities across Europe, North America, and Australia. Until vaccines became available, quarantine was the only public health defense (Fig. 10.13B).

Widespread use of inexpensive, effective poliovirus vaccines has since controlled the epidemic. As the virus has no host other than humans, it should be possible to eradicate it by vaccinating sufficient people to end the spread of the virus. Accordingly, the World Health Organization targeted the eradication of poliovirus by 2005 with a massive worldwide vaccination program. The goal was not achieved because of social, religious, and political variables that are difficult if not impossible to control. Poverty, social problems, and economic conditions often conspire to prevent vaccination of children in inner cities and in poorer countries. There is hope that poliomyelitis will soon be a disease of the past, as a consequence of the worldwide poliovirus eradication program. However, given that it may be impossible to eliminate all sources of the virus, vaccination may be part of public health programs indefinitely.

Smallpox and Measles: Diseases of Exploration and Colonization

Explosive epidemic spread may occur when a virus enters a naive population (**the evolving host-virus interaction**) (Boxes 10.15 and 10.16). The resulting infections can be frightening, often devastating, and appear to "come out of the blue." Charles Darwin was aware of this phenomenon, as he wrote in *The Voyage of the Beagle*: "Wherever the European has trod, death seems to pursue the aboriginal."

Consider the classic lethal epidemics of measles and smallpox caused by two diverse but well-known viral scourges of human populations throughout history. History records that smallpox reached Europe from the Far East in 710 A.D. and attained epidemic proportions in the 18th century as populations grew and became concentrated. The effects on society are hard to imagine today, but as an example, at least five reigning monarchs died of smallpox.

Smallpox virus continued its spread around the world when European colonists and slave traders moved to the Americas and Australia. This viral infection changed the balance of human populations in the New World. Some say that viral infections were responsible for the elimination of a myriad of native languages and the nearly exclusive use of Spanish and Portuguese in South America. This suggestion may not be hyperbole. The first recorded outbreak of smallpox in the Americas occurred among African slaves on the island of Hispaniola in 1518, and the virus rapidly spread through the Caribbean islands. Within 2 years, this tochold of smallpox in the New World enabled the conquest of the Aztecs by European colonists. When Hernán Cortez first visited the Yucatan Peninsula in 1518 and began his conquest, his soldiers infected no one with smallpox virus. However, in 1520, smallpox reached the mainland from Cuba. Within 2 years, 3.5 million Aztecs were dead, more than could be accounted for by the bullets and swords of the small band of conquistadors. Smallpox spread like wildfire in the native population (which unfortunately was highly interactive and of sufficient density for efficient virus transmission). It reached as far as the Incas in Peru before Francisco Pizarro made his initial invasion in 1533. As is true in most smallpox epidemics, some Aztecs and Incas survived, but those who did were then devastated by measles virus, probably brought in by Cortez's and Pizarro's men. Conquest occurred by a one-two virological punch rather than by military prowess. Slave traders (who were most likely immune to infection) were populating Brazil with their infected human cargo at approximately the same time, with the same horrible result. The devastation of indigenous peoples by these viruses was also recapitulated in the colonization of North America and continued into the 20th century as contaminated explorers infected isolated groups of Alaskan Inuit and native populations in New Guinea, Africa, South America, and Australia.

Hantavirus Pulmonary Syndrome: Human Disease Resulting from Changing Climate and Animal Populations

In 1993, a small but alarming epidemic of a highly lethal infectious disease occurred in the Four Corners area of New Mexico. Individuals who were in excellent health developed flu-like symptoms that were followed quickly by a variety of pulmonary disorders, including massive accumulation of fluid in the lungs, and death. Rapid action by local health officials and a prompt response by the Centers for Disease Control and Prevention were instrumental in discovering that these patients had low-level, cross-reacting antibodies to hantaviruses. These members of the family *Bunyaviridae* were previously associated with renal diseases in Europe and Asia and were well known to scientists, who associated them with viral hemorrhagic fever during the Korean War. Hantaviruses commonly infect rodents and are endemic in these populations around the world. Using PCR technology, scientists from the Centers for Disease Control and Prevention found that the patients were infected with a new hantavirus. Field biologists discovered that this virus was found in a rodent called the deer mouse (*Peromyscus maniculatus*), which is common in New Mexico. Hantavirus pulmonary syndrome has invariably been associated with the presence of this virus in New Mexico, as well as a few other isolated incidents around North America. The virus, which was given the name Sin Nombre virus (no-name virus), is an example of an emerging virus, endemic in rodents, that causes severe problems when it crosses the species barrier and infects humans.

The reason why humans became infected with Sin Nombre virus is still being debated, but one popular idea is that a dramatic increase in the deer mouse population was an important factor. In 1992 and 1993, higher than normal rainfall resulted in a bumper crop of piñon nuts, a favorite food for deer mice and local humans. Mouse populations increased in response, and contacts with humans inevitably increased as well. Hantavirus infection is asymptomatic in mice, but virions are excreted in large amounts in urine and droppings, where they are quite stable. Human contact with contaminated blankets or dust from floors or food storage areas provided ample opportunities for infection. Hantavirus syndrome is rare because humans are not the natural host, and apparently are not efficient vehicles for virus spread. However, the hantavirus zoonotic infection serves as a warning that potential human diseases lurk in the wild if we inadvertently intrude upon another host-parasite relationship.

Caliciviruses: Underappreciated but Highly Effective Zoonotic Agents

Caliciviruses are (+) strand RNA viruses and are a common causes of virus-induced vomiting and diarrhea in humans. The Centers for Disease Control and Prevention estimates that 23 million cases of acute gastroenteritis in the United States are due to members of the *Norovirus* genus in the family *Caliciviridae*. Remarkably, 50% of all food-borne outbreaks of gastroenteritis are caused by norovirus infection. Other common calicivirus-promoted symptoms include skin blistering, pneumonia, abortion, organ inflammation, and coagulation or hemorrhage. The epidemiology of calicivirus infections is poorly understood. Until recently, a common way to identify these viruses was by determining their unusual structure, visible only by electron microscopy, hardly a facile diagnostic tool. Today, diagnosis of calicivirus infection is accomplished by accurate and rapid PCR technology. Humans often become infected from water and food sources, and we now understand that the ocean and its marine animals are a major reservoir of these interesting viruses. Four genera are recognized in the family *Caliciviridae: Norovirus* (formerly known as Norwalk agent or Norwalk-like virus), *Sapovirus, Lagovirus,* and *Vesivirus.* Only noroviruses and sapoviruses cause human disease, and to date, all have resisted attempts at cultivation. Only the marine caliciviruses can be propagated *in vitro* and are known to infect terrestrial hosts, including humans.

The host range of these marine viruses is astounding. One virus serotype can infect five genera of seals, cattle, three genera of whales, donkeys, foxes, opaleye fish, horses, domestic swine, primates, and humans. The degree of exposure of land-based hosts to marine caliciviruses may be substantial, especially in confined areas such as shallow bays where aquatic mammals breed and calf. Infected whales excrete more than 10^{13} calicivirus particles daily, and the viruses remain viable for more than 2 weeks in cold seawater. Other prime sites for calicivirus encounters are aquatic theme parks and cruise ships. Human contact with contaminated water and sewage guarantees more outbreaks of calicivirus disease.

SARS: the Rise and Fall of a Highly Transmissible Zoonotic Infection

A new human coronavirus evolved in China at the end of the 20th century. As we now understand it, the disease called severe acute respiratory syndrome (SARS) first appeared in Guangdong Province in China in the fall of 2002. A Chinese doctor who treated these patients traveled to Hong Kong on 21 February 2003 and stayed on the ninth floor of the Hotel Metropole. He became ill and died in the hospital on February 22. The infection spread to 10 people staying in the hotel, who then flew to Singapore, Vietnam, Canada, and the United States before symptoms were evident. Little did they know that they were making history—the first major viral epidemic to be spread by air travel. This small number of infected people efficiently spread the new coronavirus around the world, such that about 8,000 people in 29 different countries became infected in less than a year. The fatality rate was almost 1 in 10, a chilling statistic that activated health organizations worldwide. The scientific enterprise mobilized with unprecedented speed and cooperation, such that the causative agent was identified and sequenced within a few months.

We now know that this coronavirus originated in bats and was previously known to infect occasionally only palm civets and ferret badgers (Fig. 10.9). Chance transmission to civets subsequently consumed by humans in China apparently selected for mutations that expanded the virus host range and facilitated human transmission. Crucial mutations were in the viral ligand that enable virions to bind a human receptor. Amazingly enough, the epidemic never reached pandemic proportions, despite billions of susceptible hosts and widespread seeding of infected people around the world. While public health officials did a remarkable job in quarantine and diagnosis, it appears that the virus quickly evolved to be less virulent and less transmissible. After a frightening few months, SARS all but disappeared from the human population. However, field workers continue to isolate the virus from bats, its natural host, as well as from animals that share territory with the infected bats.

Human Immunodeficiency Virus: a Pandemic from a Zoonotic Infection

The origin of human immunodeficiency virus was an enigma until two lines of research converged. The first was a remarkable study published in 1998. This study provided a time point reference for when the virus entered the human population (see Chapter 6). The second line of research required isolation and analysis of simian lentiviruses. Based on sequence homology between chimpanzee and human lentiviruses, the progenitor of human immunodeficiency virus type 1 can now be traced to a few transmissions from chimpanzees to a human in West Central Africa. This precursor of contemporary human immunodeficiency virus acquired new mutations, giving it increased tropism and propensity to spread among humans. How did the chimpanzee virus infect a human? Viral ecologists and epidemiologists now have strong evidence that humans are exposed to many zoonotic infections by the bushmeat trade in West Africa, which involves killing and consumption of wild animals, including chimpanzees, gorillas, other primates, and rodents. It is now clear that chimpanzee

retroviruses infected humans, giving rise to the M and N clades of human immunodeficiency virus type 1. Curiously, the O clade viruses first identified in Cameroon in 1994 appear to have arisen independently and may have come from gorillas, which harbor a very closely related retrovirus. Remarkably, a chance encounter of primate retroviruses with human bushmeat hunters established human immunodeficiency virus type 1 on the planet (Fig. 10.9).

The progenitor of human immunodeficiency virus type 1 was on the path to extinction because the population of wild chimpanzees had dropped to about 150,000 animals living in isolated troops. The new human host exceeds 6 billion individuals, and this niche is rapidly filling: more than 1 in 100 humans are now infected.

Shortly after the recognition of human immunodeficiency virus type 1 in the 1980s, a second related subtype, called type 2, was identified in Portugal and subsequently has been found in many countries (Chapter 6). Type 2 is not as widely distributed as type 1, being more commonly found in Africa. Interestingly, the membrane proteins of the type 2 human virus have homology to those of the sooty mangabey and strains of simian immunodeficiency virus. In fact, humans infected with human immunodeficiency virus type 2 have antibodies that cross-react with the simian virus proteins. These observations have led some scientists to speculate that the human type 2 virus was derived from transmission from the sooty mangabey into humans, most likely due to the bushmeat trade.

Host Range Can Be Expanded by Mutation, Recombination, or Reassortment

Given the large numbers of viruses and ample opportunities for infection, new host-virus interactions will certainly occur. How will they be recognized? Because the new host population will have had no experience with the virus, the infection will have unpredictable consequences (e.g., the dead-end or the evolving host-virus scenario). Such an invasion may be marked only by an immune response (i.e., production of antibodies and seroconversion) and no obvious disease. Infection could also result in mild disease, the symptoms of which may or may not be noteworthy; occasionally, astute health care workers may notice unusual symptoms in local population centers. However, sometimes the invasion can be dramatic, as in the global epidemics caused by canine parvovirus, influenza virus, and human immunodeficiency virus.

Canine Parvovirus: Cat-to-Dog Host Range Change by Two Mutations

In 1978 canine parvovirus, a member of the family *Parvoviridae*, was first identified simultaneously in several countries around the world as the cause of a new enteric and myocardial disease in dogs. Canine parvovirus apparently evolved from the feline panleukopenia virus after two or three mutations made the latter pathogenic and highly transmissible in the dog population. Feline panleukopenia virus infects cats, mink, and raccoons, but not dogs. When the new canine virus emerged, it was able to infect feline cells in culture, but did not replicate in cats. Because canine parvovirus appeared less than 30 years ago, it has been possible to analyze dog and cat tissue collected in Europe in the early 1970s to search for the progenitor canine parvovirus. Some fascinating studies have tracked and identified the virus mutants causing the epidemic. The ancestor of canine parvovirus began infecting dogs in Europe during the early 1970s, as deduced by the appearance of antibodies in dogs in Greece, The Netherlands, and Belgium. In 1978, evidence of virus infection was obvious in Japan, Australia, New Zealand, and the United States, suggesting that the new virus had spread around the world in less than 6 months. The stability of the new virus, its efficient fecal-oral transmission, and the universal susceptibility and behavior of the world's dog population were important factors in the emergence of this new virus.

We now understand the genetic changes that created a new pathogen from a well-known virus. Only two amino acid substitutions in the VP2 capsid protein were necessary to change the tropism from cats to dogs. These substitutions are both necessary and sufficient to alter host range, because changing only the feline virus VP2 sequence to the canine virus sequence enabled the feline virus to replicate in cultured canine cells and in dogs. These critical amino acids are located on a raised region of the capsid that surrounds the threefold axis of icosahedral symmetry in the $T = 1$ capsid (Fig. 10.14). These differences in the capsid structure control the tropism for cells by affecting virion binding to host transferrin receptor, the protein used to establish infection. Feline panleukopenia virions bind only to the feline transferrin receptor. In contrast, canine parvovirus virions bind to both the feline and canine transferrin receptors.

The emergence of canine parvovirus was the result of an hitherto unappreciated high rate of mutation and positive selection for mutations in the major capsid gene. The viral genome continues to evolve rapidly. Indeed, the original antigenic variant, called type 2, was only a transient player in virus evolution. It was completely replaced within 2 years by new canine parvovirus strains called type 2a and 2b, which contain additional amino acid substitutions in the VP2 protein.

The principle to be learned from canine parvovirus emergence is that host range switches by well-adapted viruses are rare, but when a switch occurs, the outcome can be severe for the new host population, New variants

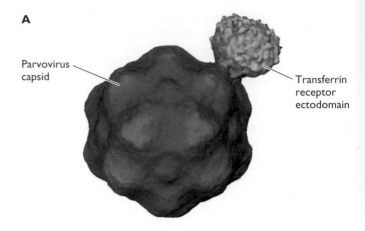

A

Parvovirus capsid

Transferrin receptor ectodomain

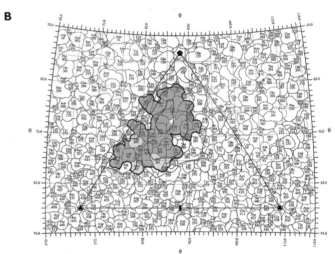

B

Figure 10.14 The transferrin receptor mediates canine and feline parvovirus host range. The transferrin receptors for feline and canine parvoviruses have a large extracellular domain (ectodomain) that is a homodimer of a single protein. The binding of the canine parvovirus virion to the ectodomain of canine transferrin receptor is determined by combinations of amino acid residues on the surface of the capsid. To examine the interactions in more detail, cryo-electron microscopy was used to determine the structure of the purified ectodomain of the feline transferin receptor bound to canine parvovirus capsids. Only a small number of transferin receptors bind to each capsid, and in the model, one such complex is shown **(A)**. The binding site (footprint) of the transferrin receptor on the surface of the canine parvovirus capsid is shown in green on a representation of the surface-exposed amino acids of the capsid, with one of the 60 asymmetric units of the icosahedron indicated **(B)**. Residues that are known to affect binding of the canine transferrin receptor or host range are indicated in yellow. Figures prepared by Susan Hafenstein and Colin Parrish, Cornell University. See S. Hafenstein et al., *Proc. Natl. Acad. Sci. USA* **104:**6585–6589, 2007.

evolve and spread widely through the nonimmune and nonadapted population (the evolving host-virus scenario). The emergence of the canine parvovirus group provided an extraordinary opportunity to study virus-host adaptation and host-range shifts in the field.

Influenza Epidemics and Pandemics: Escaping the Immune Response by Genetic Reassortment

Influenza, a disease with symptoms that have remained unchanged for centuries, is caused by a virus with a constantly changing genome (Fig. 10.15). This disease is the paradigm for the situation in which continued evolution of the virus in several host species is essential for its maintenance. We now know that all mammalian influenza viruses are of avian origin, but mammal-to-mammal transmission can occur. Considerable attention has been focused on the catastrophic influenza pandemic in 1918 that claimed over 25 million lives (Table 10.6). Sequencing data indicate that this H1N1 virus is likely to be the ancestor of all current human influenza viruses, as well as the H1N1 and H3N2 viruses circulating in the world's swine population. The 1918 pandemic was devastating, with an estimated 28% of the population of the United States being infected. Worldwide spread was facilitated by massive worldwide troop movements as a consequence of World War I. Mortality rates were over 2.5%, compared with the normal mortality of less than 0.1%. The unprecedented virulence of the 1918 influenza virus strain raised concerns that it may emerge again, as the genes from this virus still circulate in the avian influenza virus population. Reconstruction of the 1918 virus from viral nucleic acid amplified from archived tissues established that no single gene was responsible for its virulence. Rather, it appears that almost all the viral genes contributed and that the virulence was due to an overreaction of the intrinsic and innate immune response (a so-called cytokine storm, or systemic cytokine toxicity). It is likely that this virus entered the human population directly from avian sources without much adaptation.

The influenza virus serotypes in circulation are virulent, but not to the extent of the 1918 virus. Epidemiologists estimated that influenza virus kills an average of 36,000 people per year in the United States, and the number may

Figure 10.15 Phylogenetic trees for 41 influenza A virus NP genes rooted to influenza B virus NP (B/Lee/40). The nucleotide tree is shown on the left. The horizontal distance is proportional to the minimum number of nucleotide differences to join nodes and NP sequences. Vertical lines are for spacing branches and labels and have no other meaning. Blue animal symbols denote the five host-specific lineages. Pink animal symbols denote viruses in the lineage that have been transmitted to other hosts. The amino acid tree is shown on the right. The blue and pink animal symbols are the same as for the nucleic acid tree. Adapted from R. G. Webster and Y. Kawaoka, *Semin. Virol.* **5:**103–111, 1994, with permission.

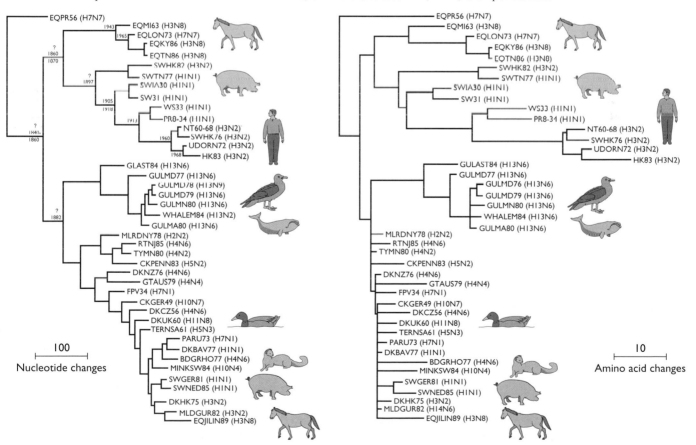

Table 10.6 The 1918–1919 influenza pandemic: one of history's most deadly events[a]

Event	No. of deaths (millions)
Influenza pandemics (1918–1919)	20–40
Black Death (1348–1350)	20–25
AIDS pandemic (through 2006)	22
World War II (1937–1945)	15.9
World War I (1914–1918)	9.2

[a] Data from *The New York Times*, 21 August 1998, and from the World Health Organization.

surge to 50,000 in severe seasons. More than 90% of the victims are 65 or older. In the 1990s, three influenza A virus serotypes predominated, with the H3N2 serotype being particularly virulent.

The life cycle of influenza virus, while comparatively well understood at the molecular level, is remarkable for its complexity in nature (Box 10.5; Fig. 10.16). New influenza viruses constantly emerge from migratory populations of aquatic birds to infect humans, pigs, horses, domestic poultry, and aquatic mammals. In birds, influenza virus replicates in the gastrointestinal tract and is excreted in large quantities, a most efficient virus distribution system. The widespread dispersal of virus in water, the facile changing of hosts, and the ease of genetic reassortment form an engine for creation of new pathogenic strains.

Figure 10.16 Emergence and transmission of H5N1 influenza virus. H5N1 influenza virus has its origins in wild waterfowl, where it was relatively nonpathogenic. Infection is thought to have spread to domestic ducks and chickens, where it evolved to be highly pathogenic in chickens. It then spread back to domestic ducks and geese, where it reassorted its genome with those of other influenza viruses of aquatic birds. The resulting new reassortants spread directly to domestic chickens, humans, and swine. These infections were facilitated by mutations in their PB2, HA, NA, and NS genes that made them more pathogenic to domestic and wild waterfowl and humans. Spread to humans without an intervening "mixing host" such as a pig raises the spectre of a pandemic in the human population.

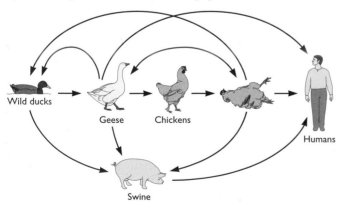

Humans are not the only hosts that suffer from infections caused by this ever-changing virus. Outbreaks of swine and avian influenza devastate operations that produce these animals for food. Despite large-scale immunization programs, virulent strains of swine influenza virus continue to emerge. When pigs are infected experimentally with an avirulent influenza virus mutant, a virulent strain can arise within a few days and cause disease. Poultry producers have similar experiences with avian influenza. A single mutation (changing a threonine to lysine in the hemagglutinin gene) in a relatively attenuated avian strain caused the 1983 epidemic that devastated commercial chicken production in Pennsylvania. The direct transfer of a virulent avian H9N1 virus to humans with lethal consequences was first documented in 1997. Similar transfer of the lethal avian H5N1 virus to humans (but with little to no human-human transmission) reminds us how quickly host-virus interactions can change (Box 10.17; Fig. 10.16).

Even in this myriad of hosts and genetic exchanges, there is some remarkable stability that is crucial to the survival of influenza virus. One surprising finding is that the avian viral genome has not changed much in more than 60 years, in contrast to the genomes of human and other nonavian viruses. Avian viral genomes exhibit mutation and reassortment rates as high as those of human and swine viruses, but only those with neutral mutations are selected and maintained in the bird population. While virulent mutants do arise occasionally, in general, birds infected with avian influenza viruses experience no overt pathogenesis. These properties indicate that influenza virus is in evolutionary stasis in birds. The avian host apparently provides the stable reservoir for influenza virus gene sequences that emerge as recombinants capable of transspecies infection.

Retrovirus Pathogenicity Change by Recombination

Retroviruses provide instructive examples of the acquisition of entirely new genes by recombination with the host genome. Such recombination can yield viruses with unexpected, and frequently lethal, pathogenic potential. A retrovirus can acquire a cellular proto-oncogene and become an acutely oncogenic virus (Chapter 7). One well-known example is feline leukemia virus, a highly conserved retrovirus that infects cats throughout the world. The virus causes lymphosarcomas and a range of degenerative diseases, including anemia and thymic atrophy. However, the sarcomas are caused not by feline leukemia virus but rather by new viruses with genomes that result from recombination of host and viral DNA sequences. These recombinant viruses are acutely oncogenic and are defective because their envelope genes have been replaced by cellular sequences. As a

BOX
10.17

DISCUSSION
H5N1 influenza virus, a global phenomenon worth understanding

Highly pathogenic H5N1 variants of influenza virus moved from Asia to India to Europe and to Africa in the space of 10 years. At least three distinct lineages can be identified by sequence analysis in Southeast Asia, Europe, northern Africa, and the Middle East. They continue to evolve, and the new variants are moving around the world in wild birds. Wild aquatic birds, notably waterfowl, are considered the natural reservoirs of low-pathogenicity avian influenza viruses and sources of infection for other species. Highly pathogenic avian influenza viruses evolve in domestic poultry from H5 or H7 subtypes of low-pathogenicity viruses and become established in these species, despite their considerably increased virulence, possibly because domestic poultry are kept at very high densities.

Of the 250 people known to have been infected with the H5N1 serotype, more than half died. The transmission of this virus to humans required close contact with bodily fluids of infected birds. Human-to-human transmission was rare and difficult to demonstrate.

Several species of cats can be infected with and can transmit the H5N1 virus. This phenomenon was first noted at zoos in Thailand when tigers and leopards died

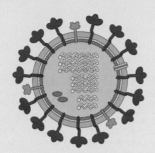

after being fed infected chickens. Although H5N1 infection moves poorly from poultry to humans, it spreads freely from poultry to cats, and then among the cats. Transmission of virus from cats to humans has not been demonstrated—yet.

Efforts to control the avian infections have led to the culling of more than 250 million domestic birds in Asia.

Migratory birds can carry the virus along their migration routes, as demonstrated by the remarkably rapid spread from Qinghai Lake in western China to Mongolia, Russia, and Turkey. However, humans spread the infection to Europe by transporting infected birds by car, truck, and railroad. After narcotics, live birds for the pet trade are the next most commonly smuggled items brought into the United States.

In over 40 years of experience with influenza, the Asian H5N1 is the most virulent virus I have encountered. If it does acquire consistent human-human transmissibility—it will likely be catastrophic.

R. G. Webster

Two Important Unanswered Questions

- Will H5N1 acquire mutations enabling efficient transmission to humans and cause a pandemic?
- How soon will H5N1 spread to the Americas?

Ito, T., H. Goto, E. Yamamoto, H. Tanaka, M. Takeuchi, M. Kuwayama, Y. Kawaoka, and K. Otsuki. 2001. Generation of a highly pathogenic avian influenza A virus from an avirulent field isolate by passaging in chickens. *J. Virol.* 75.4439–4443.
Salzberg, S. L., C. Kingsford, G. Cattoli, D. J. Spiro, D. A. Janies, et al. 2007. Genome analysis linking recent European and African influenza (H5N1) viruses. *Emerg. Infect. Dis.* 13:713–718.
Webster, R. G., Y. Guan, M. Peiris, and H. Chen. 2006. H5N1 influenza continues to circulate and change. *Microbe* 1:559–565.

consequence of unregulated expression of the oncogene from the integrated defective virus, tumors arise rapidly, often with lethal consequences. Several host genes, including c-*abl*, c-*kit*, and c-*sis*, have been incorporated into the genomes of different feline sarcoma viruses.

Some Emergent Viruses Are Truly Novel
Classically, viruses were identified by the diseases they caused; more recently, they have been identified by being grown in culture to characterize their components. Certainly, viruses exist that have never been described or grown in cultured cells. Many of these unculturable viruses cause serious, widespread disease. For example, rotaviruses infect essentially all children younger than 4 years of age in the United States, causing severe diarrhea. However, these viruses were not recognized until the 1970s, simply because the technology for finding them did not exist. Associating a disease or syndrome with a particular infectious agent used to be impossible if the agent could

not be cultured. Fortunately, with powerful recombinant DNA, hybridization, and nucleic acid amplification technology, we are now able to detect and characterize such hitherto unknown viruses with comparative ease.

Discovery of Previously Unknown Hepatitis Viruses in the Blood Supply by Recombinant DNA Technology
One of the first examples of the contemporary quantum leap in diagnostic capacity occurred with the recognition of hepatitis C virus, a member of the family *Flaviviridae*. With the development of specific diagnostic tests for hepatitis A and B viruses in the 1970s, it became clear that most cases of hepatitis that occur after blood transfusion are caused by other agents. Even after more than 10 years of research, the identity of these so-called non-A, non-B (NANB) agents remained elusive. In the late 1980s, Chiron Corporation, a California biotechnology company, isolated a DNA copy of a fragment of the hepatitis C virus

genome from a chimpanzee with NANB hepatitis. This remarkable feat was accomplished by examining DNA from about a million DNA clones for production of proteins that were recognized by serum from a patient with chronic NANB hepatitis. The sequence analysis of these DNA clones identified a (+) strand RNA genome of about 10,000 nucleotides with high homology to known flaviviral genomes. The availability of the hepatitis C virus genome sequence made possible the development of diagnostic reagents that effectively eliminated the virus from the U.S. blood supply, reducing the incidence of transfusion-derived NANB hepatitis significantly. Viral proteins produced from the DNA clones were also instrumental in establishing mechanism-based screens for antiviral drugs (Chapter 9).

These studies provide an excellent example of how a previously unknown human pathogen was identified and then analyzed by cloning technology. Interestingly, hepatitis C virus was not the only previously hidden virus hidden in the human blood supply; the same technology permitted the discovery of several viruses, including TT virus, a ubiquitous human circovirus of no known consequence.

A Revolution in Diagnostic Virology

Recombinant DNA, PCR, sequencing, and hybridization technology are causing a revolution in diagnostic virology. New nucleic acid sequences can be associated with diseases and characterized in the absence of standard virological techniques. The etiological agent of SARS was identified using a "virochip" with characteristic oligonucleotides from all the viral genomes that had been sequenced. However, because of high sensitivity and the potential for contamination by adventitious viruses, care must be taken when using any PCR-based or hybridization method. Without proper controls, these techniques have the potential to associate a particular virus incorrectly with a disease, confounding the deduction of etiology. The spurious association of multiple sclerosis with infection by human T-lymphotropic virus type 1 is an instructive example. Similarly, a variety of studies in which PCR identified human herpesviruses in the brains of patients who had died of Alzheimer's disease, implying a causal link between the disease and the virus, have been published. These conclusions are called into question by the finding of the same viruses in the tissues of similarly aged patients who died of other causes. These examples illustrate the fact that the new technologies do not circumvent the time-honored Koch's postulates, which require that the infectious agent be isolated from a diseased patient, grown, and used to cause the same disease.

While early detection and precise identification of viruses are important in public health and must be pur-

sued with vigor, the information is of limited value if no treatment or follow-up is available for the putative disease-causing agent. We do not have a robust armamentarium of antiviral drugs against even the most common pathogenic viruses. Therefore, developing treatments for previously unknown viruses discovered by these powerful techniques will be challenging.

Perceptions and Possibilities

Infectious Agents and Public Perceptions

While emerging virus infections are well known to virologists, in recent years they have become the subject of widespread public interest and concern. One reason is that less than 30 years ago, many people were eager to close the book on infectious diseases. The public perception was that wonder drugs and vaccines had microbes under control. Obviously, in a few short years, this optimistic view has changed dramatically. Announcements of new and destructive viruses and bacteria appear with increasing frequency. The reality of the human immunodeficiency virus pandemic and its effects at every level of society have attracted worldwide attention, while exotic viruses like Ebola virus capture front-page headlines. Movies and books bring viruses to the public consciousness more effectively than ever before. After the events of 11 September 2001, concern that terrorists might use infectious agents was widespread (Box 10.18). Consequently, it is important to separate fact from fiction and hyperbole. The widespread interest in potentially dangerous viral infections reminds us that a little knowledge can be a dangerous thing: scientists, as well as the public, can be misled about host-virus interactions simply by semantics and jargon.

Virus Names Can Be Misleading

Unfortunately, it is common to name a virus by the host from which it was isolated. By using the name **human immunodeficiency virus** for the virus that causes AIDS, we give short shrift to its nonhuman origins. Canine parvovirus indeed causes disease in puppies, but it is clearly a feline virus that recently changed hosts. Similarly, canine distemper virus is not confined to dogs but can cause disease in lions, seals, and dolphins. Well-known viruses can cause new diseases when they change hosts. Much is implied, and more is ignored, about the host-virus interaction when the virus is given a host-specific name.

The Importance of Pathogenic and Nonpathogenic Viruses

It is not unusual to think that disease-causing viruses are important whereas nonpathogens are uninteresting

DISCUSSION
Viral infections as agents of war and terror

Infectious agents have a documented capacity to cause harm, and can cause epidemics as well as pandemics. Fearsome and deadly viruses abound, ranging from universal scourges such as smallpox virus and influenza virus, to the less widely distributed, but no less deadly, hemorrhagic fever viruses. Any viral infection that can kill, maim, or debilitate humans, their crops, or their domesticated animals has the potential to be used as a biological weapon. Obviously, a biological attack need not cause mass destruction to be an instrument of terror, as was demonstrated by the far-reaching effects of the introduction of bacteria causing anthrax into the United States mail system. Society has only a limited set of responses to frightening outbreaks: vaccination, quarantine,

and anti-infective drugs. For example, the unintentional 1947 outbreak of smallpox in New York City originated from a single businessman who had acquired the disease in his travels. He died after infecting 12 others; to stop the epidemic, **over 6 million people** were vaccinated within a month.

Bioterrorism, like natural outbreaks, is a serious problem with no clear solutions. That fact has not stopped the U.S. government from redirecting billions of dollars to fund programs on counterterrorism and massive research programs on so-called classs A infectious agents. Some argue that this money would be better spent on public health research, or on naturally occurring common diseases. Others are concerned that publication of biological

research data could aid terrorists, and feel that measures of information control must be considered. Practically speaking, public health officials view bioterrorism as a low-probability but high-impact event, much like Hurricane Katrina. When such events occur, they are devastating. However, the hallmarks of these calamities are that they cannot be predicted with accuracy or prevented. One can only prepare for them so that appropriate actions can be taken. Preparation must be measured, practical, and functional so that damage is minimized. For biological attacks of any kind, natural or human inspired, research and free communication of results represent our greatest strengths and are the only ways to prepare.

and irrelevant. As we have seen, a virus that is stable in one host may have devastating effects when it enters a different species. Another misconception arises from human-centered thinking. How often do we think that viruses causing human diseases are more important than those that infect mammals, birds, fish, or other hosts? As noted above, naming viruses as "human" not only is inaccurate but also gives them inflated importance. Similarly, by thinking only of human needs, we are blind to the multiple networks of interactions that constitute host-virus relationships. Indeed, viral particles have access to a broad smorgasbord of hosts, and humans often represent but one stop in the evolution of a virus.

Human Economic Interests Have Significant Effects

In a material society, it is routine to direct attention to viruses that have economic consequences as opposed to those that do not. Moreover, companies and governments generally concentrate resources on viruses that affect their economies, simply because they will be rewarded monetarily for their efforts. Consequently, research is focused on antiviral drugs and therapies that offer greater economic rewards. Such a bottom-line approach is shortsighted and does not address world health problems. Furthermore, because of the mobility of society and our expanding populations, economic impulses must be balanced more prudently by a broader worldview.

What Next?

Can We Predict the Next Viral Pandemic?

The world is currently is in the midst of the AIDS pandemic, one of the most horrible in history. At the current rate of infection, more than 50 million people around the world will die within a year or two. This global devastation continues despite the availability of effective antiviral drug therapies. It is difficult to imagine an infection of greater consequence. However, it is instructive to consider the possibility that other viruses with similar potential to wreak havoc upon humans exist.

Upon reflection, the sobering reality is that some of the most serious threats do not come from the popularized, highly lethal filoviruses (e.g., Ebola virus), the hemorrhagic disease viruses (e.g., Lassa virus), or an as yet undiscovered killer virus lurking in the wild. Rather, the most dangerous viruses are likely to be the well-adapted, multihost, evolving viruses already in the human population. Influenza virus is one that fits this description perfectly. Its yearly visits show no signs of diminishing; genes promoting pandemic spread and virulence are already circulating in the virus population, and the world is ever more prone to its dissemination. Indeed, a pandemic of influenza on the scale of the 1918 to 1919 outbreak is thought by many to be the next emerging disease most likely to affect humans on an enormous scale (Table 10.6). We were reminded of this possibility in the late 1990s, with reports from China and Southeast Asia that an avian strain of influenza virus

(H5N1) had spread directly to humans. Prior to these studies, antibodies to H5 had never been isolated from humans. Obviously, once such a virus infects the naive human population, pandemic spread is possible. Scientists from around the world are now following the trajectory of the H5N1 virus as it reassorts among the myriad influenza virus populations and interacts with the many avian hosts worldwide (Box 10.17).

Many Emerging Viral Infections Illuminate Immediate Problems and Issues

Obviously, it would be a mistake to concentrate on influenza virus to the exclusion of others, because many viral infectious diseases pose immediate and urgent problems for the world. The AIDS pandemic and our experience with SARS illustrate the ease with which a new virus can enter the human population. The secondary infections that accompany AIDS have exposed the fragility of the world's health care systems and the infrastructure of developing countries. The known outbreaks of new, exotic viruses, or well-known viruses that have invaded new geographic niches, as well as the possible cross-species interactions, challenge the standard methods of diagnosis, epidemiology, treatment, and control.

Humans Constantly Provide New Venues for Meeting

To understand and control emerging virus infections, humans must recognize the scale of the problem (Fig. 10.7). The examples of poliovirus, measles virus, and smallpox virus demonstrate that viruses can suddenly cause illness and death on a catastrophic scale following a change in human behavior. Current technological advances and changing social behaviors continue to influence the spread of viruses. Some human-introduced environmental changes that are significant to virology are listed in Table 10.4. Most did not exist 50 years ago, and each one brings humans and viruses into new situations. Probably the most critical fact is that the human population is as large as it ever has been, and is still growing. As a consequence, humans are interacting among themselves and with the environment on a scale unprecedented in history.

Many changes listed in Table 10.4 have major effects upon the transmission of viruses by vectors, such as insects and rodents. When humans interfere in a natural host-virus interaction, cross-species infection is possible. Population movements, the transport of livestock and birds, the construction and use of irrigation systems, and deforestation provide not only new contacts with mosquito and tick vectors, but also mechanisms for transport of infected hosts to new geographic areas.

Humans can provide novel habitats for viruses, as demonstrated by used tires, an unexpected vehicle for movement of viruses and their hosts. Several species of tropical mosquitoes (e.g., *Aedes* species) prefer to breed in small pockets of water that accumulate in tree trunks and flowers in the tropics. The used tire has provided a perfect mimic of this breeding ground, and, as a consequence, the millions of used tires (almost all carrying a little puddle of water inside) accumulating around the world provide a mobile habitat for mosquitoes and their viruses. In the United States, such used tires are shipped all around the country for recycling, transmitting mosquito larvae to new environments literally overnight. As a result, the mosquito hosts for dengue and yellow fever viruses may be given the chance to establish a new range in New Jersey, thanks to the shipment of used tires for recycling from the South. It remains to be seen how the new insect vector will impinge on human health in the northeastern United States, but mosquito-borne viral diseases such as dengue fever clearly are of major concern.

Puerto Rico had five dengue fever epidemics in the first 75 years of the 20th century but had six recent epidemics in 10 years, with major economic costs. Simultaneously, Brazil, Nicaragua, and Cuba experienced their first dengue fever epidemics in 50 years. Brazil reported 180,000 cases in 1996. It is suspected that this viral infection is emerging because the mosquito host has new avenues for spread and many governments have poor, or no, vector control programs.

Another contemporary example of humans moving viruses to new hosts comes from transport of livestock. African swine fever virus, a member of the family *Iridoviridae*, causes a serious viral disease that is threatening the swine industries of developing and industrialized countries. Unlike any other DNA virus, it is spread by soft ticks of the genus *Ornithodoros*. African swine fever virus was spread from Africa to Portugal in 1957, to Spain in 1960, and to the Caribbean and South America in the 1960s and 1970s by long-distance transport of livestock and their resident infected arthropods. Similarly, the rapid spread of H5N1 avian influenza virus from China to Europe was probably mediated by transport of infected birds across national borders.

The construction of dams and irrigation systems can change host-virus interactions on a large scale. The creation of vast areas of standing water has introduced new sources of viruses, hosts, and vectors. The 1987 outbreak of Rift Valley fever in Mauritania along the Senegal River was associated with the new Diama Dam, which created conditions ideal for mosquito propagation. Not only do water impoundments affect insects, but also they alter the population and migration patterns of waterfowl and other animals and the viruses they carry. Previously separated hosts and viruses are brought together as a consequence.

In industrialized countries, the increasing need for day care centers has led to new opportunities for viral transmission and other types of day care-associated illness. In the United States, more than 12 million children are in day care centers for several hours a day, and the vast majority are under 3 years of age. As most parents can testify, respiratory and enteric infections are common, and infections spread easily to other children, day care workers, and the family at home.

Many of the examples discussed above involve human interactions on a large scale and reflect an aphorism called **the law of unintended consequences**: human actions often can produce unforeseen effects. Unfortunately, these unforeseen effects often have substantial negative or momentous consequences (Table 10.4). Sociologists, including the late Robert Merton, provided some insight into the concept, which has been used in developing policy and predicting the potential outcome of large-scale human activities (e.g., environmental impact statements). He stated five reasons that lead to unintended consequences: **ignorance** (we can't anticipate everything); **error** (things that worked in other situations may not apply to the current situation); **immediate interest** (long-term interests are ignored); **basic values, which may require or prohibit certain actions, even if the long-term result is unfavorable** (the long-term consequences may even change the basic values); and **self-defeating prophecy** (fear of some consequence drives people to find solutions before the problem occurs; thus nonoccurrence of the problem is unanticipated).

Preventing Emerging Virus Infections

The modernization of society and the expanding human population have led to spread of infection and selection of new virus variants. We cannot turn back the clock; however, experience and common sense can guide us. The requirements for control of new viral infections range from the obvious to the ideal. Implementation of these ideas will take money and integration of the public and private sectors at a level almost unprecedented today. Two ideas are at the forefront: safe water and better nutrition. If ways and means could be found to provide pure water and sewage disposal for the world, infection rates would drop dramatically. Better nutrition will ensure that natural human defenses are up to the task. Rodent and insect control in population centers would go far to reduce infections. Recognizing when human activities are likely to disrupt endemic disease patterns would be a major step forward. Comparative medicine, the study of disease processes across species including humans, should be fostered. Teams of medical doctors and veterinarians should work together to study zoonotic infections in the field, which

are the root cause of many emerging viral infections. It will be beneficial if government agencies that deal with animal and human health share resources and communication.

Because of the potential for rapid spread of viruses via air travel and urban development, a system of global surveillance and early warning is required for primary-care physicians and health care workers. With modern sequencing technology, it is conceivable that we could monitor all viral pathogens circulating in humans. When a new (or old) viral disease is suspected, the agent could be identified and characterized and the information could be shared widely. Methods should be developed for rapid diagnosis in the field, and primary health care workers must have access to databases. Baseline data on endemic disease and vector prevalence need to be compiled and included in this early-warning database. Active or passive vaccines could be made available, although the infrastructure for accomplishing anything more than local action is almost nonexistent and solutions are expensive to implement. Finally, responsible public and professional education is essential at all levels, as is more scientific research on host-parasite interactions. Inherent in this last proposal is the need for the scientific community to develop better mechanisms for sharing and developing the fruits of scientific research and making their work known, appreciated, and applied by the general public.

Perspectives

The relationships of viruses and their hosts are in constant flux. The perspectives of evolutionary biologists, ecologists, and epidemiologists are required to understand why this is so. At present, the interplay of environment and genes, as well as the interactions of virus and host populations, barely can be articulated, let alone studied in the laboratory. For viral infections, rapid production of huge numbers of progeny, the tolerance to amazing population fluctuations, and the capacity to produce enormous genetic diversity provide the adaptive palette that ensures survival. As Joshua Lederberg said, "In the battle by attrition, humans have a real problem competing with microorganisms. Here we are, here are the bugs, they're looking for food, we're their meal in one sense or another, how do we compete?"

One obvious reason for our success in the competition is that our intrinsic and immune defense systems are capable of recognizing and destroying invading viruses. A less obvious, but equally important, reason is that the current viral hosts represent progeny of survivors of past interactions with viruses. This experience is recorded in the vast gene pool of survivors. Indeed, heterozygosity at the many major histocompatibility complex alleles is known to confer a strong selective advantage on human populations in the battle against microbes. Viral infections have far-reaching effects, ranging from shaping of the host immune system in survivors to eliminating entire populations. Clearly, survival of host and virus is a delicate balance: once the balance is disturbed, the host, the virus, or both will be changed or even eliminated. Given the ever-changing viral populations and the drastic modifications of the ecosystem that have accompanied the current human population explosion, we are hard pressed to predict the future. Perhaps the most sobering fact for humans is that despite our intellect and ability to adapt, virus particles continue to infect and even kill us. In some senses, virus populations are, and always will be, at the top of the evolutionary ladder. The challenge is to keep them at bay and to strive to stay at least one step ahead.

References

Books

Diamond, J. 1997. *Guns, Germs and Steel: the Fates of Human Societies.* W. W. Norton and Company, New York, NY.

Domingo, E., R. W. Webster, and J. Holland (ed.). 1999. *Origin and Evolution of Viruses.* Academic Press, Inc., San Diego, CA.

Ewald, P. W. 1994. *Evolution of Infectious Disease.* Oxford University Press, Oxford, United Kingdom.

Gould, S. J. 1996. *Full House.* Three Rivers Press, Pittsburgh, PA.

Harvey, P. H., A. J. Leigh Brown, J. Maynard Smith, and S. Nee (ed.). 1996. *New Uses for New Phylogenies.* Oxford University Press, Oxford, United Kingdom.

Koprowski, H., and M. B. A. Oldstone (ed.). 1996. *Microbe Hunters Then and Now.* Medi-Ed Press, Bloomington, IL.

Krauss, H., A. Weber, M. Appel, B. Enders, H. D. Isenberg, H. Schiefer, W. Slenczka, A. von Graevenitz, and H. Zahner. 2003. *Zoonoses: Infectious Diseases Transmissible from Animals to Humans,* 3rd ed. ASM Press, Washington, DC.

Morse, S. S. (ed.). 1993. *Emerging Viruses.* Oxford University Press, Oxford, United Kingdom.

Scheld, W. M., D. Armstrong, and J. M. Hughes (ed.). 1997. *Emerging Infections 1.* ASM Press, Washington, DC.

Scheld, W. M., W. A. Craig, and J. M. Hughes (ed.). 1998. *Emerging Infections 2.* ASM Press, Washington, DC.

Reviews

Virus Evolution

Ahlquist, P. 2006. Parallels among positive-strand RNA viruses, reverse-transcribing viruses and double-stranded RNA viruses. *Nat. Rev. Microbiol.* **4:**371–382.

Best, S. P., R. LeTissier, and J. P. Stoye. 1997. Endogenous retroviruses and the evolution of resistance to retroviral infection. *Trends Microbiol.* **5:**313–318.

Claverie, J.-M. 2006. Viruses take center stage in cellular evolution. *Genome Biol.* **7:**110.1–110.5.

Domingo, E., and J. J. Holland. 1997. RNA virus mutations and fitness for survival. *Annu. Rev. Microbiol.* **51:**151–178.

Forterre, P. 2005. The two ages of the RNA world, and the transition to the DNA world: a story of viruses and cells. *Biochimie* **87:**793–803.

Gould, E. A. 2002. Evolution of the Japanese encephalitis serocomplex viruses. *Curr. Top. Microbiol. Immunol.* **267:**391–404.

Loewer, R., J. Loewer, and R. Kurth. 1996. The viruses in all of us: characteristics and biological significance of human endogenous retrovirus sequences. *Proc. Natl. Acad. Sci. USA* **93:**5177–5184.

Malim, M. H., and M. Emerman. 2001. HIV-1 sequence variation: drift, shift, and attenuation. *Cell* **104:**469–472.

Moya, A., E. Holmes, and F. Gonzalezl-Candelas. 2004. The population genetics and evolutionary epidemiology of RNA viruses. *Nat. Rev. Microbiol.* **2:**279–288.

Palumbi, S. 2001. Humans as the world's greatest evolutionary force. *Science* **293:**1786–1790.

Parrish, C., and Y. Kawaoka. 2005. The origins of new pandemic viruses: the acquisition of new host ranges by canine parvovirus and influenza A viruses. *Annu. Rev. Microbiol.* **59:**553–586.

Shadan, F. F., and L. P. Villarreal. 1996. The evolution of small DNA viruses of eukaryotes: past and present considerations. *Virus Genes* **11:**239–257.

Solomon, T., H. Ni, D. W. C. Beasley, M. Ekkelenkamp, M. J. Cardosa, and A. D. T. Barrett. 2003. Origin and evolution of Japanese encephalitis virus in southeast Asia. *J. Virol.* **77:**3091–3098.

Sturtevant, A. H. 1937. Essays on evolution. I. On the effects of selection on mutation rate. *Q. Rev. Biol.* **12:**464–467.

Suttle, C. 2007. Marine viruses—major players in the global ecosystem. *Nat. Rev. Microbiol.* **5:**801–812.

Van Regenmortel, M. H. V., D. H. L. Bishop, C. M. Fauquet, M. A. Mayo, J. Manilkoff, and C. H. Calisher. 1997. Guidelines to the demarcation of virus species. *Arch. Virol.* **142:**1505–1518.

Vignuzzi, M., J. Stone, J. Arnold, C. Cameron, and R. Andino. 2006. Quasispecies diversity determines pathogenesis through cooperative interactions in a viral population. *Nature* **439:**344–348.

Villarreal, L. P. 1997. On viruses, sex, and motherhood. *J. Virol.* **71:**859–865.

Wolfe, N., C. Dunavan, and J. Diamond. 2007. Origins of major human infectious diseases. *Nature* **447:**279–283.

Worobey, M., and E. C. Holmes. 1999. Evolutionary aspects of recombination in RNA viruses. *J. Gen. Virol.* **80:**2535–2543.

Emerging Viruses

American Association for the Advancement of Science. 2006. Special edition: Influenza. *Science* **312:**379–410.

Brinton, M. A. 2002. The molecular biology of West Nile virus: a new invader of the Western Hemisphere. *Annu. Rev. Microbiol.* **56:**371–402.

Eaton, B., C. Broder, D. Middleton, and L.F. Wang. 2006. Hendra and Nipah viruses: different and dangerous. *Nat. Rev. Microbiol.* **4:**23–35.

Nathanson, N. 1997. The emergence of infectious diseases: societal causes and consequences. *ASM News* **63:**83–88.

Reid, A. H., J. K. Taubenberger, and T. G. Fanning. 2001. The 1918 Spanish influenza: integrating history and biology. *Microbes Infect.* **3:**81–87.

Stadler, K., V. Masignani, M. Eickmann, S. Becker, S. Abarignani, H.D. Klenk, and R. Rappuoli. 2003. SARS—beginning to understand a new virus. *Nat. Rev. Microbiol.* **1:**209–218.

Steinhauer, D. A., and J. J. Skehel. 2002. Genetics of influenza viruses. *Annu. Rev. Genet.* **36:**305–332.

Strauss, J. H., and E. G. Strauss. 1997. Recombination in alphaviruses. *Semin. Virol.* **8:**85–94.

Taubenberger, J., and D. Morens. 2006. 1918 influenza: the mother of all pandemics. *Emerg. Infect. Dis.* **12:**15–22.

Weiss, R. A. 1998. Transgenic pigs and virus adaptation. *Nature* **391:**327–328.

Theory of Host-Parasite Interactions

Boots, M., and M. Mealor. 2007. Local interactions select for lower pathogen infectivity. *Science* **315**:1284–1286.

Bremerman, H., and R. Thieme. 1989. A competitive exclusion principle for pathogen virulence. *J. Math. Biol.* **27**:179–190.

Dieckmann, U., J. Metz, M. Sabelis, and K. Sigmund. 2002. *Adaptive Dynamics of Infectious Diseases.* Cambridge University Press, Cambridge, United Kingdom.

Levin, S., and D. Pimental. 1981. Selection of intermediate rates of increase in parasite-host systems. *Am. Nat.* **117**:308–315.

Matthews, L., and M. Woolhouse. 2005. New approaches to quanatify the spread of infection. *Nat. Rev. Microbiol.* **3**:529–536.

May, R., and R. Anderson. 1979. Population biology of infectious diseases II. *Nature* **280**:455–461.

May, R., and R. Anderson. 1983. Epidemiology and genetics in the coevolution of parasites and hosts. *Proc. R. Soc. Lond. Ser. B* **219**:281–313.

Selected Papers

Virus Evolution

Agostini, H. T., R. Yanagihara, V. Davis, C. F. Ryschkewitsch, and G. L. Stoner. 1997. Asian genotypes of JC virus in Native Americans and in a Pacific island population: markers of viral evolution and human migration. *Proc. Natl. Acad. Sci. USA* **94**:14542–14546.

Baric, R. S., B. Yount, L. Hensley, S. A. Peel, and W. Chen. 1997. Episodic evolution mediates interspecies transfer of a murine coronavirus. *J. Virol.* **71**:1946–1955.

Bell, P. 2001. Viral eukaryogenesis: was the ancestor of the nucleus a complex DNA virus? *J. Mol. Evol.* **53**:251–256.

Bonhoeffer, S., and M. A. Nowak. 1994. Intra-host versus inter-host selection: viral strategies of immune function impairment. *Proc. Natl. Acad. Sci. USA* **91**:8062–8066.

Chao, L. 1990. Fitness of RNA virus decreased by Muller's ratchet. *Nature* **348**:454–455.

Chao, L. 1997. Evolution of sex and the molecular clock in RNA viruses. *Gene* **205**:301–308.

Clarke, D. K., E. A. Duarte, A. Moya, S. F. Elena, E. Domingo, and J. J. Holland. 1993. Genetic bottlenecks and population passages cause profound fitness differences in RNA viruses. *J. Virol.* **67**:222–228.

Clarke, D. K., E. A. Duarte, S. Elena, A. Moya, E. Domingo, and J. J. Holland. 1994. The Red Queen reigns in the kingdom of RNA viruses. *Proc. Natl. Acad. Sci. USA* **91**:4821–4824.

Domingo, E., D. Sabo, T. Taniguchi, and C. Weissmann. 1978. Nucleotide sequence heterogeneity of an RNA phage population. *Cell* **13**:735–744.

Drake, J. W., and E. F. Allen. 1968. Antimutagenic DNA polymerases of bacteriophage T4. *Cold Spring Harbor Symp. Quant. Biol.* **33**:339–344.

Eigen, M. 1971. Self organization of matter and the evolution of biological macromolecules. *Naturwissenschaften* **58**:465–523.

Eigen, M. 1996. On the nature of viral quasispecies. *Trends Microbiol.* **4**:212–214.

Elena, S., and R. Sanjuan. 2005. Adaptive value of high mutation rates of RNA viruses: separating causes from consequences. *J. Virol.* **79**:11555–11558.

Esposito, J., S. Sammons, A. Frace, J. Osborne, M. Olsen-Rasmussen, M. Zhang, D. Govil, I. k. Damon, R. Kline, M. Laker, Y. Li, G. L. Smith, H. Meyer, J. W. Leduc, and R. M. Wohlhueter. 2006. Genome sequence diversity and clues to the evolution of variola (smallpox) virus. *Science* **313**:807–812.

Ghedin, E., N. Sengamalay, M. Shumway, et al. 2005. Large-scale sequencing of human influenza reveals the dynamic nature of viral genome evolution. *Nature* **437**:1162–1166.

Guidotti, L. G., P. Borrow, M. V. Hobbs, B. Matzke, I. Gresser, M. B. A. Oldstone, and F. V. Chisari. 1996. Viral cross talk: intracellular inactivation of the hepatitis B virus during an unrelated viral infection of the liver. *Proc. Natl. Acad. Sci. USA* **93**:4589–4594.

Hall, J. D., D. M. Coen, B. L. Fisher, M. Weisslitz, S. Randall, R. E. Almy, P. T. Gelep, and P. A. Schaffer. 1984. Generation of genetic diversity in herpes simplex virus: an antimutator phenotype maps to the DNA polymerase locus. *Virology* **132**:26–37.

Iyer, L., L. Aravind, and E. Koonin. 2001. Common origin of four diverse families of large eukaryotic DNA viruses. *J. Virol.* **75**:11720–11734.

Krish, H. 2003. The view from Les Treilles on the origins, evolution and diversity of viruses. *Res. Microbiol.* **154**:227–229.

La Scola, B., C. Desnues, I. Pagnier, C. Robert, L. Barrassi, G. Fournous, M. Merchat, M. Suzan-Monti, P. Forterre, E. Koonin, and D. Raoult. 2008. The virophage as a unique parasite of the giant mimivirus. *Nature* **455**:100–104.

Mayr, E. 1997. The objects of selection. *Proc. Natl. Acad. Sci. USA* **94**:2091–2094.

McGeoch, D. J., S. Cook, A. Dolan, F. E. Jamieson, and E. A. R. Telford. 1995. Molecular phylogeny and evolutionary timescale for the family of mammalian herpesviruses. *J. Mol. Biol.* **247**:443–458.

Mindell, D., and L. Villarreal. 2003. Don't forget about viruses. *Science* **302**:1677.

Muller, H. J. 1964. The relation of recombination to mutational advance. *Mutat. Res.* **1**:2–9.

Nichol, S. T. K., J. E. Rowe, and W. M. Fitch. 1993. Punctuated equilibrium and positive Darwinian evolution in vesicular stomatitis virus. *Proc. Natl. Acad. Sci. USA* **90**:10424–10428.

Oude Essink, B. B., N. K. T. Back, and B. Berkhout. 1997. Increased polymerase fidelity of the 3TC-resistant variants of HIV-1 reverse transcriptase. *Nucleic Acids Res.* **25**:3212–3217.

Powers, A. M., A. C. Brault, Y. Shirako, E. G. Strauss, W. Kang, J. H. Strauss, and S. C. Weaver. 2001. Evolutionary relationships and systematics of the alpha viruses. *J. Virol.* **75**:10118–10131.

Prangishvili, D., P. Forterre, and R. Garrett. 2006. Viruses of the Archaea: a unifying view. *Nat. Rev. Microbiol.* **4**:837–848.

Preston, B. D., B. J. Poiesz, and L. A. Loeb. 1988. Fidelity of HIV-1 reverse transcriptase. *Science* **242**:1168–1171.

Rambaut, A., O. Pybus, M. Nelson, C. Viboud, J. Taubenberger, and E. Holmes. 2008. The genomic and epidemiological dynamics of human influenza A virus. *Nature* **453**:615–619.

Roossinck, M. 2005. Symbiosis versus competition in plant virus evolution. *Nat. Rev. Microbiol.* **3**:917–924

Saiz, J.-C., and E. Domingo. 1996. Virulence as a positive trait in viral persistence. *J. Virol.* **70**:6410–6413.

Schrag, S. J., P. A. Rota, and W. Bellini. 1999. Spontaneous mutation rate of measles virus: direct estimation based on mutations conferring monoclonal antibody resistance. *J. Virol.* **73**:51–54.

Simmonds, P., and D. B. Smith. 1999. Structural constraints on RNA virus evolution. *J. Virol.* **73**:5787–5794

Spiegelman, S., N. R. Pace, D. R. Mills, R. Levisohn, T. S. Eikhom, M. M. Taylor, R. L. Peterson, and D. H. L. Bishop. 1968. The mechanism of RNA replication. *Cold Spring Harbor Symp. Quant. Biol.* **33**:101–124.

Takemura, M. 2001. Poxviruses and the orgin of the eukaryotic nucleus. *J. Mol. Evol.* **52**:419–425.

Wagenaar, T. R., V. T. K. Chow, C. Buranathai, P. Thawatsupha, and C. Grose. 2003. The out of Africa model of varicella-zoster virus evolution: single nucleotide polymorphisms and private alleles distinguish Asian clades from European/North American clades. *Vaccine* **21**:1072–1081.

Webster, R. G., W. G. Laver, G. M. Air, and G. C. Schild. 1982. Molecular mechanisms of variation in influenza viruses. *Nature* **296:**115–121.

Weiss, R. 2002. Virulence and pathogenesis. *Trends Microbiol.* **10:**314–317.

Zhu, T., B. T. Korber, A. J. Nahmias, E. Hooper, P. M. Sharp, and D. D. Ho. 1998. An African HIV-1 sequence from 1959 and implications for the origin of the epidemic. *Nature* **391:**594–597.

Emerging Viruses

Anderson, N. G., J. L. Gerin, and M. L. Anderson. 2003. Global screening for human viral pathogens. *Emerg. Infect. Dis.* **9:**768–773.

Chua, K. B., W. J. Bellini, P. A. Rota, B. H. Harcourt, A. Tamin, S. K. Lam, T. G. Ksiazek, P. E. Rollin, S. R. Zaki, W.-J. Shieh, C. S. Goldsmith, D. J. Gubler, J. T. Roehrig, B. Eaton, A. R. Gould, J. Olson, H. Field, P. Daniels, A. E. Ling, C. J. Peters, L. J. Anderson, and B. W. J. Mahy. 2000. Nipah virus: a recently emergent deadly paramyxovirus. *Science* **288:**1432–1435.

The Chinese SARS Molecular Epidemiology Consortium. 2004. Molecular evolution of the SARS coronavirus during the course of the SARS epidemic in China. *Science* **303:**1666–1669.

Henige, D. 1986. When did smallpox reach the New World (and why does it matter)?, p. 11–26. *In* P. E. Lovejoy (ed.), *Africans in Bondage.* University of Wisconsin Press, Madison.

Hooper, P., S. Zaki, P. Daniels. D. Middleton. 2001. Comparative pathology of the dieeases caused by Hendra and Nipah viruses. *Microbes Infect.* **3:**315–322.

Hsu, V., M. Hossain, U. Parashar, M. Ali, T. Ksiazek I. Kuzmin, M. Niezgoda, C. Rupprecht, J. Bresee, and R. F. Breiman. 2004. Nipah virus encephalitis reemergence, Bangladesh. *Emerg. Infect. Dis.* **10:**2082–2087.

Ito, T., J. N. S. S. Couceiro, S. Kelm, L. G. Baum, S. Krauss, M. R. Castrucci, I. Donatelli, Kida, H., J. C. Paulson, R. G. Webster, and Y. Kawaoka. 1998. Molecular basis for the generation in pigs of influenza A viruses with pandemic potential. *J. Virol.* **72:**7367–7373.

Jin, L., D. W. G. Brown, M. E. B. Ramsay, P. A. Rota, and W. J. Bellini. 1997. The diversity of measles virus in the United Kingdom, 1992–1995. *J. Gen. Virol.* **78:**1287–1294.

Ksiazek, T., D. Erdman, C. Goldsmith, S. Zaki, T. Peret, et al. 2003. A novel coronavirus associated with severe acute respiratory syndrome. *N. Engl. J. Med.* **348:**1953–1966.

Leitner, T., S. Kumar, and J. Albert. 1997. Tempo and mode of nucleotide substitutions in *gag* and *env* gene fragments in human immunodeficiency virus type 1 populations with a known transmission history. *J. Virol.* **71:**4761–4770.

Li, F., W. Li, M. Farzan, and S. Harrison. 2005. Structure of SARS coronavirus spike receptor binding domain complexed with receptor. *Science* **309:**1864–1868.

Marra, M. S. Jones, C. Astell, R. Host, A. Brooks-Wilson, et al. 2003. The genome sequence of the SARS associated coronavirus. *Science* **300:**1399–1404.

Morimoto, K., M. Patel, S. Corisdeo, D. C. Hooper, Z. F. Fu, C. E. Rupprecht, H. Koprowski, and B. Dietzschold. 1996. Characterization of a unique variant of bat rabies virus responsible for newly emerging human cases in North America. *Proc. Natl. Acad. Sci. USA* **93:**5653–5658.

Petsko, G. 2007. They fought the law and the law won. Genome Biol. **8:**111.

Shackelton, L., C. Parrish, and E. Holmes. 2005. High rate of viral evolution associated with emergence of carnivore parvovirus. *Proc. Natl. Acad. Sci. USA* **102:**379–384.

Sharp, G. B., Y. Kawaoka, D. J. Jones, W. J. Bean, S. P. Pryor, V. Hinshaw, and R. G. Webster. 1997. Co-infection of wild ducks by influenza A viruses: distribution patterns and biological significance. *J. Virol.* **71:**6128–6135.

Sharp, P., E. Bailes, R. Chaudhuri, C. Rodenburg, M. Santiago, and B. Mahn. 2001. The origins of acquired immune deficiency syndrome viruses: where and when? *Philos. Trans. R. Soc. Lond. Ser. B* **356:**867–876.

Smith, A. W., D. E. Skilling, N. Cherry, J. H. Mead, and D. O. Matson. 1998. Calicivirus emergence from ocean reservoirs: zoonotic and interspecies movements. *Emerg. Infect. Dis.* **4:**13–20.

Subbarao, K., A. Klimov, J. Katz, H. Regnery, W. Lim, H. Hall, M. Perdue, D. Swayne, C. Bender, J. Huang, M. Hemphill, T. Rowe, M. Shaw, X. Xu, K. Fukuda, and N. Cox. 1998. Characterization of an avian influenza A (H5N1) virus isolated from a child with a fatal respiratory illness. *Science* **279:**393–396.

Taubenberger, J. K., A. H. Reid, A. E. Krafft, K. E. Bijwaard, and T. G. Fanning. 1997. Initial genetic characterization of the 1918 "Spanish" influenza virus. *Science* **275:**1793–1795.

Tumpey, T., C. Baser, P. Aguilar, H. Zeng, A. Solorzano, D. Swayne, N. Cox, J. Katz, J. Taubenberger, P. Palese, and A. Garcia-Sastre. 2005. Characterization of the reconstructed 1918 Spanish influenza pandemic virus. *Science* **310:**77–80.

Wang, D. A. Urismajn, Y. Liu, M. Springer, T. Ksiazek, d. Erdman, E. Mardis, M. Hickenbotham, V. Magrini, J. Eldred, J. Latrelle, R. Wilson, D. Ganem, and J. DeRisi. 2003. Viral discovery and sequence recovery using DNA microarrays. *PLoS Biol.* **1:**257–260.

Webster, R., G. Yi, M. Peiris, and H. Chen. 2006. H5N1 influenza continues to circulate and change. *Microbe* **1:**559–565.

Wolfe, N., W. Switzer, J. Carr, V. Bhullar, V. Shanmugam, et al. 2004. Naturally acquired simian retrovirus infections in central African hunters. *Lancet* **363:**932–937.

Woolhouse, M. E. J. 2002. Population biology of emerging and re-emerging pathogens. *Trends Microbiol.* **10:**S3–S7.

APPENDIX A

Diseases, Epidemiology, and Disease Mechanisms of Selected Animal Viruses Discussed in This Book

This appendix presents key facts about the pathogenesis of selected animal viruses that cause human disease. Information about each virus or virus group is presented in three sections. In the first section, the viruses and associated diseases are listed. The second section, "Epidemiology," is outlined in four key areas: transmission, distribution of the virus, those at risk or risk factors, and vaccines or antiviral drugs. The third section, "Disease mechanisms," provides simple images that will enable the reader to visualize infection and the resulting pathogenesis. Each of the three sections can be made into a slide for lectures or teaching, providing a "snapshot" of the pathogenesis of a specific virus.

Adenoviruses

Virus	Disease
51 adenovirus serotypes that infect humans, classified into six subgroups	**Respiratory diseases** • Febrile upper tract infection • Pharyngoconjunctival fever • Acute disease • Pertussis-like disease • Pneumonia **Other diseases** • Acute hemorrhagic cystitis • Epidemic keratoconjunctivitis • Gastroenteritis

Epidemiology	
Transmission • Respiratory droplets, fecal matter, fomites • Close contact • Poorly sanitized swimming pools • Ophthalmologic instruments (eye infections)	**Distribution of virus** • Ubiquitous • No seasonal incidence
At risk or risk factors • Children aged <14 years • Day care centers, military camps, swimming clubs	**Vaccines or antiviral drugs** • Live, attenuated vaccine; serotypes 4 and 7 have been produced for the military

Disease mechanisms

Virus infects mucoepithelial cells of respiratory and gastrointestinal tract, conjunctiva, and cornea

Virus persists in lymphoid tissue (tonsils, adenoids, and Peyer's patches)

Antibody is essential for recovery from infection

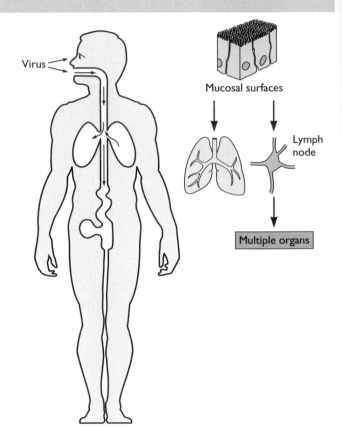

Figure 1

358

Arenaviruses

Virus	Disease
Lymphocytic choriomeningitis virus	Febrile, flu-like myalgia; meningitis
Lassa virus	Lassa fever: severe systemic illness, increased vascular permeability, shock
Junin virus	Argentine hemorrhagic fever: similar to Lassa fever but more extensive bleeding
Machupo virus	Bolivian hemorrhagic fever

Epidemiology

Transmission
- Contact with infected rodents or their secretions or body fluids

At risk or risk factors
- Lymphocytic choriomeningitis virus: contact with pet hamsters, areas with rodent infestation
- Other arenaviruses: habitat of rodents

Distribution of virus
- Lymphocytic choriomeningitis virus: hamsters and house mice in Europe, Americas, Australia, possibly Asia
- Other arenaviruses: Africa, South America, United States
- No seasonal incidence

Vaccines or antiviral drugs
- No vaccines
- Antiviral drug: ribavirin

Disease mechanisms

Persistent infection of rodents caused by neonatal infection and induction of immune tolerance

Viruses infect macrophages and release mediators of cell and vascular damage

Tissue destruction caused by T-cell immunopathology

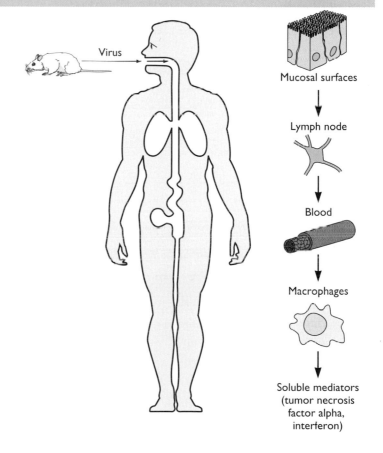

Figure 2

Bunyaviruses

Virus	Vector	Disease	Epidemiology	
Bunyavirus (49 species) • Bunyamwera virus • California encephalitis virus • La Crosse virus • Oropouche virus	Mosquito	Febrile illness, encephalitis, febrile rash	**Transmission** • Arthropod bite • Rodent excreta	**Distribution of virus** • Depends on distribution of vector or rodents • Disease more common in summer
Hantavirus (22 species) • Hantaan virus	None	Hemorrhagic fever with renal syndrome, adult respiratory distress syndrome	**At risk or risk factors** • People in area of vector, e.g., campers, forest rangers, woodspeople	**Vaccines or antiviral drugs** • None
• Sin Nombre virus	None	Hantavirus pulmonary syndrome, shock, pulmonary edema		
Nairovirus (6 species) • Crimean-Congo hemorrhagic fever virus	Tick	Sandfly fever, hemorrhagic fever, encephalitis, conjunctivitis, myositis		
Phlebovirus (9 species) • Rift Valley fever virus • Sandfly fever virus	Fly	Hemorrhagic fever		

Disease mechanisms

Primary viremia, then secondary viremia leads to virus spread to target tissues, including central nervous system, various organs, and vascular endothelium

Antibody essential for controlling viremia

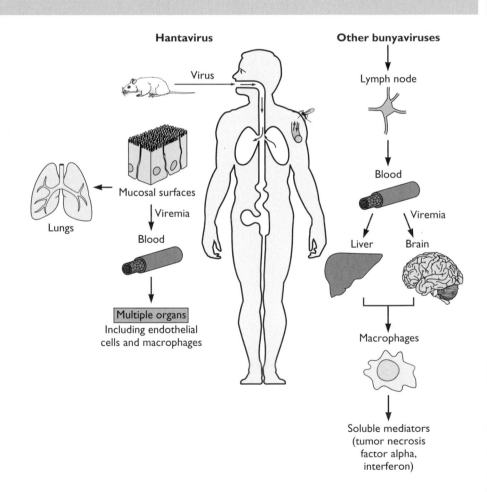

Figure 3

Caliciviruses

Virus	Disease
Logovirus	Gastroenteritis
Norovirus	
Norwalk virus	
Sapovirus	Gastroenteritis
Sapporo virus	

Epidemiology

Transmission
- Fecal-oral route from contaminated water and food
- Virions are resistant to detergents, drying, and acid

At risk or risk factors
- Children in day care centers
- Schools, resorts, hospitals, nursing homes, restaurants, cruise ships (due to infected food handlers)

Distribution of virus
- Ubiquitous
- No seasonal incidence

Vaccines or antiviral drugs
- None

Disease mechanisms

Viruses infect intestinal brush border, preventing proper absorption of water and nutrients

Viruses cause diarrhea, vomiting, abdominal cramps, nausea, headache, malaise, and fever

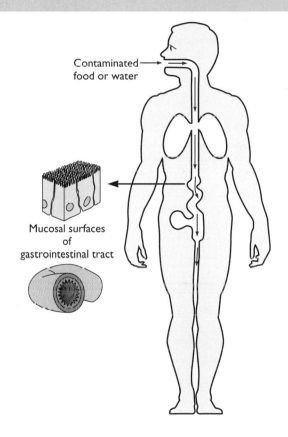

Contaminated food or water

Mucosal surfaces of gastrointestinal tract

Figure 4

Filoviruses

Virus	Disease
Marburg virus Lake Victoria marburgvirus	Hemorrhagic fever
Ebola virus Reston ebolavirus	Hemorrhagic fever

Epidemiology

Transmission
- Fruit bats are reservoirs
- Contact with infected fruit bats, monkeys, or their tissues, secretions, or body fluids
- Contact with infected humans
- Accidental injection, contaminated syringes

Distribution of virus
- Endemic in monkeys in Africa
- No seasonal incidence

At risk or risk factors
- Monkey handlers
- Health care workers attending sick persons

Vaccines or antiviral drugs
- None

Disease mechanisms

Virus replication causes necrosis in liver, spleen, lymph nodes, and lungs

Hemorrhage causes edema and shock

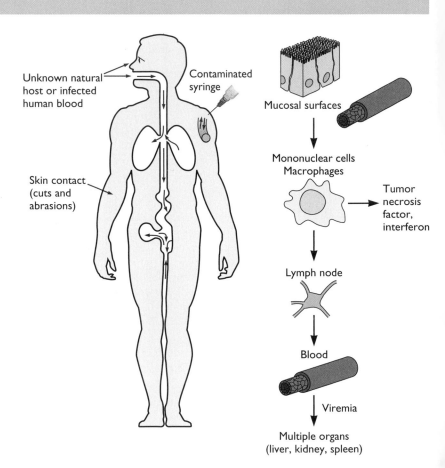

Figure 5

362

Flaviviruses

Virus	Vector	Disease
Flavivirus		
• Yellow fever virus	*Aedes* mosquitoes	Hepatitis, hemorrhagic fever
• Powassan virus	*Ixodes* ticks	Encephalitis
• Dengue virus	*Aedes* mosquitoes	Mild systemic; breakbone fever, dengue hemorrhagic fever, dengue shock syndrome
• Japanese encephalitis virus	*Culex* mosquitoes	Encephalitis
• St. Louis encephalitis virus	*Culex* mosquitoes	Encephalitis
• West Nile virus	*Culex* mosquitoes	Fever, encephalitis, hepatitis
Hepacivirus		
• Hepatitis C virus	None	Hepatitis (see "Viruses that cause hepatitis," next page)

Epidemiology

Transmission
- Mosquito or tick vectors

Distribution of virus
- Determined by habitat of vector
  - *Aedes* mosquito (urban areas)
  - *Culex* mosquito (forest, urban areas)
- More common in summer

At risk or risk factors
- People in niche of vector

Vaccines or antiviral drugs
- Live, attenuated vaccines for yellow fever and Japanese encephalitis
- No antiviral drugs

Disease mechanisms

Viruses are cytolytic

Viruses cause viremia and systemic infection

Nonneutralizing antibodies can facilitate infection of monocytes/macrophages via Fc receptors

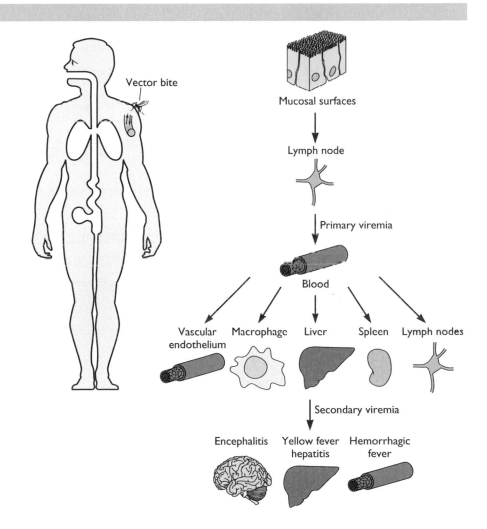

Figure 6

Viruses that cause hepatitis

Virus	Transmission	Incubation period	Mortality rate	Persistent infections	Other diseases
Hepatitis A virus (picornavirus)	Fecal-oral	15–50 days	<0.5%	No	None
Hepatitis B virus (hepadnavirus)	Parenteral, sexual	45–160 days	1–2%	Yes	Primary hepatocellular carcinoma, cirrhosis
Hepatitis C virus (flavivirus)	Parenteral, sexual	14–180 days	~4%	Yes	Primary hepatocellular carcinoma, cirrhosis
Hepatitis D virus (viroid-like; contribution to hepatitis is controversial)	Parenteral, sexual	15–64 days			
Hepatitis E virus (calicivirus)	Fecal-oral	15–50 days	1–2%; pregnant women, 20%	No	None

Epidemiology

Transmission
- Hepatitis A and E viruses
  Fecal-oral route (contaminated food, water)
  Transmitted by food handlers, day care workers, children
- Hepatitis B, C, and D viruses
  Blood, semen, vaginal secretions
  Transfusions, needle injury, drug paraphernalia, sex, breast-feeding
- Hepatitis B virus
  Saliva, mother's milk

Distribution of virus
- Ubiquitous
- No seasonal incidence

At risk or risk factors
- Hepatitis A and E viruses
  Children (mild disease)
  Adults (abrupt-onset hepatitis)
  Pregnant women (high mortality with hepatitis E virus)
- Hepatitis B, C, and D viruses
  Children (mild, chronic infection)
  Adults (insidious hepatitis)
  Adults with chronic hepatitis
- Hepatitis B or C virus
  (hepatocellular carcinoma)

Vaccines or antiviral drugs
- Hepatitis A virus: inactivated vaccine
- Hepatitis B virus: subunit vaccine
- Hepatitis C virus: ribavirin + IFN-α

Disease mechanisms

Primary replication in hepatocytes followed by viremia

Generally not cytolytic; tissue damage caused by cell-mediated immune response

All cause acute infections

Hepatitis B and C virus may cause chronic infections that can lead to hepatocellular carcinoma

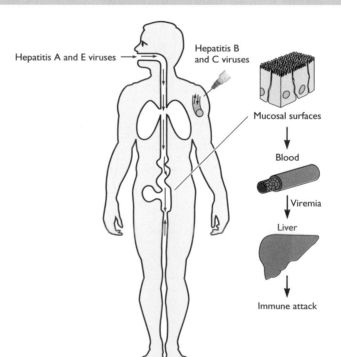

Figure 7

Herpesviruses

Herpes simplex virus

Virus	Disease
Alphaherpesviruses	
• **Herpes simplex virus types 1 and 2**	Mucosal lesions, encephalitis
• Varicella-zoster virus	Chickenpox, shingles

Epidemiology

Transmission
- Saliva, vaginal secretions, lesion fluid
- Into eyes and breaks in skin
- Herpes simplex virus type 1: mainly oral, herpes simplex virus type 2: mainly sexual

Distribution of virus
- Ubiquitous
- No seasonal incidence

At risk or risk factors
- Children (type 1) and sexually active people (type 2)
- Physicians, nurses, dentists, and those in contact with oral and genital secretions (herpetic whitlow, infection of finger)
- Immunocompromised and neonates (disseminated, life-threatening disease)

Vaccines or antiviral drugs
- No vaccine
- Antiviral drugs: acyclovir, penciclovir, valacyclovir, famciclovir, adenosine arabinoside, iododeoxyuridine, trifluridine

Disease mechanisms

Virus spreads cell to cell, not neutralized by antibody

Cell-mediated immunopathology contributes to symptoms

Cell-mediated immunity is required for resolution of infection

Virus establishes latency in neurons

Virus is reactivated from latency by stress or immune suppression

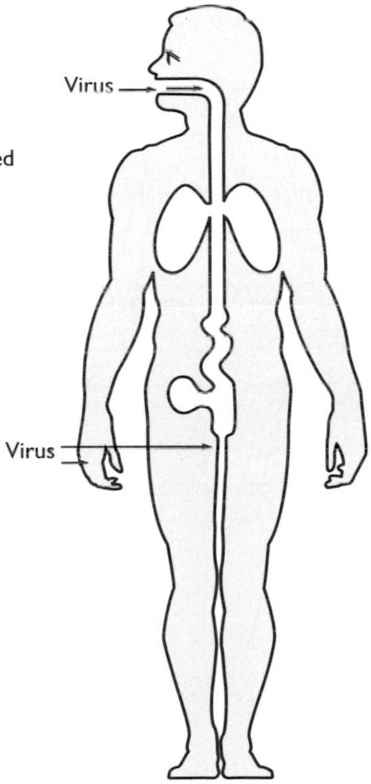

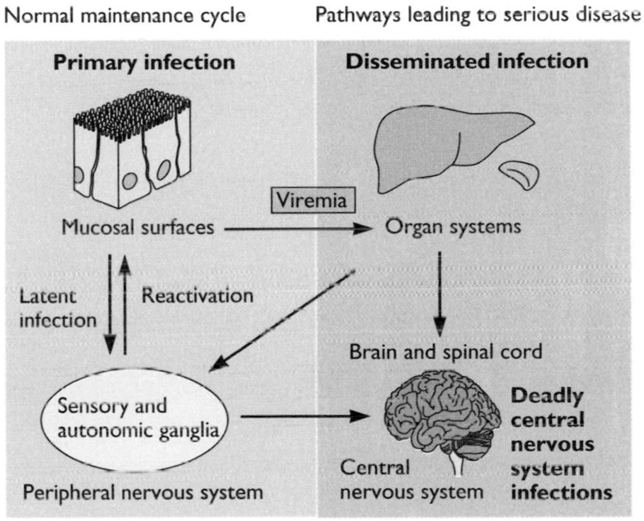

Figure 8

365

Herpesviruses

Cytomegalovirus

Virus	Disease
Betaherpesviruses	
• **Cytomegalovirus**	Congenital defects; opportunistic pathogen in immunocompromised patients
• Human herpesvirus 6	Roseola
• Human herpesvirus 7	Orphan virus

Epidemiology

Transmission
- Blood, tissue, and body secretions (urine, saliva, semen, cervical secretions, breast milk, tears)

At risk or risk factors
- Babies whose mothers become infected during pregnancy (congenital defects)
- Sexual activity
- Blood and transplant recipients
- Burn victims
- Immunocompromised (recurrent disease)

Distribution of virus
- Ubiquitous
- No seasonal incidence

Vaccines or antiviral drugs
- No vaccines
- Antiviral drugs: acyclovir, ganciclovir, valganciclovir, foscarnet, cidofovir, fomivirsen

Disease mechanisms

Infects epithelial and other cells

Mainly causes subclinical infections

Cell-mediated immunity required for resolution of infection

Latent infection in CD34$^+$ bone marrow progenitor cells, macrophages, other cells

Suppression of cell-mediated immunity leads to recurrence and severe disease

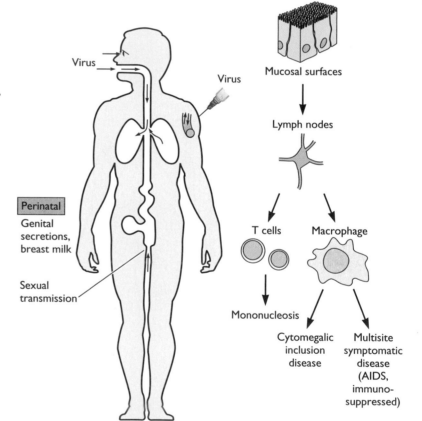

Figure 9

Herpesviruses

Epstein-Barr virus

Virus	Disease	Epidemiology		
Gammaherpesviruses • **Epstein-Barr virus**	Infectious mononucleosis; associated with a variety of lymphomas	**Transmission** • Saliva, close oral contact, or shared items (cup or toothbrush)		**Distribution of virus** • Ubiquitous • No seasonal incidence
• Human herpesvirus 8 (Kaposi's sarcoma-related virus)	Kaposi's sarcoma, rare B-cell lymphoma	**At risk or risk factors** • Children (asymptomatic or mild symptoms) • Teenagers (infectious mononucleosis) • Immunocompromised (fatal neoplastic disease) • Malaria (Burkitt's lymphoma)		**Vaccines or antiviral drugs** • None

Disease mechanisms

Infects oral epithelial cells, B cells

Immortalizes B cells

T cells are required to control infection

T cells contribute to symptoms of **infectious mononucleosis**

Associated with lymphoma in immunosuppressed patients, Burkitt's lymphoma, and nasopharyngeal carcinoma

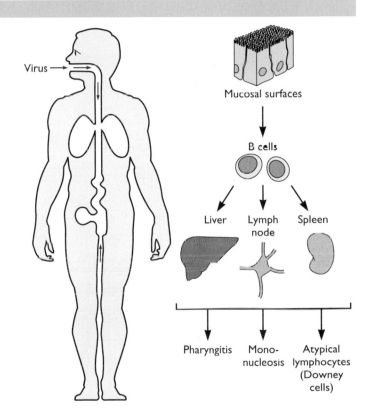

Figure 10

Herpesviruses

Varicella-zoster virus

Virus	Disease
Alphaherpesviruses	
• Herpes simplex virus types 1 and 2	Mucosal lesions, encephalitis
• Varicella-zoster virus	Chickenpox, shingles

Epidemiology

Transmission
• Virus is transmitted by respiratory droplets or contact

At risk or risk factors
• Children (ages 5–9 years) (mild disease)
• Teenagers and adults (more severe disease, possibly pneumonia)
• Immunocompromised or neonates (fatal pneumonia, encephalitis, disseminated varicella)
• Elderly, immunocompromised (recurrent zoster)

Distribution of virus
• Ubiquitous
• No seasonal incidence

Vaccines or antiviral drugs
• Live vaccine (Oka strain)
• Antiviral drugs: acyclovir, foscarnet

Disease mechanisms

Infects epithelial cells and fibroblasts, spread by viremia to skin, causes lesions of chicken pox

Cell-mediated immunopathology contributes to symptoms

Cell-mediated immunity is required for resolution of infection

Latent infection in neurons

Reactivation by immune suppression

Reactivation leads to zoster or shingles, formation of lesions over entire dermatome

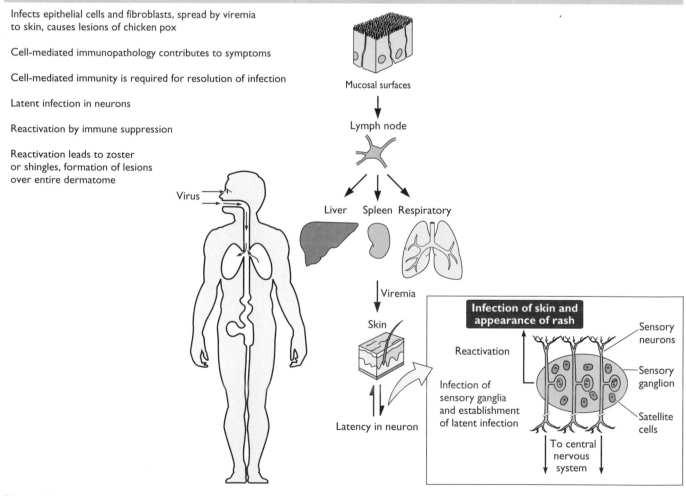

Figure 11

Orthomyxoviruses

Virus	Disease
Influenza A, B, and C viruses	Influenza • Acute febrile respiratory tract infection • Rapid onset of fever, malaise, sore throat, cough • Children may also have abdominal pain, vomiting, otitis media, myositis, croup Complications • Primary viral pneumonia • Myositis and cardiac involvement • Guillain-Barré syndrome • Encephalopathy • Encephalitis • Reye's syndrome

Epidemiology	
Transmission • Inhalation of small aerosol droplets • Widely spread by schoolchildren	**Distribution of virus** • Ubiquitous; epidemics, pandemics • More common in winter
At risk or risk factors • Adults (typical "flu" syndrome) • Children (asymptomatic to severe infections) • Elderly, immunocompromised, and those with cardiac or respiratory problems (high risk)	**Vaccines or antiviral drugs** • Killed vaccine against annual strains of influenza A and B viruses • Live, attenuated influenza A and B vaccine (nasal spray) • Antiviral drugs: amantadine, rimantadine, zanamivir, oseltamivir

Disease mechanisms

Infects upper and lower respiratory tract

Pronounced systemic symptoms caused by cytokine response to infection

Antibodies against hemagglutinin (HA) and neuraminidase (NA) are important for protection against infection

Recovery depends upon interferon and cell-mediated immune response

Susceptibility to bacterial superinfection due to loss of natural epithelial barriers

HA and NA of influenza A virus undergo major and minor antigenic changes

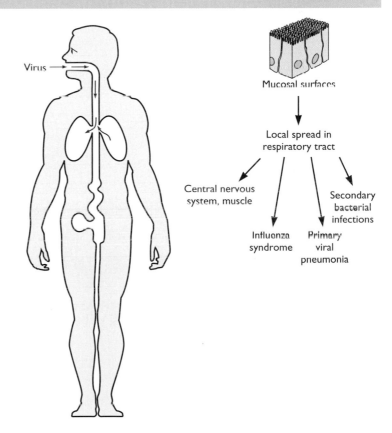

Figure 12

369

Papillomaviruses

Virus	Disease
Papillomavirus (90 genotypes)	Skin warts: plantar, common, and flat warts, epidermodysplasia verruciformis
	Benign head and neck tumors: laryngeal, oral, and conjunctival papillomas
	Anogenital warts: condyloma acuminatum, cervical intra-epithelial neoplasia, cancer

Epidemiology

Transmission
- Direct contact, sexual contact
- During birth, from infected birth canal

At risk or risk factors
- Sexual activity

Distribution of virus
- Ubiquitous
- No seasonal incidence

Vaccines or antiviral drugs
- Vaccine against types 6, 11, 16, and 18 (Gardasil)

Disease mechanisms

Infect epithelial cells of skin, mucous membranes

Replication depends on stage of epithelial cell differentiation

Cause benign outgrowth of cells into warts

Some types are associated with dysplasia that may become cancerous

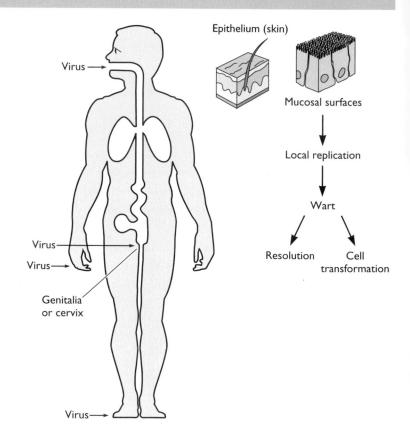

Figure 13

Paramyxoviruses

Measles virus

Virus	Disease		Epidemiology	
Morbilliviruses • Measles virus	Measles Complications: otitis media, croup, bronchopneumonia, encephalitis Subacute sclerosing panencephalitis		**Transmission** • Inhalation of aerosols • Highly contagious **At risk or risk factors** • Adults (serious disease) and children • Immunocompromised persons (more serious outcomes)	**Distribution of virus** • Ubiquitous • Endemic from autumn to spring **Vaccines or antiviral drugs** • Live, attenuated vaccine • No antiviral drugs

Disease mechanisms

Infects epithelial cells of respiratory tract, spreads in lymphocytes and by viremia

Replicates in conjunctivae, respiratory tract, urinary tract, lymphatic system, blood vessels, and central nervous system

T-cell response to virus-infected capillary endothelial cells causes rash

Cell-mediated immunity is required to control infection

Complications are due to immunopathogenesis (postinfectious measles encephalitis) or viral mutants (subacute sclerosing panencephalitis)

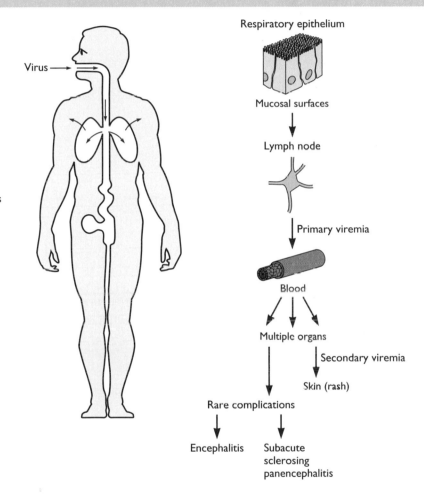

Figure 14

371

Paramyxoviruses

Mumps virus

Virus	Disease
Paramyxoviruses	
• Parainfluenza virus types 1–4	Cold-like symptoms, bronchitis, croup
• **Mumps virus**	Mumps

Epidemiology

Transmission
• Inhalation of aerosols
• Highly contagious

Distribution of virus
• Ubiquitous
• Endemic in late winter, early spring

At risk or risk factors
• Adults (serious disease) and children
• Immunocompromised persons (more serious outcomes)

Vaccines or antiviral drugs
• Live, attenuated vaccine
• No antiviral drugs

Disease mechanisms

Infects epithelial cells of respiratory tract, spreads by viremia

Replicates in salivary glands, testes, respiratory tract, and central nervous system

Cell-mediated immunity is required to control infection

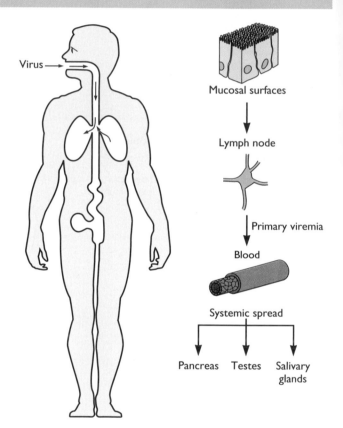

Figure 15

372

Paramyxoviruses

Respiratory syncytial virus

Virus	Disease
Pneumoviruses	
• Respiratory syncytial virus	Bronchiolitis, pneumonia, febrile rhinitis, pharyngitis, common cold

Epidemiology	
Transmission	**Distribution of virus**
• Inhalation of aerosols	• Ubiquitous
	• Incidence is seasonal
At risk or risk factors	**Vaccines or antiviral drugs**
• Infants (bronchiolitis, pneumonia)	• No vaccine
• Children (mild disease to pneumonia)	• Antiviral drug: ribavirin for infants
• Adults (mild symptoms)	

Disease mechanisms

Infects the respiratory tract, does not spread systemically

May cause bronchitis, febrile rhinitis, pharyngitis, common cold, or pneumonia

Bronchiolitis probably caused by the host immune response

In newborns, the infection may be fatal because narrow airways are blocked by virus-induced pathology

Infants are not protected from infection by maternal antibody

Reinfection may occur after a natural infection

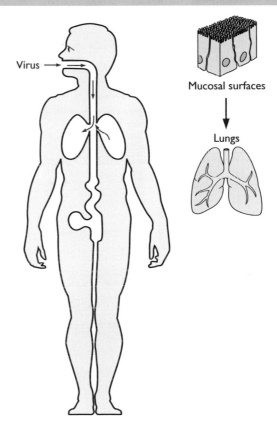

Figure 16

Parvoviruses

Virus	Disease
B19 parvovirus	Erythema infectiosum (fifth disease) Aplastic crisis in patients with chronic hemolytic anemia Acute polyarthritis Abortion
Adeno-associated virus	Commonly infects humans, not associated with illness

Epidemiology

Transmission
• Respiratory and oral droplets

Distribution of virus
• Ubiquitous
• Fifth disease most common in late winter and spring

At risk or risk factors
• Children in elementary school (fifth disease)
• Parents of infected children
• Pregnant women (fetal infection and disease)
• Patients with chronic anemia (aplastic crisis)

Vaccines or antiviral drugs
• None

Disease mechanisms

In utero infection

Virus infects mitotically active erythroid precursor cells in bone marrow

Biphasic disease
  Flu-like phase, viral shedding during viremia
  Later phase: erythematous maculopapular rash, arthralgia, and arthritis caused by circulating virus-antibody immune complexes

Aplastic crisis in patients with chronic hemolytic anemia is caused by depletion of erythroid precursors and destabilization of erythrocytes

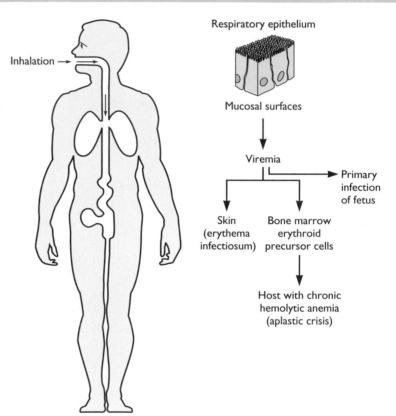

Figure 17

374

Picornaviruses

Virus	Paralytic disease	Encephalitis, meningitis	Carditis	Neonatal disease	Pleurodynia	Herpangina	Hand-foot-and-mouth disease	Rash disease	Acute hemorrhagic conjunctivitis
Poliovirus types 1–3	+	+							
Coxsackie A viruses 1–24	+	+	+			+	+	+	+
Coxsackie B viruses 1–6	+	+	+	+	+			+	
Echoviruses 1–33	+	+	+	+				+	
Enterovirus 70	+								+
Enterovirus 71	+	+					+		
Parechoviruses 1–3	+	+							
Rhinoviruses 1–100									

Virus	Respiratory tract infections	Undifferentiated fever	Diarrhea, gastrointestinal disease	Diabetes, pancreatitis	Orchitis	Disease in immunodeficient patients	Congenital anomalies
Poliovirus types 1–3	+	+				+	
Coxsackie A viruses 1–24	+	+				+	+
Coxsackie B viruses 1–6		+		+	+		+
Echoviruses 1–33	+	+	+			+	
Enterovirus 70							
Enterovirus 71							
Parechoviruses 1–3	+	+	+				
Rhinoviruses 1–100	+						

Epidemiology

Transmission
- Enteroviruses: fecal-oral
- Rhinoviruses: inhalation of droplets, contact with contaminated hands

At risk or risk factors
- Poliovirus
  - Young children (asymptomatic or mild disease)
  - Older children, adults (asymptomatic to paralytic disease)
- Coxsackievirus and enterovirus (newborns and neonates at highest risk for serious disease)
- Rhinovirus (all ages)

Distribution of virus
- Ubiquitous; poliovirus is nearly eradicated
- Enteroviruses: disease most common in summer
- Rhinovirus: disease most common in early autumn, late spring

Vaccines or antiviral drugs
- Poliovirus: live oral or inactivated polio vaccines
- No vaccines for other enteroviruses or rhinoviruses
- No licensed antiviral drugs

Figure 18

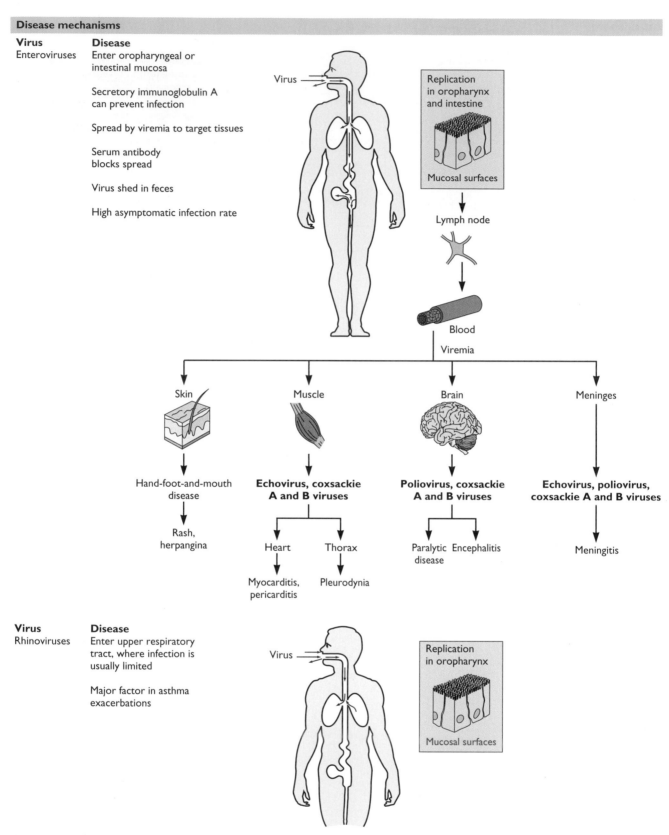

Virus
Enteroviruses

Disease
Enter oropharyngeal or intestinal mucosa

Secretory immunoglobulin A can prevent infection

Spread by viremia to target tissues

Serum antibody blocks spread

Virus shed in feces

High asymptomatic infection rate

Virus

Replication in oropharynx and intestine

Mucosal surfaces

Lymph node

Blood

Viremia

Skin

Muscle

Brain

Meninges

Hand-foot-and-mouth disease

Echovirus, coxsackie A and B viruses

Poliovirus, coxsackie A and B viruses

Echovirus, poliovirus, coxsackie A and B viruses

Rash, herpangina

Heart

Thorax

Paralytic disease

Encephalitis

Meningitis

Myocarditis, pericarditis

Pleurodynia

Virus
Rhinoviruses

Disease
Enter upper respiratory tract, where infection is usually limited

Major factor in asthma exacerbations

Virus

Replication in oropharynx

Mucosal surfaces

Figure 19

376

Polyomaviruses

Virus	Disease		Epidemiology	
Polyomavirus			**Transmission**	**Distribution of virus**
• BK virus	Renal disease in immunosuppressed patients		• Inhalation of aerosols	• Ubiquitous • No seasonal incidence
• JC virus	Progressive multifocal leukoencephalopathy in immunosuppressed patients		**At risk** • Immunocompromised persons	**Vaccines or antiviral drugs** • None

Disease mechanisms

Acquired through the respiratory route, spread by viremia to kidneys early in life

Infections are usually asymptomatic

Virus establishes persistent and latent infection in organs such as the kidneys and lungs

In immunocompromised people, JC virus is activated, spreads to the brain, and causes progressive multifocal leukoencephalopathy; oligodendrocytes are killed, causing demyelination

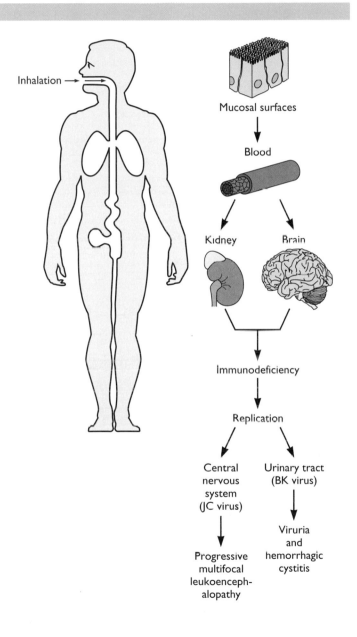

Figure 20

Poxviruses

Virus	Disease
Variola virus	Smallpox
Vaccinia virus (smallpox vaccine)	Encephalitis and vaccinia necrosum (complications of vaccination)
Orf virus	Localized lesion
Cowpox virus	Localized lesion
Pseudocowpox virus	Milker's nodule
Monkeypox virus	Generalized disease
Bovine papular stomatitis virus	Localized lesion
Tanapox virus	Localized lesion
Yaba monkey tumor virus	Localized lesion
Molluscum contagiosum virus	Disseminated skin lesions

Epidemiology

Transmission
- Smallpox: respiratory droplets, contact with virus on fomites
- Other poxviruses: direct contact or fomites

At risk or risk factors
- Molluscum contagiosum: sexual contact, wrestling
- Zoonoses: animal handlers (contact with lesion)

Distribution of virus
- Ubiquitous
- No seasonal incidence
- Natural smallpox has been eradicated

Vaccines or antiviral drugs
- Live vaccine against smallpox (vaccinia virus)
- No antiviral drugs

Disease mechanisms

Infects respiratory tract, spreads through lymphatics and blood

Molluscum contagiosum and zoonoses transmitted by contact

Sequential infection of multiple organs

Cell-mediated and humoral immunity important to resolve infection

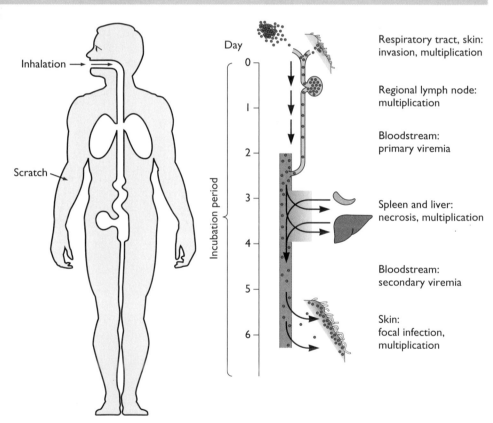

Inhalation

Scratch

Day

Incubation period

Respiratory tract, skin: invasion, multiplication

Regional lymph node: multiplication

Bloodstream: primary viremia

Spleen and liver: necrosis, multiplication

Bloodstream: secondary viremia

Skin: focal infection, multiplication

Figure 21

Reoviruses

Rotavirus

Virus	Disease
Orthoreovirus	Mild upper respiratory tract disease, gastroenteritis, biliary atresia
Coltivirus	Colorado tick fever: febrile disease, headache, myalgia (zoonosis)
Rotavirus	Gastroenteritis

Epidemiology

Transmission
- Fecal-oral route

At risk
- Rotavirus type A
  Infants <24 months of age
    (gastroenteritis, dehydration)
  Older children (mild diarrhea)
  Undernourished persons in
    underdeveloped countries
    (diarrhea, dehydration, death)

- Rotavirus type B
  Infants, older children, adults
    in China (severe gastroenteritis)

Distribution of virus
- Ubiquitous (type A)
- Less common in summer

Vaccines or antiviral drugs
- Live, attenuated, oral
  vaccines available

Disease mechanisms

nsP4 is a viral enterotoxin that causes diarrhea

Disease is serious in infants <24 months old, asymptomatic in adults

Large quantities of virions released in diarrhea

Immunity to infection depends on immunoglobulin A in gut lumen

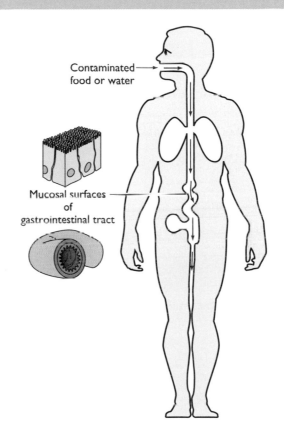

Contaminated food or water

Mucosal surfaces of gastrointestinal tract

Figure 22

Retroviruses

Human T-lymphotropic virus type 1

Virus	Disease
Deltaretrovirus	
• **Human T-lymphotropic virus type 1**	Adult T-cell leukemia Tropical spastic paraparesis
• Human T-lymphotropic virus type 2	Hairy-cell leukemia
• Human T-lymphotropic virus type 5	Malignant cutaneous lymphoma

Epidemiology

Transmission
- Virus in blood
  Transfusions, needle sharing among drug users; infected lymphocytes must be present
- Virus in semen
  Anal and vaginal intercourse
- Perinatal transmission
  Transplacental passage of infected maternal lymphocytes, infected lymphocytes in breast milk

At risk
- Intravenous drug users
- Homosexuals and hetero-sexuals with many partners
- Prostitutes
- Newborns of virus-positive mothers

Distribution of virus
- Ubiquitous
- No seasonal incidence

Vaccines or antiviral drugs
- No vaccines
- No antiviral drugs

Disease mechanisms

Infects T lymphocytes

Remains latent or replicates slowly, induces clonal outgrowth of T-cell clones

Long latency period (30 years) before onset of leukemia

Infection leads to immunosuppression

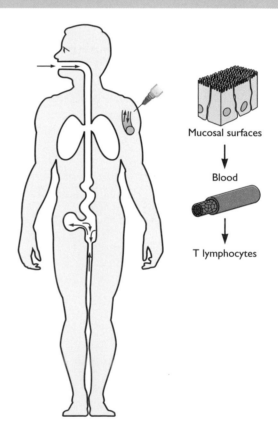

Mucosal surfaces

Blood

T lymphocytes

Figure 23

Retroviruses

Human immunodeficiency virus types 1 and 2

Virus	Disease
Lentivirus • **Human immunodeficiency virus types 1 and 2**	Acquired immune deficiency syndrome (AIDS)

Epidemiology

Transmission
- Virus in blood
  Transfusions, needle sharing among drug users, needle sticks in health care workers, tattoo needles
- Virus in semen and vaginal secretions
  Anal and vaginal intercourse
- Perinatal transmission
  Intrauterine and peripartum transmission; breast milk

Distribution of virus
- Ubiquitous
- No seasonal incidence

At risk
- Intravenous drug users
- Homosexuals and heterosexuals with many partners
- Prostitutes
- Newborns of virus-positive mothers

Vaccines or antiviral drugs
- No vaccines
- Antiviral drugs
  Nucleoside analog reverse transcriptase inhibitors (e.g., azidothymidine, dideoxycytidine)
  Nonnucleoside reverse transcriptase inhibitors (e.g., nevirapine, delavirdine)
  Protease inhibitors (e.g., saquinavir, ritonavir)
  Integrase inhibitors (e.g., raltegravir, elvitegravir)
  Fusion inhibitors (e.g., enfuvirtide, maraviroc)

Disease mechanisms

Infects mainly CD4$^+$ T cells and macrophages

Lyses CD4$^+$ T cells, persistently infects macrophages

Infection alters T-cell and macrophage function; immunosuppression leads to secondary infection and death

Infects long-lived cells, establishing reservoir for persistent infection

Infected monocytes spread to brain, causing dementia

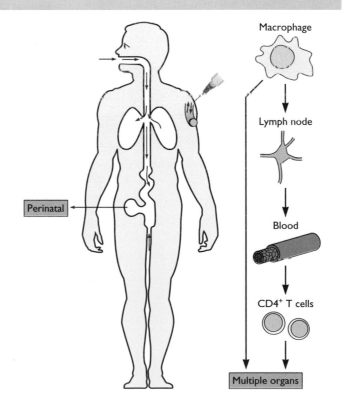

Figure 24

Rhabdoviruses

Rabies virus

Virus	Disease
Lyssavirus	
• **Rabies virus**	Rabies
• Related viruses of rodents and bats	Rarely cause rabies-like encephalitis
Vesiculovirus	
• Vesicular stomatitis virus	Flu-like illness

Epidemiology	
Transmission	**Distribution of virus**
• Reservoir: wild animals	• Ubiquitous, except certain islands
• Vectors: wild animals, unvaccinated dogs and cats	• No seasonal incidence
• Bite of rabid animal (virus in saliva), aerosols (in caves harboring rabid bats)	
At risk or risk factors	**Vaccines or antiviral drugs**
• Animal handlers, veterinarians	• Vaccines for pets and wild animals
• Those in countries with no pet vaccinations or quarantine	• Inactivated virus vaccine for at-risk personnel, postexposure prophylaxis
	• No antiviral drugs

Disease mechanisms

Replicates in muscle at bite site

Incubation period of weeks to months, depending on inoculum and distance of bite from central nervous system

Infects peripheral nerves and travels to brain

Replication in brain causes hydrophobia, seizures, hallucinations, paralysis, coma, and death

Spreads to salivary glands, from which it is transmitted

Postexposure immunization can prevent disease due to long incubation period

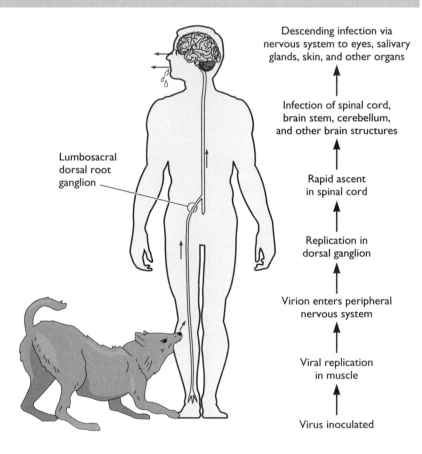

Descending infection via nervous system to eyes, salivary glands, skin, and other organs

Infection of spinal cord, brain stem, cerebellum, and other brain structures

Rapid ascent in spinal cord

Replication in dorsal ganglion

Virion enters peripheral nervous system

Viral replication in muscle

Virus inoculated

Lumbosacral dorsal root ganglion

Figure 25

Togaviruses

Virus	Vector	Disease
Alphaviruses		
• Sindbis virus	*Aedes* mosquitoes	Subclinical
• Semliki Forest virus	*Aedes* mosquitoes	Subclinical
• Venezuelan equine encephalitis virus	*Aedes, Culex* mosquitoes	Mild systemic; severe encephalitis
• Eastern equine encephalitis virus	*Aedes, Culiseta* mosquitoes	Mild systemic; encephalitis
• Western equine encephalitis virus	*Culex, Culiseta* mosquitoes	Mild systemic; encephalitis
• Chikungunya virus	*Aedes* mosquitoes	Fever, arthralgia, arthritis
Rubella virus	None	Rubella

Epidemiology

Transmission
- Alphavirus: mosquito vectors
- Rubella virus: respiratory route

At risk
- Arthropod-borne viruses: people in niche of vector
- Rubella virus: neonates <20 weeks old (congenital defects), children (mild rash), adults (more severe disease, arthritis, arthralgia)

Distribution of virus
- Arthropod-borne viruses: determined by habitat of vector
  Aedes mosquito: urban areas
  Culex mosquito: forest, urban areas
  Most common in summer
- Rubella virus: ubiquitous

Vaccines or antiviral drugs
- Live, attenuated vaccine for rubella virus
- No antiviral drugs

Disease mechanisms

Viruses are cytolytic (except rubella virus)

Cause viremia, systemic infection

Antibodies limit virus spread by viremia (e.g., to fetus in pregnant host)

Cell-mediated immunity important to resolve infection

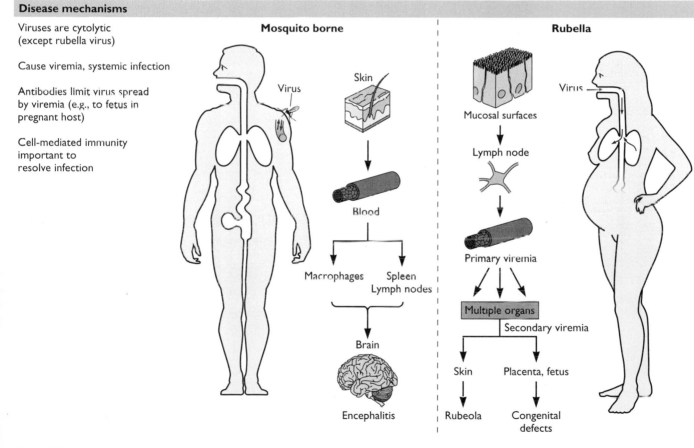

Figure 26

APPENDIX B
Unusual Infectious Agents

Introduction

In Volume I, Chapter 3, we discuss the functions specified by small and large viral genomes. A fundamental question about size is, What is the minimal genome necessary to sustain an infectious agent? Taken to the extreme, the question becomes, Could an infectious agent exist without any genome at all? The subviral agents called viroids, satellites, and prions provide answers to these questions. The adjective "subviral" was coined, in part, because these agents did not fit into the standard taxonomy schemes for viruses. The Subviral RNA Database (http://subviral.med.uottawa.ca/cgi-bin/home.cgi) boasts about 2,000 nucleotide sequences for viroids and related RNA molecules. These infectious agents are widespread, and, perhaps more important, infection by some is deadly.

Viroids

Potato spindle tuber viroid, which was discovered in 1971, is the prototype for the smallest known nucleic acid-based agents of infectious disease, the **viroids** (Fig. 1). Viroids are unencapsidated, small, circular, single-stranded RNA molecules that replicate autonomously when introduced mechanically into host plants. Infection by some viroids causes economically important diseases of crop plants, while others appear to be benign, despite their widespread presence in the plant world. Two examples of economically important viroids are coconut cadang-cadang viroid (which causes a lethal infection of coconut palms) and apple scar skin viroid (which causes an infection that results in visually unappealing apples that cannot be sold). Viroids have been classified in the families *Avsunviroidae* (group A viroids) and *Pospiviroidae* (group B viroids).

The circular single-stranded RNA molecules of all viroids range from 246 to 399 nucleotides in length and are extensively base paired, so that the RNA appears as a 50-nm rod in the electron microscope. There is no evidence that viroids encode proteins or mRNA. Unlike viruses, which are parasites of host translation machinery, **viroids are parasites of cellular transcription**

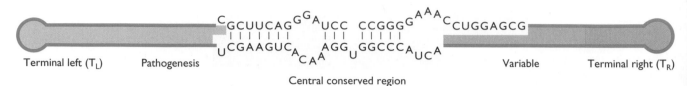

Figure 1 Model of the potato spindle tuber viroid (PSTVd), a viroid that does not form a ribozyme. The RNA strand is shown as a green closed loop. Four functional domains are indicated by different colors. The nucleotides in the upper strand of the central conserved region can form a stable stem-loop. See R. Flores, F. Di Serio, and C. Hernandez, *Semin. Virol.* **8:**65–73, 1997, for more information.

proteins: they depend on RNA polymerase II for replication. Several well-studied group A viroid RNA molecules are functional ribozymes, and this activity is essential for replication (Fig. 2). Group B viroid RNA molecules have no detectable ribozyme activity and instead require a cellular RNase for replication. In general, replication occurs in the nucleus, and most viroids can also be found in the nucleoli of infected cells.

Our current understanding is that the disease-causing viroids were transferred from wild plants used for breeding modern crops. The modern widespread prevalence of viroids can be traced to the use of genetically identical plants (monoculture), worldwide distribution of breeding lines, and mechanical transmission by contaminated farm machinery. As a consequence, these unusual pathogens now occupy niches around the planet that never before

Figure 2 The predicted secondary structure of peach latent mosaic viroid, a ribozyme-forming RNA. **(A)** The circular RNA backbone is indicated by the thick green line. Hydrogen bonds are indicated by dashes, and G · U pairs are indicated by black dots. The nucleotides involved in forming the ribozyme structures are extended to the left and are numbered arbitrarily with respect to the ribozyme sequence. When copied, the top strand forms a hammerhead ribozyme. The bottom strand also forms a hammerhead ribozyme. Conserved nucleotides in the *Avsunviroidae* are boxed and shaded, and the cleavage sites are indicated by red arrows. **(B)** The top strand complement ribozyme (the prime symbol indicates a complement of the sequence shown in panel A). **(C)** The bottom strand hammerhead ribozyme. See R. Flores, F. Di Serio and C. Hernandez, *Semin. Virol.* **8:**65–73, 1997, and M. Pelchat, D. Levesque, J. Ouellet, et al., *Virology* **271:**37–45, 2000, for more information.

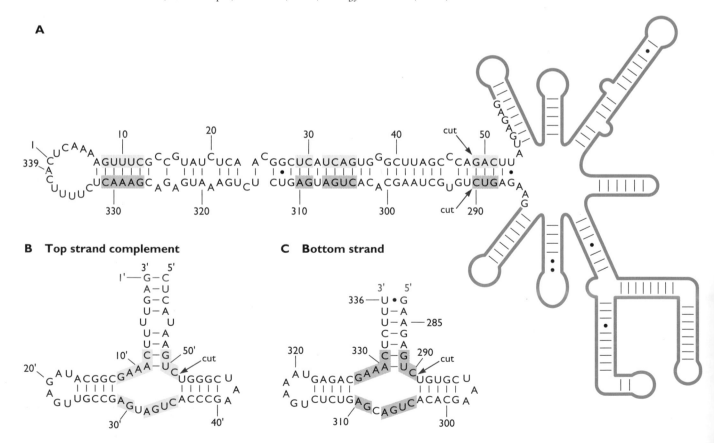

were available to them. As might be expected for an RNA replicon, mutations accumulate at high frequency. In addition, the viroid genome encodes no gene products and infection requires no host receptors, allowing viroids to evolve with amazing speed.

Satellites

Satellites are small, single-stranded RNA molecules that lack genes required for their replication but do replicate in the presence of another virus (the **helper virus**). Unlike viroids, satellite genomes encode one or two proteins. Typical satellite genomes are 500 to 2,000 nucleotides of single-stranded RNA. We recognize two classes of satellites: viruses and nucleic acids. **Satellite viruses** are distinct particles that were discovered in preparations of their helper viruses. These particles contain nucleic acid genomes that encode a structural protein that encapsidates the satellite genome. Satellite viruses are **not** defective viruses derived from the helper virus: their genomes have no homology to the helper. **Satellite nucleic acids** (sometimes called virusoids) are distinguished from satellite viruses by virtue of their packaging by a capsid protein encoded in the helper virus genome.

Most satellites are associated with plant viruses, but a well-known example of a human satellite virus is hepatitis delta satellite virus with its helper, hepatitis B virus. In plants, satellites and satellite viruses cause disease symptoms not seen with the helper virus alone. However, although hepatitis delta satellite virus is associated with human hepatitis, its contribution to disease remains controversial. Hepatitis delta satellite virus appears to be a hybrid between a viroid and a satellite. The genome is 1.7 kb of circular single-stranded RNA that folds upon itself in a tight rodlike structure (70% base paired). The RNA molecule is also a ribozyme, and this activity is required for replication. These properties resemble those of viroid genomes. On the other hand, the genome encodes a protein (delta) that encapsidates the genome, a property shared with satellite nucleic acids. Two functionally distinct forms of the delta protein are made as a result of RNA editing (see Volume I, Chapter 10). The hepatitis delta satellite virion comprises the satellite nucleocapsid packaged within an envelope that contains the surface protein of the helper, hepatitis B virus.

The world abounds with RNA replicons that share structural and functional characteristics of satellites and viroids (Table 1). Similarities to introns, transposons, and RNAs found in the signal recognition particle have been noted, but the origin of satellites and viroids remains an enigma. Moreover, we have only a modest understanding of the mechanisms by which these minimal pathogens cause disease. The speed of their divergence in the absence of obvious selection and the ease with which they can be

Table 1 Viroids and satellites

Property	Viroids		Satellites
	Group A	**Group B**	
Requires coinfection with helper virus	No	No	Yes
Encodes protein	No	No	Yes
Replication	By host RNA polymerase and viroid ribozyme	By host RNA polymerase and host RNase	By helper virus replication proteins

manipulated make them useful models to study the evolution of infection and pathogenesis.

Prions and Transmissible Spongiform Encephalopathies

Can infectious agents exist without genomes? This question challenges our definitions of what constitutes an infectious agent, but the answer seems to be yes. The question arose with the discovery and characterization of infectious agents associated with one group of slow diseases, now called **transmissible spongiform encephalopathies (TSEs)**. These diseases are rare, but always fatal, neurodegenerative disorders that afflict humans and other mammals (Table 2). Each year, thousands of individuals worldwide are diagnosed with spongiform encephalopathies, and about 1% of these cases arise by infection. There is concern that this low rate of infection may increase, because some TSEs spread when animals ingest infected tissues. At the end of 2002, 120 individuals had succumbed to variant Creutzfeldt-Jakob disease, an affliction ascribed to consumption of meat from

Table 2 Transmissible spongiform encephalopathies[a]

TSE diseases of animals

Bovine spongiform encephalopathy (BSE) (mad cow disease)
Chronic wasting disease (CWD) (deer, elk)
Exotic ungulate encephalopathy (EUE) (nyala and greater kudu)
Feline spongiform encephalopathy (FSE) (domestic and wild cats)
Scrapie in sheep and goats
Transmissible mink encephalopathy (TME)

TSE diseases of humans

Creutzfeldt-Jakob disease (CJD)
Fatal familial insomnia (FFI)
Gerstmann-Sträussler syndrome (GSS)
Kuru
Variant CJD (vCJD)

[a]From S. B. Prusiner (ed.), *Semin. Virol.* **7:**157–223, 1996.

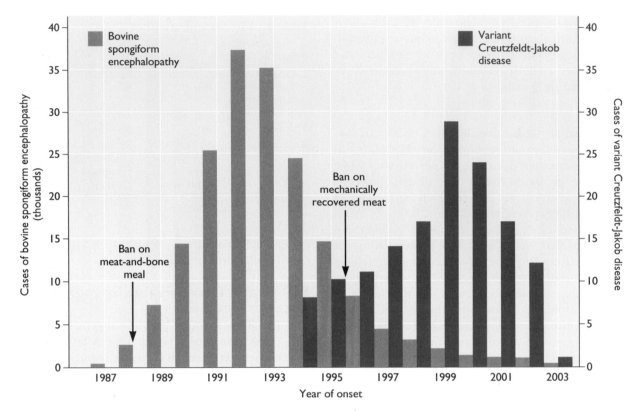

Figure 3 Time course of the reporting of bovine spongiform encephalopathy in cattle and variant Creutzfeldt-Jakob disease in humans in the United Kingdom over a period of 9 years. The peak of the bovine epidemic was in 1992, and the peak of the human disease was in 1999. The incidence of both is now rare. Data obtained from two websites (http://www.defra.gov.uk/animalh/bse/statistics/incidence.html and The 2004 Institute of Food Science and Technology Information Statement on BSE and Variant Creutzfeldt-Jakob Disease [http://www.ifst.org/hottop5.htm]).

animals with bovine spongiform encephalopathy (Fig. 3). While some investigators still contend that viruses or virus-like particles cause TSEs, most now are convinced that these diseases result from infectious proteins called **prions**.

Human TSEs

Several lines of evidence indicated that human spongiform encephalopathies might be caused by an infectious agent. Carleton Gajdusek and colleagues studied the disease **kuru**, found in the Fore people of New Guinea. Kuru spread among women and children as a result of ritual cannibalism of the brains of deceased relatives. When cannibalism stopped, so did kuru. William Hadlow made the seminal observation that lesions in the brains of humans with spongiform encephalopathy were similar to lesions in the brains of animals with scrapie. It was known that scrapie was transmissible to other animals, and experiments by others soon demonstrated that human spongiform encephalopathy can be transmitted from

humans to chimpanzees and other primates. Since 1957, the spongiform encephalopathies of animals and humans have been considered to have common features. Because of this unifying idea, experimental systems that enabled a more detailed understanding of this complex diseases were established. In humans, the spongiform encephalopathies fall into three classes, **infectious, familial,** and **sporadic,** distinguished by how the disease is acquired initially (Box 1).

Hallmarks of TSE Pathogenesis

For most TSEs, the presence of an infectious agent can be detected definitively only by injection of organ homogenates into susceptible recipient species. Clinical signs of infection commonly include cerebellar ataxia (defective motion or gait) and dementia, with death occurring after months or years. The infectious agent first accumulates in the lymphoreticular and secretory organs and then spreads to the nervous system. In model systems, spread of the disease from the site of inoculation to other organs and

TERMINOLOGY
Characteristics of the human spongiform encephalopathies

An **infectious** (or **transmissible**) spongiform encephalopathy is exemplified by kuru and iatrogenic spread of disease to healthy individuals by transplantation of infected corneas, the use of purified hormones, or transfusion with blood from patients with **Creutzfeldt-Jakob disease (CJD)**. The recent epidemic spread of bovine spongiform encephalopathy (mad cow disease) among cattle in Britain can be ascribed to the practice of feeding processed animal by-products to cattle as a protein supplement. Similarly, the new human disease, variant CJD, arose after consumption of beef from diseased cattle.

Familial spongiform encephalopathy is associated with an autosomal dominant mutation in the *prnp* gene. Familial CJD, for example, in contrast to sporadic CJD, is an inherited disease.

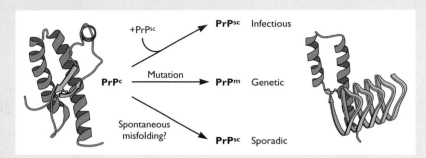

Sporadic CJD is a disease affecting fewer than a million individuals worldwide, usually late in life (from age 50 to 70). About 65% of all spongiform encephalopathies are due to sporadic CJD. As the name indicates, the disease appears with no warning or epidemiological indications. Kuru may have been originally established in the small population of

Fore people in New Guinea when the brain of an individual with sporadic CJD was eaten.

The important principle is that diseases of **all three classes** can usually be transmitted experimentally or naturally to primates by inoculation or ingestion of diseased tissue.

---

the brain requires dendritic and B cells. The disease agent then invades the peripheral nervous system and spreads from there to the spinal cord and brain. Once the infectious agent is in the central nervous system, the characteristic pathology includes severe astrocytosis, vacuolization (hence the term "spongiform"), and loss of neurons. Occasionally, dense fibrils or aggregates (sometimes called plaques) can be detected in brain tissue at autopsy. There is little indication of inflammatory, antibody, or cellular immune responses. The time course, degree, and site of cytopathology within the central nervous system are dependent upon the particular TSE agent and the genetic makeup of the host.

Identification of the First Agent Causing TSE

One of the best-studied TSE diseases is scrapie, so called because infected sheep tend to scrape their bodies on fences so much that they rub themselves raw. A second characteristic symptom, skin tremors over the flanks, led to the French name for the disease, *tremblant du mouton*. Motor disturbances then manifest as a wavering gait, staring eyes, and paralysis of the hindquarters. There is no fever, but infected sheep lose weight and die, usually within 4 to 6 weeks of the first appearance of symptoms. Scrapie has been recognized as a disease of European sheep for more than 200 years. It is endemic in some countries, for example, the United Kingdom, where it affects 0.5 to 1% of the sheep population per year.

Sheep farmers discovered that animals from diseased herds could pass the affliction to a scrapie-free herd, implicating an infectious agent. In 1939, infectivity from extracts of scrapie-affected sheep brains was shown to pass through filters with pores small enough to retain everything but viruses. In the 1970s, ultracentrifugation studies indicated that the agent was heterodisperse in size and density. Even to this day, purification of the infectious agent to homogeneity has not been achieved.

Physical Nature of the Scrapie Agent

The physical nature of the infectious agent has been a major point of contention. As early as 1966, scientists found that scrapie infectivity was considerably more resistant than that of most viruses to ultraviolet (UV) and ionizing radiation. For example, the scrapie agent is 200-fold more resistant to UV irradiation than polyomavirus and 40-fold more resistant than a mouse retrovirus. Other TSE agents exhibit similar UV resistance. On the basis of this relative resistance to UV irradiation, some investigators argued that TSE agents are viruses well shielded from irradiation whereas others claimed that TSE agents have little or no nucleic acid at all.

The infectivity of scrapie agent is also more resistant to chemicals, such as the combination of 3.7% formaldehyde and autoclaving routinely used to inactivate viruses. While it is possible to reduce infectivity by 90 to 95% after several hours of such treatment, complete elimination is exceedingly difficult. This fact has led to unfortunate infections by seemingly sterile instruments used in neurosurgery. Suffice it to say that TSE agents are not typical infectious agents.

Prions and the *prnp* Gene

The unconventional physical attributes and slow infection pattern originally prompted many to argue that TSE agents are not viruses at all. For example, in 1967 the mathematician J. S. Griffith made three suggestions as to how scrapie may be mediated by a host protein, not by a nucleic acid-carrying virus. His thoughts were among the first of the "protein-only" hypotheses to explain TSE.

An important breakthrough occurred in 1981, when characteristic fibrillar protein aggregates were visualized in infected brains. These aggregates could be concentrated by centrifugation and remained infectious. Stanley Prusiner and colleagues developed an improved bioassay, as well as a fractionation procedure that allowed the isolation of a protein with unusual properties from scrapie-infected tissue. This protein is insoluble and relatively resistant to proteases. Sequence analysis of this protein led to the cloning of the *prnp* gene, which is highly conserved in the genomes of many animals, including humans. Expression of the *prnp* gene is now known to be essential for the pathogenesis of common TSEs.

The *prnp* gene encodes a 35-kDa membrane-associated glycoprotein (PrP) found in many neurons. Mice lacking the *prnp* gene develop normally and have few obvious defects. In another study, the PrP protein was shown to be synthesized in hematopoietic stem cells and was important for their renewal. It is strongly linked to human TSE diseases. At least 18 mutations in the human *prnp* gene are associated with familial TSE diseases. Furthermore, specific *prnp* mutations appear to be associated with susceptibility to different strains of TSE (see below).

Prusiner named the scrapie infectious agent a **prion** (from the words "protein" and "infectious"). His unconventional proposal was that an altered form of the PrP protein causes the fatal encephalopathy characteristic of scrapie. Prusiner's controversial protein-only hypothesis caused a firestorm among those who study infectious disease. The hypothesis is that the essential pathogenic component **is** the host-encoded PrP protein with an altered conformation, called PrP^{sc} ("PrP-scrapie"; also called PrP^{res} for "protease-resistant form"). Furthermore, in the simplest case, PrP^{sc} is proposed to have the property of converting normal PrP protein into more copies of the pathogenic form (Fig. 4). An important finding in this regard is that mice lacking both copies of their *prnp* gene are resistant to infection. In recognition of his work on this problem, Prusiner

Figure 4 The conversion of nonpathogenic, α-helix-rich PrP^c protein to the β-sheet-rich conformation of Prp^{sc}, the pathogenic prion. (A) PrP^c is the mature normal cellular protein. The precursor is 254 amino acids long with a signal sequence that is removed. Twenty-three amino acids of the carboxy terminus also are removed as the glycosylphosphatidylinostitol (GPI) linker is added. PrP^{sc} is the β-sheet-rich, pathogenic prion. This conformation is relatively resistant to protease K digestion, in contrast to PrP^c, as indicated. This protease K-resistant PrP fragment of PrP^{sc} is diagnostic of the prion protein. H1, H2, and H3 are helical regions of PrP^c. The yellow boxes are repeats of 8 amino acids [P(Q/H)GGGWGQ]. CHO indicates two N-linked carbohydrate chains. S–S indicates disulfide bonds. **(B)**. Ribbon diagram of the PrP^c and PrP^{sc} protein backbones with α-helices in red and β-sheets in blue. From P. Chien, J. Weissman, and A. DePace, *Annu. Rev. Biochem.* **73:**617–656, 2004, with permission.

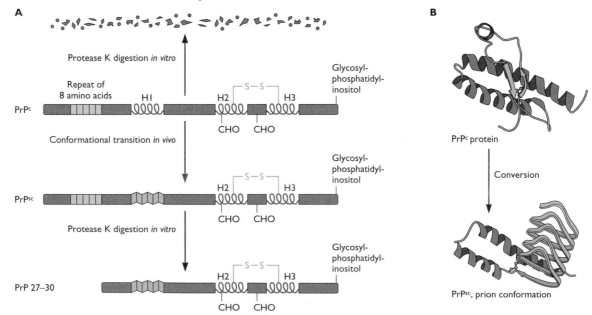

was awarded the Nobel Prize in physiology or medicine in 1997.

The conformational conversion of PrP to PrP^{sc} can be described by two models: the **refolding model** and the **seeding model.** In the former, spontaneous conversion to PrP^{sc} is virtually undetectable. However, interaction of PrP with **preformed** PrP^{sc} facilitates refolding of PrP to PrP^{sc}. PrP^{sc} can be introduced from exogenous sources or by rare mutations that lower the activation energy of conversion. In the seeding model, PrP and PrP^{sc} are in equilibrium, with the formation of PrP being highly favored. PrP^{sc} can be stabilized only when it binds to an aggregate (or seed) of PrP^{sc}. Aggregates are rare, but once formed, PrP^{sc} binds avidly and the aggregate grows rapidly by fragmentation and addition of new PrP^{sc}. Convincing evidence for the seeding model has come from studies of yeast prions.

Strains of the Scrapie Prion

As a result of many serial infections of mice and hamsters with infected sheep brain homogenates, investigators have derived distinct strains of scrapie prion. Strains are distinguished by length of incubation time before the appearance of symptoms, brain pathologies, relative abundance of various glycoforms of PrP, and electrophoretic profiles of protease-resistant PrP^{sc}. Some strains also have a different host range. For example, mouse-adapted scrapie prions cannot propagate in hamsters, but hamster-adapted scrapie prions can propagate in mice. A single amino acid substitution in the hamster protein enables it to be converted efficiently by mouse PrP^{sc} into hamster PrP^{sc}. Therefore, the barrier to interspecies transmission is in the sequence of the PrP protein. Bovine spongiform encephalopathy prions have an unusually broad host range, infecting a number of meat-eating animals, including domestic cats, wild cats, and humans. A striking finding is that different scrapie strains can be propagated in the same inbred line of mice yet maintain their original phenotypes. Stable inheritance suggests to some that a nucleic acid must be an essential component and has been used as support for a viral etiology of TSE disease. The protein-only hypothesis explains the existence of strain variation by postulating that each strain represents a unique conformation of PrP^{sc}. Each of these distinctive pathogenic conformations is then postulated to convert the normal PrP protein into a conformational image of itself. The strongest evidence supporting this hypothesis comes from studies of yeast prions.

Prions and the Food Supply

The existence of TSE in our food animals is a cause for concern, for we have only a modest understanding of the molecular nature of the infectious agent, the routes of shedding and transmission, and the processes leading to the characteristic pathogenesis (Fig. 3). Efforts are currently focused on basic research aimed at protection of the food supply and finding targets for therapeutic intervention. Remarkable progress is being made. For example, sensitive diagnostic tests for prions are now being used to screen blood and tissues. An unexpected breakthrough came when researchers discovered that the antimalarial drug quinacrine blocked accumulation of infectious prions in cultured cells. The mechanism is unknown, but this lead and others are being pursued aggressively, and limited human trials of quinacrine in patients with advanced Creutzfeldt-Jakob disease are in progress. Another promising discovery is that monoclonal antibodies specific for PrP inhibit scrapie prion replication in mice and delay the development of prion disease. The hope is that similar immunotherapy strategies will benefit humans exposed to prions.

References

Butler, D. 2001. A wolf in sheep's clothing. *Nature* **414**:576–577.

Chesebro, B., and B. Caughey. 2002. Transmissible spongiform encephalopathies (prion protein diseases), p. 1241–1262. *In* D. D. Richman, R. J. Whitley, and F. G. Hayden (ed.), *Clinical Virology*, 2nd ed. ASM Press, Washington, DC.

Chien, P., J. Weissman, and A. DePace. 2004. Emerging principles of conformation-based prion inheritance. *Annu. Rev. Biochem.* **73**:617–656.

Chiti, F., and C. Dobson. 2006. Protein misfolding, functional amyloid, and human disease. *Annu. Rev. Biochem.* **75**:333–366.

Codoner, F., J. Daros, R. Sole, and S. Elena. 2006. The fittest versus the flattest: experimental confirmation of the quasispecies effect with subviral pathogens. *PLoS Pathog.* **2**:1187–1193.

Diener, T. O. 1996. Origin and evolution of viroids and viroid-like satellite RNAs. *Virus Genes* **11**:119–131.

Diener, T. O. 2003. Discovering viroids—a personal perspective. *Nat. Rev. Microbiol.* **1**:75–80.

Flores, R., F. Di Serio, and C. Hernandez. 1997. Viroids: the noncoding genomes. *Semin. Virol.* **8**:65–73.

Pelchat, M., D. Levesque, J. Ouellet, et al. 2000. Sequencing of peach latent mosaic viroid variants from nine North American peach cultivars shows that this RNA folds into a complex secondary structure. *Virology* **271**:37–45.

Prusiner, S. B. 2006. Prions, p. 3059–3091. *In* D. M. Knipe and P. M. Howley (ed.), *Fields Virology*, 5th ed. Lippincott Wiliams & Wilkins, Philadelphia, PA.

Tessier, P., and S. Lindquist. 2007. Prion recognition elements govern nucleation, strain specificy and species barriers. *Nature* **447**:556–561.

Toyama, B., M. Kelly, J. Gross, and J. Weissman. 2007. The structural basis of yeast prion strain variants. *Nature* **449**:233–237.

Weissmann, C., M. Enari, P.-C. Klohn, D. Rossi, and E. Flechsig. 2002. Transmission of prions. *Proc. Natl. Acad. Sci. USA* **99**:16378–16383.

Zhang, C. A. Steele, L. Lindquist, and H. Lodish. 2006. Prion protein is expressed on long-term repopulating hematopoietic stem cells and is important for their self-renewal. *Proc. Natl. Acad. Sci. USA* **103**:2184–2189.

Glossary

Abortive infection An incomplete infectious cycle; virions infect a susceptible cell or host but do not complete replication, usually because an essential viral or cellular gene is not expressed. *(Chapter 5)*

Accessibility An attribute that describes the physical availability of cells to virions at the site of infection. *(Chapter 1)*

Active immunization The process of inducing an immune response by exposure to a vaccine; contrasts with passive immunization. *(Chapter 8)*

Active viremia The presence of newly synthesized virions in the blood. *(Chapter 1)*

Acute infection A common pattern of infection in which virions are produced rapidly, and the infection is resolved and cleared quickly by the immune system; survivors usually are immune to subsequent infection. *(Chapter 5)*

Adapter proteins Proteins that have no intrinsic enzymatic functions but act as linkers between functional proteins by binding to two or more at the same time. *(Chapter 6)*

Adaptive response The immune response consisting of antibody (humoral) and lymphocyte-mediated responses; unlike the innate response, the adaptive response is tailored individually to the particular foreign invader; the adaptive response has memory: subsequent infections by the same agent are met with a robust and highly specific response. Also known as the acquired immune response. *(Chapter 4)*

Adoptive transfer The transfer of cells, usually lymphocytes, from an immunized donor to a nonimmune recipient. *(Chapter 4)*

Alternative pathway One of three pathways in the complement system; activates the C3 and C5 convertases without going through the C1-C2-C4 complex. *(Chapter 4)*

Anchorage independence The ability of some cells to grow in the absence of a surface on which to adhere; often detected by the ability to form colonies in semisolid media. *(Chapter 7)*

Antigenic drift The appearance of virions with a slightly altered surface protein (antigen) structure caused by accumulation of point mutations following passage and immune selection in the natural host. *(Chapter 5)*

Antigenic shift A major change in the surface protein of a virion when genes encoding markedly different surface proteins are acquired during infection; this process occurs when viruses with segmented genomes exchange segments, or when nonsegmented viral genomes recombine after coinfection. *(Chapter 5)*

Antigenic variation The display by virions or infected cells of new protein sequences that are not recognized by antibodies or T cells which responded to previous infections. *(Chapter 5)*

Anti-inflammatory cytokines A series of immunoregulatory molecules that control the proinflammatory cytokine response; major anti-inflammatory cytokines include interleukin-1 (IL-1) receptor antagonist, IL-4, IL-6, IL-10, IL-11, and IL-13. *(Chapter 3)*

Apoptosis A sequence of tightly regulated reactions in response to external or internal stimuli that signal DNA damage or other forms of stress; characterized by chromosome degradation, nuclear degeneration and cell lysis; a natural process in development and the immune system, but also an intrinsic defense of cells to viral infection. Also called programmed cell death. *(Chapter 3)*

Attenuated   Having mild or inconsequential instead of an normally severe symptoms or pathology as an outcome of infection; having a state of reduced virulence. *(Chapter 2)*

Autocrine growth stimulation   Stimulation of cell growth by proteins produced and sensed by the same cell. *(Chapter 7)*

Autophagy   A process in which cells are induced to degrade the bulk of their cellular contents by formation of specialized membrane-bounded compartments called autophagolysosomes. *(Chapter 3)*

Auxiliary proteins   Proteins encoded in lentiviral genomes in addition to the three structural polyproteins, Gag, Pol, and Env; they include Tat, Rev, Nef, Vif, Vpr, and Vpu. *(Chapter 6)*

Avirulent virus   A virus that causes no, or mild, disease. *(Chapter 2)*

Blind screening   Screening for antiviral compounds without regard to a specific mechanism. *(Chapter 9)*

Caspases   Crucial proteases in the process called apoptosis; members of a family of <u>c</u>ysteine proteases that specifically cleave after <u>asp</u>artate residues. *(Chapter 3)*

CD markers   *See* Cluster-of-differentiation markers.

CD4+ T cells   T lymphocytes that carry the coreceptor protein CD4 on their surfaces. *(Chapter 4)*

CD8+ T cells   T lymphocytes that carry the coreceptor CD8 on their surfaces. *(Chapter 4)*

Cell cycle   The orderly and reproducible sequence in which cells increase in size, duplicate the genome, segregate duplicated chromosomes, and divide. *(Chapters 4 and 7)*

Cell-mediated response   The arm of the adaptive immune response consisting of helper and effector lymphocytes. *(Chapter 4)*

Chemokines   Small proteins that attract and stimulate cells of the immune defense system; produced by many cells in response to infection. *(Chapter 3)*

Clades   Subtypes of human immunodeficiency virus that are prevalent in different geographic areas. *(Chapter 6)*

Classical pathway   One of three complement pathways that lead to activation of C3-C5 convertases; activation occurs by direct interaction of C1q or C3b proteins with an viral protein/antibody complex on the surface of an infected cell or a virus particle. *(Chapter 4)*

Clinical latency   A state of persistent viral infection in which no clinical symptoms are manifested. *(Chapter 6)*

Cluster-of-differentiation markers   Distinct surface proteins that are recognized by specific monoclonal antibodies; these antibodies bind to various cluster-of-differentiation markers and are used to distinguish different cell types (e.g., CD4 on helper T cells). Also called CD markers. *(Chapter 4)*

Complement   A general term referring to all the components of the complement system. *(Chapter 4)*

Complement system   A set of blood plasma proteins that act in a concerted fashion to destroy extracellular pathogens and infected cells; originally defined as a heat-labile activity that lysed bacteria in the presence of antibody (it "complemented" antibody action); the activated complement pathway also stimulates phagocytosis, chemotaxis, and inflammation. *(Chapter 4)*

c-Oncogene   *See* Proto-oncogene.

Contact inhibition   Cessation of cell division when cells make physical contact, as occurs at high density in a culture dish. *(Chapter 7)*

Cytokines   Soluble proteins produced by cells in response to various stimuli, including virus infection; they affect the behavior of other cells both locally and at a distance, by binding to specific cytokine receptors. *(Chapter 4)*

Cytopathic virus   A virus that causes characteristic visible cell damage and death upon infection of cells in culture. *(Chapter 5)*

Delayed-type hypersensitivity   Cell-mediated immunity caused by CD4 T cells that recognize antigens in the skin; the reaction typically occurs hours to days after antigen is injected, hence its name; it is partially responsible for characteristic local reactions to virus infection, such as a rash. *(Chapter 4)*

Dermis   The layer of skin beneath the epidermis that supports the basement membrane or vascular network; it is composed of a dense connective tissue that provides support and elasticity to the skin. *(Chapter 1)*

Diapedesis   The process by which viruses cross the vascular endothelium, while being carried within monocytes or lymphocytes. *(Chapter 1)*

Disseminated infection   An infection that spreads beyond the primary site of infection; often includes a viremia and infection of major organs such as the liver, lungs, and kidneys. *(Chapter 1)*

DNA synthesis phase   *See* S phase.

DNA vaccine   A preparation of DNA containing the genes for one or more antigenic proteins; when the pure DNA preparation is injected into a test subject, the proteins are expressed, and an immune response to those proteins is elicited. *(Chapter 8)*

Emerging virus A viral population responsible for a marked increase in disease incidence, usually as result of changed societal, environmental, or population factors. *(Chapter 10)*

Endogenous antigen presentation The cellular process in which viral proteins are degraded inside the infected cell, and the resulting peptides are loaded onto major histocompatibility complex class I molecules that move to the cell surface. *(Chapter 4)*

Enhancing antibodies Antibodies that can facilitate viral infection by allowing virions to which they bind to enter susceptible cells. *(Chapter 6)*

Enterotropic virus A virus with a predilection to infect cells of the enteric system. *(Chapter 1)*

Epidemic A pattern of disease characterized by rapid and sudden appearance of cases spreading over a wide area. *(Chapter 1)*

Epidemiology The study of the incidence, distribution, and spread of infectious disease in populations with particular regard to identification and subsequent control. *(Chapter 2)*

Epidermis The external surface of the skin, composed of a keratinized stratified squamous epithelium. *(Chapter 1)*

Error threshold A mathematical parameter that measures the complexity of the information that must be maintained to ensure survival of a population. *(Chapter 10)*

Exogenous antigen presentation The cellular process in which viral proteins are taken up from the outside of the cell, digested, and the resulting peptides loaded onto major histocompatibility complex class II molecules that move to the cell surface. *(Chapter 4)*

Fitness The replicative adaptability of an organism to its environment. *(Chapter 10)*

Foci Clusters of cells that are derived from a single progenitor and share properties, such as unregulated growth, that cause them to pile up on one another. *(Chapter 7)*

Fomites Inanimate objects that may be contaminated with microorganisms and become vehicles for transmission. *(Chapter 2)*

Gap phases (G_1 and G_2) Phases in the cell cycle between the mitosis (M) and DNA synthesis (S) phases. *(Chapter 7)*

Genetic bottleneck A descriptive term evoking the extreme selective pressure on small populations that results in loss of diversity, accumulation of nonselected mutations, or both. *(Chapter 10)*

G_0 *See* Resting state.

Hematogenous spread Spread of virus particles through the bloodstream. *(Chapter 1)*

Hepatitis Inflammation of the liver. *(Chapter 1)*

Herd immunity The immune status of a population, rather than an individual. *(Chapter 8)*

Humoral response The arm of the adaptive immune response that produces antibodies. *(Chapter 4)*

Immortality The capacity of cells to grow and divide indefinitely. *(Chapter 7)*

Immune defenses Host defenses against pathogens comprising the innate and adaptive systems. *(Chapter 3)*

Immune memory A property provided by specialized B and T lymphocytes (memory B and T cells) that respond rapidly on reexposure to the viral infection that originally induced them. *(Chapter 8)*

Immunodominant Having the property of being recognized most efficiently by cytotoxic T lymphocytes and antibodies; said of peptides and epitopes. *(Chapter 5)*

Immunopathology Pathological changes partly or completely caused by the immune response. *(Chapter 4)*

Immunotherapy A treatment that provides an infected host with exogenous antiviral cytokines, other immunoregulatory agents, antibodies, or lymphocytes in order to reduce viral pathogenesis. *(Chapter 8)*

Incidence The frequency with which a disease appears in a particular population or area (e.g., the number of newly diagnosed cases during a specific period); distinct from the prevalence (i.e., the number of cases in a population on a certain date). *(Chapter 2)*

Inflammation A general term for the complex response that gives rise to local accumulation of white blood cells and fluid; initiated by local infection or damage; many different forms of this response, characterized by the degrees of tissue damage, capillary leakage, and cellular infiltration, occur after infection with pathogens. *(Chapter 4)*

Innate response The first line of immune defense; able to function continually in the host without prior exposure to the invading pathogen. This complex system comprises, in part, cytokines, sentinel cells, complement, and natural killer cells. *(Chapter 4)*

Insertional activation The mechanism of oncogenesis by nontransducing retroviruses; integration of a proviral promoter or enhancer in the vicinity of a c-oncogene results in inappropriate transcription of that gene. *(Chapter 7)*

Insertional mutagenesis Mutation in a genome caused by the integration of viral DNA or the DNA of a transposable element. *(Chapter 3)*

Instability elements *cis*-acting repressive sequences in lentiviral RNA molecules that respond to Rev protein to increase stability, nuclear export, and translatability. *(Chapter 6)*

Interfering antibodies   Antibodies that can bind to virions or infected cells and block interaction with neutralizing antibodies. *(Chapter 6)*

Intrinsic cellular defenses   The conserved cellular program that responds to various stresses, such as starvation, irradiation, and infection; intrinsic defenses include apoptosis, autophagy, and RNA interference; unlike immune defenses, intrinsic defenses do not include cytokines and white blood cells. *(Chapter 3)*

Killed vaccine   A vaccine made by taking an authentic disease-causing virus and treating it (e.g., with chemicals) to reduce infectivity to nondetectable levels. *(Chapter 8)*

Koch's postulates   Criteria developed by the German physician Robert Koch in the late 1800s to determine whether a given agent is the cause of a specific disease. *(Chapter 1)*

Kupffer cells   Macrophages of the liver that are part of the reticuloendothelial system. *(Chapter 1)*

Latency-associated transcript   RNA produced specifically during a latent infection by herpes simplex virus. *(Chapter 5)*

Latent infection   A class of persistent infection that lasts the life of the host; few or no virions can be detected, despite continuous presence of the viral genome. *(Chapter 5)*

Lethal mutagenesis   The elevation of mutation rates by exposure to a mutagen or an error-prone polymerase to the point at which the resulting population of genomes has lost fitness and is incapable of propagating. *(Chapter 10)*

Live, attenuated vaccine   A vaccine made from viral mutants that have reduced virulence but are competent for replication; they often also have reduced capacity for transmission. *(Chapter 8)*

Long-latency virus   A retrovirus that causes cancer in a host many years after infection; the viral genome does not encode cellular oncogenes, nor does it cause cancer by perturbing the expression of cellular oncogenes. *(Chapter 7)*

Macules   Flat, colored skin lesions caused by virus replication in the dermis. *(Chapter 1)*

Mannan-binding pathway   One of three complement pathways that lead to activation of C3-C5 convertases; mannose-binding, lectin-associated proteases cleave the C2 and C4 proteins. *(Chapter 4)*

Memory cells   A subset of B and T lymphocytes maintained after each encounter with a foreign antigen; they survive for years and are ready to respond and proliferate upon subsequent encounter with the same antigen. *(Chapter 4)*

Metastases   Secondary tumors, often at distant sites, that arise from the cells of a malignant tumor. *(Chapter 7)*

Microbicides   Creams or ointments that either inactivate or block virions before they can attach and penetrate tissues. *(Chapter 9)*

Mitogens   Extracellular signaling molecules that induce cell proliferation. *(Chapter 7)*

Mitosis   The phase of the cell cycle in which newly duplicated chromosomes are distributed to two new daughter cells as a result of cell division. Also called M phase. *(Chapter 7)*

Molecular mimicry   Sequence similarities between viral peptides and self-peptides that result in the cross-activation of autoreactive T or B cells by virus-derived peptides. *(Chapter 4)*

Monoclonal antibody-resistant mutants   Viral mutants selected to propagate in the presence of neutralizing monoclonal antibodies; often carry mutations in viral genes encoding virion protein. *(Chapter 4)*

M phase   *See* Mitosis.

Muller's ratchet   A statement positing that small, asexual populations decline in fitness over time if the mutation rate is high. *(Chapter 10)*

Natural killer cells   An abundant lymphocyte population that comprises large, granular lymphocytes; distinguished from others by the absence of B and T cell antigen receptors; these cells are part of the innate defense system. Also called NK cells. *(Chapter 4)*

Neuroinvasive virus   A virus that can enter the central nervous system (spinal cord and brain) after infection of a peripheral site. *(Chapter 1)*

Neurotropic virus   A virus that can infect neurons. *(Chapter 1)*

Neurovirulent virus   A virus that can cause disease in nervous tissue, manifested by neurological symptoms and often death. *(Chapter 1)*

NK cells   *See* Natural killer cells.

Noncytopathic virus   A virus that produces no visible signs of infection in cells. *(Chapter 5)*

Nontransducing oncogenic retroviruses   Retroviruses that do not encode cell-derived oncogene sequences but can cause cancer (at low efficiency) when their DNA becomes integrated in the vicinity of a cellular oncogene, thereby perturbing its expression. *(Chapter 7)*

Oncogene   A gene encoding a protein that causes cellular transformation or tumorigenesis. *(Chapter 7)*

Oncogenesis   The processes leading to cancer. *(Chapter 7)*

Pandemic   A worldwide epidemic. *(Chapter 1)*

Pantropic virus   A virus that replicates in many tissues and cell types. *(Chapter 1)*

Papules   Slightly raised skin lesions caused by virus replication in the dermis. *(Chapter 1)*

Passive immunization   Direct administration of the products of the immune response (e.g., antibodies or stimulated immune cells) obtained from an appropriate donor(s) to a patient; contrasts with active immunization. *(Chapter 8)*

Passive viremia   Introduction of virus particles into the blood without viral replication at the site of entry. *(Chapter 1)*

Pathogen   A disease-causing virus or other microorganism. *(Chapter 1)*

Pattern recognition receptors   Unique protein receptors of the innate immune system that bind common molecular structures on the surfaces of pathogens; they reside on the cell surfaces of sentinel cells, such as immature dendritic cells and macrophages. *(Chapter 3)*

Permissive   Able to support virus replication when viral nucleic acid is introduced; refers to cells. *(Chapter 1)*

Permissivity   A cellular environment that provides all cellular components required for viral replication. *(Chapter 1)*

Persistent infection   A viral infection that is not cleared by the combined actions of the innate and adaptive immune response. *(Chapter 5)*

Plaque-forming unit (PFU)   A measure of virus infectivity when measured as plaque-forming units per milliliter. *(Chapter 1)*

Polymorphic gene   A gene that has many allelic forms in outbred populations. *(Chapter 4)*

Power   The probability that a meaningful difference or effect can be detected, if one were to occur. *(Chapter 2)*

Prevalence   The proportion of individuals in a population having a disease; the number of cases of a disease present in a particular population at a given time. *(Chapter 2)*

Primary antibody response   The initial response of B cells when first exposed to an infection. *(Chapter 4)*

Primary viremia   Progeny virions released into the blood after initial replication at the site of entry. *(Chapter 1)*

Prions   Infectious agents comprising an abnormal isoform of a normal cellular protein but no nucleic acid; implicated as the causative agents of transmissible spongiform encephalopathies. *(Appendix B)*

Professional antigen-presenting cells   Dendritic cells, macrophages, and B cells; defined by their ability to take up antigens and present them to T lymphocytes in the groove of an major histocompatibility complex class II molecule. *(Chapter 4)*

Proinflammatory cytokines   Cytokines produced predominantly by activated immune cells; responsible for amplification of inflammatory reactions. *(Chapter 3)*

Proto-oncogene   A normal gene that, when altered by mutation, becomes an oncogene that can contribute to cancer. Also called c-oncogene. *(Chapter 7)*

Pustules   Skin lesions derived from a vesicle in which secondary infiltration of leukocytes occurs. *(Chapter 1)*

Quasispecies   Virus populations that exist as dynamic distributions of nonidentical but related replicons. *(Chapter 10)*

Recombinant vaccine   A vaccine produced by recombinant DNA technology. *(Chapter 8)*

Reservoir   The host population in which a viral population is maintained. *(Chapter 2)*

Resting state   A state in which the cell has ceased to grow and divide and has withdrawn from the cell cycle. Also called G_0. *(Chapter 7)*

Rev-responsive element (RRE)   A structural element in human immunodeficiency virus RNA that is recognized by the viral Rev protein, which mediates its export from the nucleus. *(Chapter 6)*

Satellites   Small, single-stranded RNA molecules that lack genes required for their replication, but do replicate in the presence of another virus (the **helper virus**). *(Appendix B)*

Satellite virus   A satellite with a genome that encodes one or two proteins. *(Appendix B)*

Secondary antibody response   The antibody response produced after a subsequent infection or challenge with the same antigen or virus. *(Chapter 4)*

Secondary viremia   Delayed appearance of a high concentration of infectious virus in the blood as a consequence of disseminated infections. *(Chapter 1)*

Second messengers   Small molecules, such as cyclic nucleotides and lipids, that are generated by some membrane-bound proteins in a signal transduction cascade, and that act as diffusible components in a signal relay. *(Chapter 7)*

Sentinel cells   Dendritic cells and macrophages; migratory cells that are found in the periphery of the body and can take up proteins and cell debris for presentation of peptides derived from them on major histocompatibility complex molecule. These cells respond to recognition of a pathogen by synthesizing cytokines such as interferons. *(Chapter 4)*

Signal transduction cascade   A chain of sequential physical interactions among, and biochemical modification of, membrane-bound and cytoplasmic proteins. *(Chapter 7)*

Sinusoids   Small blood vessels characterized by a discontinuous basal lamina, with no significant barrier between the blood plasma and the membranes of surrounding cells. *(Chapter 1)*

Slow infection   An extreme variant of the persistent pattern of infection; has a long incubation period (years) from the time of initial infection until the appearance of recognizable symptoms. *(Chapter 5)*

Slow viruses Viruses characterized by long incubation periods, typical for the genus lentivirus in the family *Retroviridae*. *(Chapter 6)*

Smoldering infection A low rate of viral replication equal to the rate of elimination. *(Chapter 5)*

S phase The phase of the cell cycle in which the DNA genome is replicated. *(Chapter 7)*

Subunit vaccine A vaccine formulated with purified components of virus particles, rather than intact virions. *(Chapter 8)*

Susceptibility The property of a cell that enables it to be infected by a particular virus (e.g., the presence of a viral receptor[s] on the cell surface). *(Chapter 1)*

Susceptible Producing the receptor(s) required for virus entry; refers to cells. *(Chapter 1)*

Systemic infection An infection that results in spread to many organs of the body. *(Chapter 1)*

Systemic inflammatory response syndrome Overexpression or a disproportionate host response that leads to large-scale release of inflammatory cytokines and stress mediators, resulting in severe pathogenesis or death. Also known as a cytokine storm. *(Chapter 4)*

Transcytosis A mechanism of transport in which material in the intestinal lumen is endocytosed by M cells, transported to the basolateral surface, and released to the underlying tissues. *(Chapters 1 and 4)*

Transducing oncogenic retroviruses Retroviruses that include oncogenic, cell-derived sequences in their genomes and carry these sequences to each newly infected cell; such viruses are highly oncogenic. *(Chapter 7)*

Transformed Having changed growth properties and morphology as a consequence of infection with certain oncogenic viruses, introduction of oncogenes, or exposure to chemical carcinogens. *(Chapter 7)*

Transforming infection A class of persistent infection in which cells infected by certain DNA viruses or retroviruses may exhibit altered growth properties and proliferate faster than uninfected cells. *(Chapter 5)*

Tropism The predilection of a virus to invade, and replicate, in a particular cell type. *(Chapter 1)*

Tumor A mass of cells originating from abnormal growth. *(Chapter 7)*

Tumor suppressor gene A cellular gene encoding a protein that negatively regulates cell proliferation; mutational inactivation of both copies of the genes associated with tumor development. *(Chapter 7)*

Vaccination Inoculation of healthy individuals with attenuated or related microorganisms, or their antigenic products, in order to elicit an immune response that will protect against later infection by the corresponding pathogen. *(Chapter 8)*

Vesicles Focal necroses that occur when an infection spreads from the capillaries to the superficial layers of the skin and replicates in the epidermis; usually contain inflammatory fluids. *(Chapter 1)*

Viral pathogenesis The processes by which viral infections cause disease. *(Chapters 1 and 2)*

Viremia The presence of infectious virions in the blood. *(Chapter 1)*

Virion An infectious virus particle. *(Chapter 1)*

Viroceptor A viral protein that modulates cytokine signaling or cytokine production by mimicking host cytokine receptors. *(Chapters 2 and 3)*

Viroids Unencapsidated, small, circular, single-stranded RNA molecules that replicate autonomously when introduced mechanically into host plants. *(Appendix B)*

Virokine A secreted viral protein that mimics cytokines, growth factors, or similar extracellular immune regulators. *(Chapters 2 and 3)*

Virulence The relative capacity of a viral infection to cause disease. *(Chapter 2)*

Virulent virus A virus that causes disease. *(Chapter 2)*

Viruria The presence of viruses in the urine. *(Chapter 2)*

Viruses Submicroscopic, obligate parasitic pathogens comprising genetic material (DNA or RNA) surrounded by a protective protein coat. *(Chapter 1)*

Virus evolution The constant change of a viral population in the face of selection pressures. *(Chapter 10)*

v-Oncogene An oncogene that is encoded in a viral genome. *(Chapter 7)*

Zoonoses (zoonotic infections) Diseases that are transferred from animals to humans. *(Chapters 2 and 10)*

Index

A

A18R protein, vaccinia virus, 71
A46R protein, vaccinia virus, 62
A52R protein, vaccinia virus, 62
Abacavir (ABC, Ziagen), 280, 301
ABC, see Abacavir
Abelson murine leukemia virus, 217
abl oncogene, 220
Abortive infection, 135, 160–161
Accessory proteins, human immuno-
    deficiency virus, 169, 171–174
Accidental natural infection, 341
Acid resistance, rhinovirus versus
    poliovirus, 12
Actin fiber, 222
Active immunization, 256
Active viremia, 17
Acute infection, 135–136, 161
    antigenic variation in viruses that cause,
        140–141
    course, 139
    definition, 138
    immune response, 140
    inapparent, 138
    incubation period, 138–141
    public health problems, 141–142
    requirements, 138
    r-replication strategy, 137
    uncomplicated, 139
Acute postinfectious encephalitis, 148–149
Acute-phase proteins, 59, 67
Acyclic nucleoside phosphonate, 290–291
Acyclovir, 280, 288–289, 292, 294–295, 299,
    365–366, 368
    resistance, 298
Adapter, 222
Adapter proteins, viral, 172, 226–227
Adaptive immune response, 54, 128
    antibody response, 116–120
    antigen presentation, 107–112

brain and, 120–121
cell-mediated, 110–116
cells of, 101–103, see also B cells; T cells
defined, 87–89
general features, 99–101
hygiene hypothesis, 151
integration with intrinsic cellular defense
    and innate immune response, 56
lymphocytes that carry distinct antigen
    receptors, 102–107
mucosal and cutaneous arms of immune
    system, 102–103
persistent infection and, 143–147
regulation, 99
self-limitation, 102
specificity, 102
white blood cells in, 88
ADCC, see Antibody-dependent cell-
    mediated cytotoxicity
Adefovir, 291, 294–295
Adeno-associated virus, 374
Adenosine arabinoside, 292, 365
Adenovirus
    activation of Nf-κb, 66
    antiviral drugs, 291
    disease mechanisms, 358
    diseases, 358
    E1A protein, 71, 145, 218, 220, 231–234,
        238–240, 244
    E1B protein, 77, 218, 234, 238–240, 244
    E3 protein, 77, 99, 129–130, 145
    E4 protein, 78, 218, 244
    entry into host, 10–12
    epidemiology, 358
    gene expression in virus-infected cells, 7
    gene products that suppress RNA
        interference, 79
    host alterations early in infection, 58
    immune response, 113, 115
    incubation period, 141

inflammatory response, 99
interferon response, 70–71
oncogenic, 209–210, 213
persistent infection, 142–143
regulation of MHC class I proteins, 144
regulators of apoptosis, 77
RID protein complex, 130
transformation mechanisms, 218, 244
transformed cells, 240
transmission, 36
VA-RNA I, 70–71, 79
VA-RNA II, 79
viral spread, 102
Adenovirus vaccine, 263, 358
Adenovirus vector, 121
Adjuvant, 97, 271–273
Adoptive transfer, 123, 256
Adoptive-transfer experiment, 114
    T-cell mediated delayed-type
        hypersensitivity, 114
Adrenal gland, invasion by virus, 21
Adsorption inhibitors, 290
Adult T-cell lymphocytic leukemia, 241, 243
Aerosol transmission, 9–10, 34–35
    humidity and, 40
African swine fever virus, 45, 352
Age-dependent susceptibility, 48–50
Agenerase, see Amprenavir
Agricultural animals, see Livestock
AIDS, 180–181, 352, 381, see also Human
    immunodeficiency virus
    cancer and, 191–194
    effects of virus on different tissues/organs,
        188–191
    historical aspects, 166–167, 196
    opportunistic infections, 180, 182,
        190–191
    worldwide problem, 165–166
AIDS dementia, 188
AIDS vaccine, 195–196, 271, 274–275

"Air gun," vaccine delivery, 272
Air pollution, 50
Air travel, 345
akt oncogene, 220
Akt protein kinase, 226–227, 230–231, 234–236
Alimentary tract, viral entry, 10–13
Allergy, hygiene hypothesis, 151
Alphaherpesvirus
  congenital infection, 37
  evolution, 322–324
  neural spread, 20–21
  regulation of complement cascade, 96
  tracing neuronal connections with viruses, 20
Alphavirus
  E1 protein, 326, 329
  evolution, 326, 328–329
  immune response, 120
Alternative hypothesis, 32
Aluminum hydroxide gel, 97
Aluminum salts, as adjuvants, 273
Alzheimer's disease, 350
Amantadine (Symmetrel), 280–281, 291–293, 297
AMD3100, 290
Americas, colonization by Europeans, 335, 338, 343–344
Amoeba, host of mimivirus, 324
Amprenavir (Agenerase), 280, 301
Anaphase, 207
Anaphylotoxin, 94
Anchorage-independent cells, 204
Angiogenesis, 224, 240
Animal bite, 14
Animal models, 29–30, 42
  human immunodeficiency virus infection, 195
Anogenital carcinoma, human papillomavirus, 194
Antibodies, 18, 96, 99–100, 102, 104, 128, *see also* Immunoglobulin *entries*
  as antiviral drugs, 293
  maternal, 256, 258, 342
  measurement of antiviral antibody response, 115
  "natural," 95–96
  passive immunization, 256, 258
  production by plasma cells, 116
  protective immunity, 258
  structure, 116–117
  virus neutralization, 116–120
"Antibody escape" mutant, 120
Antibody response, 116–120
  evoked by vaccination, 259–261
  primary, 116
  secondary, 116
Antibody titer, 102
Antibody-dependent cell-mediated cytotoxicity (ADCC), 120, 185
Antibody-resistant virion, 141
Antigen, 102
Antigen presentation
  endogenous, 108–110
  exogenous, 110–111
Antigenic determinant, 125
Antigenic drift, 120, 141
Antigenic shift, 141–142
Antigenic variation, 140–141
Antigen-presenting cells, 101, 103, 111–112

professional, 89, 99, 101, 108
  T cell recognition, 110–111
Anti-inflammatory cytokines, 59, 64
Antimutator, 316
Antisense oligonucleotides, 281
Antisense strand, proviral DNA, 241
Antiviral drugs, 35, 142, 279–309, *see also specific drugs*
  bioavailability, 281, 289
  costs, 287–289, 306, 351
  cross-resistance, 305–306
  cytotoxic index, 287
  delivery, 289
  discovery, 281–299, 301
  drug design, 296
    in silico drug discovery, 286–287
    structure-based, 285–286
    using genome sequencing information, 285–286
  efficacy, 279, 288
  formulation, 289
  historical aspects, 281
  human immunodeficiency virus, 187–188, 194–195, 299–307
  marketing, 287–289
  pharmacokinetics, 282
  potency, 279, 307
  proof of principle, 282
  research and development, 287–289, 351
  resistance, 298, 307
    human immunodeficiency virus, 303–306
  safety, 279, 287–289
  screening
    blind screens, 281–282
    cell-based screens, 282–284
    high-content screens, 285
    high-throughput screens, 284–285
    mechanism-based screens, 282–283
    minireplicon system, 282–283
    sources of chemicals to screen, 285
    virtual screens, 287
  side effects, 304, 306
  structure, 292
  targets, 280–281, 290–291
    entry and uncoating processes, 293
    intrinsic defense receptors, 298
    proteases, 293–297
    regulatory proteins, 295
    regulatory RNA molecules, 295
    replication of RNA viral genomes, 299
    virus-specific nucleic acid synthesis and processing, 295
  therapeutic index, 287
Antiviral state, 65–70
  produced by interferon, 63–64
Ap-1 protein, 69, 226, 229
Apaf-1 protein, 74, 76, 236
Apical surface, 14, 16, 118
Apo2/Trail receptor, 76
Apobec (apolipoprotein B editing complex), 79
  Apobec3 proteins, 173–174
Apoptosis, 72–78
  CTL-induced, 112
  defense against viral infection, 74, 77
  inhibition in transformed cells, 234–239, 244–245

integration with stimulation of proliferation, 234
  viral inhibitors, 234
  monitoring by sentinel cells, 77–78
  T-cell, Fas-mediated, 147
  viral gene products that inhibit, 74–77
  viral regulators, 77
  virus-induced, 45
Apoptosome, 76
Apoptotic body, 75
Appendix, 102
Apple scar skin viroid, 385
Arachidonic acid metabolites, 192
Arbovirus
  seasonal variation, 341
  transmission, 36
Arenavirus
  antiviral drugs, 359
  diseases, 359
  dissemination in host, 15
  emerging disease, 334
  entry into host, 12, 15
  epidemiology, 359
  evolution, 327
  immune response, 115, 120
  persistent infection, 142
  shedding from host, 15, 35
  transmission, 36
  treatment, 359
Arf proteins, 69, 237–238
Arterivirus, evolution, 328
Arthropod vector, 35, 37–38
Artificial MHC tetramer, 115
Asian tiger mosquito, 38
Asthma, hygiene hypothesis, 151
Astrocytes, 113, 120–121, 127
Atf-2 protein, 229
Atm protein, 236–237
Atripla, 306
Attachment to host cells, 281
Attenuated virus, 40, 42, 44–45
  attenuation by passage in nonhuman cell lines, 265
Attenuated (live) virus vaccine, 260, 262–269
AU-rich pentameric element, 62
Autism, MMR vaccine and, 268
Autocrine growth stimulation, 203
Autoimmune disease, 107, 124–126
Autoimmunity, HIV-infected individuals, 183–185
Autophagy, 65, 72, 78
Auxiliary proteins, human immunodeficiency virus, 169, 171
Avian erythroblastosis virus, 217
Avian influenza virus, 328, 340, 348–349
  emerging disease, 340
  H5N1, 262
  HA protein, 26
  reassortment with human viruses, 318–319
  tropism, 25–26
Avian myeloblastosis virus, 217
Avian myelocytoma virus, 217, 223
Avian reticuloendotheliosis virus, 217
Avian sarcoma and leukosis virus, 209, 217
  congenital infection, 37
  insertional activation of c-*myc*, 225
Avirulent virus, 40–41
AZT, *see* Zidovudine

B

B box, 170
B cells, 101, 104
  activated, 100
  HIV-infected individuals, 183
  immunopathological lesions caused by, 123–124
  memory, 99, 102, 104, 157
    acute infection, 140
  regulation by complement system, 95
  source/function, 88
  synthesis of MHC class II proteins, 108
  virus-infected, 148
    Epstein-Barr virus, 157, 159
B18R protein, vaccinia virus, 71
Bacteriophage, evolution, 328
Bacteriophage lambda, 321
Bacteriophage MS2, 320
Bacteriophage Qβ, 314, 320
Bad protein, 234–235
Bak protein, 74, 76
Bandicoot papillomatosis carcinomatosis virus type 1, 209
Barmah Forest complex, 329
Basement membrane, 11, 13, 16, 21–22
Basolateral surface, 14, 16, 118
Basophils, 95
  in inflammatory response, 99
  source/function, 88
Bat vector, 338–340, 345
Bax protein, 74, 76, 236
B-cell lymphoma, 191, 193–194, 214
  Epstein-Barr virus, 157–159, 193–194, 367
  human herpesvirus 8, 193, 367
B-cell receptor, 99, 101–107, 116
Bcl-2 protein, 74, 76–77, 143, 223, 234
Bcrf1 protein, Epstein-Barr virus, 71
Benign tumor, 202
Benzimidazole, 295
Benzoquinone, 293
Betaherpesvirus
  evolution, 322–324
  persistent infection, 147
Bile salts, 10
Bim protein, 234
Binary data, 32
Bioavailability, antiviral drugs, 281, 289
Biological control, 342
Biological weapon, see Bioterrorism
Bioterrorism, 43, 351
  agricultural pathogens, 256
  smallpox virus, 252
Birds
  dead-end host, 337
  West Nile virus in, 336
BK virus, 213
  disease mechanisms, 377
  diseases, 377
  epidemiology, 377
  persistent infection, 143
Blind screening, 281–282
Block copolymer, 272
Blood
  shedding of virions in, 35
  viral transmission, 36
Blood products, viral contamination, 18, 336, 349–350
  human immunodeficiency virus, 177–178

Blood-borne virus, 15
Blood-brain barrier, 120–121
Blood-tissue junction, 21–22
Body cavity-based lymphoma, 193
Body fluids, human immunodeficiency virus, 177–178
Body piercing, 14
Bone marrow, invasion by virus, 21
Booster shot, 259
Bottleneck, genetic, 315, 317–320, 331
Bovine immunodeficiency virus, 167, 169
Bovine papillomavirus, 218
  E5 protein, 228
Bovine papular stomatitis virus, 378
Bovine spongiform encephalopathy, 387–389, 391
Bovine viral diarrhea virus
  congenital infection, 37
  evolution, 320
  persistent infection, 143
Brain
  attenuated herpes simplex virus to clear brain tumors, 45
  blood-brain barrier, 120–121
  congenital viral infections, 49
  entry of viruses by olfactory routes, 19
  human immunodeficiency virus, 188–192
  immune system and, 120–121
  invasion by virus, 22
  measles virus, 148–149
  prion infections, 160
  tracing neuronal connections with viruses, 20
Breast-feeding, transmission of human immunodeficiency virus, 178–179
Brivudin, 280
Bronchiolitis, 11
Bronchitis, 11
Bunyamwera virus, 360
Bunyavirus
  disease mechanisms, 360
  diseases, 360
  emerging disease, 334
  entry into host, 12
  epidemiology, 360
  evolution, 327
  G1 glycoprotein, 47
  transmission, 36
  vector, 360
  virulence, 47
Burkitt's lymphoma, Epstein-Barr virus, 158–159, 210, 240
Bushmeat trade, 345
Bystander effect, 176
BZLF2 protein, Epstein-Barr virus, 146

C

3C protein, poliovirus, 58
C' receptor, 185
C1q, 56, 94
  pattern recognition, 96–97
C3 convertase, 93–95
C5 convertase, 93–95
CA protein, human immunodeficiency virus, 184
Caf protein, 186
Calicivirus
  disease mechanisms, 361

  diseases, 361
  emerging disease, 334
  entry into host, 10
  epidemiology, 361
  evolution, 320, 328
  marine, 344
California encephalitis virus, 360
Calnexin, 109
Calpain, 222
Canarypox virus, 270
Cancer, see also Oncogenic virus; Transformed cells; Viral transformation
  defined, 202
  genetic paradigm, 208
  human immunodeficiency virus and, 191–194
Canine distemper virus, 125
Canine parvovirus
  emerging disease, 345–346
  evolution, 314, 346
  mutation rate, 314
  VP2 protein, 346
Cap snatching, 65
Capillary
  blood-tissue junction, 21–22
  fenestrated, 21–22
Capillovirus, 328
Caprine arthritis-encephalitis virus, 167, 169
Capsid, constraints on viral evolution, 331
Carcinogenesis, 202
Carcinoma, 202, 209
CARD (caspase activation and recruitment domain), 60
Cardiac muscle, invasion by virus, 22
Cardinal data, 32
Carlavirus, 328
Carmovirus, 328
Carrier state, 150
Carrying capacity (K), 137, 313
Caspase, 72, 74–77, 97–99
  activation
    extrinsic pathway, 72
    intrinsic pathway, 72–73
Cat
  canine parvovirus in, 345–346
  influenza viruses in, 349
Cbp protein, see p300/Cbp proteins
CCr1 receptor, 65
CCr2 receptor, 65
CCr5 receptor, 65, 169, 176–177, 302
CCr6 receptor, 65
CCr7 receptor, 65
CD markers, 105
CD4, 100, 104–105, 111, 175–177, 182, 184–185
  antiviral derivatives, 300, 302–303
  autoantibodies in HIV-infected individuals, 184
  degradation, 175–176
  MHC class II-CD4 interactions, 182
CD8, 100, 104–105, 109, 116
CD19, 116
CD21, 116, 160
CD28, 90, 112, 176
CD40, 116
CD46, 96, 101, 148
CD55, 96
CD59, 96
CD80, 112

CD95, 76
CD150, 148
CD155, 24
Cdc6 protein, 233
CDId protein, 107
Cdk, *see* Cyclin-dependent kinase
Cell adhesion, 222
Cell culture, 348
　establishment, 204
　human immunodeficiency virus, 176–177
　transformation, 202–204
Cell cycle
　disruption by viral oncogene products,
　　230–234
　　abrogation of restriction point control by
　　　Rb protein, 230–234
　　inactivation of cyclin-dependent kinase
　　　inhibitors, 233–234
　　virus-specific cyclins, 233
　phases, 205–208
　regulation, 205–206
　viral transformation via control pathways,
　　230–234
Cell cycle engine, 206–208
Cell lysis, by complement system, 94–96
Cell proliferation, control
　cell cycle engine, 206–208
　regulation of cell cycle, 205–206
　sensing environment, 204–206
Cell-based screen, 282–284
Cell-mediated immune response, 88, 96,
　　99–101, 110–116, *see also* Adaptive
　　immune response
　dysfunction in HIV-infected individuals,
　　182–184, 186
　measurement, 115
Cellular defense, intrinsic, 54–81
　Apobec, 79
　apoptosis, 72–78
　autophagy, 78
　cytokines, 59–61, 64
　defined, 55
　epigenetic silencing, 78
　integration with innate and adaptive
　　immune response, 56
　interferons, 61–74
　pattern recognition receptors, 55–59
　recognition of viral invaders, 55, 57–58,
　　60–61
　RNA silencing, 78–79
　summary, 81
　Trim proteins, 79–80
Cellular origin theory, 322
Cellular transformation, 202–204
Central European tick-borne flavivirus, 339
Central nervous system
　human immunodeficiency virus, 188–192
　invasion by virus, 17, 22
　viral entry by olfactory routes, 19
Cerebrospinal fluid, human immuno-
　　deficiency virus, 177–178
Cervical carcinoma, 267–268, 370
Checkpoint, 207
Checkpoint kinase 2, 237
Chemical defenses, 54
Chemical library, 285
Chemokine(s), 48, 59, 64, 186
Chemokine receptor, 48, 65, 89–90, 176–177
Chemokine receptor antagonists, 300, 302–303

Chicken, influenza viruses in, 348–349
Chickenpox, 368
Chikungunya virus, 37–38
Children
　mother-child transmission of human
　　immunodeficiency virus, 177–178
　susceptibility to viral diseases, 48–50
Chlorella virus, 323
Chlorovirus, 323
Choroid plexus, 22
Chromatin, 153, 156
　in apoptosis, 75
　remodeling, 69
Chromium release assay, 115
Chronic infection, 142, 264
Chronic wasting disease, 387
Cidofovir, 280, 291, 294–295, 366
Cigarette smoking, 50
Cip/Kip proteins, 208, 233
Circadian rhythm, 40
Circovirus
　evolution, 314, 320
　persistent infection, 142
Cirrhosis, 246, 364
cis-acting regulatory sequences
　human immunodeficiency virus, 169–172
　sequence conservation, 316
Clade, 167, 271, 330, 345
Clara cells, 24
Class A infectious agents, 351
Climate effects, emerging disease, 344
Clonal response, 104
c-Myc protein, 220
Coconut cadang-cadang viroid, 385
Coiled-coil domain, 170
Colchicine, 19
Cold sore, 155
Collectin, 96–97
Colon, invasion by virus, 22–23
Colon carcinoma, 203, 223
Colony-stimulating factor, 59, 67
Colorado tick fever virus, 379
Colostrum, 256
Coltivirus, 379
Combinatorial chemistry, 285
Combivir, 301
Common cold, 141
　search for cure, 280
"Common host response" to infection, 7
Comovirus, 328
Comparative medicine, 353
Complement cascade, 89, 93–97
　activation of inflammation, 94–95
　activities of proteins and peptides
　　released, 95
　alternative pathway, 93–94
　cell lysis, 94–96
　classical pathway, 93–94
　inhibitor coded in variola virus
　　genome, 46
　mannan-binding pathway, 93–94
　"natural antibody," 95–96
　opsonization, 94–95
　pattern recognition by C1q, collectins, and
　　defensins, 96–97
　regulation, 94, 96
　regulation of T-cell response, 101
　solubilization of immune complexes,
　　94–95

　stimulation of inflammatory response, 98
　synthesis of complement proteins, 95
Complement receptor, 185
Complement receptor type 1 protein, 96
Complement-inhibitory proteins, 96
Computer game model, epidemics of
　　infectious disease, 34
c-oncogene, 215
Congenital infection, 37
Conjunctivitis, 13
Connective tissue, invasion by virus, 22
Contact inhibition, 203–204
Convergence model, emerging viruses, 334
Coronavirus
　entry into host, 10–11
　evolution, 328
　host susceptibility to viral disease, 49
　immune response, 115
　immunopathology, 120
Corticosteroid hormones, 50
Coughing, 10, 34, 40
Cowpox virus, 253, 378
　CrmB protein, 77
　regulators of apoptosis, 77
Coxsackievirus
　disease mechanisms, 376
　diseases, 375
　entry into host, 12
　epidemiology, 375
　immunopathology, 120, 122
　organ invasion, 22
　skin rash, 23
CpG residue, methylation, 160
CpG-containing oligonucleotide,
　　unmethylated, ligand for Toll-like
　　receptors, 57, 59, 62–63
C-reactive protein, 67
Creb protein, 229, 241
Creb-binding protein, *see* p300/Cbp proteins
Creutzfeldt-Jakob disease, 387, 389
　variant, 387–389
Crimean-Congo hemorrhagic fever
　　virus, 360
Crixivan, *see* Indinavir
crk oncogene, 220
Crk protein, 222
CrmB protein, cowpox virus, 77
Cross-presentation, 110
Cross-species infection, 333
c-SMAC, 113
c-Src protein, 228
CTL, *see* Cytotoxic T cells
Cul4A protein, 173
Cullin 5, 239
Cutaneous immune system, 102–103
CXCr1 receptor, 65
CXCr2 receptor, 65
CXCr3 receptor, 65
CXCr4 receptor, 114, 169, 176–177
Cyclin(s), 206, 208, 214, 223, 229–231, 234
　virus-specific, 233
Cyclin-dependent kinase (Cdk), 206–208,
　　229, 231, 233–234
Cyclin-dependent kinase (Cdk) inhibitors,
　　inactivation by viral proteins, 233–236
Cyclin-dependent kinase-inhibitory protein,
　　2–9, 222
β-Cyclodextrins, 281
Cysteine protease, 72

Cytidine deaminase, 173
Cytochrome *c*, 74, 76
Cytokine(s), 55, 59–61, 64, 88–91, 116, 128
    anti-inflammatory, 59, 64
    apoptosis-inducing, 77
    autoimmune disease and, 125
    "cytokine storm," 124, 298, 347
    immunotherapy, 273
    in inflammation, 67
    inhibitory, 186
    intracellular cytokine assay, 115
    NK cell-activating, 93
    proinflammatory, 59, 64, 89, 91, 97–99, 106, 186
    T-helper cells, 106
Cytokine genes, 58
Cytokine receptor, 90
    soluble, 99
Cytokinesis, 207
Cytomegalovirus
    activation of Nf-κb, 66
    antiviral drugs, 280–281, 289–291, 293, 295
    apoptosis in infected cells, 74, 77
    congenital infection, 37
    countering epigenetic silencing, 78
    dendritic cells infected with, 91
    disease mechanisms, 366
    diseases, 366
    entry into host, 12
    epidemiology, 366
    evolution, 335
    IE1 protein, 77–78
    IE2 protein, 77, 147
    immune response, 114, 125
    induction of HLA-E, 92
    interferon response, 68
    intracellular detectors of viral infection, 59
    latent infection, 144, 150
    metabolic changes in infected cells, 8
    miRNA, 79, 150
    NK modulators, 92
    persistent infection, 143–144, 146–147
    protease, 297
    regulation of MHC class I proteins, 144
    shedding of virions, 35
    "terminase" complex, 295
    transmission, 36
    treatment, 366
    US2 protein, 144–146
    US3 protein, 144–145
    US6 protein, 144–145
    US11 protein, 144–145
Cytopathic effect, 55
Cytopathic virus, 135
Cytoplasm, in apoptosis, 75
Cytosine deamination, 79
Cytoskeleton, 58
Cytotoxic index, 287
Cytotoxic T cells (CTL), 96, 100–101, 104, 107–116, 128
    adoptive-transfer experiment, 114
    assay, 115
    bypassing by mutation of immunodominant epitopes, 146–147
    cell surface proteins, 105
    control of precursor expansion, 114–115
    CTL escape mutants, 146–147
    genetic manipulation, 122

in HIV-infected individuals, 182, 186
    immunopathological lesions caused by, 121–122
    killing by infected cell, 147
    mechanisms of killing, 112
    MHC restriction, 108
    mouse model to define antiviral contribution, 122
Cytovene, *see* Ganciclovir

D
d4T, *see* Stavudine
Dai protein, 56
Dams (water control), 337, 352
Daxx protein, 78
Day care-associated illness, 353
dcc gene, 203
DC-Sign, 91, 179
Ddb1 protein, 173–174
ddC, *see* Zalcitabine
ddI, *see* Didanosine
Dead-end interaction, 37, 333–334, 336–339
"Death receptor," 77
Decay-accelerating protein, 96
Deer mouse, 344
Defective particles, 9
Defensin, 96–97, 116, 186
Deforestation, 337
Delavirdine (Rescriptor), 280, 301, 381
    resistance, 306
Delayed-type hypersensitivity, 122, 148
    T-cell mediated, 116
        adoptive-transfer assay, 114
Demyelination, 122–123
Dendritic cells, 89–91, 103
    follicular, 88, 107, 194
        HIV-infected, 188–189
    immature, 89–90, 128
        interferon production, 61–63
    interdigitating, 88
    mature, 89–91, 128
    monitoring apoptosis, 77–78
    plasmacytoid, 88
    source/function, 88
    synthesis of MHC class II proteins, 108
    synthesis of Toll-like receptors, 62
    for T-cell infection with human T-lymphotropic virus, 243
    virus-infected, 91
        human immunodeficiency virus, 179–180
Dengue hemorrhagic fever, 123
Dengue shock syndrome, 123
Dengue virus, 352, 363
    emerging disease, 334, 337
    immunopathology, 120, 123
    incubation period, 141
    replication in dendritic cells, 91
    transmission, 36–37
Dermis, 13–14
Dextran sulfate, 281, 302
Dff (DNA fragmentation factor), 75
Diagnostic virology, 350
Diapedesis, 23
Diarrhea, rotavirus-induced, 47
Didanosine (ddI, Videx), 280, 301
    resistance, 306
Dideoxycytidine, 292, 381
Dideoxyinosine, 282

DISC (death-inducing signaling complex), 76
Disease, viral
    determinants, 30
    nature, 30
    severity, 30
Disseminated infection, 9, 16, 34
DNA
    methylation, 78
    proviral, 55
    viral, integration in cellular genome, 212–215
DNA helicase-primase, antivirals targeting, 281, 288
DNA laddering, 75
DNA microarray, 6, 285–286, 288
    transcriptional profiling of host response to infection, 7
DNA polymerase
    acyclovir-resistant, 298
    viral, mutations, 44
DNA polymerase inhibitors, 281, 290
DNA priming, 274
DNA shuffling, 271
DNA vaccine, 260, 268–270
    expression vector, 270
DNA virus
    evolution, 314, 316, 322
        relationships from nucleic acid sequence analysis, 322–325
    oncogenic, 209–211
Dog, canine parvovirus in, 345–346
Dominant transforming gene, 218, 246
Dp protein, 231–232
DPT vaccine, 259
Droplet nuclei, 34
Drug holiday, 307
Dryvax, 252
Dynamin, 67
Dynein, 19, 152

E
E1 protein, alphavirus, 326, 329
E1A protein, adenovirus, 71, 145, 218, 220, 231–234, 238–240, 244
E1B protein, adenovirus, 77, 218, 234, 238–240, 244
E2 protein, hepatitis C virus, 92
E2f protein, 230, 232–233, 238
E3 protein, adenovirus, 77, 99, 129–130, 145
E3L protein, vaccinia virus, 70–71, 79
E4 protein, adenovirus, 78, 218, 244
E5 protein
    bovine papillomavirus, 228
    human papillomavirus, 228
E6 protein, human papillomavirus, 77, 194, 218, 220, 238–239, 245
E6-associated protein, 238–239
E7 protein, human papillomavirus, 71, 194, 218, 220, 232–233, 238, 245
Ear-swelling assay, 114
Eastern equine encephalitis virus
    evolution, 329
    immune response, 116
EBER-1, 151, 159
EBER-2, 151, 159
EBNA-1, 146, 151, 159, 240
EBNA-2, 151, 244
Ebna5 protein, Epstein-Barr virus, 78

Ebola virus
emerging disease, 334, 337
gene products that suppress RNA interference, 79
"geographic containment," 337
VP35 protein, 79
Echovirus
disease mechanisms, 376
diseases, 375
epidemiology, 375
transmission, 36
Economic concerns, emerging disease, 351
Ectromelia virus
NK modulators, 93
recombinant, 42–43
Efavirenz (Sustiva), 280, 301, 306
Eicosanoids, 98
eIF2α
dephosphorylation, 65
phosphorylation, 65, 71–72, 78
eIF3, 68
eIF4E, 58
eIF4G, 58
Elderly, susceptibility to viral diseases, 48–50
Elongin, 239
Elvitegravir, 381
Emerging viruses, 333–350
climate effects, 344
convergence model, 334
defined, 333
diagnostic virology, 350
diseases of exploration and colonization, 335, 338, 343–344
ecological parameters, 339–341, 352–353
escape from immune response, 347
expanding viral niches, 341–345
host range expansion, 345–349
host-virus interactions
accidental natural infections, 341
dead-end, 333–334, 336–339
encounters of virus with new host, 339–341
evolving, 333–336, 343–344
experimental analysis, 341
resistant host, 333–334, 338–339
stable, 333–335
pathogenicity change, 348–349
predicting pandemics, 351–352
prevention, 353
public perception, 350–351
sanitary practices and, 341
social parameters, 312, 339–341, 352
truly novel viruses, 349–350
zoonoses, 344–345
Emtricitabine, 306
Encephalitis, human immunodeficiency virus, 188
Encephalomyocarditis virus
detection by individual cells, 60
intracellular detectors of viral infection, 59
Endocytosis, 89, 111, 120, 146
receptor-mediated, 77
Endoneural space, 17
Endoplasmic reticulum, 145
Endosome, 57, 62, 89–90, 110–111
Endothelial cells, 21–23, 188, 190–191
Endothelin-1, 192
Endothelioma, 202, 226
Enfuvirtide (T20), 303, 381

Enhanceosome, 69
Enhancer, 24
interferon-β gene, 69
Enhancer insertion, 225
Enhancing antibody, 185–186
Enteric virus, 10–13
Enterotropic virus, 24
Enterovirus
disease mechanisms, 376
diseases, 375
dissemination in host, 15
entry into host, 12–13, 15
epidemiology, 375
incubation period, 141
shedding from host, 15
Entry inhibitors, 293, 300, 302–303
Entry into host, 9–15
alimentary tract, 10–13
eyes, 10, 12–13
respiratory tract, 9–12
site of entry establishes pathway of spread, 26
skin, 10, 13–14
urogenital tract, 10–12
Env protein, human immunodeficiency virus, 114, 271
Enzyme-linked immunospot assay (ELISPot), 115
Eosinophils, 95
in inflammatory response, 98
source/function, 88
Epidemic, 3
modeling the spread of, 138
video games model, 34
Epidemiology, 31–35
components, 35
defined, 31
incidence of disease, 32–33
prevalence of disease, 32–33
prospective studies, 33–34
retrospective studies, 33–34
video games model of epidemics, 34
Epidermal growth factor receptor, 206, 228
Epidermis, 13–14, 102
Epigenetic silencing, 78
Episome, viral, 158, 215
Epithelial cells, release of virus particles, 16
Epitope, 99
immunodominant, 146–147
Epstein-Barr virus
activation of Nf-κb, 66
acute infection, 158
antiviral drugs, 280
B-cell lymphoma, 157–159, 193–194, 367
Bcrf1 protein, 71
Burkitt's lymphoma, 158–159, 210, 240
BZLF2 protein, 146
cancer and, 208
disease mechanisms, 367
diseases, 367
EBER-1, 151, 159
EBER-2, 151, 159
EBNA-1, 146, 151, 159, 240
EBNA-2, 151, 244
Ebna5 protein, 78
entry into host, 11–12
epidemiology, 367
Hodgkin's disease, 158–159
immune response, 125–126

infectious mononucleosis, 157–159
inhibition of immune defense, 240–241
interferon response, 71
latent infection, 150–151, 156–160
phenotypes, 159
reactivation, 159–160
LMP-1 protein, 77, 151, 159, 194, 225–226, 244
LMP-2 protein, 151, 157, 159–160
miRNA, 151, 159
nasopharyngeal carcinoma, 158–159, 240
oncogenic, 215
persistent infection, 142–143, 146
primary infection, 157–158
regulation of MHC class I proteins, 144
regulators of apoptosis, 77
T-helper cell response, 106
transformation mechanisms, 244
transformed cells, 240–241
X-linked lymphoproliferative syndrome, 158
Zta protein, 160
Epstein-Barr virus receptor, 96
Equine herpesvirus, 106
Equine infectious anemia virus, 167, 169
Eradication of viral disease
agriculturally important diseases, 255–256
limitations and possibilities, 255
measles, 253
poliomyelitis, 253–255, 343
smallpox, 252–253, 255, 335
vaccination programs, 253–255
erbA oncogene, 220
erbB oncogene, 220, 228
Erk proteins, 62, 206
Error threshold, 316–317
Erythema, 97
Erythroblastosis, 228
Erythrocytes, autoantibodies in HIV-infected individuals, 184
eUI4E protein, 235
European tick-borne encephalitis, 337
Europeans, colonization of the Americas, 335, 338, 343–344
Evolution, viral, 311–333
bottlenecks, 315, 317–320, 331
constraints, 330–331
contemporary, 329–330
directed molecular, 271
DNA virus, 314, 316, 322
error threshold, 316–317
fitness, 315–316, 320
genetic shift and drift, 317
host switching, 320
host-virus interactions, 311–333
human immunodeficiency virus, 168
incorporation of cellular sequences, 320
large numbers of mutant genomes, 314
large numbers of progeny, 313–314
lethal mutagenesis, 316–317
mechanisms, 312–321
horizontal gene transfer, 321
mutation, 312–316, 331–333
nonhomologous recombination, 321
reassortment of genetic segments, 312, 318–321
recombination, 312, 320–321

selection, 312, 331
new viruses, 331–333
origin of viruses, 321–330
quasispecies concept, 314–316
rabbitpox virus in Australia, 342
replication strategies and, 330–332
RNA virus, 314, 316, 322, 326
sequence conservation in changing
    genomes, 316
two pathways, 321
virulence traits, 312
Exotic ungulate encephalopathy, 387
Experimental design, 30–34
Exponential growth, 136–137
Extinction, 316
Extracellular matrix, 204, 222
Eye
    herpes stromal keratitis, 123
    viral entry, 10, 12–13

F

F protein, paramyxovirus, 58
Fab domain, 117
Famciclovir, 280, 365
Familial adenomatous polyposis, 203
Faroe Islands, measles outbreak, 258
Fas ligand, 112, 114, 147, 176
Fas protein, 236
Fas receptor, 76, 112, 147
Fatal familial insomnia, 387
Fc domain, 117
Fc receptor, 18, 24, 95, 120, 179, 182, 185
Fecal-oral transmission, 36
Feces, shedding of virions in, 35
Feline immunodeficiency virus, 167, 169
Feline infectious peritonitis virus, 120
Feline leukemia virus, 348
Feline panleukopenia virus, 346
Feline sarcoma virus, 217, 349–350
Feline spongiform encephalopathy, 387
Fetus, viral infections, 23, 36–37, 49
Fibrinogen, 67
Fibroblasts, 202
Fibroma, 209
Fibropapilloma, 202
Filovirus
    disease mechanisms, 362
    diseases, 362
    emerging disease, 334
    epidemiology, 362
Firebreak to infection, 63
Fitness, viral evolution, 315–316, 320
Flavivirus, 363
    antiviral drugs, 281
    detection by individual cells, 60
    disease mechanisms, 363
    diseases, 363
    emerging disease, 334
    entry into host, 12
    epidemiology, 363
    evolution, 328, 335
    host susceptibility to viral disease, 48
    neural spread, 20
    NK modulators, 92
    transformation mechanisms, 244
    transmission, 36
    vectors, 363
"Flu shot," see Influenza virus vaccine
flv gene, 48

Focal adhesion kinase, 222
Foci, 203, 205
Focus-forming assay, 5
Fomite, 36
Fomivirsen, 295, 366
Foot-and-mouth disease virus
    control/eradication, 255–256
    evolution, 320
    interferon response, 71
    L protein, 71
Fortovase, see Saquinavir
fos oncogene, 220, 241
Foscarnet (Foscavir), 280, 291, 298, 366
    resistance, 306
Foscavir, see Foscarnet
Fowlpox virus, 270
Foxo protein, 235
Fractalkine, 61
Free radicals, 127
Freund's adjuvant, 97
    complete, 272
Fruit bat vector, 338–340
Fujinami sarcoma virus, 217
Fusion inhibitors, 195, 300, 302–303, see also
    Membrane fusion inhibitors
FV1 protein, 80

G

G_0 phase, 205, 230–231, 233
G_1 phase, 205, 207–208, 230–231, 233
G_2 phase, 205, 207–208
G proteins, 205
G1 protein, bunyavirus, 47
Gag protein, retrovirus, 332
α-Gal, 95–96
Galactosyl ceramide, 179
Galactosyltransferase, 95–96
Gammaherpesvirus
    constitutively active viral "receptors,"
        225–226
    evolution, 320, 322–324
    latent infections in mice, 130
Ganciclovir (Cytovene), 280, 289–292, 366
Gardasil, 370
Gardner-Arnstein feline sarcoma virus, 217
GAS element, 70
Gastrointestinal system, human
    immunodeficiency virus, 190
Gender, determinant of host susceptibility to
    viral disease, 50
Gene gun, 268–269
Gene shuffling, 269, 271
Gene therapy, 121, 160, 224
    antiviral, 295–298
Genetic drift, 317
Genetic shift, 317
Genital secretions, female, human
    immunodeficiency virus, 177–178
Genital warts, 267–268
Genome, viral
    attenuated viruses, 42
    constraints on viral evolution, 330–331
    in latent infection, 150
    packaging, antivirals targeting, 295
    sequence coherence, 322, 324
    sequence conservation, 316
    sequencing
        determining evolutionary relationships,
            322–328

information for antiviral drug discovery,
    285–286
silenced, 153–154
size, 324
"Genome-to-drug-to-lead," 287
Genomic vaccine, 269
Genomics approach, cellular metabolic paths
    in virus-infected cells, 8
Geographic variation, viral disease, 35,
    37–40
Germ line transmission, 36–37
Gerstmann-Sträussler syndrome, 387
Glioma, malignant, 45
Global travel, 37
Glomerulonephritis, 123–124
Glucocorticoid therapy, reactivation of latent
    herpesvirus, 156
Glycogen synthase, 235
Glycogen synthase kinase, 235
Glycosaminoglycan, 24
Goblet cells, 9, 11
Golgi apparatus, 145
Granulocyte(s), 88–89
Granulocyte-macrophage colony stimulating
    factor, 64
Granzyme, 91, 112, 122
Grb2 protein, 206, 222
Green fluorescent protein, 20
Growth
    mathematics of, 136–138
    transformed cells, 234–239
Growth factor(s), transformed cells, 203–204
Growth factor receptors, 228
GTPase, 48
    interferon-inducible, 65
GTPase-activating protein, 223
Guanine nucleotide exchange protein, 223
Guanine nucleotide-binding protein-coupled
    receptor, 223
Guillain-Barré syndrome, 264

H

HA protein
    avian influenza virus, 26
    influenza virus, 25, 141, 262, 293, 340
HA0 protein, influenza virus, cleavage by
    tryptase Clara, 24
HAART, see Highly active antiretroviral
    therapies
Hairy-cell leukemia, 380
Hand-foot-and-mouth disease virus, 375–376
Hantaan virus, 334, 360
Hantavirus
    antiviral drugs, 291
    diseases, 360
    dissemination in host, 15
    emerging disease, 344
    entry into host, 12, 15
    shedding from host, 15, 35
    transmission, 37
Hantavirus pulmonary syndrome, 344, 360
Harvey murine sarcoma virus, 217
Hausp protein, 237
hbz gene, human T-lymphotropic virus, 241
Heart
    human immunodeficiency virus, 190
    viral invasion of cardiac muscle, 22
Helper virus, 215, 387
Hematogenous spread, 14–19

Hemorrhagic fever, 360
Hendra virus, emerging disease, 338–339
Hepadnavirus
  immune response, 115
  oncogenic, 209
  transformation mechanisms, 244
Heparin, 281
Hepatitis, 21
Hepatitis A virus
  antiviral drugs, 281
  disease mechanisms, 364
  epidemiology, 364
  host susceptibility to viral disease, 50
  immunotherapy, 273
  incubation period, 141, 364
  mortality rate, 364
  passive immunization, 256
  transmission, 36, 364
Hepatitis A virus vaccine, 261, 263, 272,
    274, 364
Hepatitis B virus
  activation of Nf-κb, 66
  antiviral drugs, 281
  blood products contaminated with, 18
  cancer and, 208
  cellular defense against, 68
  disease mechanisms, 364
  dissemination in host, 15
  entry into host, 12, 15
  epidemiology, 364
  hepatocellular carcinoma, 242–246, 364
  host susceptibility to viral disease, 50
  immune response, 114–115
  immunopathology, 120, 122
  in vivo dynamics, 313
  incubation period, 141, 364
  interferon response, 70–71
  mortality rate, 364
  oncogenesis, 242–246
  organ invasion, 21
  persistent infection, 143
  regulators of apoptosis, 77
  shedding of virions, 15, 34–36
  studies with transgenic mice, 32
  transformation mechanisms, 244
  transmission, 36, 364
  tropism, 24
  viremia, 18
  X protein, 68, 77, 243–244
  in yellow fever vaccine, 341
Hepatitis B virus core antigen, 70
Hepatitis B virus vaccine, 263, 265, 267,
    272, 274, 364
Hepatitis C virus, 363
  antiviral drugs, 280–281, 283, 291,
    293, 295
  cancer and, 208
  CTL escape mutants, 146–147
  discovery, 349–350
  disease mechanisms, 364
  E2 protein, 92
  entry into host, 12
  epidemiology, 364
  hepatocellular carcinoma, 209, 211, 364
  immune response, 114
  immunotherapy, 273
  incubation period, 141, 364
  interferon response, 70–71
  mortality rate, 364

NK modulators, 92
NS5a protein, 71
oncogenesis, 246
persistent infection, 143, 146–147
protease, 293, 297
transformation mechanisms, 244
transmission, 36, 364
treatment, 364
viremia, 18
Hepatitis D virus, 12, 364
Hepatitis E virus, 50, 364
Hepatitis virus, discovery in blood supply,
    349–350
Hepatocellular carcinoma, 147, 202
  hepatitis B virus, 242–246, 364
  hepatitis C virus, 209, 211, 364
Herd immunity, 259
Hereditary nonpolyposis colorectal
    cancer, 203
Herpes gladiatorum, 36
Herpes labialis, 155
Herpes simplex virus
  acyclovir-resistant, 298
  antiviral drugs, 280–281, 288–289, 293,
    295–297
  attenuated, to clear brain tumors, 45
  disease mechanisms, 365
  diseases, 365
  dissemination in host, 15
  entry into host, 11–13, 15
  epidemiology, 365
  evolution, 335
  "herpes is forever," 153
  host alterations early in infection, 58
  host susceptibility, 48, 151
  ICP34.5 protein, 44, 47, 65, 71–72, 78
  ICP47 protein, 145, 155
  immune response, 65
  immunopathology, 120, 123
  incubation period, 141
  intracellular detectors of viral infection, 59
  latency-associated transcripts,
    153–154, 156
  latent infection, 136, 151–156
    establishment and maintenance,
      153–155
    reactivation, 151, 154–156
  miRNA, 153
  nervous system infection, 17
  neuroinvasiveness, 47
  neurons harboring, 151–156
  peripheral ganglia, 154–156
  persistent infection, 146
  primary infection, 150–154
  protein gB, 146
  regulation of MHC class I proteins, 144
  shedding from host, 15
  transmission, 36
  treatment, 365
  tropism, 24, 26
  virulence, 44–45, 47
Herpes simplex virus receptor, 24
Herpes simplex virus type 1
  apoptosis in infected cells, 75
  ICP0 protein, 72, 78, 153–156
  ICP27 protein, 58, 75
  interferon response, 70–72
  US11 protein, 70–72
  VP16 protein, 295–297

Herpesvirus
  adapter proteins, 226
  antiviral drugs, 280–281, 291, 295
  evolution, 322–324
  immune response, 120
  immunosuppression, 125
  NK modulators, 92
  oncogenic, 209
  reduction of T-cell receptor function, 112
  shedding of virions, 36
  transformation mechanisms, 244
  virulence, 46
Herpesvirus receptor, 293
Herpesvirus saimiri
  genes homologous to cellular genes, 223
  production of virus-specific cyclins, 233
  Tip protein, 227
Heterologous T-cell immunity, 125–126
Heterosexual contact, transmission of human
    immunodeficiency virus, 177–178
High-content screen, 285
High-throughput screen, 284–285
Highly active antiretroviral therapies
    (HAART), 188, 194, 299
Histone acetylase, 69, 239
Histone acetyltransferase, 220
Histone deacetylase, 78, 231
HIV, *see* Human immunodeficiency virus
Hivid, *see* Zalcitabine
HLA-E protein, 92
Hmg(A1) protein, 69
Hodgkin's disease, 158–159
Homing, 89
Homosexual contact, transmission of human
    immunodeficiency virus, 177–178
Hordeivirus, 328
Horizontal gene transfer, 321
Horizontal transmission, 37
Hormonal status, determinant of host
    susceptibility to viral disease, 50
Horse, West Nile virus, 336
Host
  generation time, 138
  shedding of virions, 34–36
  viral entry, 9–15, *see also* Entry into host
  viral spread, 15–21
Host defense, 29, *see also* Cellular defense,
    intrinsic; Immune *entries*
  chemical defenses, 54
  constraints on viral evolution, 331
  early actions, 53–84
  first critical moments of infection, 54
  intrinsic, 53–84
  modulation or evasion by viruses, 14–15
  physical defenses, 54
  respiratory tract, 9, 11
  viral gene products that modify, 43, 45–46
  viral life cycle and, 135–136
Host population
  age distribution, 340
  health, 312, 340–341
  population dynamics, 312, 316, 340–341
  social behavior, 312, 339–341, 352
Host range
  expansion
    mutation, 345–349
    reassortment of genetic segments,
      345–349
    recombination, 345–349

host range mutants, 43–44
host switch and evolution, 320
Host restriction, 79–80
Host susceptibility to viral disease, 48–50
  age of host and, 48–50
  epidemiological aspects, 48
  gender and, 50
  genetic determinants, 48
  herpes simplex virus, 151
  hormonal status and, 50
  nutritional status and, 50
Host-virus interactions
  accidental natural infections, 341
  dead-end, 333–334, 336–339
  emerging viruses, 333–350
  encounters of viruses with new host,
    339–341
  evolution of viruses, 311–333
  evolving, 333–336, 343–344
  expanding viral niches, 341–345
  experimental analysis, 341
  resistant host, 333–334, 338–339
  stable, 333–335
  virus names, 350
H-*ras* gene, 220
Human herpesvirus 6, 366
  antiviral drugs, 280
Human herpesvirus 7, 366
  antiviral drugs, 280
  persistent infection, 142
Human herpesvirus 8, 367
  antiviral drugs, 280
  B-cell lymphoma, 193, 367
  genes homologous to cellular genes, 223
  inhibition of immune defense,
    240–241
  K3 gene, 144–146
  K5 gene, 144–146
  K13 protein, 77
  Kaposi's sarcoma, 193, 211–212, 223
  miRNA, 79, 212
  persistent infection, 144–146
  primary effusion lymphoma, 223–224
  production of virus-specific cyclins, 233
  regulators of apoptosis, 77
  transformation of host cells, 212
  transformed cells, 240–241
Human herpesvirus 9, discovery, 211
Human immunodeficiency virus (HIV), *see
  also* AIDS; Human immunodeficiency
    virus type 1
  abortive infection, 160
  accessory proteins, 169, 171–174
  activation of Nf-κb, 66
  acute infection, 140
  animal models of infection, 195
  antibody response, 184–186
    enhancing antibodies, 185–186
    interfering antibodies, 185–186
    neutralizing antibodies, 184–185
  antigenic variation, 140, 142
  antiviral drugs, 187–188, 194–195,
    280–281, 291, 293, 299–307, 381
    challenges and lessons learned, 307
    combination therapy, 305–306
    cross-resistance, 305–306
    fusion and entry inhibitors, 300–303
    HAART therapy, 188, 194, 299
    integrase inhibitors, 300–302

nucleoside analogs, 299–301
protease inhibitors, 300–304
resistance, 303–307, 316, 333
reverse transcriptase inhibitors, 300–302
strategic treatment interruption,
  195, 307
Tat inhibitors, 300
triple-drug combination, 306
attachment to host cells, 176–177
attenuated strains, 182
autoimmunity, 183–185
auxiliary proteins, 169, 171
body fluids, 177–178
CA protein, 184
cancer and, 191–194
  anogenital carcinoma, 194
  B-cell lymphoma, 191, 193–194
  Kaposi's sarcoma, 191–193
cell culture, 176–177
cellular immune dysfunction, 182–184
  B cells, 183
  CD4+ T cells, 182–183
  cytotoxic T cells, 182
  monocytes and macrophages, 182–183
  NK cells, 183
cellular targets, 176–177
characterization, 166–169
cis-acting regulatory sequences, 169–172
clinical disorders, 167
clinical latency, 181
course of infection
  acute phase, 180–181
  asymptomatic phase, 180–181
  nonprogressors, 182
  symptomatic phase and AIDS, 180–181
  variability of response to infection,
    181–182
  virus appearance and immune cell
    indicators, 180–181
CTL escape mutants, 146
discovery, 166–169
  lessons from, 167
dissemination in host, 15
emerging disease, 334–335, 345
entry into cells, 176–177
entry into host, 15
Env protein, 271
evolution, 168, 314, 316, 320, 333, 345
  within single patient, 304–305, 331
gastrointestinal system, 190
genome, 333
heart, 190
historical aspects, 166–168, 196
host "restriction," 79–80
immune response, 116, 184–186, 321
  cellular response, 186
  humoral response, 184–186
immunosuppression, 125
in vivo dynamics, 313
incubation period, 141
instability elements, 172
integrase, 300–302
intracellular detectors of viral infection, 59
kidneys, 190
latent infection, 275
long terminal repeat, 169–170
lungs, 190
lymphoid organs, 188–189
MA protein, 184

macrophage-tropic, 176–177, 181,
  188, 331
mutations, 182, 187, 195, 304–305, 307,
  314, 317, 333
NC protein, 195
Nef protein, 92–93, 114, 145–147,
  170–172, 175–176, 182, 190–191
nervous system, 188–192
neural spread, 19
organ invasion, 22
pathogenesis, 190–192
pathological conditions associated with
  infection, 180
persistent infection, 143, 146–148,
  303–306
phylogenetic relationships among
  lentiviruses, 169
prevention of infection, 194–196
privileged sites, 195
protease, 301, 303
proteins that resemble cellular
  proteins, 184
R5 strains, 176
recombination, 142
regulatory proteins, 169, 171
replication cycle, 169–176, 300
  in AIDS, 186–188
Rev protein, 169–172, 190
reverse transcriptase, 302
"sanctuary" compartments, 187
shedding from host, 15
slow infection, 136, 160
spread in host, 179–180
structural plasticity, 141
SU protein, 47, 147, 176, 182, 184–185,
  190–191, 194, 302–303
syncytium formation, 174, 181, 184
T-cell-tropic, 176–177, 181, 331
TAR RNA, 169–171
Tat protein, 79, 114, 145, 147, 169–172,
  182, 190–191, 193, 300
  activation, 171
  interaction with TAR sequences,
    169–171
TM protein, 47, 183–185, 190–191, 194
toxic viral proteins, 47
transmission, 36
  modes, 177–179
  from mother to child, 23
  sources of infection, 177
treatment, 194–196
  antiviral drugs, *see* Human immunodefi-
    ciency virus, antiviral drugs
  immune system-based therapy,
    195, 274
tropism, 176–177
Vif protein, 79, 171–174
viral load, 304
Vpr protein, 160, 171–174
Vpu protein, 171–172, 174–175, 182
worldwide problem, 165–166
X4 strains, 176
Human immunodeficiency virus (HIV)
  receptor, 176, 293, 331
Human immunodeficiency virus (HIV)
  type 1, *see also* Human immunodeficiency virus
  binding to DC-Sign, 91
  clades, 167–168, 271, 330, 345

Human immunodeficiency virus (HIV)
type 1 *(continued)*
congenital infection, 37
disease mechanisms, 381
diseases, 381
earliest record of infection, 168
entry into host, 11–12
Env protein, 114, 271
epidemiology, 381
gene products that suppress RNA
interference, 79
group M, 167–168, 345
group N, 168, 345
group O, 167–168, 345
host susceptibility to viral disease, 48
immune response, 91, 114
immunopathology, 120
infection of memory T cells, 107
interferon response, 71
NK modulators, 92–93
protection from complement-mediated
lysis, 96
proviral DNA, 170
regulation of MHC class I proteins, 144
shedding of virions, 36
TAR RNA, 71
tropism, 24
viral spread, 102
virulence, 41
Human immunodeficiency virus (HIV)
type 2, 381
discovery, 167, 345
groups, 168
proviral DNA, 170
Vpx protein, 171–174
Human immunodeficiency virus (HIV)
vaccine, 195–196, 271,
274–275
challenges of developing, 274–275
nef deletion mutants, 176
requirements for protective immunity, 195
Human papillomavirus
anogenital carcinoma, 194
cancer and, 208
disease mechanisms, 370
diseases, 370
E5 protein, 228
E6 protein, 77, 194, 218, 220, 238–239, 245
E7 protein, 71, 194, 218, 220, 232–233,
238, 245
epidemiology, 370
evolution, 324–325
immunotherapy, 274
interferon response, 71
regulators of apoptosis, 77
transformation mechanisms, 245
transmission, 36
Human papillomavirus vaccine, 263,
267–268, 370
Human T-lymphotropic virus
activation of Nf-κb, 66
emerging disease, 334
"hand-off" system for infection, 243
persistent infection, 143
provirus organization and transcription,
241–242
slow infection, 136, 160
Tax protein, 147, 241–242, 245
transmission, 36

Human T-lymphotropic virus type 1, 245
activation of Nf-κb, 241
adult T-cell lymphocytic leukemia,
241, 243
cancer and, 208
disease mechanisms, 380
diseases, 380
epidemiology, 380
Rex protein, 241
tumorigenesis with very long latency,
241–242
Human T-lymphotropic virus type 2, 380
Human T-lymphotropic virus type 5, 380
Human-centered thinking, 351
Humoral immune response, 88, 96,
99–101, *see also* Adaptive immune
response
HIV-infected individuals, 184–186
Hybridization technology, 350
Hydroquinone, 293
5-Hydroxydeoxycytidine, 317
Hygiene hypothesis, 151
"Hypersanitized" living conditions, 151
Hypodermic needle injection, vaccine, 272
Hypothesis testing, 32

I

Iκκ, 60
Iκbα/β, 66, 226
Iatrogenic infection, 37
Icam, 24, 112, 184–185, 298
ICP0 protein, herpes simplex virus, 72, 78,
153–156
ICP27 protein, herpes simplex virus,
58, 75
ICP34.5 protein, herpes simplex virus, 44,
47, 65, 71–72, 78
ICP47 protein, herpes simplex virus, 145,
155
IE1 protein, cytomegalovirus, 77–78
IE2 protein, cytomegalovirus, 77, 147
IL, *see* Interleukin *entries*
Ileum, invasion by virus, 22–23
Immortality, transformed cells, 202
Immune complex
deposition, 123–124
solubilization, 94–95
Immune defense, 55
Immune evasion, 54
Immune globulin, 256
Immune memory, stimulation by vaccines,
256–260
Immune modulation, 54, 129–130
dysfunctional, 91
Immune response, 87
acute infection, 140
adaptive, *see* Adaptive immune response
adjuvants, 271–273
amplification step, 87
control step, 87
DNA vaccines, 268–269
host susceptibility to viral disease, 48
human immunodeficiency virus, 184–186
innate, *see* Innate immune response
mechanisms that protect transformed cells,
240–241, 244–245
mice with disruptions of genes encoding, 32
persistent infection, 143–147
"primed," 257

recognition step, 87
vaccination, 251–277
Immune senescence, 49
Immune surveillance, reduced, establish-
ment of persistent infections, 147
Immune-stimulating complex (ISCOM),
vaccine delivery, 272–273
Immunity, herd, 259
Immunization, 35, 48, *see also* Vaccination;
Vaccine
active, 256
passive, 116, 256, 258
Immunization Safety Review Committee, 268
Immunocontraception, 342
Immunodominant epitope, 146–147
Immunoglobulin A, 119
HIV-infected individuals, 183
properties, 116
secretory, 96, 102, 117–118
Immunoglobulin D
HIV-infected individuals, 183
properties, 116
Immunoglobulin E, 116
Immunoglobulin G, 96, 117, 120
HIV-infected individuals, 183
properties, 116
rate of synthesis, 116
Immunoglobulin M, 96, 117
properties, 116
Immunological synapse, 110–113
Immunomodulating agent, 274
Immunopathogenesis, 32
Immunopathology, 30, 101, 120–124
Immunosuppression, viral infection-induced,
124–125
Immunotherapy, 273–274
ex vivo approach, 273
IMP dehydrogenase inhibitor, 290–291
In silico drug discovery, 286–287
Inactivated (killed) virus vaccine, 261–262,
271–273
Inapparent acute infection, 138, 140
Incidence of disease, 32–33
Incubation period, 138–141
Independent entity theory, origin of
viruses, 322
Indinavir (Crixivan), 280, 286, 301, 304
resistance, 306
Industrialization, 342, 353
Infection, *see also specific types*
disseminated, 9, 16, 34
initiation, 6–9
localized, 16, 34
series of events, 6, 8–9
systemic, 16
Infection rate, 311
50% infectious dose (ID_{50}), 41
Infectious mononucleosis, Epstein-Barr
virus, 157–159
Inflammasome, 97–99
Inflammation
activation by complement cascade, 94–95
adjuvant-induced, 97
classical signs, 97
cytokines in, 67
Inflammatory response, 91, 97–99
in brain, 121
Influenza A virus
antiviral drugs, 280–281, 291–293, 298

evolution, 335
gene products that suppress RNA interference, 79
NP protein, 347
pathogenesis, 298
Influenza B virus
antiviral drugs, 280, 293
NP protein, 347
Influenza virus
activation of Nf-κb, 66
acute infection, 136, 138, 140–141
uncomplicated, 139
antigenic variation, 140–141, 262, 328
antiviral drugs, 296–297
cellular defense against, 67
detection by individual cells, 60
disease mechanisms, 369
diseases, 369
emerging disease, 334, 340, 347–349
entry into host, 11–12
epidemics, 347–348
epidemiology, 328, 369
evolution, 317, 327–328, 347–349
from 1889–1977, 318
genetic shift, 317
H subtypes, 318–319
H1N1, 347
H5N1, 348–349, 352
HA protein, 25, 141, 262, 293, 340
HA0 protein, cleavage by tryptase Clara, 24
historical aspects, 5
host susceptibility to viral disease, 48–50
immune response, 119, 125
incubation period, 141
interferon response, 70–71
intracellular detectors of viral infection, 59
life cycle, 348
M2 protein, 174–175, 291, 293, 297
in mice, 31
N subtypes, 318–319
NA protein, 25, 262, 296
NS1 protein, 70–71, 79
pandemics, 50, 141, 317–319, 347–348, 351–352
release of virus particles, 14
seasonality of infections, 38–40
stopping epidemics in chickens, 256
structural plasticity, 141
transmission, 36
treatment, 369
tropism, 24–26
WSN/33 strain, 25
Influenza virus receptor, 24, 31
Influenza virus vaccine, 141, 261, 263–264, 369
"split," 265
Inhibitors of apoptosis (IAP), 74–77
Ink4 proteins, 208
Innate immune response, 54
activation, setting threshold for, 130
complement system, 93–97
defined, 87–89
first line of immune defense, 87–88
general features, 89
inflammatory response, 97–99
integration with intrinsic cellular defense and adaptive immune response, 56
NK cells, 91–93
persistent infection and, 143

sentinel cells, 89–91
white blood cells in, 88
Insertional activation, nontransducing retro-virus, 224–225
Insertional mutagenesis, 55, 245
Instability element, human immunodeficiency virus, 172
Integrase, viral, 212, 328–330
human immunodeficiency virus, 300–302
Integrase inhibitors, 194–195, 300–302
Interfering antibody, human immunodeficiency virus, 185–186
Interferon, 48, 61–74, 298
animal models with defective interferon response, 68
antiviral state and, 63–70
discovery, 61
flu-like symptoms in infected individual, 64
production, 64
by immature dendritic cells, 61–63
in virus-infected cells, 61–63
signal transduction cascade, 63, 70
therapeutic applications, 64, 281
viral gene products that counter interferon response, 70–73
Interferon genes, 62
interferon-α, 68
interferon-β, 69
Interferon receptors, 63, 70, 73
Interferon regulatory factors (Irf), 57 58, 223
Irf1, 70
Irf3, 57, 60, 68
Irf7, 57, 68
Interferon-α/β, 58–59, 61, 64
inducers, 68
producer cells, 68
switching transcription on and off, 69
Interferon-γ, 58–59, 61, 64, 71, 91, 96, 98, 103, 106–108, 112–116, 269
production, 71
signal transduction pathway, 63, 71
therapeutic, 273, 364
Interferon-inducible genes, 57, 63, 65–70, 73
antiviral action of induced proteins, 65–70
Interferon-stimulated response element (ISRE), 67, 70, 73
Interleukin-1 (IL-1), 64, 67, 98, 103
Interleukin-1 (IL-1) receptor, soluble, 61
Interleukin-1β (IL-1β), 90, 99, 186, 192
Interleukin-2 (IL-2), 64, 101, 106, 110, 116, 182–183, 241
Interleukin-2 (IL-2) receptor, 182
Interleukin-4 (IL-4), 42–43, 64, 91, 106
Interleukin-5 (IL-5), 106
Interleukin-6 (IL-6), 57, 59, 64, 67, 70, 103, 106, 186, 192, 223–224, 298
Interleukin-8 (IL-8), 64–65, 298
Interleukin-10 (IL-10), 61, 64–65, 106, 186, 192
Interleukin-12 (IL-12), 57, 59, 64, 90, 106, 126, 269, 272
Interleukin-13 (IL-13), 91
Interleukin-16 (IL-16), 114
Interleukin-17 (IL-17), 61, 107, 223
Interleukin-18 (IL-18), 99

Interleukin-21 (IL-21), 107
International travel, 345
Interphase, 207
Intestine, invasion by virus, 22–23
Intracellular cytokine assay, 115
Intravenous drug use, transmission of human immunodeficiency virus, 177–178
Invirase, see Saquinavir
Iododeoxyuridine, 292, 365
Irak (IL-2 receptor-associated kinase), 62
IRES, 59, 281
Irf, see Interferon regulatory factors
Irrigation systems, 352
ISCOM, see Immune-stimulating complex
Isentress, see Raltegravir
isg56 gene, 68
Isgf3, 67, 71
ISRE (interferon-stimulated response element), 67, 70, 73

J

Jak/Stat pathway, 63, 70
Japanese encephalitis virus, 327–328, 363
Japanese encephalitis virus vaccine, 263, 363
JC virus, 213
disease mechanisms, 377
diseases, 377
enhancer, 24
epidemiology, 377
evolution, 324–325
persistent infection, 23, 143
progressive multifocal leukoencephalo-pathy, 23, 377
slow infection, 160
tropism, 24
Jenner, Edward, 252–253
Jnk pathway, 226
Jnk protein, 60, 62
jun oncogene, 220
Junin virus, 337
diseases, 359
emerging disease, 334

K

K, see Carrying capacity
K3 gene, human herpesvirus 8, 144–146
K3L protein, vaccinia virus, 71
K5 gene, human herpesvirus 8, 144–146
K13 protein, human herpesvirus 8, 77
Kaletra, see Lopinavir/ritonavir
Kaposi's sarcoma, 169, 191–193, 367
human herpesvirus 8, 193, 211–212, 223
Keratinocytes, 102–103
Keratitis, herpes, 123
Kidneys
human immunodeficiency virus, 190
invasion by virus, 22–23
Kinesin, 19
Kirs, 92
Kissing, and virus transmission, 35
kit oncogene, 220
Koch's postulates, 3–4, 350
Koplik's spots, 23, 50, 148–149
Korean hemorrhagic fever virus, 37
K-replication strategy, 137–138, 161, 313, 321
Kupffer cells, 21
Kuru, 387–389

L

L protein, foot-and-mouth disease virus, 71
L1 protein, papillomavirus, 268, 326
La Crosse virus, 42, 47, 360
Lake Victoria marburgvirus, 362
Lambdoid phage, evolution, 321
Lamin C, 304
Lamivudine (3TC), 280, 291–292, 301
  resistance, 306, 316
Langerhans cells, 90, 102–103
  human immunodeficiency virus, 179
Laryngitis, 11
Lassa virus, 337
  antiviral drugs, 291
  diseases, 359
LAT, *see* Latency-associated transcript
Latency-associated transcript (LAT), herpes
  simplex virus, 153–154, 156
Latent infection, 130, 135–136, 142, 257
  Epstein-Barr virus, 156–160
  general properties, 150
  herpes simplex virus, 151–156
  K-replication strategy, 137
Law of unintended consequences, 353
Lck protein, 113
Lectin, 96
Lentivirus, 166–167, 381
  immune deficiencies caused by, 167–169
  immune response, 115
  phylogenetic relationships, 168–169
50% lethal dose (LD$_{50}$), 41
Lethal mutagenesis, 316–317
Leukemia, 202
Leukotrienes, 98
Levovirin, 291
Leukocyte function antigen
  Lfa-1, 113, 184–185
  Lfa-3, 112
Life cycle, viral, 135–136
Limiting dilution assay, 115
Lipid A, 272
Lipid raft, disruption by drugs, 281
Lipodystrophy, 304, 306
Liver
  invasion by virus, 21
  oncogenesis by hepatitis viruses,
    242–246
Livestock
  eradication/control of viral diseases,
    255–256
  prions and the food supply, 391
  stopping epidemics by culling and
    slaughter, 255–256, 349
  transport, 352
LMP-1 protein, Epstein-Barr virus, 77, 151,
  159, 194, 225–226, 244
LMP-2 protein, Epstein-Barr virus, 151, 157,
  159–160
Localized infection, 16, 34
Logistic growth, 137
Logovirus, 361
Long terminal repeat, human
  immunodeficiency virus, 169–170
Long-latency virus, 209–210
Lopinavir, 280, 301
Lopinavir/ritonavir (Kaletra), 301
Loviride, resistance, 306
LT protein
  polyomavirus, 218, 238

simian virus 40, 77, 218, 220, 232–234,
  239, 245
Lungs, human immunodeficiency virus, 190
Lymph node, 101, 103, 188–189
Lymphatic system, 16, 128
  human immunodeficiency virus, 188–189
  pathways for viral spread, 17
  viral replication in lymphoid cells, 17
Lymphocytes, 101
  autoantibodies in HIV-infected
    individuals, 184
  intraepithelial, 102–103
  source/function, 88–89
  virus-infected, 147
Lymphocytic choriomeningitis virus
  congenital infection, 37, 49
  diseases, 359
  immune response, 114, 125
  immunopathology, 120–123
  immunosuppression, 125
  persistent infection, 136, 142, 149–150
  promotion of autoimmune disease, 126
  protection against autoimmune disease, 126
  viremia, 18
Lymphoma, 202, 209, *see also specific types*
Lysosome, 62, 111
Lyssavirus, 327, 382

M

M cells, 102–103
  human immunodeficiency virus, 179
  intestinal, 10–11, 13
M phase, *see* Mitosis
M protein, measles virus, 149
M2 protein, influenza virus, 174–175, 291,
  293, 297
MA protein, human immunodeficiency
  virus, 184
Machupo virus, 334, 359
Macrophages, 18, 21, 89–91, 96, 102, 128
  coreceptors for human immunodeficiency
    virus, 176–177
  delayed-type hypersensitivity, 116
  destruction of apoptotic bodies, 75
  interferon production, 61
  intestinal tract, 13
  monitoring apoptosis, 77–78
  respiratory tract, 9, 11
  synthesis of MHC class II proteins, 108
  synthesis of Toll-like receptors, 62
  virus-infected, 148
    human immunodeficiency virus, 180,
      182–183, 190, 192
Macule, 23
Mad cow disease, *see* Bovine spongiform
  encephalopathy
Major histocompatibility complex (MHC), 92
  artificial MHC tetramer, 115
  class I genes, 240
    polymorphism, 108
  class I proteins, 48, 92, 99, 103–105,
    107–112
    artificial MHC tetramer, 115
    in HIV-infected cells, 175–176
    in persistent infection, 143–146
    protection against human
      immunodeficiency virus, 196
    structure, 109
    viral homologs, 92

viral regulation, 144–146
class II proteins, 48, 90, 99–100, 103–112,
  116, 128
  MHC class II-CD4 interactions, 182
  structure, 111
  superantigen binding, 126–127
  synthesis, 108
  viral regulation, 146
MHC restriction, 108
Major histocompatibility complex (MHC)
  class I receptor, 92, 108
Malignant cutaneous lymphoma, 380
Malignant tumor, 202
Mammalian cells, restriction point, 230–231
Mannose-binding protein, 67
Map kinase pathway, 206, 222–223,
  226–227, 229–231, 235
MAP4 protein, 58
Maraviroc, 381
Marburg virus, 334, 337, 362
Marine viruses, 344
MASP, 94
Mast cells, 95, 98
Maternal antibody, 256, 258, 342
Maternal-neonatal transmission, 36
Mav protein, 60
McDonough feline sarcoma virus, 217
Mcp1, 65
Mcp3, 65
Mcp4, 65
Mda5 protein, 56, 58–61, 66, 68, 89, 91
Mdm-2 protein, 236–238
Measles virus
  acute infection, 141, 148–149
  acute postinfectious encephalitis, 148–149
  cellular defense against, 67
  dendritic cells infected with, 91
  detection by individual cells, 60
  disease mechanisms, 371
  diseases, 371
  dissemination in host, 15
  emerging disease, 335, 340, 343–344
  entry into host, 12, 15
  epidemics, 138
  epidemiology, 371
  eradication efforts, 253
  evolution, 328, 335
  geographic distribution, 37
  historical aspects, 343–344
  host susceptibility to viral disease, 50
  immune response, 258, 321
  immunopathology, 120
  immunosuppression, 124–125, 148–149
  incubation period, 141
  intracellular detectors of viral infection, 59
  M protein, 149
  mouse models of disease, 32
  neural spread, 19
  organ invasion, 22–23
  persistent infection, 143, 148–149
    brain, 148–149
  release of virus particles, 14
  reproduction number, 253
  shedding from host, 15
  skin rash, 23, 102
  slow infection, 136, 160
  structure, 149
  subacute sclerosing panencephalitis,
    148–149, 253, 371

transmission, 36
tropism, 148
uncomplicated measles, 148
Measles virus receptor, 96, 148
Measles virus vaccine, 253, 263, 274, 371
herd immunity threshold, 259
live attenuated, 264, 267
Mechanism-based screen, 282–283
Mek proteins, 206
Membrane attack complex, 94–95
Membrane blebbing, 74–75
Membrane cofactor protein, see CD46
Membrane fusion inhibitors, 290, 293
Membrane proteins, viral
initiation of cellular signal transduction,
225–226
in neuroinvasion, 47
Memory, adaptive immune response, 102
Memory cells, 88, 105
B cells, 99, 102, 104, 157
acute infection, 140
infected directly by viruses, 107
T cells, 99, 102, 107, 125–126, 128
acute infection, 140
HIV-infected individuals, 180, 187
Meningitis, aseptic, 149–150
Merkel cell carcinoma, 213
Merkel cell polyomavirus, 213
Mesangial cells, 124
Messenger RNA (mRNA), cellular
control by miRNA, 212
response to viral infection, 7
short-lived, 62
Messenger RNA (miRNA), viral
capping, inhibitors, 291
RNA virus, 332
Metabolites, cellular, response to viral
infection, 8
Metabolomics, 8
Metagenomic analysis, 322
Metaphase, 207
Metastasis, 202
Methylation of DNA, 78
MHC, see Major histocompatibility complex
Miasma, 3, 5
MicB protein, 79
Microarray, see DNA microarray
Microbiocide, 293
Microglia, 120, 127
β_2-Microglobulin, 108
Micro-RNA (miRNA)
antivirals targeting, 295
cellular
antiviral action, 70, 79
interferon-induced, 70, 79
cytomegalovirus, 150
Epstein-Barr virus, 151, 159
herpes simplex virus, 153
human herpesvirus 8, 212
Microspheres, vaccine delivery, 273
Microtubules, neural spread of viruses, 18
Microvilli, 10, 13
Middelburg complex, 329
Milk, shedding of virions in, 35–36
Mimivirus, 324
Minireplicon system, 282–283
Mip-1α, 64–65, 177, 186, 192, 223
Mip-1β, 64–65, 177, 186, 192, 223
Mip-3α, 65

Mip-3β, 65
miRNA, see Micro-RNA
Missing self, 92
Mitogen, 230–231
Mitosis, 205, 207–208
Mitosis-promoting factor, 206
MMR vaccine, 263
autism and, 268
MOI, see Multiplicity of infection
Molecular biology, 6
Molecular mimicry, 125, 184–185
Molluscum contagiosum virus, 378
NK modulators, 93
regulators of apoptosis, 77
Moloney murine leukemia virus, 175
proto-oncogenes targeted by, 214
Moloney murine leukemia virus vector, 224
Moloney murine sarcoma virus, 217
Monkey vector, 338
Monkeypox virus, 335, 378
Monoclonal antibody-resistant mutant, 120
Monoclonal tumor, 212
Monoculture, 386
Monocytes, 95
coreceptors for human immunodeficiency
virus, 176–177
in inflammatory response, 97–98
virus-infected, 147–148
human immunodeficiency virus, 182–183
mos oncogene, 220
Mosquito vector, 38, 336–338, 352, 360,
363, 383
myxoma virus, 342
used tires as habitat, 352
West Nile virus, 336
yellow fever virus, 338
Mouse
animal models of viral disease, 30
γHV68 latent infections, 130
genetic determinants of susceptibility, 48
influenza virus infection, 31
knockout, 32
models for studying viral pathogenesis, 32
Pkr-null mutant, 74
transgenic, 32
model to define antiviral contribution of
cytotoxic T cells, 122
Mouse hepatitis virus, regulation of MHC
class I proteins, 144
Mouse mammary tumor virus
proto-oncogenes targeted by, 214
superantigen, 126
Mousepox virus
dissemination in host, 15
entry into host, 15
pathogenesis, 6, 9
shedding from host, 15
T-helper cell response, 106
mRNA, see Messenger RNA, cellular;
Messenger RNA, viral
mT protein, polyomavirus, 226–227, 234
MT-2 protein, myxoma virus, 77
Mucosa-associated lymphoid tissue, 102–103
Mucosal immune system, 102–103
Mucosal tissue
antiviral defenses, 116–119
invasion by virus, 23
Mucus, 9
Mucus-secreting cells, respiratory tract, 9, 11

Muller's ratchet, 318–319
Multiple sclerosis, 350
Multiplicity of infection (MOI)
high-MOI infection, 137
low-MOI infection, 137
Mumps virus
disease mechanisms, 372
diseases, 372
entry into host, 12
epidemiology, 372
host susceptibility to viral disease, 50
incubation period, 141
nervous system infection, 17
neural spread, 19
organ invasion, 22
shedding of virions, 35
Mumps virus vaccine, 263, 372
Muramyl dipeptide, 272
Murine gammaherpesvirus, persistent
infection, 144–146
Murine leukemia virus, 217
congenital infection, 37
Murine sarcoma virus, 217
Murray Valley encephalitis virus, 328
Muscle, invasion by virus, 22
Mutagenesis
insertional, 55
lethal, 316–317
Mutation(s)
attenuating, 42
host range expansion, 345–349
role in viral evolution, 312–316, 331–333
Mutation rate, estimation, 314
mx genes
human, 67
mouse, 48
Mx proteins, 65–67
myc oncogene, 220, 223–225, 243
Myc protein, 235
Mycophenolic acid, 290
Myd88 protein, 57, 62
Myelin, autoantibodies in HIV-infected
individuals, 184
Myelin basic protein, autoantibodies in
HIV-infected individuals, 184
Myelomonocytes, 88
Myocarditis, 122
Myristoylation, 175
Myxoma, 209
Myxoma virus, 342
entry into host, 13
MT-2 protein, 77
persistent infection, 146
regulators of apoptosis, 77
Myxovirus, immune response, 120

N

NA protein, influenza virus, 25, 262, 296
Nairovirus, 360
Nanovirus, 320
Nasopharyngeal carcinoma, 158–159, 240
"Natural antibody," 95–96
Natural killer cells, see NK cells
NC protein, human immunodeficiency virus, 195
N-Cam proteins, 223
Ndumu complex, 329
Nef protein, human immunodeficiency
virus, 92–93, 114, 145–147, 170–172,
175–176, 182, 190–191

Nef protein, human immunodeficiency virus
(*continued*)
intracellular functions, 175
Nelfinavir (Viracept), 280, 301, 304
Neoplasm, 202
Nepovirus, 328
Nervous system, viral infection, 17
human immunodeficiency virus, 188–192
Neural spread, 17–21
anterograde, 17–18
entry into central nervous system by
olfactory routes, 19
retrograde, 17–18
tracing neuronal connections with viruses, 20
Neuraminidase inhibitors, 290, 293, 296
Neuroinvasive virus, 17, 47
Neuron
herpes simplex virus infection, 151–156
tracing neuronal connections with viruses, 20
transport of herpes simplex virus within, 152
Neurotropic virus, 17, 19, 24
Neurovirulent virus, 17, 44
Neutralization, virus, 116–120
Neutralization assay, measurement of
antiviral antibody response, 115
Neutralization index, 115
Neutralizing antibody, human
immunodeficiency virus, 184–185
Neutrophils, 89, 95
autoantibodies in HIV-infected
individuals, 184
in inflammatory response, 97–98
source/function, 88
Nevirapine (Viramune), 280, 300–301, 381
increased susceptibility, 306
resistance, 306
Newcastle disease virus
interferon response, 68
stopping epidemics in chickens, 256
Nf-κb, 57–62, 69, 90, 96, 169, 171, 175, 226
activation, 66
Nfx1-91, 220
Nipah virus
emerging disease, 338–340
interferon response, 71
V protein, 71
Nitric oxide, 67, 127, 192
Nitric oxide synthase, 67, 127
NK cell(s), 89, 91–93, 96
antibody-dependent cell-mediated
cytotoxicity, 120
HIV-infected individuals, 183
MHC class I receptors, 92
NKT cells, 107
recognition of virus-infected cells, 92
source/function, 88
viral proteins that modulate actions, 92–93
NK cell receptor, 93
NKT cells, 107
Nobel Prize, 108
Nomenclature, virus, 350
Non-A, non-B agent, *see* Hepatitis C virus
Noncytopathic virus, 135
Nonnucleoside analogs, 281
Nonoxynol, 281
Nonparametric test, 32
Nonpathogenic virus, 350–351
Nontransducing oncogenic retrovirus,
209–210, 212–214, 245

insertional activation, 224–225
Normally distributed data, 32
Norovirus, 334, 344, 361
Norvir, *see* Ritonavir
Norwalk virus, 10
Nosocomial transmission, 37
NP protein, influenza virus, 347
NS1 protein, influenza virus, 70–71, 79
NS5a protein, hepatitis C virus, 71
nsP4 protein, rotavirus, 47
Nuclear envelope, 207
Nuclear pore, 174
autoantibodies in HIV-infected
individuals, 184
Nucleoside analogs, antiviral, 281, 289, 292,
299–301
chain terminators, 294–295
mutagenicity, 317
resistance, 305–306
Nucleosome, 69, 153, 158
Nucleotide analogs, antiviral, 292
Nucleus, origin, 322, 325
Null hypothesis, 32
Nutritional status, determinant of host
susceptibility to viral disease, 50

O

Olfactory route, viral entry into CNS, 19
2'-5'-Oligo(A) synthetase, 48, 65, 71
Oncogene, 207, 210, *see also* Proto-oncogene
c-oncogene, 215
insertional activation, 224–225
transduced, 221–223
v-oncogene, 210, 215–216, 218
Oncogene probe, preparation, 216
Oncogenesis, 201–248
hepatitis B virus, 242–246
hepatitis C virus, 246
HIV-associated, 191–194
with very long latency, 241–242
Oncogenic virus, 201–248
common properties, 212
contemporary identification, 211
discovery, 208–212
DNA virus, 209–211
retrovirus, 208–210
RNA virus, 209, 211
viral genetic information in transformed
cells, 212–217
viral transforming genes
identification and properties, 215–218
origin and nature, 217–219
viral transforming proteins, 218–221
Opsonization, 94–95
Oral hairy leukoplakia, 158
Orc1 protein, 233
Orf virus, 378
Organ invasion, 21–23
Origin of viruses, 321–330
cellular theory, 322
chlorella virus, 323
independent entity theory, 322
regressive theory, 322
Orthomyxovirus
disease mechanisms, 369
diseases, 369
emerging disease, 334
epidemiology, 369
evolution, 327

Orthoreovirus, 379
Oseltamivir, 280, 290, 293, 296
Oxidative stress, 77
4-Oxo-dihydroquinolines, 295

P

P value, 32–33
p21 protein, 233–234, 236, 239
p27 protein, 220, 222, 230–231, 233–234
p38 pathway, 62
p38 protein, 62
p53 protein, 74, 76–77, 194, 203, 220, 233
alteration of activity via p300/Cbp
proteins, 238–239
inactivation, 234–239
inactivation by binding to viral proteins,
237–239
regulation of accumulation and activity,
234–237
structure, 236
p58 protein, 68
p107 protein, 232–233
p130 protein, 232–233
p300/Cbp proteins, 69, 237
alteration of p53 activity, 238–239
Pancreas, invasion by virus, 22–23
Pandemic, 5
prediction, 351–352
Pantropic virus, 24
Papilloma, 209
Papillomavirus
antiviral drugs, 281, 291
bandicoot papillomatosis carcinomatosis
virus, 209
dissemination in host, 15
entry into host, 12–13, 15
evolution, 324–325
incubation period, 141
L1 protein, 268, 326
modular nature of genomes, 326
NK modulators, 92
oncogenic, 209–210, 215
persistent infection, 143, 147
shedding from host, 15
transformation mechanisms, 245
viral transforming proteins, 218
Papule, 23
Parainfluenza virus, 372
entry into host, 11–12
50% paralytic dose (PD$_{50}$), 41
Parametric test, 32
Paramyxovirus, 371–373
detection by individual cells, 60
evolution, 327
F protein, 58
host alterations early in infection, 58
transmission, 36
Parechovirus
disease mechanisms, 376
diseases, 375
epidemiology, 375
Particle-to-PFU ratio, 9
Parvovirus
congenital infection, 37
disease mechanisms, 374
diseases, 374
epidemiology, 374
evolution, 314
skin rash, 23

Parvovirus B19, 374
Passive immunization, 116, 256, 258
Passive viremia, 16–17
Pasteur, Louis, 262
Pathogen, 3
  nonpathogen vs., 350–351
Pathogenesis, viral, *see* Viral pathogenesis
Pattern recognition receptor, 55–59, 73
Paxillin, 220, 222
PCR, *see* polymerase chain reaction technology
Pdz domain-containing proteins, 220
Peach latent mosaic viroid, 386
Penciclovir, 280, 290, 292, 365
Peptide, 99
Peptide vaccine, 265
Peptidomimetics, 281, 290, 301, 303
Perforin, 32, 91, 112, 122
Peripheral ganglia, herpes simplex virus, 154–156
Permissive cell, 6, 24, 340
Peroxynitrite, 127
Persistent infection, 36, 135–136, 161, 257, 307, 312
  definition, 142
  direct infection of cells of immune system, 147–148
  immune response, 143–147
  K-replication strategy, 137
  lymphocytic choriomeningitis virus, 149–150
  measles virus, 148–149
  mixture of *r*- and *K*-replication strategies, 137
  requirements, 142
  tissues with reduced immune surveillance, 147
Peyer's patches, 102–103
PFU, *see* Plaque-forming unit
pH sensitivity, rhinovirus versus poliovirus, 12
Phage, *see* Bacteriophage entries
Phagocytes, source/function, 88
Pharmacokinetics, 282
Pharmacophore, 291
Pharyngitis, 11
Phlebovirus, 327, 360
Phosphatase 2A, inhibition, 228–230
Phosphatidylserine, 78
Phosphoinositide-3-kinase, 226, 228, 231, 234–236
Phospholipase Cγ, 227
Phosphotyrosine, 227
Physical defenses, 54
Picornavirus
  antiviral drugs, 281
  disease mechanisms, 375–376
  diseases, 375–376
  epidemiology, 375–376
  evolution, 328
  immune response, 115
  organ invasion, 22
  transmission, 36
Picornavirus receptor, 96
Pig, influenza viruses in, 319, 348
Placental tissue, invasion by virus, 23
Plaque assay, 5
Plaque reduction assay, measurement of antiviral antibody response, 115
Plaque-forming unit (PFU), 9

Plasma cells, 100–101, 104, 116–118
Plasminogen, 25
Platelet(s), autoantibodies in HIV-infected individuals, 184
Platelet-derived activation factor, 192
Platelet-derived growth factor receptor, 206, 228
Pleconaril, 280
Pml body, 67, 78
Pneumonia, 11
Pneumovirus, 327
Polarized cells, release of virus particles, 14, 16
Poliomyelitis
  eradication efforts, 253–255, 343
  global incidence in 2008, 254
  paralytic disease, 342–343
  vaccine-associated disease, 254, 341
Poliovirus
  3C protein, 58
  circulating, vaccine-derived, 254, 341
  disease mechanisms, 376
  diseases, 375
  dissemination in host, 15
  drug-resistant variants, 299
  emerging disease, 341–343
  endemic, 342
  entry into host, 10, 15
  epidemics, 48, 341–343
  epidemiology, 375
  evolution, 316
  host alterations early in infection, 58
  host susceptibility to viral disease, 50
  immune response, 120
  inapparent acute infection, 140
  incubation period, 141
  mouse models of disease, 32
  neurovirulence, 254
  pH sensitivity, 12
  regulation of MHC class I proteins, 144
  seasonality of infections, 38–39
  serotypes, 141
  shedding from host, 15
  tropism, 24, 26
  viral spread, 16, 102
  virulence, 41, 45
Poliovirus receptor, 24
Poliovirus vaccine, 117, 141, 254–255, 261, 274, 343, 375
  circulating vaccine-derived virus, 254, 341
  live attenuated Sabin oral, 45, 117, 263, 266
  safety, 41, 254, 341
  simian virus 40 in, 211, 264–265
Poly(ADP-ribose) polymerase, 75
Polymerase chain reaction (PCR) technology, 6, 350
Polyomavirus
  adapter proteins, 226
  antiviral drugs, 291
  bandicoot papillomatosis carcinomatosis virus, 209
  disease mechanisms, 377
  diseases, 377
  epidemiology, 377
  evolution, 324–325
  LT protein, 218, 238
  Merkel cell carcinoma, 213
  mT protein, 226–227, 234

oncogenic, 209–213
  sT protein, 229
  transformation mechanisms, 245
  viral transforming proteins, 218
Polyp, colonic, 203
Polyvinylalcohol sulfate, 290
Polyvinylsulfonate, 290
Population dynamics, host population, 312, 316, 340–341
Postherpetic neuralgia, 140
Potato spindle tuber viroid, 385–386
Potyvirus, 328
Powassan virus, 363
Power (statistics), 31–33
Pox (skin lesion), 116
Poxvirus
  antiviral drugs, 291
  disease mechanisms, 378
  diseases, 378
  entry into host, 12–13
  epidemiology, 378
  evolution, 320, 325
  immune response, 321
  immunosuppression, 125
  modulation of inflammatory response, 99
  NK modulators, 92
  oncogenic, 210
  transmission, 36
Pre-B lymphoma, 214
Pregnancy, transmission of human immunodeficiency virus, 179
Prelamin A, 304
Prevalence of disease, 32–33
Primary effusion lymphoma, 193, 223–224
Prion, 160, 387–391
  in food supply, 391
  host range, 391
prnp gene, 389–391
Prodrug, 289
Programmed cell death, *see* Apoptosis
Progressive multifocal leukoencephalopathy, 23, 377
Prometaphase, 207
Promoter insertion, 225
Promyelocytic leukemia proteins, *see* Pml body
Proof of principle, 282
Prophase, 207
Prospective study, 33–34
Protease
  cellular, 25–26
  cytomegalovirus, 297
  hepatitis C virus, 293, 297
  human immunodeficiency virus, 301, 303
Protease inhibitors, 194, 281, 290, 293–297, 300–304
  cell-based screen, 284
  designer drugs, 285–286
  mechanism-based screen, 283
  resistance, 304–305
  side effects, 304, 306
Proteasome, 67, 78, 108–109, 145–146, 220
Protectin, 96
Protein(s), *see also* Membrane proteins
  cellular
    interactions with viral transforming proteins, 220
    regulation of viral transcription, 24
    virulence determinants, 47–48

Protein(s) *(continued)*
  viral
    adapter proteins, 226–227
    alteration of cellular signaling pathways,
      225–227
    apoptosis inhibitors, 74–77
    blunting of CTL response, 114
    inactivation of cyclin-dependent kinase
      inhibitors, 233–234
    inhibition of Rb protein, 231–232
    modulation of function of tumor
      necrosis factor, 97
    modulation of interferon response, 71
    modulation of NK cell actions,
      92–93
    p53 binding, 237–239
    regulation of MHC class I function,
      144–146
    regulation of MHC class II antigen, 146
    superantigens, 126–127
    toxic, 43, 47
Protein gB, herpes simplex virus, 146
Protein kinase, 205
  double-stranded-RNA-activated, 65
  mitogen-activated, 206
Protein kinase R (Pkr), 57, 66, 71–72
  Pkr-null mutant mice, 74
Protein tyrosine kinase, 204, 206, 221,
  226–227, 231, 235, 237
Proteomic technology, 286, 288
Proto-oncogene, 210, 213–214
  gain-of-function mutations, 208
Protovirus theory, origin of retroviruses,
  328–330
Provirus
  proviral DNA, 55
  viral genetic information in transformed
    cells, 212–217
PrP protein, 389–391
  refolding model, 391
  seeding model, 391
Pseudocowpox virus, 378
Pseudogene, 329–330
Pseudorabies virus, use in tracing neuronal
  connections, 20
p-SMAC, 113
Pten protein, 236
Public health problems, acute infections,
  141–142
Public perception, emerging viral diseases,
  350–351
Pustule, 23

Q
Quarantine, 35, 343
Quasispecies concept, 314–316
Quinacrine, 391
Quinolinic acid, 192

R
Rabbitpox virus, deliberate release in
  Australia, 342
Rabies virus
  disease mechanisms, 382
  diseases, 372
  emerging disease, 338
  entry into host, 14
  epidemiology, 382
  incubation period, 141

nervous system infection, 17
  neural spread, 20
  transmission, 36–37
Rabies virus vaccine, 261–262, 382
  for wild animals, 270, 273
Raccoonpox virus, 270
Raf protein, 206
Raltegravir (Isentress), 302, 381
Rantes, 65, 114, 177, 186, 298
Ras protein, 205–206, 222–223, 227,
  230–231, 234–235
Rash, *see* Skin rash
Rb protein, 77, 194, 220, 230–234
  acetylation, 238
  inhibition by viral proteins, 231–232
  restriction point control, 230–234
  viral transforming proteins that bind
    to, 238
Rb-related proteins, inhibition of negative
  regulation by, 232–233
Reassortment of genetic segments
  host range expansion, 345–349
  role in viral evolution, 312, 318–321
Receptor(s)
  sensing environment, 204
  tropism and, 24
Receptor tyrosine kinase, 228
Receptor-mediated endocytosis, 77
Recombinant DNA technology, 6, 40–43
  diagnostic virology, 350
  discovery of hepatitis viruses in blood
    supply, 349–350
  subunit vaccines, 260, 265–269
Recombinant vaccine, 260, 269–271, 274
Recombination
  host range expansion, 345–349
  human immunodeficiency virus, 142
  nonhomologous, 321
  pathogenicity change, 348–349
  role in viral evolution, 312, 320–321
  sequence-independent, 320
Refolding model, PrP protein, 391
Regressive theory, origin of viruses, 322
Regulatory proteins, human
  immunodeficiency virus, 169, 171
Release of virus particles
  directional, 14, 16
  from polarized cells, 14, 16
Renal glomerulus, invasion by virus, 22–23
Reovirus
  disease mechanisms, 379
  diseases, 379
  dissemination in host, 15
  entry into host, 10–13, 15
  epidemiology, 379
  immune response, 120
  interferon response, 70–71
  intracellular detectors of viral infection, 59
  shedding from host, 15
  σ3 protein, 70–71
  viral spread, 102
  virulence, 46–47
Replication strategy, 313, 330–332
Reporter gene, 314
Reproduction number, 253, 312
Rescriptor, *see* Delavirdine
Reservoir, 36
Respiratory syncytial virus
  antiviral drugs, 280, 291

disease mechanisms, 373
  diseases, 373
  entry into host, 11–12
  envelope protein, 61
  epidemiology, 373
  host susceptibility to viral disease, 49
  immunopathology, 120, 123
  intracellular detectors of viral infection, 59
  regulation of MHC class I proteins, 144
  treatment, 373
Respiratory tract
  influenza virus infection, 139
  shedding of virions in secretions of, 34–36
  viral entry, 9–12
Resting state, 205
Reston ebolavirus, 362
Restriction endonuclease, 214
Restriction point, 208
  control by Rb protein, 230–234
  mammalian cells, 230–231
Reticuloendothelial system, 18, 21–23
Retinoblastoma, 202
Retrogene, 330
Retrospective study, 33–34
Retrotransposon, 329–330
Retrovir, *see* Zidovudine
Retrovirus
  Apobec in infected cells, 79
  budding site, 332
  discovery, 208–210
  disease mechanisms, 380–381
  diseases, 380–381
  emerging disease, 334
  endogenous, 329
  epidemiology, 380–381
  evolution, 328–330
  Gag protein, 332
  interferon response, 68
  NK modulators, 92
  oncogene capture, 217, 219
  oncogenic, 208–210
    evolution, 320
    genome map, 215–217
    long-latency, 209–210, 245
    nontransducing, 209–210, 212–214, 245
    insertional activation, 224–225
    transducing, 209–210, 215–217, 245
  pathogenicity change by recombination,
    348–349
  protovirus theory, 328–330
  reduction of T-cell receptor function, 112
  shedding of virions, 35
  slow infection, 160
  transformation mechanisms, 245
  transforming, transduced cellular genes,
    221–223
  transmission, 36
  Trim proteins in infected cells, 79
  tumorigenesis with very long latency,
    241–242
Retrovirus vector, gene therapy, 224
Rev protein, human immunodeficiency
  virus, 169–172, 190
Reverse transcriptase, 79
  human immunodeficiency virus, 302
  mutations, 304–305
  primordial, 330
Reverse transcriptase inhibitors, 194,
  290–291, 294

nonnucleoside, 290, 300–302
nucleoside, 290, 299–301
Rev-responsive element (RRE), 171–172
Rex protein, human T-lymphotropic
  virus, 241
Rhabdovirus
  disease mechanisms, 382
  diseases, 382
  entry into host, 12
  epidemiology, 382
  evolution, 327
  neural spread, 20
  tracing neuronal connections with
    viruses, 20
Rhinitis, 11
Rhinovirus
  acute infection, 136, 138, 140
  antigenic variation, 140
  antiviral drugs, 280
  disease mechanisms, 376
  diseases, 375
  dissemination in host, 15
  entry into host, 10–12, 15
  epidemiology, 375
  host susceptibility to viral disease, 50
  immune response, 119
  pH sensitivity, 12
  seasonal variation, 341
  shedding from host, 15, 34
  structural plasticity, 141
  transmission, 36
  tropism, 24
Rhinovirus receptor, 24, 280
Ribavirin (Virazole), 280, 290–292, 359,
  364, 373
  resistance, 316
Ribonuclease L, 48
Ribonucleotide reductase, 44–45, 281
Ribozyme
  satellite viruses, 387
  viroids, 386
RID protein complex, adenovirus, 130
Rift Valley fever virus, 352, 360
  emerging disease, 334
RigI protein, 56–61, 66, 68, 89, 91
Rimantadine, 280, 291–292
Rinderpest virus, 328
RING domain, 170
RING finger proteins, 144–146
Ring-slaughter program, 255–256
Ritonavir (Norvir), 280, 301, 304, 381
RNA
  cellular, "common host response" to
    infection, 7
  double-stranded, ligand for Toll-like
    receptors, 57, 59, 62–63
  single-stranded, ligand for Toll-like
    receptors, 57, 59, 62–63
  viroid, 385–387
RNA helicase, DEXD/H box, 56, 60
RNA interference, 286
  cellular defense, 78
  hbz RNA, 241
RNA polymerase II, 69
  viroid replication, 386
RNA pseudoknot, 71
RNA silencing, antiviral defense, 78–79
RNA synthesis, viral, antivirals targeting, 295
RNA virus

double-stranded, replication, 332
evolution, 314, 316, 322, 326
  relationships from nucleic acid sequence
    analysis, 325–328
oncogenic, 209, 211
(+) strand, 326
  genome map, 328
  replication, 332
(–) strand, 326
  genome map, 327
RNA world, 322
RNA-dependent RNA polymerase, antivirals
  targeting, 295
RNase L, see Ribonuclease L
Rodent vector, 337, 344, 359
Roseola, 366
Rotavirus
  acute infection, 136
  disease, 379
  disease mechanisms, 47, 379
  dissemination in host, 15
  entry into host, 11–12, 15
  epidemiology, 379
  nsP4 protein, 47
  shedding from host, 15
  transmission, 36
  virulence, 47
Rotavirus vaccine, 263, 274
Rous sarcoma virus, 205, 215, 217
  evolution, 320
  preparation of oncogene probe, 216
  transformation mechanisms, 245
Rous-associated virus, 228, 245
Route of infection, 74
Rptk protein, 231, 235
RRE, see Rev-responsive element
r-replication strategy, 137–138, 161,
  313, 321
Rubella virus
  congenital infection, 36–37
  disease, 383
  disease mechanisms, 383
  entry into host, 12
  epidemiology, 383
  evolution, 328
  fetal infection, 23
  host susceptibility to viral disease, 49
  immunosuppression, 124–125
  incubation period, 141
  persistent infection, 143
  seasonality of infections, 38
  skin rash, 23
  transmission, 36
Rubella virus vaccine, 263, 383

S

S phase, 205, 207–208, 230–233
S-adenosylhomocysteine hydrolase inhibitor,
  290
St. Louis encephalitis virus, 36, 363
Saliva, shedding of virions in, 35–36
Sample size, 32
Sandfly fever virus, 360
Sanitation, 341
Saponins, 272
Sapovirus, 344, 361
Sapporovirus, 361
Saquinavir (Fortovase, Invirase), 280, 286,
  301, 304, 381

resistance, 306
Sarcoma, 202, 209
SARS coronavirus, 352
  emerging disease, 335, 338,
    344–345
  evolution, 314
Satellite, 387
Satellite nucleic acid, 387
Satellite virus, 387
Schwann cells, 17
Scrapie, 387–388
  physical nature of scrapie agent,
    389–390
  strains of scrapie prion, 391
Sdf-1, 186
Seasonal variation, viral disease,
  35, 37–40, 341
sec oncogene, 220
Second messenger, 205, 227
Seeding model, PrP protein, 391
Selection, role in viral evolution, 312, 331
Self antigen, 92, 125–126
Semen
  human immunodeficiency virus,
    177–178
  shedding of virions in, 35–36
Semliki Forest virus, 329
  immune response, 116
  immunopathology, 120
Sendai virus
  immune response, 119
  release of virus particles, 16
Sentinel cells, 89–91
  monitoring apoptosis, 77–78
  synthesis of Toll-like receptors, 62
  virus-infected, 91
Sequence coherence, 322, 324
Serial infections, 341
Serine protease, 25
Serine protease factor I, 46
Serine/threonine protein kinase, 206,
  214
Serum amyloid, 67
Set point, virologic, 180
Severe combined immunodeficiency, gene
  therapy, 224
Sexually transmitted disease, 10–12, 36
  human immunodeficiency virus,
    177–179
Shc protein, 206, 227
Shedding of virions, 34–36
  blood, 35
  feces, 35
  respiratory secretions, 34–36
  saliva, 35
  skin, 36
  urine, semen, and milk, 35–36
Shingles, 140, 368
Shope fibroma virus, 210
Sialic acid, 24, 31, 296, 340
Sialyltransferase, 31
SIE element, 70
σ3 protein, reovirus, 70–71
Signal 1, 112
Signal 2, 112
Signal transduction pathway, 55–58,
  204–206
  activation by viral oncogene products,
    221–230

Signal transduction pathway *(continued)*
  alteration of activity of cellular
    molecules, 228–230
  alteration of production/activity of
    cellular signal transduction proteins,
    224–230
  viral mimics of cellular signaling
    molecules, 221–224
  viral proteins that alter cellular
    signaling, 225–227
Significance level, 32
"Significant difference," 30
Sigurdsson, Bjorn, 160
Simian immunodeficiency virus, 167–169,
    174–175, 335, 345
Simian sarcoma virus, 217
Simian virus 40
  in human cancer, 210–211
  interferon response, 68
  LT protein, 77, 218, 220, 232–234,
    239, 245
  oncogenicity, 210
  in poliovirus vaccine, 211, 264–265
  regulators of apoptosis, 77
  sT protein, 229–230, 245
  temperature-sensitive mutations, 215
  transformation mechanisms, 245
  viral transforming proteins, 215, 218
Sin Nombre virus, 360
  emerging disease, 334
  immunopathology, 120
  in rodents, 337, 344
Sindbis virus
  autophagy in infected cells, 78
  host alterations early in infection, 58
  persistent infection, 143
Sinusoid, 21
  blood-tissue junction, 21
siRNA, *see* Small interfering RNA
sis oncogene, 220
Skeletal muscle, invasion by virus, 22
Skin
  barrier to infection, 54
  cutaneous immune system, 102–103
  invasion by virus, 23
  shedding of virions in skin, 36
  viral entry, 10, 13–14
Skin rash, 9, 23, 102, 116, 140, 148
Slave trade, 335, 338, 343–344
"Slim disease," 190
Slow infection, 135–136, 142, 160, 264
Slow virus, 167
Small interfering RNA (siRNA), 57, 79
Smallpox
  emerging disease, 335, 343–344
  eradication, 252–253, 255, 335
  historical aspects, 251–253, 281, 343–344
Smallpox inhibitor of complement, 46
Smallpox vaccine, 256, 263, 351, 378
  herd immunity threshold, 259
  historical aspects, 251–254
  safety, 258, 261
  U.S. stockpile, 252
Smallpox virus
  bioterrorism agent, 252
  congenital infection, 37
  destruction of laboratory stocks, 257
  evolution, 325, 335
  incubation period, 141

regulation of complement cascade, 96
reproduction number, 253
SPICE protein, 96
Smoldering infection, 135–137
Sneezing, 10, 34
Sodium-potassium ATPase, 58
Sos protein, 206, 223
Southern hybridization assay, 214
spc gene, 203
Sperm, autoantibodies in HIV-infected
    individuals, 184
SPICE protein, smallpox virus, 96
Spitting, and virus transmission, 35
Spleen, invasion by virus, 21
SPRY domain, 170
src oncogene, 205, 216
  v-src paradigm, 221–223
Src proteins, 234
SSPE, *see* Subacute sclerosing panencephalitis
sT protein
  polyomavirus, 229
  simian virus 40, 229–230, 245
Stat proteins, 63, 70–71, 73, 222
Statistics, 30–33
Stavudine (d4T, Zerit), 280, 292, 301
Strategic treatment interruption, 195, 307
Structural plasticity, 141
SU protein, human immunodeficiency virus,
    47, 147, 176, 182, 184–185, 190–191,
    194, 302–303
SU/V3 loop inhibitors, 300, 302–303
Subacute sclerosing panencephalitis (SSPE),
    148–149, 253, 371
Subtractive hybridization, 216
Subunit vaccine, 260, 265, 271–273
  recombinant DNA approaches, 260,
    265–269
Subviral RNA Database, 385
Superantigen, 126–127
Superoxide, 127
Suramin, 302
survivin gene, 237
Susceptibility, *see* Host susceptibility to viral
    disease
Susceptibility genes, 48
Susceptible cell, 6, 24, 340
Sustiva, *see* Efavirenz
Swi/Snf proteins, 69
Symmetrel, *see* Amantadine
Systemic infection, 16
Systemic inflammatory response
    syndrome, 124
Systems biology, 6

T

T cells, 99–101, 104–105
  activated, 90, 104
    killing by CTLs, 147
    superantigens, 126
  CD4, 105, 180–181
    HIV-infected, 179–182, 188,
      303–304
    HTLV-infected, 243
    immunopathological lesions caused by,
      122–123
  CD8, 105, 145, 183
  coreceptors for human immunodeficiency
    virus, 176–177
  cytotoxic, *see* Cytotoxic T cells

delayed-type hypersensitivity,
    114, 116
heterologous T-cell immunity,
    125–126
maturation, 104
memory, 99, 102, 107, 125–126, 128
  acute infection, 140
  HIV-infected individuals, 180, 187
naive, 89–90, 100, 104, 107
NKT cells, 107
noncytolytic control of infection,
    114–116
protective immunity, 258
recognition of infected cells, 108–110
recognition of professional antigen-pre-
    senting cells, 110–111
regulation by complement system, 95, 101
regulatory, 101, 105, 107
skin-tropic, 102–103
source/function, 88
Th1, 128
virus-specific, identifying and counting, 115
T20, *see* Enfuvirtide
Tab2 adapter protein, 60
TafII proteins, 236
Tak1 kinase, 60
TAK779, 290
Tanapox virus, 378
Tap proteins, 109, 145
Tar RNA, human immunodeficiency virus,
    71, 169–171
Tat protein, human immunodeficiency virus,
    79, 114, 145, 147, 169–172, 182,
    190–191, 193, 300
  activation, 171
  interaction with TAR sequences,
    169–171
Tattooing, 14
Tax protein, human T-lymphotropic virus,
    147, 241–242, 245
3TC, *see* Lamivudine
T-cell lymphoma, 214
T-cell receptor, 90, 99, 101–113, 116,
    126, 145
  structure, 105
Telomerase, 203, 220
Telophase, 207
Tenofovir (Viread), 280, 290–291, 301, 306
TfIId protein, 69, 236
TfIIIC protein, 58
Theiler's murine encephalomyelitis virus,
    120, 122
T-helper (Th) cells, 100–101, 104–105,
    110, 112
  cell surface proteins, 105
  delayed-type hypersensitivity, 116
  HIV-infected individuals, 186
  Th1, 95, 100–101, 103, 105–107, 128
    balance of Th1 and Th2 cells,
      106, 123
    cytokine production, 106
    immunopathological lesions caused by,
      122–123
  Th2, 95, 100, 103, 105–107, 116
    balance of Th1 and Th2 cells, 106, 123
    CD4, 123
    cytokine production, 106
    immunopathological lesions caused
      by, 123

Th17, 107
  virus-infected, 147–148
Therapeutic index, 287
Thimerosal, 268
Thymidine kinase, 44, 281, 289, 294–295, 298
Thymus, 104
TIBO (tetrahydroimidazobenzodiazepinone), 300–301, 306
Tick vector, 336–337, 339, 352, 360, 363
Tight junction, 13
Tip protein, herpesvirus saimiri, 227
TM protein, human immunodeficiency virus, 47, 183–185, 190–191, 194
Tnf, *see* Tumor necrosis factor
Tobamovirus, 328
Tobravirus, 328
Togavirus
  disease mechanisms, 383
  diseases, 383
  dissemination in host, 15
  entry into host, 12, 15
  epidemiology, 383
  evolution, 335
  organ invasion, 22
  shedding from host, 15
  transmission, 36
  vectors, 383
Toll receptors, discovery, 58
Toll-like receptor, 48, 56–62, 78, 89–90, 128, 269, 272
  contribution to viral pathogenesis, 62
  ligands, 57, 59, 62, 106
  recognition of microbial macromolecular patterns, 63
Toll-like receptor protein 3, antivirals targeting, 298
Tombusvirus, 328
Tonsils, 102
Tracheitis, 11
Tradd protein, 226
Traf proteins, 60, 62, 226
Transcription, viral, cellular proteins that regulate, 24
Transcriptional profiling, host response to infection, 7
Transcytosis, 10–11, 22, 118–119
Transdominant inhibitors, 295–299
Transducing oncogenic retrovirus, 209–210, 215–217, 245
Transferrin receptor, 346
Transformation, *see* Viral transformation
Transformed cells, 202
  apoptosis, 234–239
  autocrine growth stimulation, 203
  cellular transformation, 202–204
  compared to normal cells, 203–204
  detection and characterization of integrated viral DNA, 214
  escape from immune surveillance, 240
  growth parameters and behavior, 203–204, 234–239
  immortality, 202
  inhibition of immune defenses, 240–241
  integration of mitogenic and growth-promoting signals, 234–235
  mechanisms that permit survival, 234–239
  morphology, 204–205

properties, 202–204, 239–241
  viral genetic information in, 212–217
Transforming genes, viral, *see* Viral transforming genes
Transforming growth factor β, 64, 186, 192, 296
Transforming infection, 135, 161
Transforming proteins, viral, *see* Viral transforming proteins
Transgenic mouse model, cytotoxic T cell killing, 122
Translation, transformed cells, 234–235
Transmissible mink encephalopathy, 387
Transmissible spongiform encephalopathy, 387–391
  historical aspects, 389
  human, 387–388
    familial, 388–389
    sporadic, 388–389
  identification of first agent causing, 389
  pathogenesis, 388–389
Transmission of viral disease, 35–37
  germ line, 37
  horizontal, 37
  iatrogenic, 37
  nosocomial, 37
  vertical, 36–37
Tricornavirus, 328
Trif protein, 57
Trifluridine, 292, 365
Trim proteins, 79–80
Trim5α, 68, 170
Trisodium phosphonoformate, *see* Foscarnet
Trizivir, 301
Tropical spastic paraparesis, 380
Tropism, 23–26
  cellular proteases and, 25–26
  host cell proteins that regulate viral transcription, 24
  receptors for viruses and, 24
  site of entry establishes pathway of spread, 26
Tryptase Clara, 24
TT virus, 350
Tumor
  benign, 202
  defined, 202
  malignant, 202
  monoclonal, 212
Tumor necrosis factor (Tnf), 64
Tumor necrosis factor alpha (Tnf-α), 57, 59, 64, 67, 72, 76, 90–91, 98, 103, 114–115, 122, 186, 192
  viral proteins that modulate, 97
Tumor necrosis factor receptor, 77, 226
Tumor suppressor gene, 210, 220, 246
  loss-of-function mutations, 208
Tumorigenesis, properties of transformed cells, 239–241
Tymovirus, 328
Tyrosine kinase, 113
  Src family, 116

Ubiquitin ligase, 68, 79–80, 144–146, 170, 173–175, 220, 230, 237–239
Ubiquitin-proteasome pathway components, interferon-induced, 67–68

Ubiquitinylation, proteins, 67–68, 108–109, 144–146, 170, 173–175, 232, 237
Unc-93B protein, 48
Uncoating of virus, 119
  drugs targeting, 281, 293
Unintended consequences, law of, 353
Unusual infectious agents, 385–391
Uracil-DNA glycosylase, 173
Urbanization, 342
Urine, shedding of virions in, 35–36
Urogenital tract, viral entry, 10–12
US2 protein, cytomegalovirus, 144–146
US3 protein, cytomegalovirus, 144–145
US6 protein, cytomegalovirus, 144–145
US11 protein
  cytomegalovirus, 144–145
  herpes simplex virus, 70–72
Used tires, mosquito habitat, 352

V

V protein, Nipah virus, 71
Vaccination, *see also* Vaccine
  introduction of term, 253
  large-scale programs, 253–257
  schedule, 263
Vaccine, 251–277, *see also* Vaccination
  active or passive immunization, 256
  attenuated (live) virus, 260, 262–269
  booster shot, 259
  "challenge," 266–267
  clinical trials, 275
  costs, 259–260
  delivery/administration, 259–260, 272–273
  DNA, 260, 268–270
  edible plants, 273
  efficacy, 257–260
  fundamental challenge, 260–261
  genomic, 269
  historical aspects, 251–257, 262
  immunization schedule, 263
  immunotherapy, 273–274
  inactivated (killed) virus, 261–262, 271–273
  large-scale vaccination programs, 253–257
  licensed in United States, 263
  nasal spray, 264
  oral, 263–264, 272–273
  peptide, 265
  "poor take," 259
  practicality, 257–260
  production, 260–261
  protection from disease, 257
  protection from infection, 257
  public acceptance, 259, 268
  recombinant, 260, 269–271, 274
  safety, 257–261, 264, 268, 275
  side effects, 258–259
  stability, 260
  stimulation of immune memory, 256–260
  subunit, 260, 265, 271–273
  variation in responses to, 259–260
  virus-like particle, 260, 267–268
Vaccine "escape" mutant, 260, 265

Vaccine trials, 275
Vaccinia virus, 252, 378
    A18R protein, 71
    A46R protein, 62
    A52R protein, 62
    apoptotic phospholipids, 78
    B18R protein, 71
    dendritic cells infected with, 91
    E3L protein, 70–71, 79
    gene products that suppress RNA interfer-
        ence, 79
    immune response, 116, 122, 125
    interferon response, 70–71
    K3L protein, 71
    protection from complement-mediated
        lysis, 96
    recombinant vaccinia virus vaccines,
        270
    regulation of MHC class I proteins, 144
    VH1 protein, 71
Vaccinia virus vaccine, 252
Valacyclovir, 280, 289, 365
Valganciclovir, 280, 366
Varicella-zoster virus
    acute infection, 140
        inapparent, 138
    antiviral drugs, 280–281, 289
    disease mechanisms, 368
    diseases, 368
    dissemination in host, 15
    entry into host, 12, 15
    epidemiology, 368
    incubation period, 141
    interferon response, 72
    latent infection, 140, 150, 153
        reactivation, 140
    organ invasion, 23
    persistent infection, 143
    shedding from host, 15
    skin rash, 23, 102
    transmission, 36
    treatment, 368
Varicella-zoster virus vaccine, 263–264, 273
Variola virus, 378
    inhibitors of complement pathway, 46
    virulence, 46
Variolation, 252
VA-RNA I, adenovirus, 70–71, 79
VA-RNA II, adenovirus, 79
Vascular endothelial growth factor, 223–224
Vasculitis, 123
Vector boosting, 274
Vector elimination, 35
Venezuelan equine encephalitis virus
    evolution, 329
    replication in dendritic cells, 91
Venule, blood-tissue junction, 21
Vertical transmission, 36–37
Vesicle (skin rash), 23
Vesicular stomatitis virus, 382
    cellular defense against, 67
    evolution, 320
    immune response, 116, 122
    interferon response, 70
    recombinant virus vaccines, 270–271
    release of virus particles, 14
    route of infection, 74
Vesiculovirus, 327, 382
vFLIP protein, 77, 240

v-*gpcr* gene, 223–224
VH1 protein, vaccinia virus, 71
Video games model, epidemics of infectious
        disease, 34
Videx, *see* Didanosine
Vif protein, human immunodeficiency virus,
        79, 171–174
Viracept, *see* Nelfinavir
Viral pathogenesis
    defined, 29
    determinants, 30
    first human viruses, 4–5
    fundamental questions, 50
    golden age, 5–6
    history, microbes as infectious agents, 3–4
    new millennium, 6
    principles, 29–50
    studies in transgenic and knockout
        mice, 32
    views of, 8
Viral spread, 15–21
    hematogenous spread, 14–19
    human immunodeficiency virus, 179–180
    neural spread, 17–21
    organ invasion, 21–23
    site of entry establishes pathway of
        spread, 26
    virulence genes controlling, 43, 46–47
Viral transformation, 201–248, *see also*
        Transformed cells
    activation of cellular signal transduction
        pathways, 221–230
    alteration of production/activity of
        cellular signal transduction proteins,
        224–230
    viral mimics of cellular signaling
        molecules, 221–224
    disruption of cell cycle control by viral
        oncogene products, 230–234
    abrogation of restriction point control by
        Rb protein, 230–234
    inactivation of cyclin-dependent kinase
        inhibitors, 233–234
    production of virus-specific cyclins, 233
    diversity of mechanisms, 244–245
Viral transforming genes
    dominant, 218, 246
    identification and properties, 215–218
    origin and nature, 217–219
Viral transforming proteins
    functions, 218–221
    interaction with cellular proteins, 220
    simian virus 40, 215
Viral vector, 160
    defective, 121
    live attenuated, vaccine production, 260,
        269–271
Viramidine, 291
Viramune, *see* Nevirapine
Virazole, *see* Ribavirin
Viread, *see* Tenofovir
Viremia, 17–18
    active, 17
    passive, 16–17
    primary, 9, 15–17, 140
    secondary, 9, 15–17, 140
Virion, 6
Viroceptor, 61
Virochip, 322, 350

Viroid, 385–387
Virokine, 45, 61
Virologic set point, 181
Virological synapse, 179
Viroceptor, 45
Virulence, 40
    alterations, 41–42
    creation of more virulent virus, 43
    effect of inoculation route, 41–42
    genetic determinants, 41
    measurement, 40–41
        measurement of pathological lesions, 41
        measurement of survival, 41
    selection for or against, 312
Virulence genes
    cellular, 47–48
    viral, 42–47
        classes, 43
        gene products that alter host defense
            mechanisms, 43, 45–46
        gene products that alter virus
            replication, 42–44
        gene products that enable viral spread,
            43, 46–47
        noncoding sequences that affect virus
            replication, 44–45
        targets of viral virulence gene products,
            47
        toxic viral proteins, 43, 47
Viruria, 35–36
Virus
    discovery, 4
    size, 324
Virus-like particle vaccine, 260, 267–268
Virusoid, 387
Visna/maedi virus, 160, 167, 169, 175
v-oncogene, 210, 215–216, 218
VP2 protein, canine parvovirus, 346
VP16 protein, herpes simplex virus, 295–297
VP35 protein, Ebola virus, 79
Vpr protein, human immunodeficiency
        virus, 160, 171–174
Vpu protein, human immunodeficiency
        virus, 171–172, 174–175, 182
Vpx protein, human immunodeficiency
        virus, 171–174
v-*src* paradigm, 221–223
VX-478, 306

W

Warts, 36, 147, 370
West Nile virus, 363
    emerging disease, 334, 336
    host susceptibility to viral disease, 48
    intracellular detectors of viral infection, 59
    neural spread, 20
    spread to North America, 336
Western equine encephalitis virus
    evolution, 329
    immune response, 116
    transmission, 36
White blood cells, 87
Woodchuck hepatitis B virus, 242–243
Woodville, William, 253
World of Warcraft (video game), 34

X

X protein, hepatitis B virus, 68, 77, 243–244
Xanthine oxidase, 127

Xenotransplantation, 338–339
X-linked lymphoproliferative syndrome, 158

Y

Yaba monkey tumor virus, 378
Yellow fever virus, 352, 363
    emerging disease, 337–338
    historical aspects, 4
    transmission, 36–37

Yellow fever virus vaccine, 263, 363
    contaminated with hepatitis B virus, 341

Z

Zalcitabine (ddC, Hivid), 280, 301
Zanamivir, 280, 293, 296
Zap70, 113
Zerit, *see* Stavudine
Ziagen, *see* Abacavir

Zidovudine (AZT, Retrovir), 194,
    280, 289, 292–295,
    299–301, 381
    increased susceptibility, 306
    resistance, 304, 306
Zoonotic infection, 36–37, 149–150,
    328, 333–334, 336, 338–340,
    344–345, 353
Zta protein, Epstein-Barr virus, 160